MW00714482

FRACKING AMERICA

"*Fracking America* belongs on the bookshelf of everyone in the vast movement fighting the fracking scourge. It will help shape and drive this battle in the years ahead." —Bill McKibben, author and Schumann Distinguished Scholar at Middlebury College; founder, 350.org

"In this superb book, Brasch extends the focus of his previous work on Pennsylvania to America as a whole. It provides a much needed fact-based overview of the impacts on people and the environment of unconventional gas extraction and production." —Dr. Damian Short, School of Advanced Study, University of London

"Because many fracking discussions focus on environmental impact, it's satisfying to see an account that moves well beyond the usual focus to analyze some of the other reasons why fracking is an unusually dangerous pursuit. The wide-ranging discussions move from theological perspectives on fracking to connections between industry interests and political maneuvering, which have influenced politicians to create laws skewed toward industry bene-fits and against public health and environmental concerns. Dr. Brasch offers a studied, rational series of analyses centered around the mechanics of fracking and its impact on different levels. . . . Readers should anticipate the same attention to detail and facts as in his other books. Charts, graphs, news reports, scientific papers and documents support his contentions and provide authority to support every statement. . . . Discussions and assess-ments of renewable energy resources around the world, their locations, and their potentials round out what has to be the most authoritative, well-researched, rational and evidence-based discussion of fracking in America to hit the book market to date. *Fracking America* is highly recommended for anyone studying the subject at any level, whether they are newcomers to fracking or activists who have only researched environmental impact, and need to fill in the blanks on political processes and impacts that hold important questions about American freedoms and political maneuvering." —Diane Donovan, *Midwest Book Review*

"Walter Brasch has presented the many complicated issues surrounding shale gas drilling in a clear, easy to understand manner without oversimplifying them. This book is packed with information everyone living in any area being drilled or likely to be drilled by the gas industry needs to know."
—Karen Feridun, founder, Berks Gas Truth

ii

"*Fracking America* is an indispensable book for anyone who wants to understand natural gas fracking and the environmental risks it presents. But it is the politics of fracking that will cause the greatest outrage. Walter M. Brasch has done a public service by creating a devastating and coherent account of how fracking destroys the environment and contributes to significant health problems."
—DAVID DeKOK, reporter, Reuters; and author of *Unseen Danger: A Tragedy of People, Government, and Centralia*; and *The Epidemic: A Collusion of Power, Privilege, and Public Health*

"Dr. Brasch meticulously researched the issues behind fracking in all its complexity: geologic, environmental, social, economic, and political. Shale gas being touted more and more as the greener energy of our future makes this book essential reading for anyone who wants to safeguard their health and the environment."
—Carol Terracina Hartman, Ph.D., environmental journalist

"A remarkable piece of work, and a great resource for anyone working on the issue."
—Joe Uehlein, labor/environmental activist and musician.

"[*This book*] is woven together in a cohesive way that takes the reader through the complexities of issues surrounding shale gas drilling. If you are not alarmed and fearful while reading Brasch's book, you are not sufficiently engaging its content. If I were teaching a course on environmental ethics, Brasch's books would be on the reading list." —The Rev. Dr. Leah Schade, *EcoWatch*

"A journalistic coup in the vein of Rachel Carson's *Silent Spring*."
—Carolyn Howard-Johnson, *HoJo News*

"The best thing on fracking I have seen. The fact that Dr. Brasch is careful in his fact-checking does not mean that he lacks aggressiveness in the pursuit of wrong-doing. On the contrary, he does not take his eye off the quarry." —Chuck Brown, editor, *Common Sense 2*.

"This is one of the best books to really drill down into what fracking is all about, wherever you live. One really great thing is Dr. Brasch's painstaking research into the many social impacts of fracking. Recommended for anyone facing fracking, and everyone concerned about this wrong-turn industry."
—Alex Smith, EcoShock Radio Network

BOOKS BY WALTER M. BRASCH

Fracking America:
Sacrificing Health
and the Environment for
Short-Term Economic Benefit

by Walter M. Brasch

Greeley & Stone, Publishers, LLC
Sacramento, California 95825

© Copyright 2015 Walter M. Brasch
All rights reserved

No part of this book may be reproduced or transmitted in any
form or by any means, electronic or mechanical, including
photocopying, or within any commercial information retrieval and
storage system, without permission in writing from the author.

Publisher's Cataloging-in-Publication
(Provided by Quality Books, Inc.)

Brasch, Walter M., 1945- author.
 Fracking America : sacrificing health and the
environment for short-term economic benefits /
by Walter M. Brasch.
  pages cm
  Includes bibliographical references and index.
  LCCN 2015910920
  ISBN 978-0-942991-27-7 (trade paper)
  ISBN 978-0-942991-28-4 (case bound)

  1. Hydraulic fracturing--Environmental aspects--
United States. 2. Hydraulic fracturing--Health aspects
--United States. 3. Hydraulic fracturing--Political
aspects--United States.   I. Title.

TD195.G3B737 2015   333.8'2330973
      QBI15-600157

Design: MaryJayne Reibsome

PRINTED IN THE UNITED STATES OF AMERICA
∞ recyclable paper

Greeley & Stone, Publishers, LLC
2350 Wyda Way, suite 1113
Sacramento Calif. 95825
www.greeleyandstone.com

PHOTO: Gary F. Clark

Rig near Warrendale, Pa.

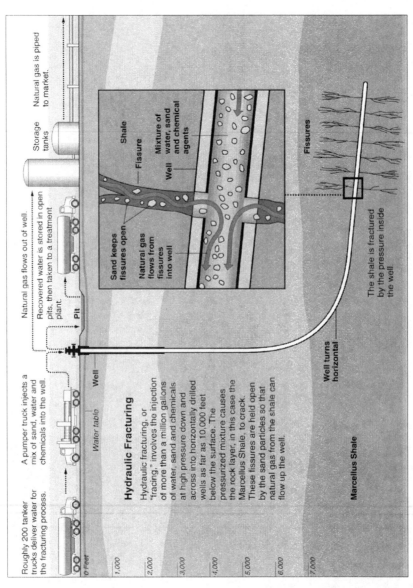

Roughly 200 tanker trucks deliver water for the fracturing process.

A pumper truck injects a mix of sand, water and chemicals into the well.

Natural gas flows out of well.

Recovered water is stored in open pits, then taken to a treatment plant.

Storage tanks

Natural gas is piped to market.

Water table

Well

Pit

Hydraulic Fracturing

Hydraulic fracturing, or "fracing," involves the injection of more than a million gallons of water, sand and chemicals at high pressure down and across into horizontally drilled wells as far as 10,000 feet below the surface. The pressurized mixture causes the rock layer, in this case the Marcellus Shale, to crack. These fissures are held open by the sand particles so that natural gas from the shale can flow up the well.

Shale

Fissure

Well

Mixture of water, sand and chemical agents

Sand keeps fissures open

Natural gas flows from fissures into well

Fissures

Well turns horizontal

The shale is fractured by the pressure inside the well

Marcellus Shale

GRAPHIC: Al Granberg

0 Feet
1,000
2,000
3,000
4,000
5,000
6,000
7,000

Contents

PHOTO: Doug Duncan/USGS

Well site in the Marcellus Shale

Lower 48 states shale plays

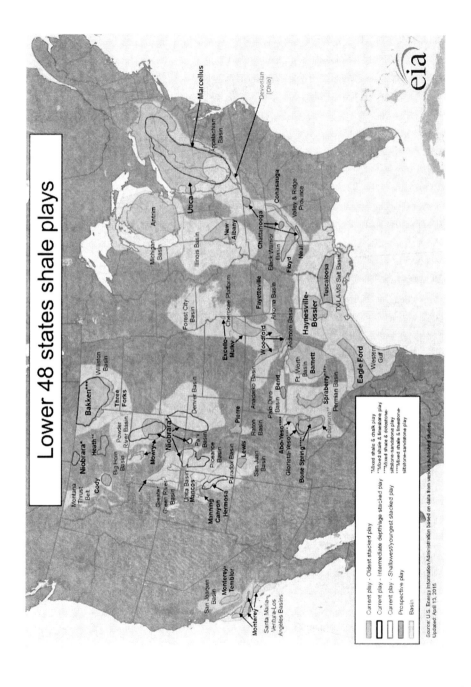

PHOTO: Bill Cunningham/USGS

Wellheads in the Fayetteville Shale in Arkansas

Preface and Acknowledgements

Fracking America is my third book about fracking.

I didn't want to write any.

Sometime early in 2008, Rob Kall, a friend and editor of *OpEdNews*, had a couple of conversations with me about the gas extraction process known as fracking. He said the public had little information about what it was and its effects; what information there was came from the oil and gas industry; he strongly urged me to study, investigate, and fill "a huge hole" in coverage of this relatively new process.

I did some cursory reading, agreed this was a subject that needed to be covered, but didn't want to take the time and spend the energy to learn about the geology and engineering of a process that was relatively new, nor to take the time to study the complexities of the economic, political, environmental, and health issues.

A couple of years later, David Young, a friend and environmental activist, sent me an email. He asked if I thought there were First Amendment violations within the proposed Pennsylvania law that regulated the natural gas industry.

I knew the proposed law, pushed by the newly-elected Republican governor and the Republican majorities in both the House and Senate, was controversial. Although seeming to regulate natural gas drilling, it was being written by pro-industry lobbyists and legislators.

By now, I had been aware of the issue for more than two years, but I was still overloaded with other work. I was writing a syndicated newspaper column, doing weekly radio commentaries, and promoting my recently-released book, an historic novel, *Before the First Snow,* which has an undercurrent of environment and energy activism. The main characters are environmental activists who became involved in a battle in the early 1990s to stop the building of a nuclear waste plant by a corrupt corporation that took safety shortcuts. Not only do the activists have to prove the proposed plant is dangerous to the environment, they have to overcome the lure of jobs in a depressed economy, and show that the temporary boost in economic prosperity is not enough to overcome the decades of health and environmental problems.

I also knew that all the issues and problems in the building and operating of nuclear power plants in the 1980s and early 1990s were almost the same as in the natural gas fracking process more than two decades later. The government and Big Energy were again collaborating, holding out jobs as the lure to get the people to overlook the hazards of construction and continued operation of one source of energy. Substitute names, and the issues raised decades earlier are the same today.

Frankly, after an intense time of writing and research to once again expose problems of nuclear energy, which I acknowledge has significant controls and employs some of the brightest minds in the energy field although there are health risks, I didn't want to write about another environmental campaign so soon. I also knew that I didn't have much of a choice—the health and environmental risks of fracking were becoming too great to ignore.

There was one part of Pennsylvania's proposed law that led me to believe there was a bigger story than just a controversial law. Buried deep in the 174-page bill was a section that allowed any natural gas company to withhold disclosing some of the composition of its fracking fluids if the company could establish that it was proprietary information. This would forbid health care professionals from getting the knowledge they might need to treat persons with idiopathic illnesses who came into the emergency rooms or physician offices. If the physician suspected a link between symptoms and the toxic chemicals and compounds in the fracking fluid, the companies would allow the physician to learn the composition of those "trade secrets." But the physician would have to sign an agreement not to disclose the composition or the cause of illness to anyone, including the patient, medical specialist, or public health worker.

The cavalier attitude of protecting "corporate secrets" at the expense of public health led me to push aside other writing, and begin a month-long investigation, assisted by health care professionals and lawyers.

The article led to many reader responses, but I was still

unwilling to do other articles. Many activists were already protesting, and I didn't know what else I could do to add to the public's knowledge. But, I kept getting reader responses, and suggestions for other stories. After writing two more articles, I finally realized my next book had to be about fracking, a process which most Americans had little knowledge about and which had barely registered on any search engine prior to 2008. Even three years later when I began *Fracking Pennsylvania*, most Americans didn't know what fracking is, and even more didn't understand the issues. A survey published in September 2012 by the Center for Climate Change Communication at George Mason University revealed that 39 percent of Americans hadn't heard of fracking, and an additional 16 percent only knew "a little." About 22 percent knew "some," 13 percent didn't know whether they knew about it or not, and only 9 percent said they knew "a lot." Part of the problem was a lack of media coverage. In the three years beginning 2009, there were only nine news stories that focused on fracking that were broadcast by ABC, CBS, NBC, CNN, and Fox News, according to an investigation conducted by *Extra!* Of the nine stories, six were favorable to fracking, four of them from Fox News, according to *Extra!*'s data search.

I began by trying to balance all sides, to be objective, to allow the facts to direct my work. But as I accumulated mounds of evidence, I realized that fracking, even under the best of conditions, is a problem.

One of the first persons I contacted was Dr. Helen Podgainy Bitaxis, who had actively spoken out about the state's gag order. A pediatrician in Coraopolis, Pa., she gave me good information, which allowed me to better understand that those who care about our children also care that no harm comes to them because of the state's aggressive push to accede to corporate wishes that could bring jobs, but at a cost that would impact the health and environment.

I also received excellent advice and information about fracking and its effects from psychologists Diane Siegmund and Kathryn Vennie, and environmental activist Eileen Fay. Others who are active in the battle for the environment also assisted, some just by answering a question, others by explaining and answering myriad questions over several months.

Among thousands of activists who have been vigorous in pursuing the truth about fracking, and have provided me with solid information, are Barbara Arrindell, David Braun, Tracy Carluccio, Robbie Cross, Bob Donnan, Julie Ann Edgar, Joanne Fiorito, Gloria Forouzan, Dr. Harvey M. Katz, Ralph Kisberg, Debbie Ziegler Lambert, Dr. Wendy Lee, Jan Milburn, Elizabeth Nordstrom, Liz Rosenbaum, Paul R. Roden, Patti Rose, Rebecca Roter, the Rev. Dr. Leah Schade, Vera Scroggins, Briget and Doug Shields, Nathan Richard Sooy, Michele Novak Thomas, Maya K. van Rossum, and Jon Vogle.

Dory Hippauf's "Connecting the Dots" series exposes innumerable connections between government and the energy industry, as well as the industry's shadow and front groups, and connected the "dots" of the mountains of money and influence that went into political campaigns.

Karen Feridun's weblog, "Berks Gas Truth," is an excellent place to get information from an environmental activist with significant knowledge and a willingness to stand up for what's right.

Judy Morrash Muskauski is the nation's best aggregator of fracking information. She is the administrator of "Fracking in Northeast Pennsylvania," a website that began as an information site about fracking in Luzerne County, but is now a hub for almost every story about fracking. Her attention to accuracy and the search for information make it easy to learn about fracking throughout the country. She and Feridun are also compilers of the largest database of websites and blogs about the fracking industry.

Feridan and Muskauski are former librarians; they and Hippauf reviewed my manuscripts and made numerous suggestions that improved them for publication. They have become a mini-brain trust; if they don't know the answers, they know who to contact.

Working within a department that had become politically charged, several career staff and scientists in the Pennsylvania Department of Environmental Protection responded to my innumerable questions; off the record, they often gave information and suggestions that opened new areas of investigation. Similarly, persons working in the natural gas industry also provided assistance, with the understanding I would verify their information and protect their identities.

Although I had discussed fracking in other states, most of *Fracking Pennsylvania* was focused upon the one state at the center of the biggest gas-rush in American history. The success of *Fracking Pennsylvania*—as Rob Kall had predicted more than three years earlier—led to a realization that I needed to expand the book and now focus upon the critical issues that enveloped most of the country, not just the 100,000 square mile Marcellus Shale region. The result is *Fracking America.* (Some of *Fracking Pennsylvania* was re-verified and, where necessary, modified and updated for *Fracking America.*)

Those who helped me with *Fracking Pennsylvania* were there to help me with this book. In addition to these valuable friends and resources, hundreds of others in every state that has fracking have come forward with information—scientists, health professionals, environmentalists, economists, numerous persons who worked in the oil and gas industry, and individuals who have been affected by fracking; without them, *Fracking America* would not be possible.

In addition to interviews with more than 350 people, many willing to provide information, some reluctant, all of whose information I had to verify, I have read thousands of pages of government and court documents, academic research studies, corporate reports, and SEC filings.

As a journalist and media analyst, I also immersed myself in how reporters, "fractivists," and the industry use the media to present their message.

As always there have been some very special people who have made this book even better than I could do individually. Corey Ellen, MaryJayne Reibsome, Diana Saavedra, and Morris Stone of Greeley & Stone Publishers constantly amaze me with their abilities and willingness to make sure that excellence is the only standard acceptable.

A special thanks also goes to my wife, Rosemary R. Brasch, who read the manuscripts, commented on them, and made innumerable valuable suggestions. As always, I am indebted to my parents, Milton and Helen Haskin Brasch, for advice, wisdom, and love.

Fracking America is a fact-based overview of the issues surrounding the natural gas industry and fracking. As with

Fracking Pennsylvania, Fracking America is not meant to be an extensive analysis of the science and engineering of the process to extract natural gas nor a comprehensive discussion of the economic, health, environmental, and political issues. It is meant as a basic reference to acquaint people with the issues, with the hope they will dig deeper into areas that directly concern them and rally their friends and neighbors to help protect the health and environment of the people, wildlife, and natural vegetation.

—WALTER M. BRASCH

PHOTO: Gary F. Clark

Rig Platform in the Marcellus Shale

"The people have a right to clean air, pure water, and to the preservation of the natural, scenic, historic and esthetic values of the environment. Pennsylvania's public natural resources are the common property of all the people, including generations yet to come. As trustee of these resources, the Commonwealth shall conserve and maintain them for the benefit of all the people."
—*The Pennsylvania Constitution (Article I, section 27, 1972)*

"By any responsible account, the exploitation of the Marcellus Shale Formation will produce a detrimental effect on the environment, on the people, their children, and future generations, and potentially on the public purse, perhaps rivaling the environmental effects of coal extraction."
—Pennsylvania Supreme Court (December 2013)

Extracting water for a drill site in the Fayetteville Shale in Arkansas

PHOTOS: Bill Cunningham/USGS

Water storage tanks in the Fayetteville Shale

Introduction

There are two kinds of natural gas—biogenic and thermogenic. Biogenic gas, lying closer to the surface, is produced by methanogenic organisms as a metabolic byproduct. Thermogenic gas, deeper in the earth, is composed of organic decomposition that is affected by heat and pressure. The differences, depth of the gas, and the geology of the area all affect the fracking process.

Available Natural Gas

At the beginning of the 21st century, deep shale produced only about one percent of all natural gas; however, within a decade, it was producing about 30 percent.[1] About 45 percent of all natural gas produced in America is expected to be produced by fracking and horizontal drilling by 2035.[2]

The U.S. Energy Information Administration (U.S. EIA) estimates there may be as much as 7.3 trillion cubic feet of shale gas worldwide.[3] However, this includes gas that may not be profitable to be mined. Potential natural gas in China, which is at the verge of developing the technology, may be about 1,115 trillion cubic feet. The U.S. is fourth—behind China, Argentina, and Algeria and ahead of Canada and Mexico—in available shale gas. The first estimates of available shale gas in the United States suggested there were 2,560 trillion cubic feet, enough for a 100 year supply;[4] however, the high estimates often came from the industry itself and "include every bit of natural gas known or imagined, whether or not it is economically or technologically viable to recover," says Karen Feridun, founder of Berks Gas Truth. Revised estimates by the U.S. Geological Survey in 2013 revealed only 481 trillion cubic feet of recoverable gas, enough for less than a two decade supply, assuming full production.[5]

Nevertheless, natural gas could temporarily replace coal and oil, and serve as a transition to wind, solar, and water as primary energy sources, releasing the United States from dependency upon fossil fuel energy and allowing it to be more self-sufficient—but only if the federal government decreases its subsidy to the fossil fuel industry and increases support to industry and scientists working to increase renewable energy. By 2014, the United States became the world's largest producer of crude oil, shipping an average of about 8.5 billion barrels a day—up from 4.8 billion barrels a day in 2008—to refineries and export terminals.

Several shales in the United States contain natural gas—the Antrim Shale in Michigan, which in the 1980s became one of the first major shale plays in the United States;[6] Barnett and Barnett–Woodford shales in Texas and Oklahoma; Bakken shale in North Dakota and Saskatchewan; Fayetteville shale in Arkansas; Haynesville–Bossier shale in eastern Texas and northwest Louisiana; the Eagle Ford shale in southern Texas; the Pearsall shale, located below the Eagle Ford shale, at depths of about 8,000–11,000 feet, but which has minimal drilling; Mancos and Lewis shales in the San Juan Basin in New Mexico, Colorado, and parts of Utah; the Niobara shale in Colorado, Kansas, Nebraska, and Wyoming; the Monterey Shale in California's San Joaquin Basin, which yields primarily oil and some gas through conventional drilling methods; the South Georgia Basin in South Carolina, Georgia, and northern Florida; the New Albany Shale in the Illinois basin; the Woodford Shale in Oklahoma; the Chattanooga Shale in Tennessee; several smaller shales along the East Coast from Delaware into South Carolina; the South Newark Basin, which extends through parts of New Jersey and southern Pennsylvania; the Utica Shale, a deeper shale which ranges from parts of Ontario and Quebec in Canada, through parts of New York, Pennsylvania, Ohio, and West Virginia; and the Marcellus Shale, centered in Pennsylvania, but extending into New York, Ohio, and West Virginia.[7] The top three plays—Marcellus, Haynesville, and Barnett—account for about two-thirds of all gas production.[8]

The Marcellus Shale, which became the largest natural gas play in the United States in 2013,[9] extends beneath the Allegheny Plateau, through southern New York, much of Pennsylvania, east Ohio, West Virginia, and parts of Maryland

and Virginia. The Marcellus Shale, which lies on top of the Devonian and Utica shales, was created during the Middle Devonian epoch, about 400 million years ago.

The Marcellus Shale, also known as the Marcellus Formation, was named by geologist James Hall in 1839 for Marcellus, N.Y., a small village near Syracuse, where an outcrop of the shale was first discovered.[10] (An outcrop is a geological formation that is usually covered by soil and vegetation but is exposed by erosion.)

Recoverable gas in the Marcellus Shale is trapped between impervious layers of limestone between 40[11] and 900[12] feet thick, and between 4,000 and 8,500 feet deep.

In 2002, the U.S. Geological Survey (USGS) estimated there were about 1.9 trillion cubic feet of natural gas in the Marcellus Shale,[13] not substantial enough to allow profitable extraction by energy companies. However, the following year, Range Resources drilled the first well in Washington County, Pa., and other companies soon drilled exploratory wells.[14] The gas rush boom began about 2008 when Dr. Terry Engelder, professor of geosciences at Penn State, figured there might be about 500 trillion cubic feet of gas in the shale. That figure was probably artificially inflated. The Marcellus Shale may contain only about 141 trillion cubic feet of natural gas, about 30 percent of all available shale gas, according to the U.S. Energy Information Administration.[15] The USGS revised figure is about 84 trillion cubic feet of available natural gas.[16] Other estimates vary, but none are even half that provided by Dr. Engelder.

Contrary to some industry claims that the natural gas in all shales could provide energy for more than a century, the reality is that it may provide only enough for a couple of decades. The high-end estimates do not take into consideration that at present only 10–15 percent of all natural gas in the Marcellus Shale is viable; the costs to go after the rest of the available gas may not be financially justified if gas prices to consumers increase significantly.

In 2014, natural gas accounted for about 27 percent of the nation's electricity generation, behind coal (39 percent), and ahead of nuclear power (19 percent), hydropower (6 percent), other renewable (7 percent), and petroleum (1 percent).[17] Several states receive most of their electricity generation from coal; Ohio receives about two-thirds of its energy from coal.[18] However, petroleum-based products still provide considerably more fuel for transportation than do natural gas or solar energy.

The first commercial gas well was drilled near Fredonia, N.Y., in 1821, but it wasn't until the 1920s in Kentucky that shale gas production became economically feasible on a large scale.[19]

The first commercial well to use vertical fracking was drilled in the Hugoton gas field in Kansas by Halliburton in 1949.[20] Vertical fracking, which shatters shale rock to get methane gas, usually goes no more than 3,000 feet into the earth, and uses 20,000–80,000 gallons of water per conventional well.[21] To extract coalbed methane, about 50,000–350,000 gallons of water are required.[22]

Even with the newer vertical fracking technology, known as conventional drilling, vast resources of gas and oil are left in the earth because it isn't economically feasible to extract it. The development of high-pressure high-volume horizontal fracking, a combination of horizontal drilling and hydraulic fracturing, known as unconventional drilling, allows the fossil fuel industry to recover more gas and oil. Although conventional drilling brings up about 30–35 percent of the gas, unconventional drilling can effectively trap only about 4–6 percent of the gas, but results in greater volume.

The process of horizontal fracking extracts natural gas from low permeability sandstone/shale (the gas is also known as "tight gas") or from coal seams (the gas is known as coalbed methane.) About 65 percent of all gas extraction is done by horizontal fracturing.[23] The process requires a sequence of 10–32 separate stages.[24]

The procedure was first developed in the Barnett Formation in Texas by Mitchell Energy. George P. Mitchell, a petrochemical engineer, told *Forbes Magazine* in 2009, "My engineers kept telling me, 'You are wasting your money' . . . And I said, 'Well damn it, let's figure this thing out because there is no question there is a tremendous source bed that's about 250 feet thick.'"[25]

Dan Steward, geologist and former vice-president of Mitchell Energy, was one of the pioneers of the process that even the harshest critics acknowledge is the result of creative engineering. "Some people thought it [fracking] was stupid," Steward told the Associated Press in September 2012; even in his own company, "probably 90 percent of the people" didn't believe fracking could be possible or, if so, would not be profitable.[26] In an interview with the Breakthrough Institute, a liberal think tank with a focus on global ecological problems, Steward discussed the origins of fracking:

"In the seventies we started looking at running out of gas, and that's when the DOE [Department of Energy's] started looking for more.

"The DOE Eastern Gas Shales Project [in the Appalachia basin in 1976] determined there was a hell of a lot of gas in shales. It was the biggest accumulation of data and knowledge to date. It set the stage for people to have the basic background and caused people to start asking questions, and that's always important.

"They did a hell of a lot of work, and I can't give them enough credit for that. DOE started it, and other people took the ball and ran with it. You cannot diminish DOE's involvement. . . .

"Mitchell got involved in the Cotton Valley limestone looking for gas, but it was tight rock, and George [Mitchell] said, 'I want to frack it.' But he had a hard time to get his people to go along.

"Mitchell was interested in Barnett and his geophysicist said, 'It looks similar to the Devonian, and the government's already done all this work on the Devonian.' . . .

"In the 1990s they [the Department of Energy] helped us to evaluate how much gas was there, and evaluate the critical properties as compared to Devonian shale of Appalachia basin. They helped us with our first horizontal well. They helped us with pressure build-ups. And we worked with them on crack mapping. In 1999 we started working with GTI (formerly GRI) on re-fracks of shale wells. . . .

"[At] the time we started trying the Barnett, the thinking was we had to have open natural fractures. And so as we moved along we drilled wells and built the database.

"There was trial and error. Frequently that's what has to happen. You have to take best science and trial and error things. That's how Barnett [Shale] got started. . . .

"We tried on two wells one time and a third time three years later where we were trying this microseismic, and we thought

5

when they get the bugs worked out this is going to be [a] break. Until 1997 the bugs hadn't been worked out. . . .

"By the year 2000, Mitchell Energy had proven shale as workable and viable. The energy industry recognized it, but financial markets didn't recognize [it] until 2002, and politicians only realized it in 2006.

"In 2002 we actually started drilling horizontals in the Barnett as part of Devon Energy, which had bought Mitchell. They started with single fracks in the horizontals. And then from frack mapping they realized that we're not connecting with much of the rock. . . .

"It wasn't until 2006 and 2007 that we finally found the porosity, thanks to work done by the Bureau of Economic Geology, which I think gets state and federal funds. We knew there was porosity in the rock but couldn't see it with available technology. They experimented with ways to look at the rock at a higher resolution. And applied argon milling used in metallurgy where you could mill a surface on a piece of metal that was so clean that you could look at surface with high resolution."[27]

Kevin Begos of the AP reported:

"Energy Department researchers processed drilling data on supercomputers at a federal lab. Later, technology created to track sounds of Russian submarines during the Cold War was repurposed to help the industry use sound to get a 3-D picture of shale deposits and track exactly where a drill bit was, thousands of feet underground."[28]

Alex Crawley, former associate director of the Department of Energy's National Petroleum Technology Office, and one of the most important persons to have pushed to develop the new technology, told the AP, "It was a lot of pieces of technology that the industry thought would help them. Some worked out, some didn't."[29] But the technology carried risks, some of which the industry ignored or discarded. Less than a year before he died, George Mitchell told the *Economist* magazine:

"We know that there are significant impacts on air quality, water consumption, water contamination, and local communities. We need to ensure that the vast renewable resources in the United States are also part of the clean energy future, especially since natural gas and renewables are such great partners to jointly fuel our power production. Energy efficiency is also a critical part of the overall energy strategy that our nation needs to adopt."[30]

Mitchell, who had sold his company in 2002 for $3.5 billion, died in July 2013, having spent much of his life as an innovator and philanthropist.

At the University of London, Dr. Damien Short created the Extreme Energy Initiative (EEI) in 2013 to concentrate on the effects of unconventional fossil fuel extraction. The term, "extreme energy," was created in 2010 by Dr. Michael Klare, professor of peace and world security studies at Hampshire College. Extreme Energy is defined, according to the EEI, "as a category into which can be consigned all the new, more intensive and environmentally destructive energy extraction methods that have been spreading across the globe in recent years—methods such as 'tar sands' open-cast mining (also called the 'oil sands'), mountaintop removal, deep water drilling, shale gas extraction by hydraulic fracturing ('fracking') and coal bed methane (also called coal seam gasification), found at shallower depths near aquifers. Biofuels can also be considered 'extreme energy' as large-scale biofuel agricultural operations can re-move the potential for land to grow food."[31]

The EEI hosts conferences, workshops, seminars, and short courses; it will also initiate and develop peer-reviewed publish-able research about extreme energy, "bringing together scholars, practitioners, policy makers, and activists working on issues related to extreme energy production." says Dr. Short.

More than 15 million people in 11 states live within one mile of a gas well completed since 2000, according to a survey by the *Wall Street Journal*.[32]

Natural gas can be used to heat houses and retail businesses; it is a primary fuel in manufacturing industries, including metals refining, glass production, and pulp and paper pro-cessing.[33] Natural gas provides energy for a number of household appliances, including clothes dryers and ovens. About 54 percent of all single-family residences in the U.S. are heated by natural gas.[34] About 17 million vehicles world-wide, mostly buses and fleet vehicles, run on compressed natural gas (CNG),[35] which burns cleaner than gasoline, but has a lower fuel efficiency than hybrids. The cost to convert a car from gasoline to CNG can be $6,500–$12,000. About 250,000 cars, trucks, and buses in the United States have been adapted to run on CNG,[36] which can be about half the cost of gasoline derived from oil, depending up fluctuations in natural gas and oil markets. The U.S. has about 1,000 high-volume fueling stations, most of which are in California, New York, and Texas. A home compressor system, which can fill an average car in 10–20 hours, costs about $4,500.[37] However, research published in the February 14, 2014, issue of *Science*, cautions although natural gas as a fuel gives off fewer carbon dioxide emissions than oil or diesel fuel, mining natural gas by horizontal fracturing under current procedures negates any benefits.[38] Dr. Adam R. Brandt of Stanford University, who led a team of 16 scientists from Stanford, MIT, and the U.S. Department of Energy, told the *New York Times*, "Switching from diesel to natural gas [is] not a good policy from a climate perspective."[39]

The natural gas boom of the past half-decade, propped up by industry-favorable laws pushed by Republican-dominated legislatures, is primarily because the U.S. has determined it cannot be dependent upon foreign oil, natural gas burns cleaner and is less expensive than other fossil fuels, and the development of horizontal fracking.

Well-paying jobs became plentiful; however, most are temporary, ending when the gas companies declare a site no longer profitable. But, high volume hydraulic horizontal fracturing has left in its wake health and environmental issues that are as serious as those that surrounded the coal and oil industries.

The natural gas industry defends fracking as safe and efficient,[40] claiming fracking has been used more than 65 years. It's just one of the industry's half-truths. Fracking has been relatively safe—but it's vertical fracking, not horizontal fracking, which uses significantly more water, chemicals, sand, and elements, and poses significantly more dangers to the public health and environment than does vertical fracking.

Fracking doesn't necessarily have to be into the earth. Israel discovered vast natural gas fields in the Mediterranean. The

Tamar field is believed to be able to produce about 8.5 trillion cubic feet of gas; the Leviathan field may hold as much as 18 trillion cubic feet of gas. However, as the AP's Tia Goldenberg reports, "Selling this gas overseas will require Israel to navigate a geo-political quagmire that risks angering allies and enemies alike."[41]

Off the coast of California in the Santa Barbara channel, large oil platforms have been used as a base for fracking. However, such drilling is illegal; California banned fracking in the ocean after a disaster in 1969 spread more than three million gallons of crude oil, and may have triggered the beginning of the environmental movement. An Associated Press investigation, based upon documents obtained under the Freedom of Information Act, revealed that fracking was employed at least a dozen times in two decades since it was banned.[42] In August 2013, the California Coastal Commission, which protects the coast and marine life, even if drilling occurs in federal waters more than three miles off the coast, initiated an investigation to determine how the drilling could have occurred for more than two decades without oversight, and why it hadn't been notified of potential danger to the coast. The platforms have been "grandfathered," but EPA rules, in effect on new drilling as of March 1, 2014, require oil and gas companies to report all chemicals discharged into the ocean. In July 2014, Govs. Jay Inslee (Washington), John Kitzhaber (Oregon), and Jerry Brown (California) strongly urged the federal government to ban drilling off the West Coast. The governors told Secretary of the Interior Sally Jewell:

> "While new technology reduces the risk of a catastrophic event such as the 1969 Santa Barbara oil spill, a sizable spill anywhere along our shared coast would have a devastating impact on our population, recreation, natural resources, and our ocean and coastal dependent economies."[43]

Ten months after the three governors urged an end of off-shore drilling, a 24-inch onshore oil pipe burst, sending more than 100,000 gallons of oil onto 8.7 miles of beaches near Santa Barbara, Calif.;[44] about 20,000 gallons spilled into the Pacific Ocean, causing a 9.5 square mile slick that threatened marine life, shore birds, and a migrating seal population.[45] It took more than three hours for Plains All American Pipeline to shut off the flow.[46] The company had 175 federal safety and maintenance violations between 2006 and 2015, according to the Pipeline and Hazardous Materials Administration. In 2010, the EPA and the Department of Justice had ordered Plains to pay about $40 million to resolve 10 separate oil spills in Kansas, Louisiana, Oklahoma and Texas between 2004 and 2007.[47]

The Process of Horizontal Fracturing

The process George Mitchell had developed—and which he had warned of its health and environmental effects—requires drilling a bore hole into the earth's crust as deep as 12,000 feet; at that depth, temperatures can be as much as 250 degrees Fahrenheit.

After drilling down vertically past the cap rock, which traps oil and gas beneath it, the company sends fracking tubing, which has small explosive charges in it, to create a perforated lateral borehole, about 90 degrees from the vertical hole, which fractures the shale, a sedimentary bedrock, for up to about 6,000 feet to open channels and force out natural gas and fossil fuels. However, shale can be broken and gas released, but not brought to the surface, as much as 2,000 feet from the end of the pipeline.

A mixture of chemical additives, radioactive isotopes, proppants, and water is put into the tubes at a pressure of up to 15,000 pounds per square inch.

A typical well pad includes up to 30 separate wells. There are more than 500,000 active wells in the United States.[48] For the first few weeks, while the companies drill through the shale, a 120-foot tall derrick, known as a rig, identifies the location; after that, the rig comes down; the wells remain, now connected to pipelines and compressor stations. How long the rig remains for each well on the pad is determined by the geology of the shale formation; for Marcellus wells, the rig remains between 15 and 30 days,[49] moved to another well site on the pad, and then removed from the pad entirely prior to the beginning of the actual fracking process.

"Fracking turns solid bedrock into broken shards whose cracks become potential pathways for contamination, some of it radioactive. Broken shale is not reparable by any known technology," says Dr. Sandra Steingraber, a biologist and distinguished scholar in residence at Ithaca College.[50]

CHEMICAL ADDITIVES AND PROPPANTS

Chemical additives, most of them toxic and labeled as carcinogens "are used to prevent pipe corrosion, kill bacteria, and assist in forcing the water and sand down-hole to fracture the targeted formation," explains Thomas J. Pyle, president of the Institute for Energy Research.[51] To lubricate the drill, drillers use a chemical-laden solution they call "Mud," which often lies unprotected on large mats around the site, primarily to level the ground for the trucks.

Proppants, as much as 10,000 tons of silica sand[52] or bauxite, keep the fractures open to allow the gas to flow from the shale into the well bore.

The industry claims most of the ingredients in the fracking fluids are harmless, not unlike many food products. Dr. William Stringfellow, director of the Ecological Engineering Research Program at the University of the Pacific, disagrees. "You can't take a truckload of ice cream and dump it down the storm drain," Dr. Stringfellow said, pointing out, "Even ice cream manufacturers have to treat dairy wastes, which are natural and biodegradable. They must break them down rather than releasing them directly into the environment."[53]

About 650 of the 750 chemicals used in fracking operations are known carcinogens, according to a report filed with the U.S. House of Representatives in April 2011.[54] Almost two years later, the EPA listed more than 1,000 chemicals and elements—including acids, biocides, breakers, corrosion inhibitors, friction reducers, gelling agents, iron control chemicals, scale inhibitors, and surfactants[55]—that could be used in fracking fluid.[56] Among the most common chemicals reported in fracking fluids are benzene, which can lead to leukemia and several cancers, reduce white blood cell production in bones, and cause genetic mutation; formaldehyde, which irritates the eyes, ears, nose, and throat, and can lead to leukemia and genetic and birth defects; hydrogen sulfide, which attacks neurological and respiratory systems; nitrogen oxide and sulfur

11

dioxide, which can cause pulmonary edema and heart disease; radon, which has strong links to lung cancer; and hydrofluoric acid, which can cause genetic mutation and chronic lung disease; if it touches the skin, it can cause third degree burns, and affect bone structure, the central nervous system, and cause cardiac arrest.

A subsequent EPA analysis, based solely upon what the industry reported to the database FracFocus, revealed that in gas production fracking, hydrochloric acid was the most prevalent chemical in the additives, used in 73 percent of all disclosures. Among the other chemicals used were methanol (72 percent of disclosures), various distillates (70 percent), Isopropanol (47 percent), Ethanol (37 percent), Propargyl alcohol (34 percent), Glutaraldehyde (33 percent), and Ethylene Glycol (32 percent.) For oil production, Methanol was used in 72 percent of all fracking, followed by distillates (61 percent), Peroxydisulfuric acid (60 percent), Ethylene glycol (59 percent), Hydrochloric acid (58 percent), Guar gum (52 percent), and Sodium hydroxide (50 percent).[57]

Halliburton claims its CleanStim Hydraulic Fracturing Fluid System can replace many of the toxic chemicals;[58] however, the food-based product hasn't been used enough, or tested enough, to make a significant difference.

CASINGS

Drilling companies place concentric steel rings (casings) within the well bore and then pour cement into the bore to try to keep toxic chemicals deep in the earth from polluting the water sources they must drill through to get to the gas, and to reduce subsequent air and water pollution.

Cement well casings in Pennsylvania between 2010 and 2012 had an immediate failure rate of 6.95 percent, leading to pollution of ground water,[59] according to Dr. Anthony Ingraffea, professor of civil and structural engineering at Cornell University.

Pennsylvania in 2011 and Texas in 2013[60] improved regulations for well casing protection, but even with the tighter regulations there still are problems.

A study by Stephen Bachu and Theresa Watson found there

was as much as a 60 percent chance of gas leakage as wells age.[61] There is "no such thing as a perfect well seal," says Paul Hetzler, an environmental engineering technician for the New York Department of Environmental Conservation. "Occasionally sooner, often later, well seals can and do fail," he says.[62] Erosion caused by heat, pressure, and fluids are a primary reason why there can never be a perfect seal. The steel casings will rust out; the cement protection, subjected to stress, will begin to deteriorate within 10 years. Even Halliburton, which developed corrosion resistant cement systems "to endure high steam temperatures and low pH and still provide effective zonal isolation for longer than conventional cements," acknowledges it can extend the life of a well only five to ten years.[63]

About half of "squeeze jobs," attempts to repair well casings, fail. "Based on recent statistical evidence, one could expect at least 10,000 new wells with compromised structural integrity," Dr. Ingraffea wrote in a "white paper" about fluid migration and structural integrity of the boreholes.[64] "No matter what the industry tells you, their own data proves conclusively to any reasonable scientist or engineer that it is impossible to design any well so it will never leak," Dr. Ingraffea told a rally of Artists Against Fracking in August 2012.[65] Subsequent research conducted by Dr. Ingraffea and three others on data from about 75,000 wells concluded that unconventional wells in the Marcellus Shale "are at a 2.7-fold higher risk relative to the conventional wells in the same area."[66]

"The gas industry tells the public it's safe because they're extracting the gas so far below the ground surface that any contaminants they use will never make their way up to an aquifer," Dr. Patricia Culligan, professor of civil engineering and engineering mechanics at Columbia University, told *Columbia Magazine*.[67] If gas wells aren't properly prepared and sealed, "contaminants have a pathway to migrate into upper aquifers [and] once you've polluted an aquifer, it's almost impossible to undo the damage," says Dr. Culligan.[68]

Energy in Depth (EID)—the propaganda and disinformation organization formed in 2009 and funded by energy companies,[69] and whose mission includes opposing "new environmental regulations, especially with regard to hydraulic fracturing"[70]— claims evidence of a high failure rate of cement well casings is

13

faulty, and there were only 184 failures in more than 34,000 wells.[71] But, EID's data was for 1983 to 2007, before horizontal fracking began in the Marcellus Shale.

SACRIFICING PUBLIC INTEREST TO GET WATER

Most wells drilled since 2005 require between three and nine million gallons of fresh water for the first frack; a well may be fracked several times (known as "restimulation"), but most fracking after the first one is usually not economical. Between 2005 and 2013, shale gas companies used about 250 billion gallons of fresh water; Texas used about 110 billion gallons, followed by Pennsylvania (about 30 billion gallons), and Arkansas and Colorado (26 billion gallons each.)[72] In a few of the smaller, shallower shales, such as the Chattanooga Shale, as few as 200,000 gallons of water could be used. However, Seneca resources needed almost 19 million gallons of water to frack a well in northeastern Pennsylvania in 2012,[73] and Encana Oil & Gas USA used more than 21 million gallons of water to frack one well in Michigan the following year.[74] Between 2005 and 2014, gas extraction companies used about 250 billion gallons of water, according to a Duke University study.[75]

About 11 million gallons of water a day for fracking operations in Pennsylvania are taken daily from the Susquehanna River, which provides water for about 6.2 million people; about three times the current water withdrawal would occur if fracking could reach the peak the industry claims.[76] Between 2005 and 2013, Pennsylvania shale gas companies used about 30 billion gallons of water.[77] In contrast, the Delaware River Basin Commission (DRBC) has a moratorium on taking water from that river until health and environmental studies have been completed. The Delaware River Basin provides water to about 17 million people in a 14,000 square mile area. About half of all water for New York City is drawn from the river, the largest unfiltered water supply east of the Mississippi River.

Water is provided by companies that draw up to three million gallons a day from rivers and lakes, by individuals who sell water from their ponds, and by municipalities and water districts. Because of the chemicals added to the water, the methane that can't be extracted, and the chemicals and elements released from the earth during fracking, the water can't be used again, except to frack other wells.

A 2011 analysis showed that companies were using an average of 2.4–7.8 million gallons of water per unconventional (fracked) well;[78] the amount is based upon the geology of the basin. Pennsylvania energy companies use an average of 4.0–5.6 million gallons of water per frack, higher than wells in the Barnett Shale (2.3–3.8 million gallons), Fayetteville Shale (2.9–4.2 million gallons), and Haynesville Shale (2.7–5.0 million gallons), according to data compiled by the Groundwater Protection Council and ALL Consulting.[79]

Beginning about 2008, the water in the nation's aquifers has been decreasing significantly. The depletion of the rivers, lakes, and aquifers is because of population growth, higher usage, climate change, and a severe drought that has spread throughout California, the Southwest, and the Midwest. This depletion, according to Leonard Konikow, a research hydrologist at the U.S. Geological Survey, is about three times the rate as between 1900 through 2008.[80] About 47 percent of all fracking is in high or extremely high water stress areas.[81]

Significant reductions in water availability are now common for the 1,450 mile long Colorado River, which provides water to about 40 million people in California and the southwest, including the agriculture-rich Imperial Desert of southeastern California. The basin lost about 53 million acre feet of freshwater between December 2004 and November 2013, according to research conducted by NASA and scientists at the University of California at Irvine.[82] Lake Mead, a part of the Colorado system, provides water to Las Vegas and the Nevada desert communities; its water level dropped about 140 feet, forcing marinas to move and allowing the desert to reclaim some of its land; the U.S. Bureau of Reclamation expects the water level to drop another 20 feet by Summer 2016.[83] The drop in water is close to the point where the U.S. Department of the Interior may declare a water shortage and impose strict water-use regulation.[84]

The drought in California became so severe that Gov. Jerry Brown, by executive order in April 2015, ordered the state's 400 water suppliers to reduce urban water distribution by 25 percent. Among those affected were colleges, golf courses, cemeteries, and residents using water for lawns. Exempt were the state's agriculture industry, which uses about 80 percent of all water, and oil/gas companies using water for fracking.[85] The oil and gas industry in California uses almost 800 million gallons of water a year.[86] However, the industry is now selling recycled chemical-laden wastewater to farmers.[87]

Water is so critical to fracking that during the drought in the Midwest, oil and gas companies bought water from farmers and municipalities and trucked it in from as far away as Pennsylvania.[88] Agriculture fields and, sometimes, livestock suffered because the industry needed water, and was paying premium prices, as much as $1,000–$2,000 for about 326,000 gallons (an acre foot);[89] the normal price is about $30–$100 for the same amount. Energy companies are "going to pay what they need to pay," said Dr. Reagan Waskom, director of the Colorado Water Institute at Colorado State University.[90] In most cases, the natural gas industry can outbid farmers for water supplies.

The need for water has also led to significant fish kills. In testimony before a Senate committee in October 2011, Katy Dunlap of Trout Unlimited revealed how little the shale gas industry cares about aquatic life:

"Horton Run, a tributary of the East Fork of Sinnamahoning Creek and classified as an "Exceptional Value" trout stream, was virtually de-watered by water withdrawals for gas well development. Fish kills have occurred as a result of water withdrawals that de-watered Cross Creek and Sugarcamp Creek in Washington County, Pa. Four gas companies have paid a total of $1.7 million to settle charges of illegal water withdrawals from Pennsylvania trout streams, including Chief Oil & Gas, which took 3.5 million gallons from a tributary of Larry's Creek, and Range Resources, which took 2.2 million from Big Sandy Run. Additionally, water withdrawals have damaged Meshoppen, Pine and Sugar creeks. These examples clearly demonstrate the risk that water withdrawals from small headwater streams pose to aquatic habitat."[91]

The Coalition for Environmentally Responsible Economies (CERES), basing its analysis upon more than 25,000 wells, reports almost 47 percent of wells that use fracking technology were developed in areas with high or extremely high water stress levels; 92 percent of all gas wells in Colorado are in extremely high-stressed regions; In Texas, 51 percent are in high or extremely high stress water regions.[92]

Energy companies drilling in Texas use the most water, followed by Pennsylvania, Colorado and Arkansas.[93]

Suzanne Goldenberg reported the problem in August 2013 for readers of the *Manchester Guardian*'s U.S. edition:

> "Beverly McGuire saw the warning signs before the well went dry: sand in the toilet bowl, the sputter of air in the tap, a pump working overtime to no effect. But it still did not prepare her for the night last month when she turned on the tap and discovered the tiny town [of Barnhart, Texas] where she had made her home for 35 years was out of water. . . .
>
> "Even as the drought bore down, even as the water levels declined, the oil industry continued to demand water and those with water on their land were willing to sell it. The road west of town was lined with signs advertising 'fresh water', where tankers can take on a box-car-sized load of water laced with industrial chemicals."[94]

An additional problem occurs in Wisconsin, Minnesota, and the Midwestern states where silica sand is mined to send to the fracking sites. Mike Ludwig of *TruthOut* reports that a large sand processing plant near Maiden Rock, Wisc., "can suck up to 1.3 million gallons of water a day out of a local aquifer to clean frack sand before it's loaded onto waiting trains."[95] He contrasts that with the Prairie Island nuclear generating plant in nearby Red Wing, Minn., which uses about 1.6 million gallons of water a day; the plant can provide energy for about one million homes.[96] Water for nuclear plants is mostly recyclable; water to wash sand isn't.

The drought has had one beneficial consequence. Several companies, including Water Rescue Services and Pure Stream, are developing technologies to recycle wastewater for additional use by separating the toxins and radioactive elements.

A newer significantly more expensive technique is being developed that could reduce water usage. Propane Hydraulic Fracturing uses large volumes of liquid propane and smaller quantities of butane mixed with phosphoric acid to create a gel, which is then mixed with sand and several compounds. Gary P. Hoffman, of Applied Thermodynamics, explains:

"This gelled hydraulic fluid is then forced down the perforated well bore exactly in the same manner as the slick water [low viscosity water, with less sand and chemicals] is under the conventional method. Once the rock is fractured, the high pressure is released. This causes the gel to flash back into gaseous propane. The propane gas, and the methane gas, now come roaring back up the well bore as a mixed, high explosive gas. Here they are controlled, it is hoped, by a blow-out preventer, and are gradually separated into the two gasses."[97]

Lineal and crosslinked gels reduce water usage, but some of the same problems of slick water drilling remain.

Another method to extract gas is known as an "acid job," which uses high volumes of chemicals that are poured into a previously dug well to dissolve rocks. Much of the drilling in the Monterey Shale in central California, which underlies one of the most productive agricultural regions in the country uses acid jobs.[98]

Although "acid jobs" require only a few thousand gallons of water, as opposed to millions of gallons in horizontal fracking, there is the potential for significantly greater health problems. "These are super-hazardous poisonous chemicals and we have no idea what they are doing out there with it—how deep it is going, the volumes—nothing," Bill Allayaud of the Environmental Working Group, told Reuters in May 2013.[99]

POST-DRILLING

About half of all water and various elements and chemicals—known variously as wastewater, flowback, blow-back, or brine, which is saltier than seawater—is brought to the surface and must be disposed. The Government Accountability Office reports that the wastewater "drilled at depths generally ranging from 5,000 to 8,000 feet have salt and mineral levels 20 times higher than produced water from coalbed methane wells drilled at depths of 1,000 to 2,000 feet."[100]

In most drilling operations, this flowback, as well as solid waste, liquid waste, and sludge can be stored in ponds; Earthworks measured ponds as much as 200 feet by 120 feet by 14 feet deep, with a capacity of about 2.5 million gallons of waste.[101] This waste is then pumped into tanker trucks for disposal at sewage facilities or injected into deep wells. In some cases, it just lies in uncovered open ponds. The other half of the mixture stays within the ground, sometimes affecting water supplies; however, that half, contaminated with fracking chemicals, will eventually come to the surface. Unlike water used in agriculture, fracking wastewater can never be recycled or reused.

Compressor stations prepare natural gas to be pressurized to move through pipes, some extending up to 100 miles, to the next station.

The mined gas is transported by truck and by more than 1.5 million miles of pipelines.[102] Extensive truck traffic to bring water, fracking fluids, and equipment to the drilling sites and take gas and wastewater away have clogged public roads and highways, often blocking local residents from using those roads to leave or enter their own properties. Environmentalist Robert F. Kennedy Jr. points out that nationally, "The Industry now acknowledges that it absolutely cannot afford to pay localities the costs of roads damaged from the thousands of truck trips per wellhead, leaving those ruinous costs to local taxpayers."[103]

Because the term "fracking" is commonly used to refer to the entire process rather than one part, the natural gas industry has become adept at trying to convince the public that fracking isn't as dangerous as people believe, conveniently leaving out other parts of the process that includes not just drilling, but extracting, refining, transporting, and storing.

Also not included in the industry's tight and self-serving definition of "fracking" are worker accidents, health and environmental effects.

WHEN A WELL RUNS DRY

Within two to three years of drilling a well, the cost to mine natural gas by horizontal fracking becomes less profitable. The solution is to "cap" the well. There may be as many as one million abandoned wells in the United States, according to the Interstate Oil and Gas Commission.[104] However, not all wells are capped. An Associated Press investigation revealed there were about 3,200 uncapped wells in the Gulf of Mexico.[105]

Texas has about 8,400 known uncapped wells.[106] In New York, which has only vertical wells, about 5,000 are uncapped.[107] Uncapped wells can allow brine, which can include naturally-occurring toxic and radioactive elements, to come to the surface and pollute groundwater systems. Even wells that are capped (also known as plugged) can be a problem. Most are capped by cement, which deteriorates.

Pennsylvania charges companies $10,000 per well to assure the company caps the well when it ceases to be productive. However, "even $50,000 is a very optimistic number," says Dr. Austin Mitchell, an engineer, who has analyzed costs of capping wells.[108] Many of the companies that drilled the land took immediate profits and later declared bankruptcy, avoiding the costs to cap the wells or clean up toxic spills. Most forfeited the bond, knowing the clean-up and capping costs would far exceed what they had paid as assurance they would cap the wells.

Underground Storage

The U.S. currently has about two trillion cubic feet of natural gas in about 400 underground storage sites.[109] About 80 percent of all underground storage is in depleted gas or oil fields; the rest are in aquifers or salt caverns. However, in the past two decades, the number of salt cavern storage sites has grown steadily, mostly in the Gulf Coast areas.[110]

A sinkhole near Bayou Corne, La., about 80 miles west of New Orleans, is one of the reasons the residents of the Finger Lakes may be worried. The walls of an underground three square mile salt cavern,[111] which Texas Brine had been mining for the fracking industry until 2011, collapsed in August 2012. The U.S. Geological Survey determined that the mining pro-bably caused several micro-earthquakes that had weakened the walls.[112] The collapse and release of methane gas forced the state to issue a mandatory evacuation for about 300 residents living in about 150 homes. The depth of the sinkhole was originally 449 feet at its deepest part.[113] Although the depth decreased because of collapsing walls and earth slides, the surface dimensions continued to expand; by mid-2015, it was 32.5 acres.[114] To rid the area of the gas and its escape into local aquifers and swamps, Texas Brine built several vents to burn off the escaping gas, leading to additional air pollution.[115] Among other health problems are the presence of the toxic benzene and diesel fuels in the release of oil.[116]

Polluting the Environment

The United Nations Global Environmental Alert Services (GEAS) concluded in November 2012 that horizontal fracking "may result in unavoidable environmental impacts even if UG [underground gas] is extracted properly, and more so if done inadequately."[117] The report further concluded:

"Even if risk can be reduced theoretically, . . . in practise many accidents from leaky or malfunctioning equipment as well as from bad practises are regularly occurring. This may be due to high pressure to lower the costs or to improper staff training, or to undetected leaks leading to contamination of the ground water. . . .

"Existing laws and regulations of the mining activities often do not address specific aspects of hydraulic fracturing. For governments who choose this path, UG will require dedicated regulations. . . .

"The debate on UG exploitation cannot be disassociated from a 'comeback' of fossil fuels. UG is and will be produced by the same actors. Although only very recent, the history of UG exploitation already includes instances of water contamination, leakages to soil, wide-scale land clearing and negative health impacts. Furthermore, increased extraction and use of UG is likely to be detrimental to efforts to curb climate change. Given the increased demand for fossil energy, the UG may be used in addition to coal, rather than being a sub-stitute. Even under the optimistic assumption of the sub-stitution of coal by UG, UG will likely have a limited reduction impact on 21st century global warming. The claim that UG can reduce GHG emissions is conditional on whether UG, over its entire life-cycle, is demonstrated to have a much lower GWP than coal.

"Given the ever-increasing demand for energy, UG use is likely to grow. With large gas reserves and the comparative advantage of using existing infrastructures, equipments and networks from the oil and gas industry (drilling equipment, pipelines, thermal power stations, etc) UG will remain a tempting power source for the industry and for some governments who want to decrease their foreign dependency on energy. However it will face strong opposition given low public acceptance in certain places. As a non-renewable source of energy, UG remains a stop-gap measure in the transition to a low carbon future. In order to develop energy plans that maximize benefits and minimize harm, other forms of energy will also be needed. . . .

"New technologies and or energy supplies, such as biofuels or UG, are often greeted as a panacea, but under further investigation are revealed to be less ideal than originally thought. Further research and appropriate, transparent and well-enforced regulation are all critical to possible development of the unconventional gas industry."[118]

Dr. Stephen Cleghorn, a sociologist and farmer in Jefferson County, Pa., says the industry is still "figuring it out." He says "they have models [but] they don't really know. . . . It's being invented as they go." Even some industry executives now acknowledge fracking is not safe. The first years of fracking in Pennsylvania "was not the industry's finest hour," said Mark Bolling, executive vice-president of Southwest Energy.[119] The CEO of Schlumberger—identified by Bloomberg News as "the world's largest oilfield-services provider"—agrees. During a conference call with investors in February 2010, Andrew Gould acknowledged, "At the moment, we're doing it by brute force and ignorance."[120]

The industry is still experimenting.

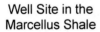

Well Site in the Marcellus Shale

PHOTO: Ken Skipper/USGS

PART I:
Historical, Political, and Economic Issues

The forests that once dominated Pennsylvania
were destroyed during the lumber boom era
of the second half of the nineteenth century.

PHOTO: Department of Energy

The Shippingport Atomic Power Station,
about 35 miles northwest of Pittsburgh, Pa.,
was the nation's first nuclear power plant
developed solely for peaceful use.

CHAPTER 1

A Brief History of America's Energy Policies

The history of energy exploration, mining, and delivery is best understood in a range from benevolent exploitation to worker and public oppression. A company comes into an area, leases or buys land in rural and agricultural areas for mineral rights, increases employment, usually in a depressed economy, strips the land of its resources, creates health problems for its workers and those in the immediate area, and then leaves.

It makes no difference if it's timber, oil, coal, nuclear, or natural gas. All energy sources are developed to move mankind into a new era; all energy sources are developed to bring as much profit to corporations as quickly as possible, often by exploiting the workers.

Before the settlement of Pennsylvania in the 1680s, more than 20 million acres of forests covered almost all of the land. During the latter half of the nineteenth century, the lumber industry had clear-cut several million acres, leading Pennsylvania into an era that almost rivaled the Gold Rush in California. By World War I, the companies had stripped the land, taken their profit, and then moved on, leaving devastation in their wake. Only when the people finally realized that destroying the forests led to widespread erosion and flooding did they begin to reforest the state. Almost a century after the lumber companies denuded the forests, the natural gas industry, with encouragement from the state, leased more than 150,000 acres of forests for wells, pipelines, and roads.[121]

In 1859, Edwin Drake successfully drilled the world's first commercially-successful oil well near Titusville, Pa. By 1881, the region in northwestern Pennsylvania was producing more than three-fourths of the world's oil.[122] In the early days of oil exploration, John D. Rockefeller established Standard Oil, which would become the largest monopoly in the world. However, the boom began a decline that year, ending about 1901;[123] by 1907, California, Texas, and Oklahoma had surpassed Pennsylvania in oil production.

Between 1859 and 1933, the beginning of Franklin D. Roosevelt's "New Deal," Pennsylvania, under almost continual Republican administration, was among the nation's most corrupt states.[124] The robber barons of the timber, oil, coal, steel, and transportation industries, enjoying and contributing to the Industrial Age of the 19th century, bought their right to be unregulated. In addition to widespread bribery, the energy industries, especially coal, assured the election of preferred candidates by giving pre-marked ballots to workers, many of whom were immigrants and couldn't read English.

When the coal companies determined underground mining was no longer profitable, they began strip mining, shearing the tops of hills and mountains to expose coal, causing environmental damage that could never be repaired even by the most aggressive reforestation programs. Pennsylvania is the only state producing anthracite coal; it is fifth in the nation in production of all coal, behind Wyoming, West Virginia, Kentucky, and Illinois.[125]

John Wilmer, an attorney who formerly worked in the Pennsylvania Department of Environmental Protection (DEP), in a letter to the editor of *The New York Times* in March 2011, explained that "Pennsylvania's shameful legacy of corruption and mismanagement caused 2,500 miles of streams to be totally dead from acid mine drainage; left many miles of scarred landscape; enriched the coal barons; and impoverished the local citizens."[126] His words are a warning about what is happening in the oil and natural gas fields throughout the country.

Three years later, the Pennsylvania Supreme Court, in a decision that reaffirmed the rights of the people and the protection of their environment over the greed of politicians and the oil and gas industry, in a ruling that could be applied to any state, declared:

"Pennsylvania has a notable history of what appears retrospectively to have been a shortsighted exploitation of its bounteous environment, affecting its minerals, its water, its air, its flora and fauna, and its people. The lessons learned from that history led directly to the Environmental Rights Amendment [to the Pennsylvania Constitution], a measure which received overwhelming support from legislators and the voters alike. When coal was "King," there was no Environmental Rights Amendment to constrain exploitation of the resource, to protect the people and the environment, or to impose the sort of specific duty as trustee upon the Commonwealth as is found in the Amendment. Pennsylvania's very real and mixed past is visible today to anyone travelling across Pennsylvania's spectacular, rolling, varied terrain. The forests may not be primordial, but they have returned and are beautiful nonetheless; the mountains and valleys remain; the riverways remain, too, not as pure as when William Penn first laid eyes upon his colonial charter, but cleaner and better than they were in a relatively recent past, when the citizenry was less attuned to the environmental effects of the exploitation of subsurface natural resources. But, the landscape bears visible scars, too, as reminders of the past efforts of man to exploit Pennsylvania's natural assets. Pennsylvania's past is the necessary prologue here: the reserved rights, and the concomitant duties and constraints, embraced by the Environmental Rights Amendment, are a product of our unique history.[127]

The nation's first commercial nuclear power plant to develop peaceful uses of energy was the Shippingport Atomic Power Station, along the Ohio River, about 35 miles northwest of Pittsburgh. The plant went online in December 1957 and stayed in production through October 1982. Pennsylvania is second in the nation, behind Illinois, in production of electricity from nuclear reactors.[128] During the last four decades of the twentieth century, the nation built 132 nuclear plants, with politicians and Industry claiming nuclear energy was clean, safe, efficient, and would lessen the nation's ties to oil. Chernobyl, Three Mile Island, Fukushima Daiichi, and thousands of violations issued by the Nuclear Regulatory Commission, have shown that even with strict operating guidelines, nuclear energy isn't as clean, safe, and as efficient as claimed. Like all other energy industries, nuclear power isn't infinite. Most plants have a 40–50 year life cycle. After that, the plant becomes so radioactive that it must be sealed.

Nuclear energy has several advantages over fossil fuels—it uses relatively little fuel to create considerable energy, the water a plant uses is mostly recyclable, and the new Westinghouse AP100 model reactor has significant safety features most current plants don't have. One other advantage: nuclear power plants don't contribute to global warming. One major disadvantage: Adequate nuclear waste disposal is still a problem.

In the early 21st century, the natural gas industry follows the model of the other energy corporations, and uses the same rhetoric. The Heartland Institute, a politically-conservative think tank that says it exists to "promote free-market solutions to social and economic problems," claims, "Shale extraction has proven remarkably safe for the environment and the newfound abundance of domestic natural gas reserves promises unprecedented energy prosperity and security."[129] Its claims are largely inaccurate.

Laws to Protect Health and Environment

THE NIXON AND FORD ADMINISTRATIONS

The National Environmental Policy Act (1970),[130] Clean Air Act (1970),[131] the Clean Water Act (1972),[132] and the Safe Water Drinking Act (1974)[133] were all enacted under the Richard M. Nixon administration.

The Resource Conservation and Recovery Act (RCRA),[134] passed in 1976 during the Gerald Ford administration (1974–1977), established a strict protocol for hazardous waste management. Four years later, Congress specifically exempted wastes from natural gas and oil production, requiring the EPA to prove that such waste was detrimental to health. The EPA gave the states authority to regulate the control of waste. In essence, interstate transportation of waste went unregulated.

In 1980, Congress passed two major bills to specifically benefit the development of natural gas industry. First, it established significant tax breaks to encourage drilling of unconventional wells; second, it passed the Comprehensive Environmental Response, Compensation, and Liability Act (CERCLA)[135] that created the "superfund" for cleaning hazardous waste, and held the energy industry financially accountable for hazardous waste that could affect health. However, Congress specifically exempted oil and natural gas industries from CERCLA.

THE REAGAN ADMINISTRATION

In 1988, the last year of the Reagan–Bush administration, Congress exempted oil and gas hazardous waste from oversight by the RCRA.

A *New York Times* article in March 2011 summarized what had happened during the Reagan years:

> "When Congress considered whether to regulate more closely the handling of wastes from oil and gas drilling in the 1980s, it turned to the Environmental Protection Agency to research the matter. E.P.A. researchers concluded that some of the drillers' waste was hazardous and should be tightly controlled.
>
> "But that is not what Congress heard. Some of the recommendations concerning oil and gas waste were eliminated in the final report handed to lawmakers in 1987. 'It was like the science didn't matter,' Carla Greathouse, the author of the study, said in a recent interview. 'The industry was going to get what it wanted, and we were not supposed to stand in the way.'
>
> "E.P.A. officials told her, she said, that her findings were altered because of pressure from the Office of Legal Counsel of the White House under Ronald Reagan."[136]

THE CLINTON ADMINISTRATION

Under the Bill Clinton administration (1993–2001), "national environmental targets were made more stringent, and environmental quality improved," according to an 84-page report by Sheila M. Cavanagh, Robert W. Hahn, and Robert N. Stavins, and published in September 2001.[137] According to the authors:

> "Most important among the new targets were the National Ambient Air Quality Standards (NAAQS) for ambient ozone and particulate matter, issued by EPA in July 1997, which could turn out to be one of the Clinton Administration's most

enduring environmental legacies. Also, natural resource policy during the Clinton years was heavily weighted toward environmental protection. Environmental quality improved overall during the decade, continuing a trend that began in the 1970s, although improvements were much less than during the previous two decades.

"Second, the use of benefit-cost analysis for assessing environmental regulation was controversial in the Clinton Administration, while economic efficiency emerged as a central goal of the regulatory reform movement in the Congress during the 1990s. When attention was given to increased efficiency, the locus of that attention during the Clinton years was the Congress in the case of environmental policies and the Administration in the case of natural resource policies. Ironically, the increased attention given to benefit-cost analysis may not have had a marked effect on the economic efficiency of environmental regulations.

"Third, cost-effectiveness achieved a much more prominent position in public discourse regarding environmental policy during the 1990s. From the Bush Administration through the Clinton Administration, interest and activity regarding market-based instruments for environmental protection—particularly tradeable permit systems—continued to increase.

"Fourth, the Clinton Administration put much greater emphasis than previous administrations on expanding the role of environmental information disclosure and voluntary programs. While such programs can provide cost-effective ways of reaching environmental policy goals, little is known about their actual costs or effectiveness.

"Fifth and finally, the Environmental Protection Agency placed much less emphasis on economic analysis during the 1990s. EPA leadership was more hostile to economic analysis than it had been under the prior Bush Administration, and it made organizational changes to reflect this change in priorities."[138]

THE BUSH–CHENEY ADMINISTRATION

The energy policy during the eight years of the George W. Bush–Dick Cheney administration (2001–2009) gave favored status to the Industry, often at the expense of the environment. In addition to negating Bill Clinton's support for the Kyoto Protocol to reduce greenhouse-gas emissions, an act signed by 191 countries, former oil company executives Bush and Cheney pushed to open significant federal land to drilling. Included in the proposal was the 19 million acre Arctic National Wildlife Refuge (ANWR); drilling in the ANWR would disrupt the ecological balance in one of the nation's most pristine areas. The Bureau of Land Management also approved about one-fourth of all corporate requests to avoid any required public comment for land development in the western United States.[139]

An EPA study published in 2004 concluded that fracking was of little or no risk to human health.[140] However, Wes Wilson, a 30-year EPA environmental engineer, in a letter to members of Congress and the EPA inspector general, called that study "scientifically unsound," and questioned the bias of the panel, noting that five of the seven members had significant ties to the industry. "EPA's failure to regulate [fracking] appears to be improper under the Safe Water Drinking Act and may result in danger to public health and safety," he wrote.[141]

The following year, the Energy Policy Act of 2005[142] (P.L. 109-58)—by a 249–183 vote in the House and an 85–12 vote in the Senate—exempted the oil and natural gas industry from the Safe Water Drinking Act, the Clean Water Act, and the National Environmental Policy Act.[143] That exemption applied to new well pads and the accompanying new roads and pipelines, allowing the newly-developed method of horizontal fracking to proceed, and opening significant investment by individuals, corporations, and financial institutions. The Natural Resources Defense Council noted that the EPA interpreted the exemption "as allowing unlimited discharges of sediment into the nation's streams, even where those discharges contribute to a violation of state water quality standards."[144] Vice-President Dick Cheney, whose promotion of Big Business and opposition to environmental policies is well-documented, had pushed for that exemption. His hand-picked "energy task force," composed primarily of industry representatives, had concluded fracking was a safe procedure. Cheney had been CEO of Halliburton, one of the world's largest energy companies, with headquarters in Houston and Dubai; the exemption became known derisively as the Halliburton Loophole. That legislation, says former Vice-President Al Gore, "put the whole industry in such a privileged position [that] it disadvantages the advocates of the public interest, which was the intention."[145]

Robert F. Kennedy Jr., an environmental lawyer and activist for more than three decades, was blunt in his assessment of the natural gas industry's lobbying and propaganda machine: "The industry's worst actors have successfully battled reasonable regulation, stifled public disclosure while bending compliant government regulators to engineer exceptions to existing environmental rules."[146]

George P. Mitchell, the innovator whose company developed horizontal fracking, agreed with Kennedy. "The [Obama] administration is trying to tighten up controls," Mitchell told *Forbes Magazine*, and emphasized, "They should have very strict controls. The Department of Energy should do it."[147] In an exclusive interview with *Economist* magazine, Mitchell later strengthened his belief for regulation:

"A strong federal role is also necessary, starting with the Environmental Protection Agency's new rules calling for more controls over the most dangerous air pollution associated with hydraulic fracturing. The rules will also mitigate methane leakage during the drilling process."[148]

An editorial in the Oct. 19, 2011, issue of *Scientific American* pointed out the reality of the lack of adequate regulation, and the willingness of states to embrace the natural gas industry:

"Fracking is already widespread in Wyoming, Colorado, Texas and Pennsylvania. All these states are flying blind. A long list of technical questions remains unanswered about the ways the practice could contaminate drinking water, the extent to which it already has, and what the industry could do to reduce the risks."[149]

Scientific American recommended the federal government establish "common standards" and that "[S]tates should put the brakes on the drillers" until the EPA completes a preliminary study of the effects of the environmental effects of fracking.[150] The agency won't be looking at health impacts.

Bills introduced in the U.S. House (H.R. 2766[151]) by Reps. Diana DeGette (D-Colo.), John Salazar (D-Colo.), and Maurice Hinchey (D-N.Y.); and the U.S. Senate (S. 1215[152]) by Sens. Robert P. Casey Jr. (D-Pa.) and Chuck Schumer (D-N.Y.) in

June 2009 would have given federal authority under the Safe Water Drinking Act to regulate hydraulic fracturing; however, the bills languished. New bills (H.R. 1084[153] and S. 587[154]), introduced in March 2011 in the 112th Congress, also died without a vote. In May 2013, Reps. DeGette and Christopher Gibson (R-NY) reintroduced legislation (HR 1921[155]), known as the Fracturing Responsibility and Awareness of Chemicals Act (FRAC Act). That bill was referred to the Subcommittee on Environment and the Economy, where it died.[156]

OBAMA ENVIRONMENTAL POLICIES

President Barack Obama has repeatedly spoken against the heavy use and dependence upon fossil fuels, and once saw the use of natural gas as a transition fuel to expanded use of wind, solar, and water energy. In his first month in office, he pushed through an $831 billion stimulus package to help bring the country out of the deep recession of the previous three years; in that bill was $80 billion for renewable energy projects.[157] However, political realities may have changed some of his pre-presidential beliefs.

Hillary Clinton, his secretary of state, promoted the use of natural gas within foreign countries. In 2010, Clinton told a meeting of foreign ministers, "Natural gas is the cleanest fossil fuel available for power generation today."[158] One reason for the Obama/Clinton push for natural gas exploration and distribution in overseas countries is because geopolitics plays "a significant role in whether a number of gas projects are realized and come online and where pipelines are built. . . . Individual country decisions about natural gas resources can have dramatic impacts on responses in international discourse," according to a research analysis published by the Kennedy School of Government at Harvard University.[159] Amy Myers Jaffe and Dr. Meghan L. O'Sullivan, co-editors of the study, also pointed out, "The relative fortunes of the United States, Russia and China—and their ability to exert influence in the world—are tied in no small measure to global gas developments and vice versa."[160] Gazprom, Russia's state-owned energy corporation, and the world's largest exporter of natural gas, had a $38 billion profit in 2012.[161] Two years later, Gazprom signed a $400 billion agreement to deliver 38 billion cubic meters of natural gas annually to China, beginning in 2018.[162]

In his January 2012 State of the Union, the President advocated use of all available energy sources and the development of natural gas exploration, which "will create jobs and power trucks and factories that are cleaner and cheaper, proving that we don't have to choose between our environment and our economy."[163]

In April, he issued an executive order to coordinate oversight of the natural gas industry. The order affected White House offices and cabinet departments, including the EPA. The natural gas industry and several Republican leaders had previously criticized the President for what they believed was his opposition to natural gas development, and had lobbied heavily for such an order to reduce what they saw as overlapping jurisdictions and over-regulation of the Industry.

"We are pleased to see this action today, which will help promote consistency between the Administration and policies that are put in place [and will improve] government communication and coordination that will help our members continue to safely deliver this foundation fuel to 177 million Americans every day,"[164] said Dave McCurdy, president of the American Gas Association. He also noted that the President "has been promoting responsible production and broader use of this domestic, abundant, affordable, clean and reliable energy source." Favorable statements also came from the Marcellus Shale Coalition, America's Natural Gas Alliance, American Petroleum Institute, Independent Petroleum Association of America, American Chemistry Council, and the Dow Chemical Co.[165] Even with an industry-favorable decision by the President, there was still sniping from the right-wing. "We don't need another working group, or any more bureaucracy," a spokesman for House Speaker John Boehner (R-Ohio) said in a prepared statement.[166]

President Obama later shunted aside his earlier belief that natural gas could be a bridge fuel, a temporary solution to the energy crisis. In his 2013 State of the Union address, while calling for better research and technology, he had declared:

> "[T]he natural gas boom has led to cleaner power and greater energy independence. We need to encourage that. And

that's why my administration will keep cutting red tape and speeding up new oil and gas permits."[167]

President Obama's declaration would be the prelude for his natural gas policy as part of the Climate Change speech delivered five months later:

> "Now, even as we're producing more domestic oil, we're also producing more cleaner-burning natural gas than any other country on Earth. And, again, sometimes there are disputes about natural gas, but let me say this: We should strengthen our position as the top natural gas producer because, in the medium term at least, it not only can provide safe, cheap power, but it can also help reduce our carbon emissions.
>
> "Federally supported technology has helped our businesses drill more effectively and extract more gas. And now, we'll keep working with the industry to make drilling safer and cleaner, to make sure that we're not seeing methane emissions, and to put people to work modernizing our natural gas infrastructure so that we can power more homes and businsses with cleaner energy.
>
> "The bottom line is natural gas is creating jobs. It's lowering many families' heat and power bills. And it's the transition fuel that can power our economy with less carbon pollution even as our businesses work to develop and then deploy more of the technology required for the even cleaner energy economy of the future."[168]

The President, perhaps relying more upon pro-fracking proponents than scientific literature, was inaccurate. While natural gas in commercial and transportation may burn cleaner than oil energy, the process to get that gas still fouls the air and water. Further, the U.S., by increased fracking, will not become less energy dependent upon foreign countries; the U.S., with an abundance of natural gas, has no dependence upon foreign countries.

Heather White, chief of staff of the Environmental Working group, which calls itself "the nation's leading environmental health research and advocacy organization,"[169] issued a strong opposition statement to the policy:

> "People are worried about their water, their health and the value of their property after drilling. They are beset by

frenzied leasing requests from natural gas 'land men' and in some cases, experiencing drilling-related earthquakes. These communities have deep, long-term concerns about the environmental and financial impacts of natural gas drilling in key battleground states like Ohio, Pennsylvania, Virginia, Colorado and North Carolina. Yet it seems to us the White House has missed this political reality in its fervor over natural gas drilling."[170]

In March 2015, the Department of Interior finally issued new rules to cover the 100,000 gas wells already drilled and those that will be drilled on public land. Those rules, which took effect in June 2015, require drillers, within 30 days of completing fracking on a well, to disclose the chemicals used, and to improve safety standards on storage of chemicals, well construction, and use of wastewater. The new rules also require companies to provide detailed geological information about each well.

The oil and gas industry wasn't pleased with the new safety requirements. In a prepared statement, Barry Russell, president of the Independent Petroleum Association of America (IPAA), which filed a lawsuit to challenge the regulations, emphasized the economic benefits of gas drilling, and included within his statement the industry-wide half-truths about safety:

> "At a time when the oil and natural gas industry faces incredible cost uncertainties, these so-called baseline standards will threaten America's economic upturn, while further deterring energy development on federal lands. . . .
> "Hydraulic fracturing has been conducted safely and responsibly in the United States for over sixty years. These new federal mandates will add burdensome new costs on our independent producers, taking investments away from developing new American-made energy, much-needed job creation, and economic growth. The agency needs to further engage stakeholders, adequately assess the costs, and compare the proposal to existing safeguards under state law."[171]

The President's policy also eliminated funding by the United States of new coal plants throughout the world; by executive order, the President also directed the EPA to "put an end to the limitless dumping of carbon pollution from our power plants, and complete new pollution standards for both new and existing power plants."

Most environmentalists praised the President's Climate Change policy, but there was a sharp division by those who opposed the use of horizontal fracking to extract natural gas.

The Green Party praised the President's call for the EPA "to strengthen regulation of carbon emissions from existing power plants [as] a long overdue step in the right direction," but that his "'all of the above' approach to energy is still a disaster for the climate."[172]

Dr. Jill Stein, the Green Party's 2012 presidential nominee, argued:

> "You can't give your child an 'all of the above diet' with toxic lead and arsenic, and think that adding some spinach and blueberries is going to make it OK. Likewise, reducing carbon pollution from coal does not make fracking, tar sands oil, deep water and Arctic drilling OK. The climate is spiraling into runaway warming. Obama's promotion of cheap dirty fossil fuels makes coal regulations just window dressing on a disastrous policy. . . .
>
> "The world already has five times as much oil, coal and gas available as climate scientists say the atmosphere can tolerate if global temperature rise is to stay below the internationally accepted limit of two degrees Celsius. So rather than drill for more fossil fuels, we must keep 80 percent of those reserves locked away safely underground to avoid a climate disaster. In order to achieve this, and protect the climate for everyone rather than corporate profits for a few, we need to consider making energy a democratically-controlled public utility, with a mission to take us from fossil fuels to renewables at least cost."[173]

As expected, the Republicans complained about the President's policy.

The President's policy "is absolutely crazy," said House Speaker John Boehner (R-Ohio), who protested what he erroneously believed would be a loss of jobs in the fossil fuel industry—"Why would you want to increase the cost of energy and kill more American jobs at a time when the American people are still asking the question, where are the jobs?"[174] But it was Boehner and his fellow Republicans who had blocked several jobs bills that were not connected to the fossil fuel industry. The American Society of Civil Engineers had estimated that to bring the nation's infrastructure just to the level of "good," the cost would be $3.6 trillion by 2020.[175] The society issued a grade of D+ for the quality of the nation's infrastructure, ranging from a B– for solid waste management to a D– for the integrity of inland waterways and levees.[176] Sen. Bernie Sanders (I-Vt.) pointed out in 2014 that if the federal, state, and local governments spend just $1 trillion of the need to upgrade roads, bridges, dams, wastewater plants, and other parts of the nation's infrastructure, about 13 million jobs would be created.[177] Thus, the reality is that "jobs" isn't the primary focus of those who support fracking, but that they're really supporting the oil/gas industry and, especially, its generous financial support to politicians.

One year later, President Obama laid out stricter rules governing fossil fuel development and generation, demanding coal producers to reduce carbon dioxide emissions from their power plants by 30 percent from 2005 levels by 2030;[178] the result would be five billion fewer tons of carbon dioxide in the air in the decade beginning 2020. Coal-fired power plants accounted for almost 30 percent of all greenhouse gas.[179] The President ordered states to submit full plans for compliance no later than June 2018.

As expected, politicians in coal-producing states saw the President's policy as a continued war upon coal, rather than an attempt to reduce greenhouse gas emissions and help protect the Earth's climate. "The administration has set out to kill coal and its 800,000 jobs," Sen. Mike Enzi (R-Wyo.) said even before the President's address to the nation.[180]

From the floor of the U.S. Senate, Minority Leader Mitch McConnell (R-Ky.), falsely argued:

> "Declaring a 'War on Coal' is tantamount to declaring a 'War on Jobs.' It's tantamount to kicking the ladder out from beneath the feet of any Americans struggling in today's economy."[181]

McConnell, who would become Senate majority leader in January 2015, declared his top priority would be "to do whatever I can to get the EPA reined in."[182] What Boehner and McConnell didn't say was that there were as many workers employed in wind and solar energy as in coal,[183] and employment projections showed only an increase in renewable energy employment. Neither man addressed the issues of global warming or the necessity to improve the health and environment of their American constituents.

Pennsylvania state Sens. Tim Solobay and Gene Yaw founded the Coal Caucus in September 2013. First praising the gas drillers, Yaw gushed, "Since the Industrial Revolution, coal has also fueled our economy having created hundreds of thousands of jobs. Collaboratively, we can change the dynamic of coal as an energy resource. . . . This Coal Caucus will serve as a champion for increased investment in coal and coal-driven technology."[184]

Solobay was equally effusive in praising the coal industry: "Coal is critically important to our effort to reduce dependence on foreign fuels. In addition to being a major employer in Pennsylvania, the industry provides consumers with protection from energy shortages and price spikes."[185] Substitute "gas" for "coal" and you have almost a duplicate of Solobay's views pre-caucus. Dory Hippauf says Americans have been shoved onto a Fossil Fuel Carousel and are now supposed to believe that the "dirty" coal energy that politicians said would be replaced by "clean" natural gas energy isn't so dirty any more. One effect that concerned environmentalists, however, was that reducing coal energy would increase the use of natural gas.

When not attacking what they believed was the President's attack on coal, conservatives objected to the rules for economic reasons. The greenhouse gas rules "have the potential of crippling the United States economy . . . [and] have the potential of placing a $100 [billion] additional fixed cost on the system by requiring the construction of new power-generation facilities," said Tom Lubnau, Wyoming's speaker of the house.[186]

In addition to significant opposition from Republican politicians and the business community, the American Legislative Exchange Council (ALEC), which has significant financial support from major oil/gas companies and trade associations,[187]

39

began a behind-the-scenes campaign to block the President's policy. Documents obtained by *The* [London] *Guardian* in May 2014 revealed that ALEC had begun a campaign to lobby politicians to convince state attorneys general, especially in Republican-controlled states, to sue the EPA to block implementation of the new rules to reduce carbon emissions. ALEC also helped write about a dozen anti-EPA bills, according to the *Guardian*.[188]

First oil well (Titusville, Pa.), about 1860

CHAPTER 2
The Economics of Fracking

Drilling for natural gas is not because energy companies are good citizens who want to improve the local economy and hire a robust work force. The companies form limited liability corporations (LLCs), which can be created and folded quickly, and have every intention of showing a profit while giving their investors a good return on investment.

It takes about $5 million to bring a Marcellus Shale well into production, says Patrick Creighton of the industry-sponsored Marcellus Shale Coalition.[189] The cost per well ranges from $3.5 million in the Barnett Shale to $10 million in the Bakken Shale. However, financial analyst/economist Deborah Rogers says, "Only 20 percent of wells drilled will actually make money," noting "Eighty-five percent of wells are abandoned in the first five years."[190]

Each well is expected to generate about $16 million during its lifetime,[191] which can be as few as ten years, according to the Pennsylvania Budget and Policy Center. Many cease to be profitable after three to five years.

"With these new shale gas wells, you get peak production in one or two years, when the pressure is strongest, then a steep fall-off," says geologist Sally Odland of Columbia University's Lamont–Doherty Earth Observatory.[192] Gas well production in the Marcellus Shale drops 32 percent in the second year, and 45 percent in the third year, according to data compiled by Marcellusgas.org. Nationally, oil and gas production drops about 50–75 percent after the first year, according to geologist/shale analyst Pete Stark.[193] For 2013, the average production value of a Marcellus Shale well was about $2.1 million; at the end of 2014, the production value was about $1.2 million.[194]

Estimates of 30–40 years of well production are "unproven

41

[and] has significantly stressed the financials of the industry as payback based on long productive output is not yet a proven part of shale gas dynamics," says Robert Magyar, one of the nation's leading experts in energy production.[195]

The Lure of Jobs

A 73-page report from IHS, which says it's "the leading provider of diverse global market and economic information," claims "Unconventional oil and natural gas activity is reshaping America's energy future and bringing significant benefits to the US economy in terms of jobs, government revenues, and GDP." The report, released in September 2013, concluded:

> "The full economic contribution from the unconventional oil and natural gas value chain and energy related chemical manufacturing has added 2.1 million jobs in 2012, and that contribution will increase to almost 3.1 million by the end of the decade and almost 3.9 million in 2025.
> "The value chain's annual contributions to GDP will nearly double, from almost $284 billion in 2012 to almost $533 billion in 2025. Government revenues will average $115 billion annually and will cumulatively grow by a total of more than $1.6 trillion from 2012 to 2025."[196]

"For an organization like the American Petroleum Institute (API) being able to cite the findings and reputation of IHS goes a long way toward making its point to government officials," said Kyle Isakower, API vice president.[197] Steve Forde of the Marcellus Shale Coalition called the report, "an important advocacy tool and in the initial years in the Marcellus development, they really helped to build the case that this is real."[198]

What the report didn't say was that funding was provided by the API, Natural Gas Alliance, the Natural Gas Supply Association, and other industry groups, which then used the report to justify the industry's economic impact. The report, which optimistically inflated the economic benefits of shale gas drilling, also didn't look at the significant health and environmental impacts from shale gas drilling.

The U.S. Chamber of Commerce and local Chambers, which support business and conservative candidates for political office, have also touted the economic impacts,[199] while skirting the health and environmental effects.

In July 2012, the Chambers announced a multi-million dollar "advocacy and education campaign" to lobby Pennsylvania, West Virginia, Ohio, and other states with large shale deposits about what they see as the benefits of fracking and natural gas drilling.[200] The U.S. Chamber, which spends more in lobbying expenses than any company or organization, spent about $1 billion between 1998 and 2014, according to the Center for Responsive Politics.[201]

President Barack Obama specifically noted that the development of natural gas as an energy source to replace fossil fuels could generate 600,000 jobs, a claim that had previously been advanced by the natural gas industry. "[T]he industry is working hard to move forward responsibly and will continue to create good paying, family sustaining jobs that will make our economy more competitive," said Gene Barr, president of the Pennsylvania Chamber of Commerce.[202]

The Marcellus Shale Coalition claimed 88,000 new jobs were created in Pennsylvania in 2010 because of Marcellus Shale drilling. [203] In an OpEd column for the *Philadelphia Inquirer* in September 2012, Kathryn Z. Klaber, Marcellus Shale Coalition president, argued, "Marcellus development is creating tens of thousands of jobs at a time when they're most needed," pointing out that according to "state data, nearly 239,000 jobs across the commonwealth are tied to our industry."[204]

When Gov. Tom Corbett proudly puffed that more than 240,000 jobs were created in Pennsylvania,[205] he included every temporary job, every worker from Texas and Oklahoma who doesn't pay Pennsylvania taxes and sends much of his income to his family out of state, every independent truck driver who gave up another job to get a fracking job, every clerk and every prostitute who came to the fracking fields to also make money from the artificial economic boom. Dr. Tim Kelsey, Penn State professor of agricultural economics, reports that at most fewer than 35,000 jobs were created, only about half in the core industry.[206]

In 2010, Pennsylvania was ranked seventh in the nation in job creation; at the end of 2014, it was ranked 50th.[207] Of 6.4 million employed Pennsylvanians in mid-2014, only about

37,500 were employed in fossil fuel extraction (including oil, gas, coal mining, and logging), according to the Bureau of Labor Statistics (BLS).[208]

In Texas, which is dominated by oil and gas extraction, fewer than 10 percent of the state's workers are employed in mining and logging, according to the BLS.[209] Even in North Dakota, which has been undergoing a boom because of oil extraction in the Bakken Shale, jobs in the mining and logging industry are less than 8 percent of all employment.[210]

"[It] is very dangerous when we hype things, because it sets expectation which perhaps can't be fulfilled to the degree that you would like," Matthias Bischel, projects and technology director at Royal Dutch Shell, told the *Financial Post*.[211]

"Those who lobby [for fracking] tend to overstate the benefits and understate the costs," said Pennsylvania State Rep. Greg Vitali.[212] Research studies by economists Deborah Rogers, Dr. Jannette M. Barth, and others agree with Vitali and debunk the idea of significant job creation.

Rogers used data provided by the Department of Labor's Bureau of Labor Statistics to discredit the industry's inflated jobs creation estimates. The Perryman Group, commissioned by the U.S. Chamber of Commerce, claimed there were 111,131 jobs created by drillers in northern Texas during 2008.[213] However, BLS statistics reveal there were only 166,500 jobs created in the entire United States in the production sector that year. Chesapeake Energy claimed 53,200 jobs for the Fort Worth/ Arlington area in 2010;[214] Rogers questioned that claim since there were only 93,800 jobs created in the production sector for the entire country in 2010.

"The grandiose job creation claimed by the industry is not at all consistent with data from unbiased, publicly available sources," said Dr. Barth in testimony before the New York State Committee on Environmental Conservation in October 2011.[215]

In an article that critiqued an industry-funded study, Dr. Barth pointed out an economic reality:

> "There will be short-term jobs for truck drivers and some workers in related heavy industrial and business service sectors; there will be some other local businesses who benefit in the short-term boom as well, such as hotels and restau-

rants [because of] the transient workforce that comes in from other states. . . .

"The natural gas industry has attracted workers by offering relatively higher wages, causing some workers to leave their jobs with existing businesses who cannot afford to compete with the wages offered by the gas industry. These smaller businesses close and leave the area, and when the short-term boom created by the gas industry ends, the communities may become almost ghost towns with empty buildings and high levels of poverty and unemployment."[216]

The gas industry," says Dr. Barth, "has an incentive to mislead the public in order to gain public support for gas drilling."[217]

The Lure of Immediate Gratification

Because the oil and gas culture is a part of the lifestyle of the people of several states, "people have fully embraced fracking [to gain imme-diate financial benefit] without knowing its impact," says Elisabeth Radow, an attorney specializing in the environment and real estate.

When the natural gas industry first comes into a region to lease land, the sales people hold out the lure of a lump sum payment plus a royalty on gas extracted. For families faced by the recession, leasing mineral rights for unused land or converting portions of non-producing farm land to the cor-porations might have been a way to get out of debt and even have a decent bank account to prepare for future emergencies. Some owners with dozens of acres to lease became rich; most didn't.

During the first years of exploration in the Marcellus Shale, energy companies paid as little as $5–$25 an acre for a five year lease, plus a 12.5 percent royalty. By 2009, bonuses increased to $1,000–$7,000 and royalties sometimes exceeded the legal minimum of 12.5 percent as landowners became more knowledgeable, says Dave Messersmith of the Penn State Exten-sion office. By 2010, some of the large landowners became—as *60 Minutes* dubbed them—"shaleionaires."[218] Landowners in the Marcellus Shale occasionally negotiated additional benefits, including clean-up costs and a certain amount of free natural gas for home heating use. However, several provisions do not benefit landowners, leaving the newly-minted "shaleionaires" significantly less money than promised.

LEGALLY REDUCING OWNER COMPENSATION

Landowners soon realized that the corporations were legally deducting from royalty payments post-production and marketing costs, often more than half the expected royalties, shattering the dreams of numerous people who believed leasing mineral rights would make them rich. Throughout the country, property owners averaged less than $500 a month from land they leased to the gas companies.[219] When landowners figured out they weren't getting what they thought they should get, and then were told that what the gas companies deducted was legal and part of the basic contracts, they became even more upset. Not wanting a public fight, most companies often settle.

Shortly after several landowners sued Chesapeake Appalachia in August 2013, the company settled for about $7.5 million, with potential distribution to more than 1,000 landowners. The settlement allowed Chesapeake to still charge landowners for 72.5 percent of the total costs, including all of the costs of gas transportation to compressors and terminals.[220]

In June 2013, Range Resources settled a class action suit for $87.5 million. The suit was filed in 2010 by landowners who claimed Range had underpaid royalties.[221] Range Resources acknowledged it had improperly calculated royalties in Pennsylvania, and agreed to pay about 2,000 landowners at least $1.75 million, up to a maximum $20 million, depending upon market prices over the next few years.

Energy companies can also tie up land indefinitely, even with non-producing wells, if there is no expiration date; contracts can specify the royalties are only paid for working wells on a property. As landowners became more financially savvy, they demanded specific expiration dates. However, significant mineral rights are owned by few people. Research by Dr. Timothy W. Kelsey, Dr. Alex Metcalf, and Rodrigo Salcedo for the Center for Economic & Community Development, reveals:

"[O]wnership of the land in the Pennsylvania counties with the most Marcellus drilling activity is concentrated in a relatively small share of residents, and in owners from outside the county. The majority of residents of these counties together own little of the total land area, and so have relatively little 'voice' in the critical leasing decisions which affect whether and how Marcellus shale drilling will occur in their county. Half of the resident landowners in the counties together only control 1.1 percent of the land area, and renters have no 'voice' at all. Rather it is the top 10 percent of resident landowners, plus outside landowners (both public and private), who are able to make the major leasing decisions that affect the rest of the community. . . .

"[A] majority of lease and royalty income from Marcellus shale development will go to a relatively small share of the resident population in these counties, with much of the remainder going to others outside the counties. A little less than half (48.9 percent) of the lease and royalty dollars . . . go to the top ten percent of local land-owners, while 39.8 percent will go to the public sector or non-resident landowners. The remaining 11.3 percent of lease and royalty income will be divided between the bottom 90 percent."[222]

ADDITIONAL CONTRACTUAL REALITIES

In addition to the lack of voice and income, there are additional realities not addressed by proponents of fracking.

"[T]he potential for groundwater contamination [is] one of the most high-profile risks associated with drilling [and leads] to a large and significant reduction in property values," according to Drs. Elisheba Spiller and Lucija Muehlenbachs of Resources for the Future, and Christopher Timmins of Duke University's Department of Economics. They evaluated 19,000 properties in Washington County, Pa., and concluded, "By itself, groundwater risk reduces property values by up to 24 percent" of homes using well water within 1.25 miles of a shale gas well.[223] Whether or not there is actual pollution caused by the fracking process, the perception of the potential of pollution is enough to cause a decline in property values that exceeds any economic benefit from signing gas leases.

"In general, homes that are in close proximity of fracking may have a decreased home value of about 20 percent as a result of construction, noise, traffic, airborne dust, and road damage," according to the 360 Mortgage Group in Austin, Texas.[224] In many cases, the value drops by as much as half; in some cases, it's nearly impossible to sell houses near natural gas wells.

"The most important thing to many people is their homes, [which] is an emotional safe place," says Elisabeth Radow. But when contaminated water makes it unsafe, "and fracking makes it impossible for you to sell your home or that you may only get a fraction of its value, people become upset," Radow says.

Richard Plunz, director of the Urban Design Lab at Columbia University, told *Columbia Magazine*:

> "If there's a lease on your land, your property is devalued. People didn't understand that initially. They were told by the gas-company landsmen, 'You're going to make a fortune, and you won't even see a well.' But even without a well, nobody is going to buy property that has a lease. The value of the neighbors' property probably decreases, too. No one wants to buy a house in an industrializing landscape.
>
> "The long-term economic prospects for these towns are diminished. The land will be undesirable, scarred with roads and well pads and possibly contaminated. The owners will have collected their proceeds from the production as long as possible, but when the profits end they can simply walk away. With that, the town's tax revenue fades."[225]

After 36 years as a Realtor, Jennifer Canfield of Damascus, Pa., says she got out of real estate because many clients didn't wish to buy homes or land when they learned it was near property that was leased to the gas companies. "Many clients," says Canfield, "were from the metropolitan areas, and they wanted to avoid areas that had heavy industrial use." The effects of fracking upon home sales became personal for Canfield and her husband, whose house on six acres in Wayne County in northeastern Pennsylvania was on the market for four years. It finally sold in Summer 2013 at a price significantly below value. The sale occurred only after Newfield Appalachia and Hess Co. terminated leases in the county, and decided not to move from exploration to production.

A further problem is that in what the natural gas industry derisively calls the "Sacrifice Zone," many home owners can't get second mortgages, home equity loans, or insurance. Both the Federal Housing Administration and the Department of

Housing and Urban Development have refused to provide mortgages if the home owner has leased gas rights within 300 feet of the house.[226]

Another issue is that wells and rigs on residential property often leads to a determination that the entire property is no longer residential but industrial. This can result in higher taxes and financial institutions to deny mortgages because they are unable to do adequate risk analyses.

Among major lending institutions that refuse to issue mortgages are Bank of America, Santandr, and Wells Fargo. In July 2013, Rabobank announced it would not lend money to farmers who rent or lease their land to shale gas extraction companies or to any company involved in shale gas extraction.[227] The multinational financial institution, one of the largest in the world, is headquartered in The Netherlands.

The North Carolina State Employees Credit Union issued a statement in April 2012 that further clarified the reality about purchasing houses from owners who have already leased mineral rights or from future owners who have secured both surface and mineral rights but plan to lease rights, hoping the royalties would offset the mortgage payments:

> "The standard Fannie Mae/Freddie Mac Deed of Trust document recorded for most real estate liens prohibits the homeowner from selling or transferring any part of the property during the term of the loan without obtaining prior written approval from an official of the financial institution holding the mortgage. This includes the oil, gas and minerals found on the property. Any property financed with a State Employees' Credit Union mortgage falls under the aforementioned restriction. Approval of exceptions from State Employees' Credit Union would not be granted due to heightened risk concerns associated with extraction of these natural resources, including hydraulic fracturing technology (other-wise known as fracking or horizontal drilling)."[228]

Homeowners with Fannie Mae mortgages can't lease mineral rights, according to Callie Dosberg, a Fannie Mae executive. She told the *American Banker* that Fannie Mae doesn't allow "surface instruments," which includes rigs and platforms. She says homeowners could request an exception, "but generally the agency does not approve such requests."[229]

Homeowners are also finding it difficult to get insurance. Nationwide Insurance no longer insures homes and property in the heavily industrialized areas of natural gas drilling. A company news release in July 2012, stated:

> "Fracking-related losses have never been a covered loss under personal or commercial lines policies. Nationwide's personal and commercial lines insurance policies were not designed to provide coverage for any fracking-related risks . . .
> "From an underwriting standpoint, we do not have a comfort level with the unique risks associated with the fracking process to provide coverage at a reasonable price."[230]

What they're saying "is they won't cancel your insurance but if there's drilling on your property, they won't renew the insurance next year," says Elisabeth Radow. Other insurance companies are likely to not insure or renew policies of homeowners who allow fracking on their property.

Under federal law, persons living in a flood plain are required to have flood insurance.[231] If an insured house is flooded and uninhabitable, the Federal Emergency Management Agency (FEMA) could buy that house at fair market value and then forbid construction on the site. However, FEMA is rejecting buy-outs for homeowners who have signed leases to natural gas companies. A review of public documents by Sharon Kelly of DeSmogblog.com revealed "at least 18 homeowners in Pennsylvania were denied access to the Hazard Mitigation program because of oil and gas leases or pipeline rights-of-ways on their properties."[232] Part of the reason is that by signing away mineral rights, "The landowner and the oil company now co-own the rights to use the surface of the property and the oil company owns all of the oil and gas under the property," FEMA attorney Michael C. Hill wrote to the Pennsylvania Emergency Management Agency in June 2011.[233] "Thus the landowner alone cannot transfer all of the surface rights nor the oil and gas rights," he pointed out. If FEMA doesn't buy the affected property, the consequences can be significant, says Kelly:

"If their land is not acquired, the affected landowners may be caught in a bind—unable to sell their land, they may face sharply higher federal flood insurance rates, which are hiked after repeated flooding, partly as an incentive for homeowners to participate in the HGMP [Hazard Mitigation Grant Program]. If FEMA moves to acquire the land, the government may be legally required to conduct a review under the National Environmental Policy Act, some legal experts say, but this process is famously slow and can stretch on for decades.[234]

In May 2014, FEMA officially banned from hazard mitigation protection following natural disasters properties that are in the fracking zone and which the owners have leased mineral rights or have allowed wells, rigs, and other oil/gas excavating machinery and roads on their property. The logic for FEMA reimbursement for properties damaged or destroyed by floods, earthquakes, and other natural disasters is that the homeowner can't be responsible of what is known as "acts of God." Allowing oil and gas companies to drill on one's property can compromise the geological integrity of the property and of the building foundation and, equally important, doesn't give the homeowner a clear title to the land. Rep. Lou Barletta (R-Pa.) called the policy "anti-fracking," and chaired a public hearing in August 2014 to advance a pro-fracking agenda and oppose FEMA's mitigation policy.[235]

It isn't unreasonable to expect farmers who have struggled in the recession to snap at the lure of thousands of dollars from an energy company that will lease mineral rights and some of their land and which will also pay monthly royalties that could help pay mortgages, repair barns, get better supplies for crops, or more care for farm animals. It isn't unreasonable for those who are unemployed or underemployed to grasp for anything to help themselves and their families, nor is it unreasonable to expect that persons—roustabouts, clerks, truck drivers, safety inspectors, helicopter pilots, and engineers among several hundred thousand in dozens of job classifications—will take better paid jobs, even if it often means 60–80 hour work weeks under hazardous conditions. What is unreasonable is that state governments allow continued destruction of agricultural and forest lands in their rush for economic gain. "We gave up the green of natural resources for the green of money," says Dr. Harvey M. Katz, whose specialty is freshwater biology.

Years from now, thousands of landowners who allowed drilling on their property may wonder if the immediate gratification of a few dollars or even sudden wealth was worth the cost of what happened to the health and lifestyle of the people and their environment.

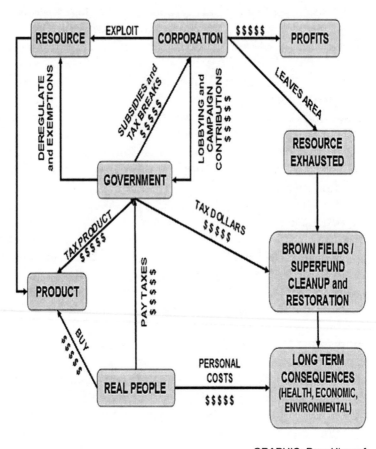

GRAPHIC: Dory Hippauf

CHAPTER 3
Seizing Private Property

Robert Donnan of Washington County, Pa., south of Pittsburgh, opposed fracking since late 2008, not long after the gas companies moved into the state and began dumping drilling wastewater into the Monongahela River, source of his tapwater. His daily e-mail newsletter is a well-read compilation of some of the more important stories in the shale gas industry and of the campaigns to end fracking.

A Vietnam War combat veteran, who became a respected landscape contractor in a 35 year career, Donnan was about to fight a different kind of battle, one that would cause him to do what he never thought he would ever do—sign a lease with a gas company.

In 2012, he refused to lease his two percent share of 74 acres, part of a larger 296 acre parcel owned by more than 20 members of his extended family. Later that year, in a routine DEP file review, he learned that Range Resources was in the process of drilling into the family's parcel. "They said they had signed two family heirs to a lease," says Donnan, "and that a two percent ownership interest was enough for Range to drill." Donnan says his attorney had said that companies probably wouldn't drill with less than 50 percent agreement with the gas rights owners. But, Range Resources didn't have even the two percent it thought it had. "Neither person who signed a Range lease was related to our family," says Donnan.

He was ready to fight, but no attorney he approached would take the case on a contingency basis. "They all wanted a high hourly fee," he says.

Without the money to hire lawyers and fight Range Resources, Donnan, who owns less than 1.5 acres of the 74 acre parcel gas rights, and other members of his family faced a reality they had not foreseen:

"These latest developments were enough to make my cousin, who owns that surface property above part of our gas rights, very nervous, and rightly so. What if one or two heirs signed a 'bad lease' giving Range the right to install roads, impoundments, pipelines and drilling pads on her property? That was the last thing she wanted. So at that point she knew she had to protect her property and hired a skilled attorney to craft an extremely protective lease, one that would prevent any surface activity on her land. A couple other family heirs signed a lease at the same time.

"It was too late to change what had already happened because Range completed the drilling, and the wells would start producing in March 2013. A senior land agent at Range told the heirs representing the rest of our family that any heir who did not sign a lease by May 31 would not get royalties, not even the minimum payment required by law."

It was now a question, says Donnan, of "Range stealing our gas and not paying us, or signing a lease." Torn by the principle of not dealing with a gas company and losing all income and dealing with Range, even if reluctantly, and getting some funds to help scrutinize the gas companies, continue to try to educate the public about the problems of fracking, and try to protect the environment, Donnan signed. This prompted Range's Matt Pitzarella to tell the Associated Press that Donnan's signing of a lease was probably "an endorsement" of drilling.[236] "It was no such endorsement, but a ridiculous spin," says Donnan, who refused to be interviewed by the AP because of what he believed was "the pro-industry slant of that particular reporter's previous articles."

About three months after Donnan received an upfront check from Range, he and his wife donated $500 to help build a "smart home" for a severely wounded Afghanistan war veteran and his wife. The day after he received his first royalty check from Range in October 2013, Donnan and his wife donated $1,000 to Clean Water Action to continue the fight for clean water as it relates to Marcellus Shale drilling in southwestern Pennsylvania.

Robert Donnan had a hard choice that he thinks about almost every day, but realizes that since the drilling was nearly complete and his cousins had signed a lease, the best thing he could have done was to take the money and use some of it to address drilling issues, like clean streams and rivers.

Others don't have a choice.

Split Estate

The law of Split Estate, dating to the English common law belief that the king owned all the land, allows two different owners to split ownership of surface rights and underground rights. Thus, it is possible that a person who purchases a house, farm, or even a garage, may not have the underground rights, those rights having been held by the original owner or sold to someone else.

According to the U.S. Bureau of Land Management (BLM):

> "[M]ineral rights are considered the dominant estate, meaning they take precedence over other rights associated with the property, including those associated with owning the surface. However, the mineral owner must show due regard for the interests of the surface estate owner and occupy only those portions of the surface that are reasonably necessary to develop the mineral estate."[237]

Eminent Domain

Julia Trigg Crawford of Direct, Texas, is the manager of a 650-acre farm that her grandfather first bought in 1948. The farm produces mostly corn, wheat, and soy. On its north border is the Red River; to the west is the Bois d'Arc Creek.

TransCanada is an Alberta-based corporation that is building the controversial Keystone Pipeline that will carry bitumen— thicker, more corrosive and toxic than crude oil—through 36-inch diameter pipes from the Alberta tar sands to refineries on the Gulf Coast, almost all of it to be exported. If the pipeline was at full capacity, it would add about 240 billion tons of carbon dioxide into the atmosphere, according to Environment America.[238] The $2.3 billion southern segment, about 485 miles from Cushing, Okla., to the Gulf Coast was completed at the end of 2013 after President Obama in February 2012 directed his administration "to cut through the red tape, break through the bureaucratic hurdles and make this project a priority, to go ahead and get it done."[239] However, the northern leg of the $7 billion 1,959 mile pipeline, from Hardisty, Alberta, to Steele City, Neb., was held up until President Obama either succumbed to corporate and business pressures or blocked the construction because of environmental and health issues.

When TransCanada first approached Crawford's father, Dick, in 2008, and offered to pay about $7,000 for easement rights, he refused, telling the company, "We don't want you here." He said the corporation could reroute the line, just as other pipeline companies in oil-rich Texas had done for decades. TransCanada increased the offer the following years, but the family still refused. In August 2012, with Julia now managing the farm, TransCanada offered $21,626 for an easement—and a threat. "We were given three days to accept their offer," she says, "and if we didn't, they would condemn the land and seize it anyway." She still refused.

And so TransCanada, a foreign corporation, exercised the right of eminent domain to seize two acres of an American farm so it could build a pipeline. In May 2013, heavy machinery rolled onto Crawford's farm. Crossing an easement and past a barbed wire enclosure that separates the land TransCanada seized from the rest of the farm, the bulldozers and graders peeled away the topsoil of a 1,200 foot strip. For three months, the workers and their equipment crossed the Crawford farm, putting the last segment of the Southern pipeline in place.

Governments may seize private property if that property is for public use; the Fifth Amendment's "taking clause" requires the landowner to be given fair compensation. Although exercising eminent domain to seize land for the public good is commonly believed to be restricted to the government, a 5–4 Supreme Court ruling in 2005 (*Kelo v. City of New London, Conn.*[240]) expanded the "takings" clause to allow government to seize private property and then transfer it to private companies for "economic development." To get that "right," all TransCanada had to do was fill out a one-page form and check a box that the corporation declared itself to be a "common carrier." The Railroad Commission of Texas, which regulates oil and gas but has no responsibility over railroads, merely processes the paper, rather than investigates the claim; it has admitted it never denied "common carrier" status.[241] In the contorted logic that is often spun by corporations, TransCanada declared itself to be a common carrier because the Railroad Commission said it was, even though the Commission's jurisdiction applies only to intrastate, not interstate, carriers.

On Aug. 21, 2012, the day before Judge Bill Harris of Lamar County rendered his decision on Crawford's complaint, the sheriff, with the judge's signature, issued a writ of possession giving TransCanada the right to seize the land. The next day, Harris issued a 15-word decision, transmitted by his iPhone, that upheld TransCanada's rights. In Texas, as in most states, the landowner can only challenge the settlement not the action.

Crawford's refusal to sell is based upon a mixture of reasons. The Crawford farm is home to one of the most recognized Caddo Nation Indian burial sites in Texas, and the 30 acre pasture that TransCanada wanted to trench represented the southern-most boundary of this archeological site. Both the Texas Historical Commission and TransCanada's archeological firm concurred that the vast majority of the 30 acre pasture in question qualifies for the National Registry of Historic Places. An archeological dig undertaken after TransCanada showed up to seize the land recovered 145 artifacts in just a three-foot deep, 1,200 foot by 20 foot section. But the executive director of the Texas Historical Commission sent a letter stating that no new artifacts had been found in the slice of land TransCanada planned to build.

Another reason Crawford refused to be bought out was that she didn't want TransCanada to drill under the Bois d'Arc Creek "where we have state-given water rights." That creek irrigates about 400 acres of her land. "Any pipeline leak, she says, "would contaminate our equipment, and then our crops in minutes." It isn't unreasonable to expect there will be an incident that could pollute the water, air, and soil for several miles.

Between 2000 and October 2013, there were 8,132 pipeline incidents, resulting in 218 deaths, 832 injuries, and about $5.4 billion in property damage, according to the federal Pipeline and Hazardous Materials Safety Administration.[242] Cornell University's Global Labor Institute concluded that economic damage caused by potential spills from the Keystone XL pipeline could outweigh the benefits of jobs created by the project.[243] Between 2010 and 2013, there were 14 spills on the operational parts of the Keystone Pipeline.[244]

Crawford and her attorney, Wendi Hammond, challenged TransCanada's right to seize public property, arguing not only is TransCanada, which had net earnings of $1.3 billion in 2012, a foreign corporation but it also doesn't qualify as a "common carrier" since the benefit is primarily to itself. The state appeals court ruled against her in August 2013. "They have far more lawyers and funds than we have," says Crawford, who held a music festival to help raise funds. Additional donations have come from around the world, many from those who weren't immediately affected by oil and gas exploration, transportation, and processing, but who understand the need to fight a battle that could, at some time, affect them.

"The company basically goes to court, files condemnation petitions, says, 'We are common carrier, have the power of eminent domain, we are taking this property.' And that's all there is to it," says Debra Medina,[245] of WeTexans, a grassroots organization opposed to the seizure of private land by private companies.

At least 89 Texas landowners have had their properties condemned and then seized by TransCanada.[246]

Eleanor Fairchild, a 78-year-old great-grandmother living on a 300-acre farm near Winnsboro, Texas, also protested the seizure of her land. She and her husband, a retired oil company geologist now deceased, bought the land in 1983. TransCanada planned to bisect her farm, which includes wetlands, natural springs, and woods. In October 2012, Fairchild and activist/ actor Daryl Hannah raised their arms and stood before an excavator that would begin to clear her land and dig out trees. "This is not just about my land," said Fairchild, "It's about all of our country. It needs to be stopped."[247] Both women were arrested and charged with criminal trespass. Hannah was also charged with resisting arrest. TransCanada's response was, "It is unfortunate Ms. Hannah and other out-of-state activists have chosen to break the law by illegally trespassing on private property."[248] Apparently, the corporation didn't want anyone to know that Hannah had a long history of environmental activism, including numerous protests against the creation of the pipeline, and that the private property Hannah and Fairchild stood upon was originally Fairchild's land that TransCanada had seized.

TransCanada isn't the only oil and gas company that uses and bends eminent domain laws.

Chuck Paul, who lost about 30 of his 64 acre horse farm because of required easements by the natural gas Industry, told the *Fort Worth Weekly:*

> "The gas companies pay a one-time fee for your land, but you lose the right to utilize it as anything more than grassland forever. . . . You can never build on those easements. They took my retirement away by eminent domain.
> "They just screwed me. And they might want to come back and take more. And there's nothing you can do to stop them. If they'd have paid me fair market value, I would have sold it to them. But with the right of eminent domain they have, well, they can do anything they want."[249]

In Arlington, Texas, Dr. Ranjana Bhandari, an economist, and her husband, Dr. Kaushik De, professor of physics at the University of Texas, refused to grant Chesapeake Energy the right to take gas beneath their home, although Chesapeake promised $18,000 an acre signing bonus and royalties.[250] "We decided not to sign because we didn't think it was safe, but the Railroad Commission doesn't seem to care about whose property is taken," Dr. Bhandari told Reuters.[251] Chesapeake seized the mineral rights to capture natural gas beneath the family's home. Between January 2005 and October 2012, the Railroad Commission approved all but five of Chesapeake's 1,628 requests to seize mineral rights, according to the Reuters investigation. Arlington, a Dallas suburb, has Chesapeake wells in public parks, schools, the University of Texas at Arlington, and even the athletic field at Oakridge School.[252]

The Texas Supreme Court, in *Texas Rice Land Partners and Mike Latta v. Denbury Green Pipeline–Texas* (2012),[253] had previously ruled, "Even when the Legislature grants certain private entities 'the right and power of eminent domain,' the overarching constitutional rule controls: no taking of property for private use." In that same opinion, the Court also ruled, "A private enterprise cannot acquire unchallenged-able condemnation power . . . merely by checking boxes on a one-page form and self-declaring its common-carrier status."

Texas isn't the only state that has a broad eminent domain policy that allows Big Energy to seize private property.

Most states' new laws that "regulate" fracking were written by conservatives who traditionally object to "Big Government" and say they are the defenders of individual property rights. But, these laws allow oil and gas corporations to use the power of eminent domain to seize private property if the corporations can't get the landowner to agree to an easement, lease, or sale.

In Pennsylvania, Act 13 allows the natural gas industry to "appropriate an interest in real property [for] injection, storage and removal" of natural gas.

Sandra McDaniel, of Clearville, Pa., was forced to lease five of her 154 acres to Spectra Energy Corp., which planned to build a drilling pad. The government, says McDaniel, "took it away, and they have destroyed it." According to Reuters, "McDaniel watched from the perimeter of the installation as three pipes spewed metallic gray water into plastic-lined pits, one of which was partially covered in a gray crust. As a sulfurous smell wafted from the rig, two tanker trucks marked 'residual waste' drove from the site."[254]

By 2012, the energy companies "took all the land they needed" in Pennsylvania, says Penn State's Dave Messersmith, "so now they're buying just smaller patches."

In Tyrone Twp., Mich., Debora Hense returned from work in August 2012 to find that Enbridge workers had created a 200 yard path on her property and destroyed 80 trees in order to run a pipeline. Because of an easement created in 1968 next to Hense's property, Joe Martucci of Enbridge Energy Partners said his company had a legal right to "to use property adjacent to the pipeline." Martucci says his company offered Hense $40,000 prior to tearing up her land, but she refused. Hense says she had a legal document to prevent Enbridge from destroying her property; Enbridge says it had permission from the Michigan Public Service Commission.[255]

Near the banks of the Delaware River outside Montague, N.J., contractors for Kinder Morgan Energy Partnership in February 2013 cut down dozens of white pines to allow construction of a loop of the Tennessee Gas Pipeline. The trees were on property owned by Ruth and George Feighner, and condemned by the Federal Energy Regulatory Commission through eminent domain laws. "My wife and I had always wanted a place in the country, so we ended up buying a 200-year-old farmhouse and spent a decade gutting and rebuilding it," before moving into it in 1992, George Feighner, 86, a retired oil company chemist, told KWWL-TV. "For as long as they can remember, it has been a nightly summertime ritual for the couple [to] sit on a swing on their back porch and gaze upon the nearby mountain and trees framed by the backdrop of the waning sun," wrote reporter Eric Obernauer.[256] As chainsaws began clear-cutting what was the Feighners' property, George began to cry. The company had offered $17,000 for the land; the Feighners refused it.[257] "There is no amount of money that Tennessee Gas could offer me that would make me happy to look at a wide barren strip of land that I would have to see every day," said Ruth Feighner.[258]

The rush to grab as much land as possible was noticed as early as May 2010 when Marcus C. Rowland, Chesapeake's chief financial officer, acknowledged, "at least half and probably two-thirds or three quarters of our gas drilling is what I would call involuntary."[259] However, not all states accede to requests by private energy companies to enforce eminent domain rules. The Georgia Department of Transportation denied Kinder Morgan's request to impose eminent domain to build a part of the $1 billion Palmetto Pipeline from Belton, S.C., to Jacksonville, Fla. The decision was announced in May 2015 after the state held seven public meetings, two public hearings, and received more than 3,000 public comments.[260] Against public opposition, Kinder Morgan declared the project was "in the best interests of Georgia's consumers," and that it planned "to pursue all available options to move forward with the project." Those options would include lengthy and costly court actions.[261]

The use of eminent domain to assist private corporations

isn't just a United States reality. The United Kingdom, in October 2014, disregarded its own public survey and decided that oil and gas companies could have access to shale 300 meters (about 984 feet) below homeowners' residences. The government survey had shown that about 99 percent of the 40,647 residents who responded to the survey objected.[262] The government response was simply, "[N]o issues have been identified that would mean that our overall policy approach is not the best available solution."[263] Matt Hancock, energy minister and a member of the ruling Conservative party, told the *Guardian,* "These new rules will help Britain to explore the great potential of our national shale gas and geothermal resources, as we work towards a greener future—and open up thousands of new jobs in doing so."[264] However, Fergus Ewing, Scotland's energy minister, issued a strong objection to the seizure of mineral rights:

> "UK Government proposals to remove the right of Scottish householders to object to drilling under their homes, without so much as debate in the Scottish Parliament, flies in the face of Scotland's cautious, considered and evidence based approach on this issue. It is also fundamentally an issue affecting land ownership rights."[265]

In the United States, on Julia Trigg Crawford's farm is an old and creaky windmill, ravaged by time and a few shotgun shells. "But it's still standing there," says Crawford who may be a bit like that windmill. She's a 6-foot tall former star basketball player for Texas A&M who is standing tall and proud in a fight she says "began as a fight for my family," but became one "for the people, for the landowners who wanted to stand up and fight for their rights but didn't think they could."

The U.S. House of Representatives voted, 353–65, in February 2014 to withhold economic development funds for two years to governmental agencies that allow the use of eminent domain by private enterprise to further private development. There was one exception—TransCanada could continue to use eminent domain rules to seize property for the Keystone XL pipeline.[266] The following month, the Texas Supreme Court refused to review Crawford's petition to set aside lower court rulings that gave TransCanada the right to seize her land.[267]

Forced Pooling

Related to Eminent Domain is Forced Pooling, which requires landowners to surrender mineral rights to private energy companies. Thirty-nine states allow forced pooling.

The concept of forced pooling, dating to the 1930s, allows energy companies to ask the state to allow them to "pool" mineral rights on large areas (often at least 640 acres, depending upon the state), and only when a majority of landowners in the projected parcel (usually 65–90 percent) have already signed over their mineral rights. Forced pooling is not a threat to owners of larger properties who may object and not sign leases, but to the owners of smaller properties, sometimes as little as a quarter acre, who are in the larger parcel.

Under all state laws, energy companies must notify landowners in the prospective parcel; the landowners have a right to protest at a public hearing. If the state approves, then all landowners holding mineral rights are forced to sell or lease their rights to the energy company.

Forced pooling proponents say it is an improvement upon the "Rule of Capture" that allows energy companies with wells on adjacent properties to "capture" the oil or gas beneath the property that is not owned or leased by the company. Forced pooling, say the energy companies, allows them to be more efficient in natural gas extraction by preventing a patchwork of land that is leased and unleased, reduce the number of wells, and benefit all landowners by allowing them to receive a share of royalties, usually 12.5 percent for the entire parcel. Thus, a person who owns 10 percent of the land in the parcel would collect one-tenth (1.25 percent) of the 12.5 percent paid to all owners of the mineral rights. The energy industry also claims it protects landowners from not being able to be compensated when natural gas is taken from their land by the horizontal pipes that run beneath that land. Opponents say forced pooling is nothing less than a government seizure of private property rights to benefit private industry.

In Windsor, Colo., several schools could be sitting over the horizontal tubes of the fracking process. "The district has little control over whether the mineral rights owner can drill for oil and gas beneath the district's schools," according to the *Fort Collins Coloradoan.*[268]

A study by Western Resource Advocates revealed 26 wells within 1,000 feet of a public school in four counties.[269] The non-profit advocacy group pointed out, "It is illegal in Colorado to idle a vehicle for more than 5 minutes within 1,000 feet of a school—but you can drill for oil and gas, spewing potentially toxic chemicals into the air, as long as you aren't closer than 350 feet."[270] Although the wells must be at least 350 feet away, the underground horizontal drilling could extend as much as a mile. Hundreds of schools in Colorado may be forced to enter into lease agreements with gas drillers to protect the surface ground, Stephanie Watson, assistant superintendent for the Windsor district, told the *Coloradoan.*[271]

Prior to the development of horizontal fracturing, Idaho wasn't a good place to look for gas or oil. Until 2013, the state didn't even have a regulatory commission to oversee gas and oil extraction and transportation. That need became apparent when drillers began their search for gas and oil in the south-western part of the state near the Washington border. Among industry-friendly regulations is a setback of only 200 feet from homes for drilling equipment and forced pooling with the approval of 55 percent of owners, significantly fewer than many states require. "If you're going to force someone to have stuff on their property when they own their mineral rights, then they need to have a lot more say so over what happens there," Alma Hasse, a member of the recently-formed Idaho Residents Against Gas Extraction (IRAGE), told a meeting of the Idaho Oil and Gas Commission.[272]

Pennsylvania's forced pooling law had applied only to the deeper Utica Shale, and did not apply to the Marcellus Shale, which is one reason why the industry was once looking to drill into the Utica Shale. A bill (HB 977[273]) to allow forced pooling and to create a state office to regulate its implementation was introduced in the Pennsylvania House of Representatives in

2009. Gov. Ed Rendell gave lukewarm approval, but said he would not sign the bill unless a minimum distance between wells was required and that landowners received "full, fair" compensation.[274] That bill died after three co-sponsors withdrew their support.[275]

A subsequent bill (SB 447[276]), which addressed many of the problems of the previous bill, never left committee. Opposing the bills were 31 environmental and outdoors organizations and Gov. Tom Corbett[277] who declared in April 2011 he would not sign the bill because, "I do not believe in private eminent domain, and forced pooling would be exactly that."[278]

In July 2013, Corbett signed a bill (SB 259[279]) to allow forced pooling.

"If you wanted to address pooling, we should have been doing it in a stand alone bill we could debate, not hiding it in here [a bill clarifying royalty rates] and fast-tracking it through," said Trevor Walczak, president of the Pennsylvania chapter of the National Association of Royalty Owners.[280]

The first test of the new law came in September 2013 when Hilcorp Energy filed court papers to mine gas in the Utica Shale in Lawrence and Mercer counties in the northwest part of the state. The company had previously signed mineral rights for all but 35 of the 3,267 acres it wanted to drill. Martin Clingan, owner of a 200-acre golf course, told the Pittsburgh *Tribune-Review* he believed it wasn't right for owners of a few small properties to block his right to get maximum benefits from his property. He said he had received a signing bonus of $650,000 in 2010, and would receive 18 percent royalties. He explained why he took the money: "Maybe the ground owes me a little bit of something—moneywise—back. If anybody tells you it's not the money, they're crazy." Suzanne Matteo doesn't agree with Clingan. "It's not about the money for us," she said, pointing out, "We want peace; we want clean air."[281]

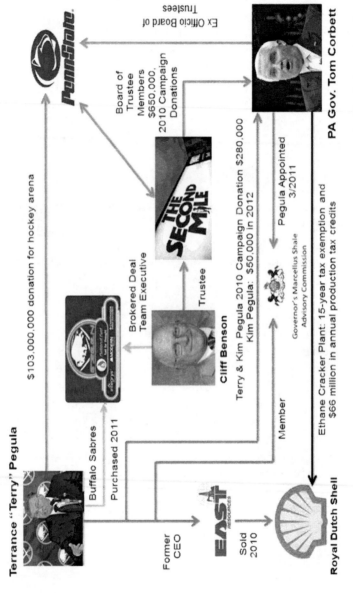

Terrance "Terry" Pegula

$103,000,000 donation for hockey arena

Buffalo Sabres
Purchased 2011

Former CEO

Sold 2010

Royal Dutch Shell

Ethane Cracker Plant: 15-year tax exemption and $66 milllion in annual production tax credits

Member

Governor's Marcellus Shale Advisory Commission

Pegula Appointed 3/2011

Brokered Deal Team Executive

Cliff Benson

Trustee

Terry & Kim Pegula 2010 Campaign Donation $280,000
Kim Pegula: $50,000 in 2012

Board of Trustee Members $650,000, 2010 Campaign Donations

Ex Officio Board of Trustees

PA Gov. Tom Corbett

GRAPHIC: Dory Hippauf

66

CHAPTER 4
Following the Money

The oil and natural gas industry has a history of effective lobbying at the state and national levels.

For the 2008 election cycle (Jan. 1, 2007–Dec. 31, 2008), contributions were about $37.9 million, but dropped during the off-year election cycle of 2009–2010 to $32.1 million. Total contributions for the 2011–2012 election cycle were $63.6 million, with almost 90 percent of it going to Republicans.[282] Total contributions between 1990 and 2013 were $238.7 million.[283]

Contributions in the 2011–2012 cycle by PACs and individuals associated with the oil and gas industry to Mitt Romney, the Republican presidential nominee, as of the FEC report filed Nov. 12, 2012, the week after the election, was $4,763,934.[284] During one of the televised debates, Romney stated the EPA was "a tool in the hands of the President to crush the private enterprise system," and suggested that regulation of fracking should be left solely to the individual states.[285] His faulty reasoning was, "The EPA and those extreme voices in the environmental community and in the President's own party are just frustrated beyond belief that the states have the regulatory authority over fracking."[286]

Other contributions to Republican presidential candidates went to Rick Perry ($969,824), Ron Paul ($178,121), Newt Gingrich ($139,410), Herman Cain ($122,147), Rick Santorum ($91,925), Tim Pawlenty ($52,050), Michele Bachman ($47,140), and Jon Huntsman ($32,000). Gary Johnson, the Libertarian candidate, received $10,700; Dr. Jill Stein, the Green Party candidate, received $750 from the oil and gas industry.[287] President Obama received $710,277. During the 2008 presidential campaign, Obama had received $952,900 from the oil and gas industry.[288] His opponent, Sen. John McCain, accepted $2,599,892.[289]

Oil and gas corporations spent about $144 million in lobbying expenses in the 2012 presidential year and 2014 off-election year, according to the Center for Responsive Politics.[290] All 20 of the top recipients of oil and gas donations in both 2010 and 2012 election cycles were Republicans. Almost 80 percent of all oil and gas industry political contributions went to Republican candidates. Only six of the top 50 candidates who received Industry funds were Democrats, according to data analysis by Citizens for Responsibility and Ethics in Washington (CREW).[291]

The top recipients of oil and gas industry contributions in the 2013–2014 election cycle were Sen. John Cornyn (R-Texas), $865,206; House Speaker John Boehner (R-Ohio), $674,389; Sen. Cory Gardner (R-Colo.), member of Energy and Natural Resources committee, $640,049; Sen. Mary Landrieu (D-La.), chair of the Energy and Natural Resources committee who lost in her re-election campaign, $615,596; Senate Majority Leader Mitch McConnell (R-Ky.), $573,158; Sen. Bill Cassidy (R-La.), member of Energy and Natural Resources committee, $551,553; Sen. Tom Cotton (R-Ark.), $438,905; Sen. James Lankford (R-Okla.), $428,325); Sen. Steven Daines (R-Mont.), member of the Energy and Natural Resources committee, $424,469); and Rep. Mike Pompeo (R-Kansas), member of the Committee on Energy, $357,950. Nineteen of the top 20 recipients of oil and gas funds were Republicans.[292]

The average donation by the oil and gas industry in 2014 to Republican House candidates was about $70,000; for Democrats, the average was about $13,500. For Republican senate candidates, the average donation in 2014 was about $90,000; the average donation to Democrats was about $30,000.[293]

President Obama in August 2013 pointed to a reason why fracking has been allowed in more than 30 states:

> "[W]hat we've seen too often in Congress is that the fossil fuel industries tend to be very influential . . . on the energy committees in Congress. And they tend not to be particularly sympathetic to alternative energy strategies. And, in some cases, we've actually been criticized that it's a socialist plot that's restricting your freedom for us to encourage energy-efficient light bulbs, for example."[294]

The natural gas industry contributed $3.7 million in campaign contributions to current members of the House Energy and Commerce Committee, according to CRP.

The top recipients of oil and gas contributions during the 2012 election for House or Senate, according to the CRP, were Rep. Richard "Rick" Berg (R-N.D.), a member of the Ways and Means Committee, former chair of the Energy and Commerce Committee, and recipient of the Petroleum Council's Legislator of the Year award in 2009 ($433,949); Rep. John Boehner (R-Ohio), speaker of the House ($407,699); Senate Majority Leader Mitch McConnell (R-Ky.) ($395,950); Sen. Orin Hatch (R-Utah), a member of the Subcommittee on Energy, Natural Resources, and Infrastructure ($370,650); Rep. Denny Rehberg (R-Mont.) ($373,469); Sen. John Barrasso (R-Wyo.), a member of the Energy and Natural Resources Committee ($339,716); Rep. Mike Pompeo (R-Kans.), a member of the Energy and Commerce committee ($286,300); Rep. Eric Cantor (R-Va.), House majority leader ($245,250); Rep. Kevin McCarthy (R-Calif.), majority whip ($235,700); Rep. Dave Camp (R-Mich.), chair of the Ways and Means committee ($222,250); and Sen. Dean Heller (R-Nev.), a member of the Energy and Natural Resources Committee ($208,500).[295]

For the 2013–2014 cycle, top recipients were Sen. John Cornyn (R-Texas), $334,609; House Speaker John Boehner (R-Ohio), $236,689; Sen. Mary Landrieu (D-La.), $199,500; Rep. Bill Cassidy (R-La.), $134,750; and Rep. Mike Conaway (R-Texas), $121,900.[296]

The top lobbyists in 2014 were Koch Industries ($13.8 million), ExxonMobil ($12.7 million), Occidental Petroleum (9.2 million), American Petroleum Institute ($9.1 million), and Royal Dutch Shell ($8.4 million).[297]

The top contributors to candidates and outside spending groups in 2013–2014, an off-election cycle, were Koch Industries ($9.5 million), Ken Davis Finance ($2.3 million), Chevron Corp. ($2.1 million), ExxonMobil ($2.1 million) Marathon Petroleum ($1.3 million), Western Refining ($1.3 million), Occidental Petroleum ($1.1 million), Devon Energy ($820,000), Valero Energy ($732,000), Chesapeake Energy ($719,000), Bonanza Energy ($667,000), Energy Transfer ($658,000), and Halliburton Co. ($621,000).[298] More than 98 percent of all donations went to Republicans.

The politician who received the most campaign contributions from 2004 to 2012 from companies and front groups that were identified as specifically involved in fracking, a subset of the oil and gas industry, was Rep. Joe Barton (R-Texas), who received $509,447, according to the Citizens for Responsibility and Ethics in Washington (CREW).[299] Barton, who had previously chaired the Committee on Energy and Commerce, was principal sponsor of legislation to exempt the oil and gas industry from the Safe Water Drinking Act.

Rep. Fred Upton (R-Mich.), chair of the House Energy and Commerce committee, received $704,150 from the oil and gas sector between 1998 and 2015. He is a cousin of Kathleen McClendon, wife of former Chesapeake Energy CEO Aubrey McClendon. The McClendons contributed $12,400 to Upton in 2010 and 2011.[300] Upton acknowledged owning Chesapeake stock valued between $150,000 and $350,000 in 2010. Once a moderate on environmental regulation, Upton now believes that regulation of carbon emissions "threatens to drive energy prices higher, destroy jobs and hamstring our economic recovery."[301]

Sen. Ted Cruz (R-Texas), the first Republican to declare in 2015 he was running for the Presidency, received $294,950 from the oil and gas industry during his first election for senate in 2012. In July 2015, his Super-PAC, Keep the Promise, received a $15 million contribution from billionaire brothers Farris and Dan Wilks, whose fortune was largely from the fracking boom.[302] Toby Neugebauer, a Houston investor, donated $10 million to the Super PAC.[303] Cruz said one of his priorities if elected president would be to dismantle the Environmental Protection Agency.

Influenced by Money, Exercising Their Power

The Republicans' determination to support the natural gas industry was apparent in how they treated President Obama's nomination of Rebecca Wodder, a biologist and environmental scientist, to be assistant secretary of the Department of the Interior. Wodder had previously stated that fracking "has a nasty track record of creating a toxic chemical soup that pollutes groundwater and streams." Shortly after her nomination in June 2011, Sen. James Inhofe (R-Okla.), ranking member of the Senate Committee on Environment and Public Works, and the recipient of $576,250 in donations from 2009–2014 from the oil and gas industry PACs and individuals associated with the industry,[304] declared he had:

> "serious concerns about her nomination: she is the CEO of a far-left environmental organization [American Rivers] and was a staunch supporter of the Clean Water Restoration Act, a bill that would have given the federal government authority over practically every body of water in the country, no matter how small. She is also an active proponent of federal regulation of hydraulic fracturing—a practice that is efficiently and effectively regulated by States."[305]

Several weeks later, before a Senate Energy and Natural Resources hearing, Wodder refused to retract her statement about fracking. This led Inhofe to declare Wodder was "beholden to an extremist environmental agenda,"[306] and lead a campaign against her nomination. Although the Senate Environmental and Public Works Committee, on a strict party vote, later approved her nomination, there were not enough votes for her recommendation by the Energy and Natural Resources Committee.[307] In December 2011, the nomination was sent back to the White House; a month later, the Obama administration reluctantly did not renominate her.

In triumph, Inhofe boasted, "I am pleased that through our rigorous oversight, we succeeded in preventing Rebecca Wodder, another member of President Obama's job-destroying 'green team,' from assuming an influential position within the administration."[308] An Obama official, who asked not to be identified, had a different opinion. He said, "The fact that Rebecca Wodder, a highly qualified nominee, could not get confirmed is another reflection of how some Republicans have ground the Congress to a halt. If the nomination process wasn't so politically supercharged, Rebecca would have been confirmed months ago."[309] Wodder was then appointed senior advisor to the Secretary of the Interior, a position that did not need Senate confirmation.

Inhofe was also involved in the Republican boycott of a vote for Gina McCarthy, President Obama's nominee for EPA director to replace Lisa Jackson. In May 2013, the eight

Republican minority members of the Senate Committee on Environment and Public Works boycotted a confirmation vote.

McCarthy, who had a distinguished career as an environmental advisor for Republican and Democratic administrations in Massachusetts and Connecticut, was the EPA's administrator for air and radiation. The minority had previously stated they would not block McCarthy's nomination if she would provide written answers to a cumbrous and unprecedented 1,079 written questions; many of the questions had nothing to do with environmental policy and everything to do with politics. McCarthy answered those questions, but the senators still blocked her nomination.

The Republican reason for their political action was a poorly-disguised smokescreen of their philosophy of minimal regulation and maximum profit, often at the expense of the environment, worker safety, and public health:

McCarthy, said Sen. John Barrasso (R-Wyo.):

> "has been extremely unresponsive with the information we requested. We're not asking to amend any bedrock environmental laws. We're asking for access to the scientific data and reasoning behind the justification for expensive new rules and regulations that continue to cause high unemployment."[310]

Sen. Mike Crapo (R-Idaho) agreed: "Far too many federal regulations, many from the EPA, make it harder to do business and maintain jobs while providing little benefit to their intended purposes."[311]

"This latest attempt by Republicans in the Senate to derail progress to protect our air, our water and the health of our families is nothing short of cowardly," said Michael Brune, executive director of Sierra Club. Sen. David Vitter (R-La.), the committee's minority chair, and the other Republicans, said Brune, "have tried everything to push the agenda of big polluters, constantly and grossly abusing the Senate rules with filibusters, holds and delays. Now, they are simply shirking their duties to show up and legislate."[312]

Barrasso, Crapo, and Vitter are among the top recipients in Congress of campaign contributions from persons and political action committees associated with Big Energy. Common Cause reported Vitter received about $1.1 million, Barrasso received about $527,000, and Crapo received about $329,000.[313]

Robert F. Kennedy Jr. was equally angry about the Republican obstructionism and their support of Big Energy at the expense of the environment, health, and safety:

> "Vitter is an unabashed mouthpiece for the petroleum industry and record breaking receptacle for petrodollars. . . . With cash gushers of oily money cascading down their open gullets, the Republican leadership's mercenary devotion to Big Oil shouldn't shock us. However, the boldness of the party's most recent assault on the public interest might cause us to ponder how GOP's honchos' knee jerk slavishness to petroleum interest has infected its rank and file."[314]

President Obama, several months after the Republicans first refused to allow a vote on McCarthy's nomination, condemned Republican obstructionism. McCarthy, said the President, has been "forced to jump through hoops no Cabinet nominee should ever have to—not because she lacks qualifications, but because there are too many in the Republican Party right now who think that the Environmental Protection Agency has no business protecting our environment from carbon pollution. The Senate should confirm her without any further obstruction or delay."[315]

McCarthy was finally confirmed by the Senate, 59–40, in July 2013, but only after Majority Leader Harry Reid agreed he would not change the rules on filibusters in exchange for the Republicans approving up-or-down simple majority votes on several of President Obama's nominees.

The same month the Republicans unloaded a barrage of written questions at McCarthy and then walked out of a committee meeting to vote on moving the nomination to the full Senate, a peer-reviewed research study showed an interesting correlation between conservatives and attitudes about energy conservation and the environment. Writing in the *Proceedings of the National Academy of Sciences*, Drs. Dena M. Gromet, Howard Kunreuther, and Richard P. Larrick summarized their research as showing:

"[M]ore politically conservative individuals were less in favor of investment in energy-efficient technology than were those who were more politically liberal. This finding was driven primarily by the lessened psychological value that more conservative individuals placed on reducing carbon emissions. [Further], in a real-choice context, more conservative individuals were less likely to purchase a more expensive energy-efficient light bulb when it was labeled with an environmental message than when it was unlabeled."[316]

It would not be unreasonable to assume that whoever President Obama nominated to become the EPA director would have met resistance by elected officials of a political party that not only distrusts environmental protectors but also generally opposes the science that reveals humans in industrialized nations have been a primary cause of global warming.

EFFECTS OF *CITIZENS UNITED* AND CORPORATE INFLUENCE

Political influence through lobbying increased significantly because of the Supreme Court's 5–4 decision in *Citizens United v. Federal Election Commission*.[317] That decision, handed down in January 2010, gave corporations the same First Amendment rights as individuals. Thus, one of the political effects of that decision was to allow corporations to create significant campaign assistance for political candidates, including producing campaign ads for print and electronic media. The decision was supported by the U.S. Chamber of Commerce, the National Rifle Association, and most conservative organizations. In response, Barack Obama said that the *Citizens United* decision "gives the special interests and their lobbyists even more power in Washington—while undermining the influence of average Americans who make small contributions to support their preferred candidates."[318] Later that month, in his State of the Union address, the President was even more forceful, arguing, "The Supreme Court reversed a century of law to open the floodgates for special interests—including foreign corporations—to spend without limit in our elections. Well I don't think American elections should be bankrolled by America's most powerful interests, or worse, by foreign entities."[319]

Chevron, the third largest U.S. corporation, took advantage of the *Citizens* decision to contribute $2.5 million in October 2012 to the Congressional Leadership Fund,[320] which supports conservative politicians; almost all of those they support are in favor of expanded oil and natural gas drilling, and have no objection to the use of fracking. Chevron also acknowledged spending $9.5 million solely in lobbying costs for members of Congress in 2011.[321]

CONFLICTS OF INTEREST WITH THE KEYSTONE XL

Against strong public opposition, most of Congress aligned itself with the oil and gas industry in the battle to construct the Keystone XL pipeline from Alberta to the Gulf Coast. However, there have been notable exceptions.

"Americans are realizing that transporting large amounts of this corrosive and polluting fuel is a bad deal for American taxpayers and for our environment," Rep. Ed Markey (D-Mass.) said in March 2013.[322] That same month, the U.S. Senate voted, 62–37, for a non-binding resolution supporting the development of the pipeline; two months later, the House, by a 241–175 vote, largely along party lines, voted to approve the pipeline in a quixotic attempt to bypass the President's authority, and exempt the pipeline from further environmental review.[323] In November 2014, the House again voted, this time 252–161, to support completion of the Keystone XL pipeline. It was the ninth time the House voted to support the pipeline or to force the President to approve it.[324] A similar bill in the Senate failed by one vote. In January 2015, with Republicans now in control of the Senate, a bill to support construction of the pipeline passed, 62–36. However, President Obama vetoed that proposed legislation, only the third time in six years that he exercised his right of veto. The Senate did not have enough votes to override that veto.

House Speaker John Boehner, one of the strongest pro-ponents for completing the Keystone XL pipeline, declared the President "is standing with a bunch of left-fringe extremists and anarchists."[325] What was not recorded in the public record was that Boehner in 2010 had invested $15,000–$50,000 in each of seven companies that would benefit if the pipeline was completed.[326]

Those who voted for the proposed legislation to complete the Keystone XL had received about $13 million in campaign funds in 2014; those opposed had received about $800,000, according to data bases of Oil Change International and the Center for Responsive Politics.[327] The average campaign contribution of those voting for the pipeline was about $50,000; those opposed had received less than $6,000.[328]

The financial contributions were entirely legal because of the *Citizens United* decision. "This circle of legalized corruption is the best possible investment for the fossil fuel industry today, as the polluters confront a rising tide of climate science and citizen activism. With the heat on from fossil fueled global warming, the price of bribery is going up," said Stephen Kretzmann and Matthew Maiorana, writing for Oil Change International.[329]

While Americans were debating the Keystone XL pipeline, the Department of State and the Federal Energy Regulatory Commission quietly approved development of a combination pipeline–rail transportation system to move tar sands oil through middle America. The Department of State permit allows Enbridge to accept and move about 350,000 barrels of oil per day from Alberta, Canada. The FERC permit allows Enbridge to develop a rail facility with a capacity to unload tar sands oil from Alberta, Canada, and move the oil from Flanagan, Ill., to Port Arthur, Texas in the Flanagan South and Seaway Twin pipelines.[330] The rail facility is scheduled to be completed in 2016.

BRIBES AS ACCEPTABLE BUSINESS PRACTICES

To help assure both a market and higher prices overseas, the energy exploration and distribution corporations have a plan to reduce government oversight. The American Petroleum Institute, the Independent Petroleum Institute, the U.S. Chamber of Commerce, and the National Foreign Trade Council sued the Securities and Exchange Commission in October 2012 to reduce oversight of corporate finances and practices.[331] The plaintiffs wanted the federal courts to declare unconstitutional, under the

First Amendment, a regulation within the Dodd–Frank Act of 2010 that requires energy companies to disclose on their annual reports any payments made to foreign governments and their leaders. The purpose of that regulation, as well as the Foreign Corrupt Practices Act of 1977, which the plaintiffs also protested, is to eliminate bribes paid to get and keep contracts. The API claimed the SEC rule would make "American firms less competitive against state-owned oil companies."[332] The U.S. District Court for the District of Columbia in July 2013 vacated the SEC regulations and ruled the regulatory agency "misread the statute to mandate public disclosure of the reports."[333]

"Transparency is . . . taking a beating as U.S. firms fight for the right to bribe foreign governments and hide their activities from American shareholders and the citizens of the nations where they do business," observed Sara Jerving and Mary Bottari, writing in *PR Watch*, a web-based newsletter published by the Center for Media and Democracy.[334]

Following the Money into Louisiana

Companies directly involved with the release of 210 million gallons of oil and the deaths of 11 workers on a BP oil platform in the Gulf Coast in 2010 could be exempt from lawsuits filed in Louisiana. The Republican-controlled state legislature passed a bill, quickly signed by Gov. Bobby Jindall, that effectively killed a lawsuit filed by the Southeast Louisiana Food Protection Authority–East[335] against 97 oil and gas companies. The Authority demanded the companies either fix damage caused by dredging waterways or pay the Authority to have the damage fixed.

State Sen. Robert Adley, who sponsored the bill,[336] called the Authority a "rogue agency." In signing the legislation, Gov. Jindal said the bill "will help stop frivolous lawsuits and create a more fair and predictable legal environment."[337] Attorney General Buddy Caldwell had warned the new law could have unintended consequences, including the dismissal of lawsuits aimed at BP, Halliburton, and others involved in the Gulf Oil spill.[338] The new law "is a huge victory for the oil and gas industry as well as the economy for the state of Louisiana," Don Briggs, president of the Louisiana Oil and Gas Association, told the New Orleans *Times–Picayune*.[339] Environmentalists had a different opinion. "If there is a pipeline rupture in the coastal zone, the ability of local governments to litigate will be limited," said John Barry, former vice-president of the Authority, who Jindal removed after the suit was filed.[340]

A coalition of environmental organizations had previously documented that Jindal had received, over a 10-year period, about $1 million in campaign donations from the oil and gas industry.[341] The largest contributors to Sen. Adley's campaigns were oil and gas industry lobbyists and front groups. Adley, who owned ABCO Petroleum for two decades, is currently the owner of Pelican Gas Management Co.[342]

Following the Money into Ohio

Marathon Petroleum received about $365 million in local, state, and federal subsidies since 1998, according to Good Jobs First database. In 2011, Marathon received tax credits worth about $78.5 million for retaining 1,650 jobs and creating 100 more in Ohio.[343] In fiscal year 2014, Marathon had a net income of $2.7 billion.[344]

In his announcement about the subsidies, Gov. John Kasich, a conservative Republican, told the public, "I think Marathon always wanted to be here. All we're doing is helping them."[345]

Nevertheless, although he received fossil fuel campaign donations of $213,519 during the previous decade, Gov. Kasich twice tried but failed to convince his own party to pass legislation that would increase the severance tax on the shale gas industry. When asked by Jim Siegel of the *Columbus Dispatch* if the tax would ever be increased to what Kasich believes would be a fair level, the governor replied, "I think we wait for Batchelder to retire." Batchelder is William Batchelder, Ohio speaker of the House. A *Dispatch* investigation revealed he was Ohio's top recipient of oil and gas money. The investigation also revealed that during the 2010–2012 election cycle, oil and gas interests contributed about $830,000 into Republican legislative campaigns. This included $255,000 from the Ohio Oil and Gas Association and $131,000 from Chesapeake Energy.[346]

Between July 1, 2011 and June 30, 2013, the natural gas

industry donated about $1.8 million to candidates, officials, and parties, according to Common Cause of Ohio.[347] In 2014, the oil and gas industry donated about $1.5 million to political and court candidates, including the Senate President, the incoming and outgoing House speakers, the Senate chair of the Energy Committee, and $8,000 to one Supreme Court justice and $7,200 to another, according to an analysis of campaign financing by the *Columbus Dispatch*.[348]

"What the drilling industry has bought and paid for in campaign contributions they shall receive," Ohio Supreme Court Justice William M. O'Neill wrote in a sarcastic dissenting opinion to a 4–3 Court decision to deny local governments a right to regulate fracking.[349]

"It seems like the law only changes when this industry allows the law to be changed," Jack Shaner of the Ohio Environmental Council told Columbus *Dispatch* reporter Jim Siegel in August 2013.[350]

Ohio is not an isolated example of corporate influence upon state legislatures.

Following the Money into New York

Hoping to overturn the state's moratorium on fracking—which became a ban in 2015—the natural gas industry seeded the state with political contributions.

Between 2007 and July 2013, the fracking industry contributed $1.1 million to candidates and spent about $15.6 million for advertising and lobbying, according to Common Cause of New York. Support industries (including General Electric and several advertising/PR agencies) spent $9.6 million on candidates and $17.9 million for lobbying; business interests (including the New York Farm Bureau and the Business Council of New York State) spent $3.2 million in contributions and $13.9 million in lobbying, according to Common Cause.[351]

As fracking became more prevalent in Pennsylvania, and more contentious, numerous corporations and trade associations increased their contributions and lobbying in N ew York. The American Petroleum Institute had spent $416,000 on lobbying between 2007 and 2012 but increased its campaign to $1.2 million between January 2012 and August 2013, according to Common Cause of New York. ExxonMobil, which had spent an average of about $194,000 a year in lobbying expenses between 2007 and 2012, increased its lobbying expense to $2.2 million in the 18 months beginning January 2012. America's Natural Gas Alliance, which did no active lobbying before January 2012, spent $290,000 in the 18 months after January 2012, according to Common Cause.[352]

Although fracking corporations and special interests donated about $1 million to Gov. Andrew Cuomo and $61,264 to House Speaker Sheldon Silver, both Democrats, most of their contributions went to Republicans, donating $3.1 million to state senate Republican candidates and $1.3 million to the state's Assembly Democrats since 2012, according to Common Cause of New York.[353]

During the 2012 elections, the oil/gas industry contributed more than $400,000 to candidates in the 10 southern tier counties, eight of which border Pennsylvania. Most of the contributions were to Republican incumbents from Broome County, likely to be the first county fracked if the state lifts the moratorium.

The industry and its allies contributed $82,428 to Debra A. Preston, the Republican incumbent executive of Broome County, according to data compiled by Common Cause of New York. Oil/gas donations were 22 percent of the $373,858.12 total she had raised.[354] Preston's opponent, Tarik Abdelazim, who ran a strong anti-fracking campaign, received no gas industry support.

In the race in Senate District 52 (Broome, Chenango, and Tioga counties), the industry gave $190,700 to State Sen. Thomas Libous, an incumbent Republican. With campaign contributions of about $1.3 million, Libous easily outspent and defeated John Orzel, who questioned fracking, raised only $12,500, and took 33 percent of the vote.[355] (Between 2007 and March 2013, Libous received $353,205 in contributions from the shale gas industry, according to Common Cause.[356] In July 2015 he was convicted of lying to FBI agents and forced to resign from the Senate.[357]) Why the industry gave so much to Libous in a race his opponent had almost no hope of winning could be because in addition to being a resident of Binghamton, the county seat of Broome County, Libous is the Senate's deputy majority leader, and someone who would likely have had a major influence on whether the state went ahead with plans to frack the southern tier. In 2013, Libous successfully blocked a moratorium bill from getting to a vote.[358] The senator's official website says one of his top priorities is "fighting for stronger ethical standards in state government."[359]

The oil and gas industry also contributed $70,345 to State Sen. Catharine M. Young,[360] and $51,375 to State Sen. Thomas F. O'Mara.[361] Both Republican incumbents had no opposition in their 2012 races, and had been leaders while members of the state Assembly. Young is also assistant majority whip and is a member of the Environmental Conservation committee.[362]

In races for Assembly seats in the southern tier counties, the oil/gas industry contributed $45,217 in six races, including three races where the incumbents were unopposed. With one excep-tion, the donations to the winning candidates, five Republicans and a Democrat, were 9.4 to 14.7 percent of their fund-raising totals.[363]

For Congress, Rep. Tom Reed (R-Corning) defeated Nate Shinagawa, who was endorsed by the anti-fracking movement. Reed received $89,521 from individuals and PACs associated with the oil/gas industry.[364]

Money may have also influenced the head of New York's Republican party. Gov. Andrew Cuomo is "caving in to environ-mental luddites," state Republican chair Ed Cox said at an annual dinner of the Independent Oil and Gas Association of New York a day after elections in November 2013. "Any reasonably knowledgeable person in New York's political world knows that Cuomo's official position is pure political posturing," said Cox who demanded Gov. Cuomo "end the moratorium immediately."[365] At the time, Cox and his wife held about $4.25 million in Noble Energy stock; he collected $250,000–$350,000 in 2012 as a member of the company's board of directors, according to the Associated Press.[366]

Against the pro-fracking interests, anti-fracking groups spent only $5.4 million in lobbying and $1.9 in campaign contributions between 2007 and July 2013,[367] but increased their influence upon voters by extensive use of social media, including Twitter and Facebook, and by strong grassroots organizing.

Following the Money into Pennsylvania

Mixed into Pennsylvania's energy production is a symbiotic relationship of business and government and a history of corruption and influence-peddling in all of its 67 counties.

The Keystone State, with no limits on political contributions, is also the target of energy company generosity to politicians. The natural gas industry, through PACs and individuals associated with the natural gas industry, contributed about $8 million to Pennsylvania candidates and their PACs between 2007 and November 2014, according to Common Cause.[368]

Total lobbying expenditures, in addition to direct financial contributions, between 2007 and the end of 2014 were $41 million.[369] Almost $31 million of the $41 million for lobbying was spent by 10 corporations and industry groups: Marcellus Shale Coalition ($11.31 million), PPL, ($5.04 million); Range Resources ($4.36 million), UGI ($3.25 million), Chesapeake ($2.10 million), EQT ($1.97 million), Shell ($1.20 million), Dominion ($1.04 million), Exco Resources ($956,000), and Alpha ($752,733).[370]

Of the top 15 members of Congress from Pennsylvania who accepted campaign funds from PACs and individuals affiliated with energy companies, 11 were Republicans, according to Common Cause. Rep. Tim Murphy (R-Pa.), accepted $477,034 from individuals and PACs associated with the oil/gas industry from 2001 through 2014, according to Common Cause.[371] He was tenth among all Congressional recipients. Murphy is chair of the oversight and investigations subcommittee of the Energy and Commerce Committee. Sen. Pat Toomey (R-Pa.) received $477,034 between 2009 and 2015 from individuals and PACs associated with the oil/gas industry.[372]

Rep. Lou Barletta (R-Pa.) is on the list of the top third of House members who received campaign donations from individuals and PACs associated with the gas industry. Barletta received $31,725 in donations in the 2009–2010 election cycle.[373] For the 2011–2012 cycle, now a freshman representative, Barletta received $29,450.[374] For the 2013–2014 cycle, he received $32,800 from individuals and PACs associated with the oil/gas industry, according to Common Cause.[375] Barletta is a member of the Congressional Marcellus Shale Coalition—

and, according to reporting by Rick Dandes of the Sunbury (Pa.) *Daily Item*, "owns stock in [eight] gas industry companies, including several that are actively drilling in Pennsylvania and beyond."[376] Depending upon market fluctuation, value of the stock in 2012 was worth more than $75,000. However, as reported by the *Daily Item*:

> "[T]here are no rules in place to bar him from buying stock. The practice is both legal and permitted under the ethics rules that Congress has written for itself, which allow law-makers to take actions that benefit themselves or their families except when they are the lone beneficiaries.
> "The financial disclosure system Congress has implemented also does not require the lawmakers to identify potential conflicts when they take official actions that intersect or overlap with their investments."[377]

Barletta's communications director told the *Daily Item* that Barletta "has an incredibly diversified portfolio" that includes non-energy stocks.[378]

At the state level, the natural gas industry contributed about $3.1 million to Tom Corbett's political campaigns between 2000 and April 2012;[379] about $2.6 million of that was for his campaign for governor in 2010.[380] An additional $6 million may have been funneled through the Republican Governor's Association (RGA), most of it cloaked by multiple transfers from the RGA office in Washington, D.C., and untraceable to specific individuals or corporations.[381] The first influx of energy company money came in Corbett's first run for attorney general. Will Bunch, columnist for the *Philadelphia Daily News*, revealed the long connection between Corbett and the industry:

> "The $450,000 in campaign checks that energy mogul Aubrey McClendon [CEO of Chesapeake Energy] wrote that fall [2004] helped elect a man he said he'd never even met—a relatively obscure GOP candidate for Pennsylvania attorney general, Tom Corbett. . . .
> "That investment arguably changed not just the history but also the political direction of the state. The influx of cash helped Corbett narrowly win the closest attorney general's

race in Pennsylvania history and propelled him toward the governor's mansion. . . .

"Did Oklahoma gas driller McClendon see the coming boom in drilling in the gas-rich, Pennsylvania-centered formation known as the Marcellus Shale back in 2004? And did he see his massive campaign contributions—filtered through an obscure GOP committee—as a shrewd down payment on future political access and influence?

"Or was it merely a case of what McClendon and Chesapeake officials have maintained all along—that the energy millionaire was simply writing so many checks for conservative causes that year, including $250,000 for the notorious John Kerry-bashing Swift Boat Veterans for the Truth, that he wasn't even aware that his cash was going to the state's future top prosecutor?"[382]

Chesapeake Energy, which once owned drilling rights to about 15 million acres in the U.S., and about 1.5 million acres in the Marcellus Shale,[383] with 1,935 permits[384] became the state's leading producer of natural gas,[385] and also a company with a spotty history of environmental concerns.[386] By the end of December 2015, Chesapeake had the second most violations (610) of all companies drilling in Pennsylvania, behind Cabot (618), and ahead of Swepi (475), Talisman (394), and Range Resources (370).[387]

Not long after the November 2010 elections, Gov.-elect Tom Corbett and a swarm of Republican legislators began developing plans to further open the state to fracking.

Rep. Brian L. Ellis, sponsor of HB 1950, which would become the base for Act 13, received $23,300 from PACs and individuals associated with the oil and gas Industries. Sen. Joseph Scarnati, the senate president *pro-tempore* who sponsored the companion Senate bill (SB 1100), received $500,870, as of the end of 2014, according to Marcellus Money.[388] Rep. Dave Reed, chair of the majority policy committee, received $178,950; Rep. Mike Turzai, majority floor leader, received $272,100; Rep. Samuel H. Smith, minority leader, received $102,500; Sen. Don White, a member of the Environmental Resources and Energy committee, received $82,100; Rep. Timothy Solobay received $138,942; Sen. Jake Corman, chair of the Appropriations Committee, received $89,650. Of the 20 Pennsylvania legislators who received the most money from the industry between 2001 and 2012, 16 are Republicans, according to Common Cause.[389]

Pennsylvania State Sen. Gene Yaw, chair of the Environmental Resource and Energy Committee, in his first run for the Senate in 2008 accepted only $3,700 in campaign contributions from energy companies; the largest were $1,000 donations from Anadarko Petroleum and Chesapeake Energy.[390] In his first re-election campaign in 2012, he received no contributions from the shale gas industry. He didn't need it. Yaw leased 148 acres in Lycoming County to Anadarko, and had received a signing bonus. In March 2013, now in his second term, Yaw introduced two bills to expand natural gas usage in the state. When asked by WNEP-TV about a possible conflict-of-interest, Yaw replied he had signed his lease with Anadarko in 2006 before he was elected to the senate; but in 2011, he renewed that lease for an additional five years. "Conflict of interest is the most easily-thrown-about concept when you can't think of anything else to say," said Yaw.[391] However, Eric Epstein of Rock the Capital, a watchdog on state government, countered Yaw's cavalier attitude. "You can not simultaneously promote and regulate an industry, if you have the ability to pass and sponsor legislation that will increase your quarterly dividend."[392] A week later, outside a meeting closed to the public to discuss issues related to Anadarko's request to drill in Loyalsock State Forest, Jim Hamill of WNEP-TV again asked Yaw to discuss his ties to the shale drilling industry and about his possible conflict of interest. "No, and I don't know what part of N-O you don't understand. Last week we talked and you turned it into a totally negative article, something that should have never been done," Yaw testily replied.[393]

State Reps. Timothy J. Solobay and H. William DeWeese were the two Democrats who received the most money from Big Energy. Solobay, who voted for Act 13, received $60,325, according to Common Cause; DeWeese, who opposed the bill, received $60,250.[394]

Rep. DeWeese, first elected in 1976, had been Speaker of the House and Democratic leader. In April 2012, DeWeese was sentenced to 30 to 60 months in prison for theft, conspiracy, and conflict of interest, all related to the use of legislative staff and public resources for campaign work.[395] He was also ordered to pay $25,000 in fines and about $117,000 in restitution. DeWeese maintained that the prosecution, begun while Tom Corbett was attorney general, was political.[396] DeWeese had charged that Corbett's prosecution of state legislators, at first primarily Democrats, was to set a base for Corbett's run for governor. However, DeWeese also charged that Corbett targeted him because he opposed Corbett's belief that natural gas companies should not pay an extraction fee.[397] DeWeese, whose district includes about 2,000 wells, increased his criticism after Corbett became governor.

Political donations may also have bought access to state regulators. An investigation by Allegheny Front determined that in a one-year period before the state passed Act 13, Corbett's energy executive, Patrick Henderson, had more than 130 meetings with fossil fuel executives and lobbyists, and only 29 meetings with environmental groups.[398] Allegheny Front also determined that in his 21-month tenure as secretary of the Department of Conservation and Natural Resources, Richard J. Allan had at least 54 meetings with fossil fuel executives and lobbyist, and no meetings with the state's two largest environmental groups.[399]

Following the Money into North Carolina

Several North Carolina politicians with close ties to the energy industry pushed pro-Industry bills into law that allowed the companies to begin fracking operations in July 2015.

The *Asheville Citizen–Times* pointed out, "There are a lot of problems with that law. There was little public input, the chemicals used are secret, local governments are shut out of the process and the moratorium on fracking was lifted before the Mining and Energy Commission had completed its work.[400]

The *Fayetteville News–Observer* suggested a continuation of a temporary moratorium:

"Use the period to let oil and gas prices settle to assess

whether it's even worth drilling into North Carolina. And use it to allow more health and environmental data to accumulate in fracking states.

"In a few years, if the oil and gas markets are strong and the health and environmental issues known, North Carolina could decide whether to allow fracking." [Dec. 20, 2014][401]

The Legislature wasn't convinced, and fast-tracked the bill into law. Gov. Patrick L. McCrory, who had worked for Duke Energy for 28 years, received $308,836 in campaign donations from Duke Energy, Progress Energy, and persons associated with Duke and Progress during his 2008 and 2012 campaigns.[402] He lost to Bev Perdue in 2008, and then won the election in 2012; during that campaign, he received $173,905 from the oil and gas industry.[403]

Among the legislators is State Sen. Bob Rucho, co-chair of the North Carolina Energy Policy Commission. Rucho received $29,650 in campaign funds from Duke Energy and $25,000 from Piedmont Natural Gas.[404] McGuireWoods, a 1,000-member law and lobbying firm that represents the American Petroleum Institute, Halliburton, and Koch Industries, was behind a $2,897 per plate fund-raising dinner for Rucho in 2013, according to an investigation by Greepeace.[405]

George Howard, vice-chair of the state's Mining and Energy Commission, has a pro-Industry perspective, claiming it is "physically impossible" for fracking to affect aquifers. Howard, as Greenpeace pointed out in 2013, has a significant financial stake in the fracking industry, including a multi-million dollar shale play project in Pennsylvania [and has] also invested in the area of North Carolina most likely to be leased by fracking companies."[406]

The Department of Environment and Natural Resources "has become not the guardian of sustainability, but the second-best, most-genuflecting Department of Commerce in North Carolina," said State Rep. Rick Glazer.

Corporate Welfare for Gas Drillers

The extent of energy company influence is easily seen by a vote in the U.S. Senate in March 2012. President Barack Obama had called for an end to the $4 billion direct subsidy to oil companies.[407] Overall, state and federal governments provide more than $21 billion a year in direct and indirect subsidies, according to research conducted by Shakuntala Makhijani for the *Oil Change Report*.[408] The subsidies include more than $5 billion a year for continued development of fossil fuel resources.

Over the past decade, the five largest oil companies had made $1 trillion in profits. "We have been subsidizing oil companies for a century. That's long enough," the President said in 2012.[409] The Senate disagreed. Forty-three Republicans and four Democrats blocked the elimination of subsidies. Although the final vote was 51–47 to end the subsidies, a simple majority was not enough because the Republicans threatened a filibuster that would have required 60 votes to pass the bill. A Think Progress financial analysis revealed that the 47 senators who voted to continue subsidies received $23,582,500 in career contributions from the oil and gas industry. In contrast, the 51 senators who had voted to repeal the subsidies received $5,873,600.[410] Slightly more than a year after the Senate blocked the elimination of subsidies, President Obama again called for a repeal of the subsidies, charging, "[B]ecause billions of your tax dollars continue to still subsidize some of the most profitable corporations in the history of the world, my budget once again calls for Congress to end the tax breaks for big oil companies, and invest in the clean-energy companies that will fuel our future."[411]

Numerous reporters and columnists, taking the word of the oil and gas industry, have stated that eliminating the subsidies would raise gasoline and home heating oil prices. However, Media Matters[412] and Factcheck.org,[413] citing several energy analysts and economists, pointed out that eliminating the subsidy would have negligible effect upon prices to consumers.

A cracker plant takes natural gas and breaks it up to create ethylene, primarily used in plastics. Several oil giants are building ethane cracker plants in Louisiana and Texas, the first construction since 2001. The largest plants are those that Chevron Phillips and Exxon Mobil are building in Baytown, Texas. Each plant, which should be operational in 2017, is expected to cost $5–6 billion.

In Pennsylvania, Tom Corbett extended benefits to a foreign corporation, which was thinking about building an ethane cracker plant about 30 miles northwest of Pittsburgh. Shell Chemical, a division of Royal Dutch Shell, which owns or leases about 900,000 acres in the Marcellus Shale basin,[414] considered placing the plant beside the Ohio River in Pennsylvania, Ohio, or West Virginia. All three states were interested, but Pennsylvania held out the most lucrative corporate welfare check for the company, which had a profit of $26.8 billion in 2012, and which spent $14.5 million in lobbying during that year, about 10 percent of all lobbying costs for all gas and oil corporations.[415] The Pennsylvania legislature handed over a 15 year exemption from local and state taxes, apparently without consulting local officials.[416] Tom Corbett then approved a $1.65 billion tax credit over 25 years, tweeting, "A crackerplant would create up to 20,000 permanent jobs in Southwest PA."[417] However, the reality is considerably lower. Shell stated it planned to hire only 400 to 600 persons; because of the location, many new employees would probably be Ohio and West Virginia residents. Even if all possible indirect jobs—including more low-wage clerks at local fast food restaurants—were added, the most would be about 6,000–7,000 employees.

Pennsylvania may have been able to attract the plant without giving up so much corporate welfare. A Shell news release stated the company "looked at various factors to select the preferred site, including good access to liquids rich natural gas resources, water, road and rail transportation infrastructure, power grids, economics, and sufficient acreage to accommodate facilities for a world scale petrochemical complex and potential future expansions."[418] Even then, Shell said it could be "several years" before construction would begin.

Corbett may have believed that extending corporate welfare to Royal Dutch Shell was just good business, and would spur job creation and the economy. But, there is another probability for his generosity, and it's both personal and political. Dory Hippauf's "Connecting the Dots" series[419] explains why Corbett may have been so generous with extending tax credits and subsidies, and it begins with billionaire Terrence (Terry) Pegula, who sold East Resources to Royal Dutch Shell in 2010 for $4.7 billion.[420] East Resources, according to reporting in the *Buffalo*

News, had "a less-than-stellar track record in the environmental dicey business of drilling for natural gas."[421] Terry and Kim Pegula donated $280,000, [422] and Shell donated about $358,000, to Corbett's political campaign for governor.[423] Kim Pegula added another $100,000 to Corbett's campaign in February 2011, a month after Corbett was inaugurated.[424] One month later, Corbett appointed Terry Pegula to the newly-formed Marcellus Shale Advisory Commission, which was loaded with pro-fracking energy company executives before being disbanded in 2011 after delivering its pro-industry report.

Compromising the Opposition

Although Big Energy has been relatively effective in seeding Congress with campaign donations that would sprout whenever additional regulations were proposed or environmental proponents for national office were brought to various committees for review, a financial relationship with the Sierra Club would appear to be bizarre. Yet, the 1.4 million member Sierra Club—the nation's largest pro-environmental organization and one which actively opposes the development of coal as an energy source—had received $27 million between 2007 and 2010 from Chesapeake Energy, but may have deliberately withheld that information from its members.[425] However, since August 2010, the Sierra Club has refused to accept donations from the natural gas industry. By 2010, "our view of natural gas [and fracking] had changed [and we] stopped the funding relationship between the Club and the gas industry, and all fossil fuel companies or executives," said Michael Brune, who became Sierra's executive director in March 2010, succeeding Carl Pope who had accepted the Chesapeake donations for the Sierra Club.[426] The Sierra Club philosophy on energy has evolved from believing natural gas can be a "bridge fuel" to eliminating the use of fossil fuels "as soon as possible." Its new policy, established in January 2015, affirms a ban:

> "We must replace all fossil fuels with clean renewable energy, efficiency and conservation. Fracking poses unacceptable risks to our communities, our environment and our climate. . . . Fracking has negative impacts on air and water quality and frequently necessitates unacceptable drawdowns

on surface water and groundwater. . . . Sierra Club Chapters are the best judge of the most strategic way at the state and local level to end fracking and limit the damage from fracking until it can be ended. Chapters are authorized to decide whether the best course of action at the state and local levels is to advocate for bans, moratoria and/or stronger regulations on fracking."[427]

Conclusion

Lisa Sumi, an environmental scientist who had been research director of Earthworks: the Oil and Gas Accountability Project (OGAP), and the person who coined the term "fracking" in 2004, believes:

"This betrayal of the public interest also severely weakens state claims that they can protect the public from the impacts of the shale boom. A rule–even an improved rule–on the books means little if an oil or gas company knows that it can be ignored with little or no consequence."[428]

"When state agencies say they will 'regulate' or 'monitor' hydraulic fracturing, we should not accept this as a guarantee of any kind," says Eileen Fay, an animal rights/environmental writer. Fay argues that because of legislative corruption, it is a responsibility of citizens to protect their own health and environment by "putting pressure on our legislators."

It's possible that significant campaign contributions and potential employment in the private sector didn't influence politicians or staff of environmental regulatory agencies to rush to embrace the natural gas industry and its controversial use of high volume horizontal hydraulic fracturing. It's possible these politicians had always believed in fracking, and the natural gas industry was merely contributing to the campaigns of those who believed as they do. However, with the heavy amount of money spent by the natural gas lobby and willingly accepted by certain politicians, there is no way to know how they might have voted or how strong their support of fracking would be had there been no lobbying and campaign contributions.

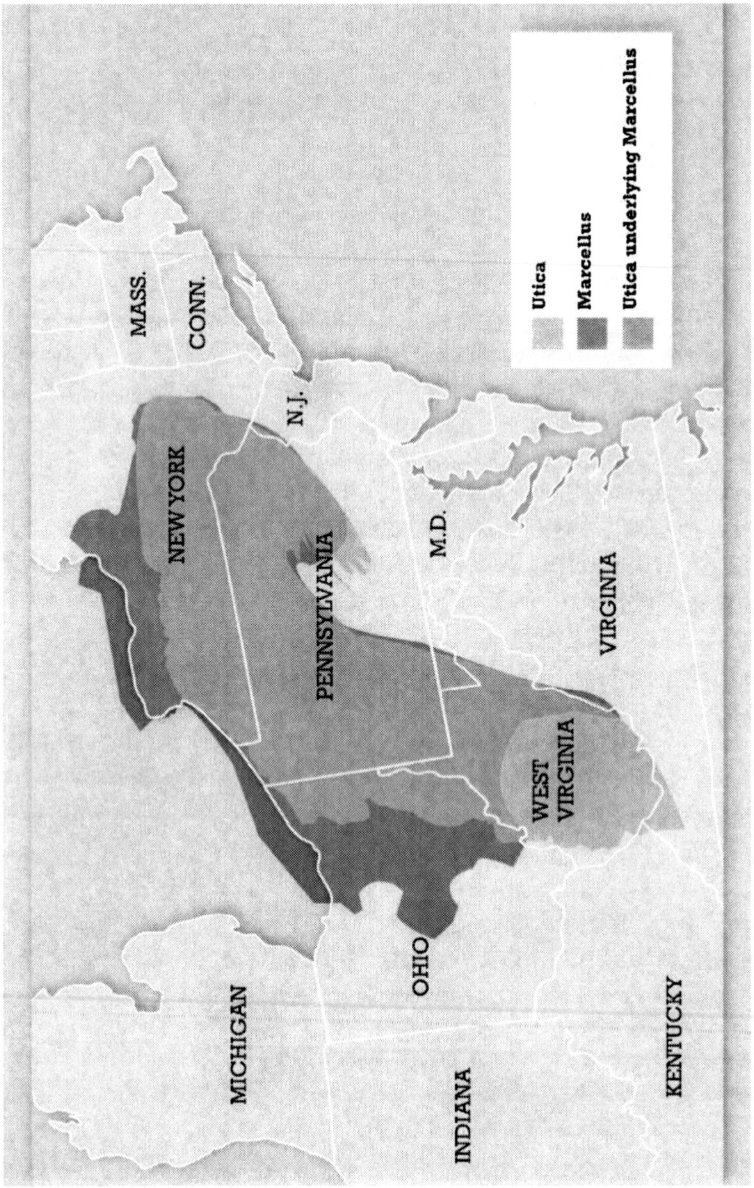

Legend:
- Utica
- Marcellus
- Utica underlying Marcellus

GRAPHIC: Marcellus Shale Coalition

CHAPTER 5
Pennsylvania's
All-Gas Administration

With the recently-closed Sunoco oil refinery in suburban Philadelphia in the background, a Pennsylvania government official declared his agency "will work hand in glove, very closely, cooperatively and spiritedly with anyone who would want to participate in activities at this facility."[429] Employees, he said, will "work night and day" to help any industry get necessary permits to reopen the facility, perhaps as a natural gas processing refinery. The official continually touted the economic benefits the natural gas industry was bringing to the state.

If that government official was the head of the Department of Community and Economic Development, Labor and Industry, or even the governor's press secretary, it might be just another political speech. But this official was Michael Krancer, and his agency was the Department of Environmental Protection.

On this fifth Friday in June 2012, the public realized it was now official—multi-billion dollar energy companies and the state agency that is charged with protecting the environment were having a public love fest.

The DEP headquarters is located in the Rachel Carson state office building in Harrisburg, named for the woman whose book, *Silent Spring* (1962), was a call for environmental awareness and protection. The irony of an emasculated regulatory agency was not lost upon those who believe energy companies and the government are in collusion.

Almost as an afterthought, Krancer, speaking on behalf of the potential owners of the refinery, claimed they weren't interested "in any short circuiting of environmental protections because they would need to live in this community."[430]

The words sounded good, but the constant noise and odors,

93

products of turning crude oil into gas, were just some of the problems the residents endured in order to benefit from a vigorous job market and the economic benefits that accrued to small businesses, such as cafés and restaurants. The 110 year history that Sunoco Marcus Hook was in the community also included numerous incidents of air and water pollution. The latest, before the plant closed, was toxic air emissions in May, June, and December 2008 that resulted in the DEP, under Gov. Ed Rendell, fining Sunoco $173,000.[431] The Environmental Protection Agency (EPA) had listed the refinery as a regulated site,[432] having had serious violations in the three years prior to its closing in December 2011.

In July, less than two weeks after Krancer said the state was willing to work with energy companies, Gov. Tom Corbett, who took office in January 2011, announced the state awarded a $15 million grant to Brazilian petrochemical company Braskem,[433] which had recently bought a part of the refinery to split propylene, an oil byproduct, into polypropylene, a polymer that is used in dozens of products, including carpets, cups, and chairs.

There are more than 515,000 gas wells in the country, with Texas having the most (98,000), followed by Pennsylvania (70,000), West Virginia (60,000), and Oklahoma (40,000).[434] The last two years of the Ed Rendell administration and the four years of the Tom Corbett administration (2011–2015) are not unusual or unique, but a microcosm of what is happening in any state that has experienced a sudden "boom" in oil and natural gas production.

At the epicenter of the Marcellus Shale exploration, Pennsylvania rushed to embrace the natural gas industry and its use of horizontal fracking, apparently disregarding its own Constitution that requires the state to recognize, "The people have a right to clean air, pure water, and to the preservation of the natural, scenic, historic and esthetic values of the environment."[435] In contrast to the words in the Constitution, each of the state's well pads takes up about 8.8 acres;[436] access roads, pipelines, and staging areas take up an additional 21 acres.[437]

Gov. Corbett's public announcements in March 2011, two

months after his inauguration as Pennsylvania governor, established the direction for gas drilling in Pennsylvania. In his first budget address, Corbett declared he wanted to "make Pennsylvania the hub of this [drilling] boom. Just as the oil companies decided to headquarter in one of a dozen states with oil, let's make Pennsylvania the Texas of the natural gas boom. I'm determined that Pennsylvania not lose this moment."[438] (By the end of 2014, Pennsylvania was second to Texas in production of natural gas.[439])

Enthusiastic about fracking Pennsylvania, Corbett told an industry-sponsored conference in Philadelphia in September 2012, "I am convinced that we are at the beginning of a new industrial revolution and you are at the tip of the spear."[440]

Like a five-year-old running from house to house on Halloween to gather as much candy as possible, Pennsylvania's elected leaders were wide-eyed ecstatic about natural gas drilling. Between Jan. 1, 2005, and the end of 2014, the Pennsylvania Department of Environmental Protection issued 15,775 permits to 124 companies,[441] and denied fewer than 400 requests to construct unconventional gas wells.[442]

By executive order in March 2011, Gov. Corbett created the Marcellus Shale Advisory Commission (MSAC),[443] which he loaded with persons from business and industry.[444] The MSAC had begun as the Energy & Environment Committee (EEC) during the transition from election to inauguration. Dory Hippauf's "Connect the Dots" series noted that EEC members included representatives of four major lobbying groups that represented 29 clients involved in fracking.[445] Not one member of the MSAC was from the health professions; of the seven state agencies represented, not one member was from the Department of Health. Lt. Gov. Jim Cawley, who chaired the Commission, boasted, "The Marcellus [Shale] is revitalizing our main streets in downtowns."[446] Like Corbett, he didn't say anything about protecting the people's health and their environment.

Corbett's first major political appointment after his election in November 2010 was to name C. Alan Walker, an energy company executive, to head the Department of Community and

Economic Development. The *Pennsylvania Progressive* identified Walker as "an ardent anti-environmentalist and someone who hates regulation of his industry."[447] A *ProPublica* investigation revealed not only had Walker given $184,000 to Corbett's political campaign, but that:

> "In 2002, three of Walker's coal companies notified Pennsylvania's Department of Environmental Protection that they had run out of money and were going to stop treating the 173 million gallons of polluted water they produced each year and released into tributaries of the Susquehanna River. The state eventually got a court injunction to force them to continue treating the wastewater as required by state and federal law."[448]

In his first budget bill, Corbett authorized Walker to "expedite any permit or action pending in any agency where the creation of jobs may be impacted."[449] This unprecedented reach apparently applied to all energy industries.

Corbett selected Richard J. Allan, who has a B.S. in environmental sciences/biology to be director of the Department of Conservation and Natural Resources, Although he volunteered with non-profit environmental organizations, most of his professional career was as a scrap metal dealer; he had no experience in public lands management.[450]

To head the Department of Environmental Protection, Corbett selected Michael Krancer, whose wife and father had donated a combined $306,500 to Corbett's 2010 political campaign;[451] Krancer himself donated $15,447 in gifts to Corbett since 2007.[452] Krancer was an attorney and partner in law firms that represented energy clients. From 1999 to 2007, and then from 2009 to 2011, he was a judge and then chief judge of the Pennsylvania Environmental Hearing Board (EHB); by most accounts, he was fair, concerned, and tireless. Inbetween times when he was on the EHB, he was assistant general counsel for the Excelon Corp., a national energy company.

Krancer willingly bought into the political philosophy that fracking would bring jobs to the state and improve the economic position that, like the rest of the country, had fallen into the Great Recession during the last years of the George W.

Bush–Dick Cheney administration. However, on Pennsylvania's western border, Gov. John Kasich (R-Ohio), as much a proponent of natural drill fracking as Corbett and Krancer, took a more realistic view: "You could have a situation where we are not getting the jobs, they're [oil/gas industry] taking the resources, and all their profits and they're heading home."[453]

Within months of Corbett's inauguration, Krancer took personal control over DEP's issuance of any violations. By Krancer's decree, every inspector could no longer cite any well owner in the Marcellus Shale development without first getting the approval of Krancer or John T. Hines, the executive deputy secretary. Hines later left the DEP to become government relations advisor for Shell Oil Co.[454]

"It's an extraordinary directive [that] represents a break from how business has been done" and politicizes the process, John Hanger told *ProPublica*.[455] Hanger, DEP secretary in the Ed Rendell administration, said the new rules "will cause the public to lose confidence entirely in the inspection process."[456] The new policy was the equivalent of every trooper having to get permission from the state police commissioner before issuing a traffic citation, Hanger said.[457] Because the new policy is so unusual and broad "it's impossible for something like this to be issued without the direction and knowledge of the governor's office," said Hanger. Corbett denied he was responsible for the decision. Five weeks after Krancer's decision was leaked to the media, and following a strong negative response from the public, environmental groups and the media, the DEP rescinded the policy—which Krancer now claimed was only a three-month "pilot program."[458]

Nevertheless, the DEP was still going to issue fewer fines than previously. Part of it was a lack of field personnel to cover the burgeoning gas industry; much of it was a political decision. "The idea of an issuance of a notice of violation is not to issue a fine," Michael Krancer told the state Senate Appropriations Committee in February 2012. The purpose of DEP inspections, said Krancer, "is to bring conduct which is potentially volatile to the attention of the operator so the operator can do something about it."[459] Andrea Ryder, DEP district supervisor for air quality assurance, said the department's job "is to educate, not to penalize; we're trying to get people to do it right the next time." DEP data reveals there were many "next times" where fines weren't issued or were negligible, compared to the companies' profits. Fewer than one-third of all violations result in penalties.[460] Auditor General Eugene DePasquale, elected in November 2012, skewered the DEP for that philosophy. "When DEP does not take a formal, documented action against a well operator who has contaminated a water supply, the agency loses credibility as a regulator and is not fully accountable to the public,"[461] he wrote as part of an investigation that was published in July 2014. The Auditor General's report also noted, "While a 'cooperative' approach may bring about instances of operator compliance, such collaboration raises concerns that DEP chooses to play the role of a mediator instead of a regulator. [Mediation] should not be the substitute for enforcement actions required by law."[462]

Pennsylvania isn't the only state that is reluctant to fine oil/gas companies. The Railroad Commission of Texas is "more concerned about bringing people into compliance than we are in punishment after the fact," said Commissioner David Porter. Only about two percent of all violations result in fines.[463] Most violations are repeated, without financial consequences.

Corbett's core belief that political loyalty was more important than environmental oversight and regulation became more obvious when he created a deputy secretary for oil and gas management, severed oil and gas staff from reporting to regional directors, and replaced regional managers who were Democrats.

Historically, new administrations did not normally replace DEP executives who were of the "wrong" political party. One of those replaced was George Juguvic Jr., in charge of the Southwest District that included Pittsburgh. Jugovic, now general counsel for PennFuture, an environmental advocacy non-profit, has a B.S. in environmental science, a law degree, taught classes in environmental criminal law at the University of Pittsburgh, and had been at DEP several years. He was replaced by Susan Malone, a Corbett political ally who had worked with him in the Office of Attorney General, but had no experience in environmental science or management.

FRACKING THE STATE PARKS AND FORESTS

In October 2012, the Corbett administration ordered John Norbeck, the widely-respected head of the state's park system, to resign. Norbett told the *Pittsburgh Post-Gazette* he was told he was terminated because of "philosophical differences," but was not told the reason for his forced resignation.[464] The *Harrisburg Patriot-News* reported:

> "Environmentalists and advocates for the state parks fear Norbeck's abrupt departure could be a sign that the Corbett administration is preparing to open the park gates to drilling rigs to tap natural gas under the Commonwealth's popular park system; Norbeck is known to be an opponent of drilling in the state parks."[465]

In March 2010, the *Pocono Record* (Stroudsburg, Pa.) had called for a moratorium on drilling in state parks and forests:

> "Pennsylvania should not risk the integrity of the beautiful forests that gave the state its name. Yet that's a possibility, given the gas-drilling free-for-all over the Marcellus shale deposit that underlies a huge swath of Penn's Woods. Energy companies are clamoring for expanded leases on state forest land to drill and extract the natural gas. . . .
> "[Gas] drilling presents a tempting potential economic bonanza, bringing millions of dollars in lease revenue to the state and jobs and revenues to counties and communities with large deposits. Responsible legislators also must weigh its substantial risks, which include chemical spills, water pollution, the incursion of new roads in pristine remote land and heavier traffic in rural areas. . . .
> "Our legislators have a sworn duty to protect Pennsylvania's natural resources. . . . Don't let our historic forests become a cash cow for drillers."[466]

Sixty-one of the 120 state parks sit above the Marcellus Shale.[467] The Nature Conservancy of Pennsylvania argued that deforestation from drilling, even if not on public forests and parks, would alter and destroy the ecological balance.[468]

Seven months after the *Pocono Record* editorial, Gov. Ed Rendell, who had authorized leasing mineral rights to 74,000 acres of public land in 2008,[469] by executive order imposed a moratorium on drilling on public lands.[470] However, shortly after taking office, Corbett repealed environmental assessments of gas wells in state parks, and decided that about half of the 1.5 million acres of state forest lands over the Marcellus Shale could be leased for mineral rights.[471]

In 2013, Corbett officially issued his own executive order that rescinded Rendell's executive order. Corbett claimed drilling adjacent to state forests and parks, with pipelines running beneath public land, could add about $75 million to the forthcoming year's budget.[472] By then, there had already been about 1,400 wells drilled in state forests before the beginning of horizontal fracturing in 2008; about half were still in production.[473]

"Pennsylvania politicians sold gas companies the right to pollute Pennsylvania's land, air, and water for bargain basement prices," said Josh McNeil, executive director of Conservation Voters of Pennsylvania. "For their $23 million political investment," said McNeil, "gas companies avoided hundreds of millions in taxes that could have paid for thousands of teachers, roads and desperately needed environmental protections."[474]

Within two weeks after becoming governor in January 2015, Tom Wolf rescinded Corbett's executive order, stating that the purpose of state parks and forests "is to allow people to enjoy nature at its best."[475] But whatever the new governor could do was mitigated by what had already been done the previous four years.

SACRIFICING THE PEOPLE

In his first budget message Tom Corbett had pushed for draconian cuts in education, health care for the poor, child care, and services to the disabled. The 2012–2013 budget, readily passed by the Republican-controlled legislature, reflected even deeper cuts to health and human services.

In his first two annual budgets, Tom Corbett had cut funding to the Departments of Environmental Protection and Conservation and Natural Resources. In his first year, he cut about $2 million from the DEP and $27 million of the $82 million budget for the Department of Conservation and Natural Resources.

The following year, Corbett cut DEP funding about $10.8 million, a 7.8 percent decrease, reducing the state appropriation to about $124 million.[476] However, the DEP also received a $40 million cut in its $268 federal appropriation from the previous year, leaving it with about $51 million less to work with. Michael Krancer did not publically complain about the reductions, although significant increases in drilling left overworked staff who were having difficulty conducting regular field inspections.

In a commentary published in the Harrisburg (Pa.) *Patriot-News*, State Rep. Greg Vitali pointed out, "Nonunion [DEP] staff has not received a salary increase in four years. Noncompetitive salaries combined with increasing workloads due to these staffing cuts have made it difficult for the DEP to attract and retain quality people."[477] Vitali says he was told by a former senior staff member of the DEP, "We are hemorrhaging jobs to the oil and gas industry."[478]

Michael Krancer claimed the decreased funding would not affect the department's well inspections. He told the House Appropriations Committee in March 2012 that the inspections program "has been a function of the permit fees," and there was increased staff to deal with the increased number of wells.[479] Eighty inspectors conducted 5,000 field inspections in 2011. What Krancer did not say is that each inspector had increased workloads, and there was a paperwork backlog in processing permit fees. (Nationally, between 2010 and 2011, the number of drilling rigs increased by 22 percent, but the number of inspections fell by 12 percent, according to an analysis of 50,000 inspection reports by the *New York Times*.[480]) A scathing report by the state's auditor general three years after Corbett became governor charged that the DEP is "underfunded, understaffed, and does not have the infrastructure in place to meet the continuing demands placed upon the agency by expanded gas development."[481]

PUTTING 'PUBLIC HEALTH AND SAFETY AT RISK'

Instead of adding staff to take care of that backlog, Tom Corbett issued an industry-friendly executive order in July 2012. That order directed the DEP to "establish performance standards for staff engaged in permit reviews and consider compliance with the review deadlines a factor in any job performance evaluations."[482] That requirement, said George Jugovic Jr., is "going to put the public health and safety at risk." Jugovic told the Associated Press, "I think the message is clear. Issuing the permit has a higher priority than doing a fair and thorough job of insuring that the application complies with the law." The order, said Jugovic, "does not recognize any of the complexities of what the agency is required to do [except to] beat down an already demoralized staff."[483] Statements made in a closed-door legal proceeding the previous year revealed that because of political and administrative pressures, DEP staff spent as few as 35 minutes per application, with supervisors spending as few as two minutes to review each application.[484] "Such a cursory review leaves little time to consider and include necessary permit provisions or technical requirements to protect public health and the environment," Lisa Sumi wrote in *Breaking All the Rules,* a 124-page summary of extensive field research conducted by Earthworks.[485] Sumi, an environmental consultant and author of several major research studies on energy, noted, "In Pennsylvania, citizens have conducted research and file reviews that have exposed deficiencies in permits [but] citizens do not have the resources to review all permits, nor should they be doing the work that agencies are charged to do."[486]

CLOUDING TRANSPARENCY

In keeping with its developing trend to cloud transparency, the DEP established a policy to delay notifying the public of water contamination. The DEP had previously issued "notices of contamination" to the public as soon as DEP scientists had made the determination. The new policy restricted notification until after DEP executives were notified and approved the release of information. "This change in procedure is unnecessary and a dangerous policy," said Karen Feridun of Berks Gas Truth. "With this secretive change in policy," without giving the public time to comment, said Thomas Au of the Sierra Club, "the DEP has violated fundamental democratic values of transparency and public participation."[487]

A CNNMoney investigation revealed the DEP "does not have to notify landowners if a violation is discovered," even if there are significant health, safety, and environmental problems.[488] Reporter Erica Fink learned there were 62 safety violations in four years on property owned by four Lycoming County families; there were 26 natural gas wells on their properties. None of the families were notified of the violations. The DEP posts violations online but "the digital records are short on specifics—most importantly whether a violation poses a health risk."[489] When Fink tried to do a hard copy file review, she found additional problems. DEP refused several requests for interviews, said Fink. Additionally, the process to review specific public records "required a visit to the regional DEP office [in Williamsport, Pa.], which had to be scheduled weeks in advance" and the information was "largely in legal and technical language." It was from a meticulous review of paper records that Fink learned there had been a spill of 294 gallons of frac fluid at one of the wells, but "[t]here was no mention of this spill in DEP's online records, and the paper records did not clearly indicate whether the ground water was tested after the spill."[490]

The Auditor's General's investigation, released in July 2014, confirmed Fink's observations: "DEP's documentation was, and continues to be, egregiously poor."[491]

A Valentine's Day Gift

On Tuesday, Feb. 14, 2012, Tom Corbett, surrounded by Republicans, gave the natural gas industry a Valentine's Day gift, and proudly signed Act 13 of 2012, an amendment to Title 58 (Oil and Gas Act; 58 P.S. §601.101 *et seq.*)[492] of the Pennsylvania Consolidated Statutes. Pennsylvania's new law that regulates and gives favorable treatment to the natural gas industry was initiated and passed by the Republican-controlled General Assembly. The House had voted 101–90 for passage,[493] the Senate voted 29–20.[494] Both votes were mostly along party lines. Environmental and conservation groups spent under $50,000 to lobby against the proposed act; the natural gas industry spent about $1.3 million for it, according to data compiled by Dory Hippauf.[495]

Act 13 is generally believed to be "payback" by Corbett and the Republican legislators for campaign contributions. "The industry has largely had its way in Pennsylvania and has spent millions to put their friends in the state legislature and the Governor's mansion," said James Browning, Common Cause regional director of state operations.[496] The focus for the oil and gas industry, says Browning, is on protecting these investments and maintaining access to key elected officials."

Much of Act 13 was modeled after legislation prepared by the American Legislative Exchange Council (ALEC),[497] which promotes a right-wing agenda. A month before the bill passed, Tom Corbett and all members of the state's House of Representatives received a three page letter from Kathryn Z. Klaber, president of the Marcellus Shale Coalition, and Stephanie Catarino Wiseman, executive director of the Associated Petroleum Industries of America. In that letter, Klaber and Wiseman emphasized the economic benefits of shale gas drilling, and issued a veiled threat that any attempt to increase safety and environmental standards "will result in an overly burdensome regulatory environment which will be factored in when investment decisions are made."[498] Attached to that letter was a five page list of suggested modifications to the current bill. Almost all of those suggestions were incorporated into the final bill.

"Thanks to this legislation," said Corbett upon signing the bill, "this natural resource will safely and fairly fuel our generating plants and heat our homes while creating jobs and powering our state's economic engine for generations to come."[499]

State Sen. Charles T. McIlhinney was effusive in his praise:

> "While not perfect, it is a balanced and thoughtful approach to protecting our environment and regulating an industry that is here to stay in Pennsylvania."[500]

McIlhinney may have believed what he said, but he was wrong on almost every point.

Act 13, says State Sen. Jim Ferlo, "is completely inadequate in regulatory oversight." Most of the Legislature's Democrats agreed.

State Rep. Mark Cohen, the state's longest-serving legislator, said the new law "produced far too little revenue for local communities, gives the local communities local taxing power which most of them do not want, because it pits one community against the other, and gives no revenue at all to other areas of the state."

State Sen. Daylin Leach, a Democrat from suburban Philadelphia, agrees. "At a time when we are closing our schools and eliminating vital human services, to leave billions on the table as a gift to industry that is already going to be making billions is obscene."

THE LAW

Act 13 allows companies to place drilling wells 500 feet from houses, and 1,000 feet from public drinking supplies.[501] Well pads, which are areas where trucks can park and technicians can mix the chemicals for fracking, can be 300 feet from the residential buildings. The law also allows compressor stations to be placed in residential districts 200 feet from a homeowner's property line and 750 feet from houses. Drillers can also place wells 300 feet from streams, creeks, rivers, ponds, and wetlands.

The Pennsylvania law also requires companies to provide fresh water, which can be bottled water, to areas in which they contaminate the water supply, but doesn't require the companies to clean up the pollution or even to track transportation and deposit of contaminated wastewater.

The law doesn't allow for local health and environmental regulation, and forbids municipalities from appealing state decisions about well permits. As a result of numerous concessions, the natural gas industry is given special considerations not given any other business or industry in Pennsylvania.

Numerous sections of Act 13 call for tax waivers or subsidies. Waivers of state sales tax on the purchase of large items or hotel rooms for out-of-state workers were in place before Act 13 was passed. The new law also provides subsidies to the natural gas industry, but provides no incentives or tax credits to companies to hire Pennsylvania workers. (The law didn't close the "Delaware Loophole," which allows businesses headquartered or incorporated in other states to avoid paying corporate taxes on net income in the state they actually do business.) The law also allows subsidies for purchase of trucks that weigh at least 14,000 pounds that run on natural gas. The program, which began December 2012, pays companies $25,000 or 50 percent of the cost, whichever is less, for purchase or to reimburse costs to convert trucks that run on diesel or gasoline. Trucks that use compressed natural gas (CNG) usually cost 10–20 percent more than trucks that use gas or low-sulfur diesel. The three year program is expected to cost $20 million.[502]

THE SEVERANCE TAX

Pennsylvania is the only state among the 16 major gas-production states that does not have a severance tax. Gov. Ed Rendell had wanted a 5 percent severance tax on the value of gas produced, plus a fee of 4.7 cents for every 1,000 cubic feet in order to create additional revenue for the general fund. However, he couldn't get the support of the Legislature, and decided not to pursue that source of revenue. If Pennsylvania imposed a severance tax, the industry had a plan. According to Karen Feridun of Berks Gas Truth:

> "They were already saying that they were too poor to set up operations *and* pay a severance tax, so they planned to lobby for getting a 3–5-year window to set up before starting to pay the tax. Drilling is like wringing a sponge. You get the most liquid out of a sponge the first few times. You get more out in subsequent wrings, but nothing like what you got the first couple of times. The drillers planned to drill like crazy in those first 3–5 years, knowing that they'd be paying next to nothing when the tax finally took effect."

As early as 2008, the *Pocono Record* called for a tax.[503] More than two years later, with Tom Corbett now governor, it said the state was foolish for not having a severance tax:

> "[G]as is a sought-after resource in this energy-hungry world, and a tax could produce much-needed revenues to help host communities deal with the effects of drilling and to

create a pool of funds to remediate the inevitable environmental damage drilling causes.

"It would be great if Corbett showed some guts and spearheaded an effort to bring Pennsylvania in line with other states by—at least—imposing a tax on that resource.

"Then again, selling out is all too common, while true political courage is rare indeed."[504]

The natural gas industry didn't need to worry. Tom Corbett, pushing hard for the gas industry, had originally wanted no tax or impact fees placed upon natural gas drilling.[505] At the time Corbett signed Act 13, Pennsylvania had a $4 billion deficit,[506] which could have been significantly reduced had Pennsylvania imposed fees and taxes in line with other states. However, even *if* there was an adequate severance tax, there was debate as to whether it should have been allocated to the general fund or used primarily to repair the damage caused by the natural gas industry.

As public discontent increased, Corbett suggested a 1 percent tax, which was in the original House bill before being deleted. In contrast, other states that allow natural gas fracking have tax rates as high as 7.5 percent of market value (Texas) and 25–50 percent of net income (Alaska).[507] The Pennsylvania rate can vary, based upon the price of natural gas and inflation, but is still among the five lowest of all states that currently have gas fracking. Over the lifetime of a well, Pennsylvania will collect about $190,000–$350,000, while West Virginia will collect about $993,700, Texas will collect about $878,500, and Arkansas will collect about $555,700, according to data compiled and analyzed by the non-partisan Pennsylvania Budget and Policy Center.[508]

The additional revenue for 2013–2014, according to the Center would be $573 million if Pennsylvania used the rate assessed by Texas to the oil and gas industry; if it used the West Virginia rate, it would have been an additional $303 million for 2013–2014.[509]

In September 2012, the Pennsylvania Public Utility Commission announced it billed energy companies $206 million for impact fees from 4,453 wells for 2011.[510] The energy industry is believed to have earned about $3.5 billion that year from wells in Pennsylvania, according to a previous AP analysis.[511] The PUC reported that the state was taking $25 million; 40 percent of the remainder was to go to 37 counties and 1,500 municipalities that had wells, and 60 percent split among state agencies that dealt with drilling. Had state law allowed taxes upon oil and gas property, Pennsylvania counties in the Marcellus Shale could have received about $600 million revenue for 2014.

The industry also received a huge gift in 2002, long before the Marcellus Shale drilling was begun. The state's Independent Oil and Gas Association found a loophole in the state law which specifically mentions coal as being taxable, but does not mention oil and gas. A state Supreme Court decision in 2002 [*IOGA et al. v. Board of Assessment Appeals of Fayette County, Pennsylvania, and County of Fayette, Pennsylvania*] determined that because state law did not specifically include oil and gas drilling and property in the law, county assessors could not tax oil and gas production. Attempts by Democrats to change the law in 2012 never came to a vote in the Legislature. The failure to tax oil and gas drilling was "a huge travesty," said James Hercik, president of the Assessor's Association of Pennsylvania.[512]

With the election of Dr. Tom Wolf as governor in November 2015, the creation of a severance tax again became a critical issue. Gov. Wolf's proposal, similar to what Gov. Ed Rendell had proposed six years earlier, was to tax production at 5 percent, with an additional tax of 4.7 cents per thousand cubic feet. That additional tax, still beneath most taxes in other gas-production states, would be primarily used to restore, and then improve, funding to the state's educational systems and to improve environmental regulation. Opposing the new tax were drillers, suppliers, lobbying groups, and the Pennsylvania Chamber of Business and Industry, all of whom conveniently overlooked health and environmental issues to incorrectly claim such a tax not only "would cost Pennsylvania more than $20 billion in lost gross domestic product and up to 18,000 jobs over the next decade,"[513] but would also be the highest tax in the country.[514] An analysis by StateImpact Pennsylvania reveals the tax rate would be the highest in the country, but only if the price of natural gas remained near historic lows; as the price rebounds to previous levels, the effective rate would be well within the range of that of other states.[515] In a blistering response to the Chamber, Gov. Wolf said the attack on the proposed tax "is based less on the facts and our shared policy goals, and more on your need to appease oil and gas special interests. [The] facts you outlined in your letter are simply talking points from the oil and gas drillers. . . . It is bogus rhetoric, and it does nothing to change Pennsylvania, fix our schools, or create jobs."[516] Nevertheless, within six weeks of the new budget being due, the Chamber placed 30-second TV ads, all reiterating gas/oil talking points, in the Pittsburgh, Harrisburg, and Wilkes-Barre/Scranton markets, and uploaded onto YouTube longer videos that attacked the proposals,[517] and deftly claimed higher taxes would force companies to leave the state. However, the companies already paid little or no corporate income taxes, according to an analysis by the Pennsylvania Budget and Policy Center.[518]

CONSTITUTIONAL ISSUES

Two separate New York courts upheld the rights of local governments to create and enforce zoning ordinances to ban fracking and fracking-related activity. [*Cooperstown Holstein Corporation v. Town of Middlefield* and *Anschutz Exploration Corp v. Town of Dryden*.] Both rulings came in the same week in February 2012. A state appellate court in May 2013 affirmed the towns' right to ban fracking.[519] Slightly more than a year later, the state's highest court, the Court of Appeals, consolidating both cases, affirmed the right of a municipality to limit or ban fracking.[520]

The Dryden case, which set a statewide precident and had ramifications in other gas-drilling states, began with the formation of the Dryden Resource Awareness Coalition (DRAC) in 2009, with Marie McRae and Deborah Cipolla-Dennis as the Coalition's first leaders. Attorney Todd M. Mathes established a base for the possibility that municipalities could restrict fracking operations by legally regulating traffic and road use.[521] Attorneys David and Helen Slottje devoted their time and energy to help develop a legal foundation. They advised the Coalition to get enough signatures to convince the town supervisors to take action that led them to pass an ordinance in August 2011 to uphold its rights to determine zoning and ban fracking. (Helen Slottje, who helped other New York communities defend themselves against the power of the oil and gas industry, earned the Goldman Environmental Prize in 2014, the highest honor an environmentalist can receive. The response by *Marcellus Drilling News*, an industry front publication, was that the award "was started years ago by a husband and wife . . . who both attended University of California at Berkeley. Which should tell you all you need to know about the prize. Created by wackos for wackos. So it's no surprise the foundation awarded this year's $150,000 prize to a New York wacko . . . [who] is in wacko mecca—San Francisco— . . . to pick up her check."[522]

"For too long the oil and gas industry has intimidated and abused people, expecting to get away with it. That behavior is finally coming back to haunt them, as communities across the country stand up and say 'no more,'" said Deborah Goldberg, managing attorney with EarthJustice, which represented the Town of Dryden.[523]

In Pennsylvania, Act 13 guts local governments' rights of zoning and long-term planning, doesn't allow local governments to impose restrictions on gas well drilling stricter than the state law, and doesn't allow municipalities to deny any gas company the state-mandated right to drill. The state legislature specifically inserted those clauses into Act 13 to negate a Pennsylvania Supreme Court decision in 2009 that affirmed the right of municipalities to determine zoning and planning regulations. [*Huntley Huntley v. Oakmont* 600 Pa. 207, 964 A.2d 855[524]].

The law also gives the Public Utilities Commission (PUC) the right to overturn local zoning, and to act as a *de facto* court of appeals for individuals and the gas companies.[525] The PUC is composed of appointed officials and, thus, is likely to have a predetermined political bias.

In April 2012, seven municipalities, the Delaware River-keeper Network, and physician Mehernosh Khan filed suit against the Commonwealth to block enforcement of parts of the law that revoked local government authority to create and enforce zoning ordinances, and charged that Act 13 was unconstitutional under several sections of the state's constitution and the 14th Amendment of the U.S. Constitution. The suit charged that Act 13 was created "to elevate the interests of out-of-state oil and gas companies," and that by overriding local zoning authority, "municipalities can expect hundreds of wells, numerous impoundments, miles of pipelines, several compressor and processing plants, all within [their] borders, they will be left to plan around rather than for orderly growth."

By mixing residential living with industrial activity, "The state has surrendered over 2,000 municipalities to the industry," said Ben Price of the Community Environmental Legal Defense Fund. He called the law "a complete capitulation of the rights of the people and their right to self-government."[526]

In securing support for Act 13, the Corbett Administration had claimed the support of the Pennsylvania State Association of Township Supervisors (PSATS), an association of 1,455 local municipalities and townships. "PSATS has been cited again and again by the Governor and the Majority leadership to show how their legislation had the 'participation' and 'support' of local elected officials," *Marcellus Monthly* reported.[527] However, it wasn't the members who supported Act 13 but the association's paid executives, a split that become apparent with the subsequent suit.

David Ball, councilman for Peters Twp. in the southwestern part of the state, argued that the association's reporting of the suit in its April 2012 newsletter was "deliberately misleading." In a letter dated May 7, 2012, to David Sanko, PSATS executive director, Ball attacked Sanko's failure to defend the townships:

> "Where was PSATS whose job it is to advocate to ensure our protections? The court confirmed with its injunction that the legislation posed real danger to municipalities. By its silence, PSATS exposed municipalities to danger that needed to be enjoined. . . .
> "PSATS gives all appearances of having sold out to the gas industry

Sanko was an appointed member of Tom Corbett's Marcellus Shale Advisory Commission.

Several Pennsylvania residents went to municipal meetings to speak against Act 13 and to advise the local governments of the impact the Act had against their communities. However, PSATS "was actively working against us," says Karen Feridun of Berks Gas Truth. She says PSATS "was telling municipalities that it was a waste of their time and that it was a state, not a local, issue."

At the PSATS annual meeting in May 2012, with Gov. Tom Corbett as the featured guest speaker, the membership passed two resolutions opposing the state prempting local authority from regulating zoning regarding the oil and gas industry.

By speaking on behalf of Act 13, while staying silent on certain parts that allowed the state to take jurisdiction over local zoning in order to make gas drilling easier for energy companies, PSATS' professional staff abrogated the rights of municipalities to protect the health, safety, and welfare of residents, as was brought out in the legal challenge to the entire Act. Like David Ball, many of the supervisors and council members who opposed Act 13 are conservative Republicans. They didn't object to the drilling, as long as public health and safety are protected, but they did object when the state usurped local authority in order to benefit one industry.

The Commonwealth Court ruled against the state, and upheld the rights of municipal government to establish adequate zoning. In his majority opinion issued in July 2012, President Judge Dan Pellegrini, wrote that the zoning provision of Act 13:

> "violates substantive due process because it does not protect the interests of neighboring property owners from harm, alters the character of neighborhoods and makes irrational classifications—irrational because it requires municipalities to allow all zones, drilling operations and impoundments, gas compressor stations, storage and use of explosives in all zoning districts, and applies industrial criteria to restrictions on height of structures, screening and fencing, lighting and noise."[528]

In a separate order, Judge Pellegrini declared the zoning provision of Act 13 "unconstitutional and null and void." As expected, Gov. Corbett appealed the decision to the state

Supreme Court. In a public statement, Corbett continued to hammer on the economic benefits of natural gas drilling, while ignoring the health and pollution effects from fracking. In a subsequent ruling, Judge Pellegrini informed the Corbett administration, "Jobs do not justify the violation of the constitution."[529] The state's PUC, supporting the Corbett administration, decided the ruling of the Commonwealth Court that vacated parts of Act 13 apparently didn't apply to it. The PUC withheld payments of about $1 million from drilling fees to four of the townships that were part of the suit against the Corbett administration and the PUC. State Rep. Jesse White called the decision to withhold funds "political extortion."[530] The Commonwealth Court agreed. In October 2012, Senior Judge Keith Quigley issued a "cease and desist" order to the PUC, ruling that the agency did not have authority to determine local zoning ordinances or to withhold drilling impact fees.[531]

Gov. Corbett continued to claim the full support of the Pennsylvania State Association of Township Supervisors. "It's frustrating," says Dave Ball, "that even after the membership passed the two resolutions that strongly opposed the zoning provisions of Act 13 that executives of PSATS have refused to publically announce those votes, and that the Governor [who knows of those resolutions] continues to claim township support."

The petitioners filed a cross-complaint to the Supreme Court, which would allow that Court not only to uphold the lower court's decision but also to rule on the other eight complaints. The state filed a 45-page brief, arguing that the Commonwealth Court "failed to acknowledge and uphold the supreme authority of the Legislature."[532] It was an unusual position since the Constitution, not what the Legislature does, is the supreme authority. To overturn the Commonwealth Court's ruling, the Supreme Court would have to declare its own recent decision in *Huntley, Huntley v. Oakmont* to be null and void, a highly unlikely possibility.

Frustrated by PSATS' failure to defend municipalities while allowing the Corbett administration to continue to claim PSATS support, members had begun to suggest they might not renew their membership. Finally, at the end of August 2012, more than three months after members passed resolutions opposing parts of Act 13, and more than a month after the court determined that part of Act 13 was unconstitutional, David Sanko finally issued a lame statement. "Now that the courts have made their determination, we're standing with that," the PSATS executive director said, and now claimed that Act 13 "was a compromise for everybody [and] there was something ugly in that bill for everyone."[533]

The arrogance of PSATS executives, who now reluctantly supported the will of the membership, was evident in its impudent claim, and abrupt dismissal of requests for information about the resolutions and detailed requests for the organization's budget, much of it from public tax funds, and communications between it and the governor's office that might suggest collusion. PSATS claimed it was a private organization, and not bound by public records disclosures. In September 2012, the state's Office of Open Records ruled in *Walter Brasch v. PSATS [Docket 2012-1184]* that PSATS "has not met its burden of proof to withhold responsive records," and that the organization, created by statute, "exists solely as an extension of, and to serve, township governments in the Commonwealth that are otherwise subject to the RTKL [right-to-know law]."

PSATS appealed to the Court of Common Pleas. Terrence Barna, the attorney representing Brasch, said there were "two primary arguments before President Judge Kevin Hess—First, that PSATS was a 'Local agency' under the RTKL by virtue of being a 'Local authority,' as that term is defined in the Pennsylvania Statutory Construction Act (1 Pa. C.S.A. §1991); Second, that PSATS also fit the definition of 'Local agency' as a 'similar governmental entity.'" However, the Court ruled against both arguments, stating that PSATS did not meet the requirements of a public agency. The Court did not address the issue of the association taking membership dues entirely from public funds.

THE SUPREME COURT TRUMPS THE LEGISLATURE

The Corbett administration had claimed the Governor's signature on legislation passed by the General Assembly had given Act 13 the force of absolute law. The Pennsylvania

Supreme Court, by a 4–2 vote in December 2013, dismissed that argument:

> "[T]he General Assembly's police power is not absolute; this distinction matters. Legislative power is subject to restrictions enumerated in the Constitution and to limitations inherent in the form of government chosen by the people of this Commonwealth. [p. 65]

In a 162-page opinion, written by Chief Justice Ronald Castille, a Republican, the Supreme Court declared:

> "To describe this case simply as a zoning or agency discretion matter would not capture the essence of the parties' fundamental dispute regarding Act 13. Rather, at its core, this dispute centers upon an asserted vindication of citizens' rights to quality of life on their properties and in their home-towns, insofar as Act 13 threatens degradation of air and water, and of natural, scenic, and esthetic values of the environment, with attendant effects on health, safety, and the owners' continued enjoyment of their private property." [p. 57] . . .
>
> "The displacement of prior planning, and derivative expectations, regarding land use, zoning, and enjoyment of property is unprecedented. [p. 111]
>
> "For communities and property owners affected by Act 13, however, the General Assembly has effectively disposed of the regulatory structures upon which citizens and communities made significant financial and quality of life decisions, and has sanctioned a direct and harmful degradation of the environmental quality of life in these communities and zoning districts. [p. 125] . . .
>
> "[Another] difficulty arising from Section 3304's requirement that local government permit industrial uses in all zoning districts is that some properties and communities will carry much heavier environmental and habitability burdens than others. [p. 125] . . .
>
> "Act 13's blunt approach . . . exacerbates the problem by offering minimal statewide protections while disabling local government from mitigating the impact of oil and gas development at a local level. Remarkably, [it] prohibits local government from tailoring protections for water and air quality. . . . and for the natural, scenic, and esthetic characteristics of the environment." [p. 126]

In a concurring opinion, Justice Max Baer said Act 13 forced "municipalities to enact zoning ordinances, which violate the substantive due process rights of their citizenries."[534]

The Court also ruled that the law, which allowed the DEP to waive setback requirements, already minimal, was unconstitutional. The Supreme Court also remanded to the Commonwealth Court to determine if the entirety of Act 13 violated Article III, section 32 of the state constitution that forbids the creation of special laws to benefit one industry.

The Supreme Court decision didn't ban horizontal fracking; it merely established that the state doesn't have the authority to impose zoning conditions upon local municipalities. Thus, a local municipality may forbid drilling in certain zones, and allow it in other (e.g., industrial) zones. The Court also left intact the responsibility of the Department of Environmental Protection, not the local municipalities, to regulate all technical aspects of horizontal fracking.

The Court also looked at the history of energy production and exploitation in the state, at current horizontal fracking practices, and at the state constitution to issue a broad condemnation of the intent of Act 13. That ruling could have been the basis of court rulings in any state:

> "The type of constitutional challenge presented today is as unprecedented in Pennsylvania as is the legislation that engendered it. But, the challenge is in response to history seeming to repeat itself: an industry, offering the very real prospect of jobs and other important economic benefits, seeks to exploit a Pennsylvania resource, to supply an energy source much in demand. The political branches have responded with a comprehensive scheme that accommodates the recovery of the resource. By any responsible account, the exploitation of the Marcellus Shale Formation will produce a detrimental effect on the environment, on the people, their children, and future generations, and potentially on the public purse, perhaps rivaling the environmental effects of coal extraction. The litigation response was not available in the nineteenth century, since there was no Environmental Rights Amendment. The response is available now.
>
> "The challenge here is premised upon that part of our organic charter that now explicitly guarantees the people's right to an environment of quality and the concomitant

expressed reservation of a right to benefit from the Commonwealth's duty of management of our public natural resources. The challengers here are citizens—just like the citizenry that reserved the right in our charter. They are residents or members of local legislative and executive bodies, and several localities directly affected by natural gas development and extraction in the Marcellus Shale Formation. Contrary to the Commonwealth's characterization of the dispute, the citizens seek not to expand the authority of local government but to vindicate fundamental constitutional rights that, they say, have been compromised by a legislative determination that violates a public trust. The Commonwealth's efforts to minimize the import of this litigation by suggesting it is simply a dispute over public policy voiced by a disappointed minority requires a blindness to the reality here and to Pennsylvania history, including Pennsylvania constitutional history; and, the position ignores the reality that Act 13 has the potential to affect the reserved rights of every citizen of this Commonwealth now, and in the future." [pp. 117-119][535]

The Supreme Court also reversed the Commonwealth Court's opinion that Dr. Meherosh Khan, a plaintiff in the original suit, did not have standing to sue for full disclosure of all chemicals used in fracking:

"In light of Dr. Khan's unpalatable professional choices in the wake of Act 13, the interest he asserts is substantial and direct. Moreover, Dr. Khan's interest is not remote. A decision in this matter may well affect whether Dr. Khan, and other medical professionals similarly situated, will accept patients and may affect subsequent medical decisions in treating patients—events which may occur well before the doctor is in a position to request information regarding the chemical composition of fracking fluid from a particular Marcellus Shale industrial operation. Additional factual development that would result from awaiting an actual request for information on behalf of a patient is not likely to shed more light upon the constitutional question of law presented by what is essentially a facial challenge to Section 3222.1(b)." [p. 26]

In 2014, unwilling to endure the risks of fracking and the state's refusal to protect the health of the people, even with a court opinion, Dr. Khan and his wife moved to New York.[536]

Maya Van Rossum, the Delaware Riverkeeper, and one of the plaintiffs in the suit, said the Court's ruling:

"has vindicated the public's right to a clean environment and our right to fight for it when it is being trampled on. . . . [T]he environment and the people of Pennsylvania have won and special interests and their advocates in Harrisburg have lost. This proves the Constitution still rules, despite the greedy pursuits of the gas and oil industry. With this huge win we will move ahead to further undo the industry's grip of our state government."[537]

Tracy Carluccio, deputy director of the Delaware Riverkeeper Network, accurately noted:

"The gas industry tried to take over every inch of every municipality in Pennsylvania for drilling, regardless of the zoning rights of local governments and the residents they represent. The industry and their backers in Harrisburg overreached when they thought they could literally take over the state, turning it into one big drilling and gas infrastructure site. We fought this law because it was illegal and because it spelled ruin for public health and the environment, even though we, as plaintiffs, didn't have nearly the resources our powerful and well-funded opponents had. This proves, when you have the law and environmental rights on your side, it's worth fighting and you can win."[538]

The Court's decision, said Rep. Jesse White, sends "a clear message . . . to Gov. Corbett and his friends in the energy Industry: Our fundamental constitutional principles cannot be auctioned off to wealthy special interests in exchange for campaign dollars."[539]

The Harrisburg *Patriot-News* editorialized, "As you're being schmoozed by lobbyists and lavished with campaign contributions from powerful industries that want special treatment, there's a limit on how far you can go to please them, because the Pennsylvania Constitution has an Environmental Rights Amendment."[540] [Dec. 20, 2013]

However, the Supreme Court's message was not heard by the gas industry and politicians.

Disregarding environmental and health effects, Dave Spigelmyer, Marcellus Shale Coalition president, argued, "If we are to remain competitive and our focus is truly more job creation and economic prosperity, we must commit to working together toward common sense proposals that encourage—rather than discourage—investment into the Commonwealth."[541]

In a joint statement, Sen. Joe Scarnati, Senate president *pro temp*, and House Speaker Sam Smith said they were "stunned" by the decision that "will so harshly impact the economic welfare of Pennsylvanians."[542] The two leaders of the General Assembly also referred to the Pennsylvania State Association of Township Supervisors, which they said "was fully engaged in numerous legislative discussions and supported the zoning language during passage, until they later decided to oppose it in court. The language in dispute was crafted with their full input."[543] Neither Scarnati nor Smith referred to the reality that while the executives of PSATS supported the language of Act 13, the membership did not.

Within two weeks of the Supreme Court's ruling, the Corbett Administration, represented by the Department of Environmental Protection and the Public Utilities Commission, filed a Request for Reconsideration. Corbett's office of general counsel argued in its petition:

> "In announcing a never-before-employed balancing test against which the constitutional validity of the law is to be judged, the Pennsylvania Supreme Court made its own sweeping factual findings regarding the impact of Act 13, none of which finds any support in the sparse and uneven factual record that was made before Commonwealth Court. The Supreme Court's decision is a stunning departure from the historical practice of that Court, and an unrestrained venture into a fact-finding role that the Court always has insisted is not its proper place in the judicial system.
>
> "Accordingly, today's request for reconsideration seeks to give Act 13 its fair day in court, as every law of this Commonwealth deserves when challenged. We are asking the Supreme Court to follow its own established precedent and remand the case to Commonwealth Court for the development of an evidentiary record (through a fair and thorough process in which all parties have a real and equal opportunity to participate), application of the Court's newly-pronounced stan-

dards to the facts as found by the Commonwealth Court (with legal briefs from all parties), and finally a fair and final determination as to whether Act 13 violates Article I, Section 27 based on a full record and formal findings."[544]

Johnathan Kamin and John Smith, co-counsels for the plaintiffs in the suit against the DEP and PUC, called the petition by the Corbett administration "bizarre." In a news release, they said "The PUC's and DEP's Request For Reconsideration so that it can implement and utilize unconstitutional legislation is an affront to the Citizens of Pennsylvania, and truly demonstrates the Executive Agencies' inappropriate stake in this legislation."[545]

Rep. White, whose district in western Pennsylvania is in the Marcellus Shale, told the *Marcellus Monitor* that Corbett's efforts for reconsideration "has 'sore loser' written all over it. It's mind-boggling to me that the DEP is taking such extraordinary steps to advocate for less environmental and constitutional protections."[546]

The Corbett administration chose to spend taxpayer funds to hire an outside law firm to pursue its Request for Reconsideration. One of the partners was Christopher Carusone, who had been Corbett's chief of staff. The decision to hire the Conrad O'Brien law firm of Philadelphia was apparently a no-bid contract.

Although the Pennsylvania Supreme Court's decision doesn't apply outside the state, it has forced other states to reconsider their own legislation, some of it written by the same persons who wrote much of Pennsylvania's Act 13.

EXEMPTING THE AFFLUENT SUBURBS

Pennsylvania politics continued to play out in how the Republicans separated one of the wealthiest and more high tech/industrial areas of the state from the rural areas. Less than a week before the 2011–2012 fiscal year budget was scheduled to expire, the majority party had slipped an amendment into the 2012–2013 proposed budget, (SB1263[547]), to ban drilling in a portion of southeastern Pennsylvania for up to six years in the South Newark Basin, a rift basin that includes Philadelphia and parts of five suburban counties. That basin could provide at least

360 billion cubic feet of natural gas.[548]

The Republican legislators who drew up that amendment claim the amendment was needed to better study the effects of fracking. "We basically said we didn't know [the South Newark Basin] was there before when we did Act 13," said State Sen. Charles T. McIlhinney Jr.,[549] who had enthusiastically backed Act 13 but now desperately sponsored the budget amendment when he realized that his comments several months earlier that Act 13 not only protected the environment and local zoning control but would protect his own district from drilling were inaccurate, and that drilling could occur in the affluent Philadelphia suburbs. McIlhinney now said, "We need to slow this down until we can do a study on it—see what's there, see where it is, see how deep it is, study the impact, get the local supervisor's [sic] thoughts on it."[550]

"Where was *our* study?" demanded Rep. White. "We were here [when Act 13 was passed] under the guise of, we had to have uniformity, we had to have consistency, we n e e d e d to be fair," said Rep. White, "and now . . . we're saying, 'Maybe, for whatever reason, we're going to give a few people a pass.'"[551]

Karen Feridun pointed out, "Studies are not being conducted before drilling begins anywhere else in the state, nor are studies being conducted on the potential impacts of the pipeline operations already coming here."

David Meiser, chair of the Bucks County Sierra Club, said the Pennsylvania Legislature "should either exempt all counties from Act 13 and not just try to get special treatment from Sen. McIlhinney's core area, or repeal the law entirely."[552]

Tracy Carluccio, deputy director of the Delaware Riverkeeper Network, called the amendment "outrageous," arguing:

> "This really smacks of cronyism and self-preservation in order to protect a few elite areas in this state that have more power than others . . .
>
> "This is absolutely wrong. This kind of favoritism where a legislator thinks he can protect himself and his turf at the expense of everybody else is what people cannot tolerate and what they complain about in politics today. We should be making decisions about gas drilling based on sound public policy—not based on what some legislator thinks he can slip in in order to save his own neck."[553]

Sen. McIlhinney proudly claimed the last-minute legislation "makes good on my promise that Act 13 was not intended to apply to Bucks County."[554] And, once again, McIlhinney was wrong. While the toothless moratorium protects the southeastern counties from drilling, it doesn't protect those counties from drilling-related activities, including construction of pipelines and compressor stations.

Significant questions need to be raised why a state law discriminated against the rural counties of the Marcellus Shale while protecting the health and welfare of the more affluent suburban counties that are home for many of the state's most powerful and wealthiest constituents, which at that time included the head of the DEP and the lieutenant governor.

PHOTO: Vera Scroggins

Anti-fracking protestors attend inaugural of Pennsylvania Gov. Tom Wolf, Jan. 20, 2015. Although significantly more environmentally conscious than his predecessor, and a strong advocate for increased taxes on unconventional drilling and more rigorous enforcement of regulation upon the gas industry, Gov. Wolf does not believe in a ban.

PART II:
Health and Environmental Issues

Flaring near Hickory, Pa.

PHOTO: Robert Donnan

124

CHAPTER 6

Methane as a Greenhouse Gas

On Oct. 23, 2015, Southern California Gas technicians discovered a leak of methane from a failed casing on one of the pipes in its Alisa Canyon storage facility, about 30 miles northwest of Los Angeles.[555] It would take about five months to stop the leak from the pipe 8,000 feet underground.

At its peak a month after discovery, between 101,000–132,000 pounds of methane was going into the air every hour, according to the California Air Resources Board.[556] Residents complained of nausea, dizziness, headaches and nosebleeds[557] from the sulfur-like odor that is put into natural gas to identify it. By the end of the year, the L.A. Unified School District closed two schools until March 2016,[558] the Federal Aviation Administration banned flights under 2,000 feet from going within a half-mile of the leak,[559] and about 2,900 households were relocated for several months.

It was the worst methane leak in U.S. history, amounting to about 15 percent of all hourly greenhouse gas emissions in the U.S., Dr. Robert W. Howarth, a geochemist, told Herman K. Trabish of Utility Dive.[560]

About 78 percent of unprocessed natural gas is methane (CH_4),[561] the simplest hydrocarbon and one of the major contributors to ozone layer depletion. (Other major gases in methane are butane, ethane, and propane; lesser gases are hexane, nitrogen, pentane, hydrogen sulfide, and carbon dioxide, which can be as much as 8 percent of natural gas.[562])

In March 2014, National Oceanic and Atmosphere Administration (NOAA) scientists at the Mauna Loa Observatory in Hawaii measured the concentration of carbon dioxide in the atmosphere as 402 parts per million, believed to be the highest level in more than 800,000 years.[563] (Scientists measure trapped air particles in glaciers to determine historical atmospheric levels.) The increasing rate "is consistent with rising fossil fuel emissions," said Dr. James Butler of NOAA.[564]

If current and projected levels of carbon dioxide and methane enter the system within the next two decades, it could raise the earth's overall temperature by two degrees Centigrade; this would be a "point of no return" on climate change.

Advocates of fracking argue natural gas is "greener" than coal and oil energy, with significantly fewer carbon, nitrogen, and sulfur emissions and, thus, better for the environment.

Most scientists—and the person whose company pioneered horizontal fracking—disagree. "[M]ethane is a powerful greenhouse gas pollutant, and uncontrolled leakages call into question whether natural gas is cleaner than coal from a global climate perspective," said George P. Mitchell.[565]

About 100 billion cubic feet of natural gas are released into the atmosphere each year.[566]

Escaped methane from natural gas into the air and water increases problems with public health and the environment;[567] production and distribution errors and problems mining natural gas make it the equivalent of coal and oil; when released directly into the atmosphere, methane is "more than 20 times more potent than carbon dioxide" over a 100 year period, said Lisa Jackson, a chemical engineer and EPA administrator, 2009–2013.[568] However, over a 20 year period, "one pound of it traps as much heat as at least 72 pounds of carbon dioxide," says Dr. Anthony R. Ingraffea, professor of civil and environmental engineering at Cornell University.[569] His studies reveal:

> "When burned, natural gas emits half the carbon dioxide of coal, but methane leakage eviscerates this advantage because of its heat-trapping power. . . .
> "[R]ecent measurements by [NOAA] at gas and oil fields in California, Colorado and Utah found leakage rates of 2.3 percent to 17 percent of annual production."[570]

A team of researchers from Cornell University determined that leaked methane gas into the air from fracking operations could have a greater negative impact upon the environment than either oil or coal. Geochemist Dr. Robert W. Howarth, engineer Dr. Tony Ingraffea, and ecology researcher Renee

Santoro, concluded:

> "The footprint for shale gas is greater than that for conventional gas or oil when viewed on any time horizon, but particularly so over 20 years. Compared to coal, the footprint of shale gas is at least 20% greater and perhaps more than twice as great on the 20-year horizon and is comparable when compared over 100 years. . . . The GHG [greenhouse gas] footprint of shale gas is significantly larger than that from conventional gas, due to methane emissions with flow-back fluids and from drill out of wells during well completion. . . . The large GHG footprint of shale gas undercuts the logic of its use as a bridging fuel over coming decades, if the goal is to reduce global warming."[571]

Dr. Ingraffea and three other scientists associated with Cornell University, in a separate study, also determined that unconventional wells drilled before 2010 had a methane leakage rate of about 10 percent; after 2010, about 6 percent of all unconventional wells had methane leaks, almost three times higher than conventional wells. The study, published in 2014 in the *Proceedings of the National Academy of Sciences*, was based upon data analysis of 75,000 wells in Pennsylvania between Jan. 1, 2000 and Dec. 31, 2012.[572]

Natural gas producers in the Denver area are losing about 4 percent of their gas production into the atmosphere, about twice the earlier estimate.[573] NOAA scientists, led by Dr. Gabrielle Petron, recorded considerably higher atmospheric levels of propane and butane than in Los Angeles or Houston, both of which have severe air pollution problems. Dr. Petron's team found that significantly increased levels of methane in the air came from oil and gas production. She also found significant levels of the carcinogen benzene. Analyzing thousands of data readings, the team found benzene in the atmosphere was between 385 and 2,055 metric tons per year, significantly higher than the previously reported estimates of 60 to 145 tons released each year.[574]

Infrared videography revealed air pollution near 11 well sites in Pennsylvania, Maryland, and West Virginia. The Chesapeake Bay Foundation, which hired Dr. Howarth to document the presence of airborne pollution, said the video "establishes that the industry is not sufficiently limiting the amount of leaks from drilling and processing operations."[575] Dr. Howarth filmed 15 plants, four of which had no evidence of airborne pollution. While conceding that natural gas is cleaner than coal and oil, the Foundation also told the Associated Press in November 2011, "The alarming rate at which extraction activities have increased in the bay watershed gives us great pause as we attempt to understand the full implications."[576]

The natural gas industry countered the scientists by claiming that a four percent decline in U.S. emissions in 2012 from 2011 is evidence that the increased use of the "greener" natural gas and the decline in coal usage is the reason. "Everybody started to believe that shale gas is driving all these CO_2 reductions," said Shakeb Afsah of the CO2 Scoreboard Group. Most of the reduction, says Afsah, was because a milder winter reduced the need for heat, and because of "energy-efficiency and conservation measures in the transportation, residential and commercial sectors."[577]

There are no records of exact numbers of abandoned wells.[578] However, even without knowing the number, it's possible to determine that abandoned wells have contributed to climate change. Research completed in 2014 by Mary Kang of Princeton University reveals that methane leaks from the wells, which are not monitored, could cause between 4 and 13 percent of all human-caused emissions just in Pennsylvania.[579] In the first decades of oil exploration in Pennsylvania, the state had no regulations for cleaning, capping, and monitoring wells; however, legislation passed in 2014 required companies to identify abandoned wells near their own sites and to monitor them for leaks, beginning in 2016.[580]

FLARING AND VENTING

All natural gas companies use flaring and venting to get rid of excess gas that can't be immediately sent to a processor and to reduce possibilities of explosions. Flaring is the controlled burning of natural gas, often with the production of oil. Venting is the controlled release of unburned gas directly into the atmosphere.

Flaring, also known as burnoffs, can produce as much as two

million tons of carbon dioxide a year, according to a *New York Times* compilation of data.[581] During 2013, oil and gas drilling companies flared about 89.9 bcf of gas in the United States.[582] Much of the flaring is the result of fossil fuel corporations rushing to drill for oil, having excess natural gas from fracking and no infrastructure or financial means in place to move it into pipelines and compressors.

World-wide, about 460 bcf of gas is flared each year,[583] releasing as much as 360 million tons of carbon dioxide a year, according to data analysis by General Electric.[584] Venting accounted for about 170 billion cubic feet of gas released in the United States during 2013.[585]

Between 2005 and the end of 2012, nonconventional wells put more than 110 million tons of carbon dioxide equivalent into the air, according to data analyzed by the Research and Policy Center of Environment America. Companies drilling in Texas released about 44 tons of carbon dioxide into the air, followed by Colorado (25.4 tons), and Pennsylvania (8.8 tons of emissions).[586] The three states accounted for about 71 percent of all emissions from unconventional wells. An investigation by Earthworks and SkyTruth revealed that flaring in the Bakken shale, about 26 bcf in 2013, resulted in the equivalent of carbon emissions from 1.1 million cars and light trucks; flaring in the Eagle Ford Shale resulted in the equivalent of carbon emissions from about 350,000 cars and light trucks, according to an investigation by a reporting team from the *San Antonio Express–News*.[587] That four-part series revealed that state regulators didn't have any idea how much natural gas is being flared or vented. However, the reporters determined that in Texas, about 33 bcf of natural gas was lost between 2009 and 2012, about two-thirds in the Eagle Ford Shale. The reporters also determined that seven operations didn't have necessary permits, despite assures from the Railroad Commission of Texas that it maintains strict control over all operations.[588]

New EPA Regulations

In January 2009, WildEarth Guardians and the San Juan Citizens Alliance had sued the EPA to force compliance with the Clean Air Act of 1970. That Act requires the EPA every eight years to review standards for industrial categories that could cause or contribute to air pollution and endanger public health and the environment. The last time the EPA conducted a study was in 1985. In issuing a consent decree in February 2010, the U.S. District Court for the District of Columbia ordered the EPA to present a final set of regulations no later than Jan. 31, 2011.[589] The Court later granted EPA requests for extensions.

On April 17, 2012, with strong input from the natural gas industry and having consulted environmental groups, the EPA filed the regulations on the last date allowed by the Court.[590] The regulations require natural gas drillers to install special equipment to separate gas and hydrocarbons in the flowback, reduce flaring, and install new valves to reduce emissions. The EPA says the rules "are expected to yield a nearly 95 percent reduction in VOC [volatile organic compounds] from more than 11,000 new hydraulically fractured gas wells each year" and by 25 percent the methane gas released into the air.[591] The EPA believes the new regulations will reduce VOCs, which can cause ground level smog, by 190,000–290,000 tons, air toxins by 12,000–20,000 tons, and methane by 1.0–1.7 tons.[592]

However, in a major concession to the natural gas industry, the rules did not go into effect until January 2015, six years from the time the suit was first filed.

The Politics of Climate Change

The U.S. House of Representatives, largely divided by partisan political beliefs, voted 231–192 in May 2014[593] to deny the Department of Defense public funds to assess and respond to threats of climate change. That vote scuttled military analysis of the effects of rising sea levels and food shortages caused by drought.

Climate change "constitutes a serious threat to global security, an immediate risk to our national security, and, make no mistake, it will impact how our military defends our country," the President said at a commencement address at the U.S. Coast Guard Acdemy in May 2015.[594]

Among those who had previously voted to deny funds were Reps. John Shimkus (R-Ill.) and Joe Barton (R-Texas).

Shimkus, senior member of the House Energy and Commerce Committee, claimed the world would not end with rising sea levels because he believed God said he would never again destroy the earth like he did in Noah's time. "I believe that is the infallible word of God, and that's the way it is going to be for his creation," said Shimkus, adding, "Man will not destroy this earth."[595]

Shimkus' reasons echo those of the fundamental Religious Right. Several evangelical organizations have convinced conservative statehouse legislatures to require teachers to present not only evolution and creationism as equal theories, but to present "evidence" that global warming is not caused by mankind.[596]

Rep. Barton, tossing around junk science, claimed carbon dioxide is not a pollutant "in any normal definition of the term" because it is in soda and Perrier water, and is "necessary for human life [and] does not cause cancer, does not cause asthma." The reason why regulation of greenhouse gasses is bad, claimed Barton, is because, "You can't regulate God. Not even the Democratic majority in the U.S. Congress can regulate God."[597] *The Wall Street Journal*, which has an editorial philosophy that climate change is not caused by humans, claimed Barton was the "House GOP's leading expert on energy policy."[598] Barton, one of the top congressional recipients of oil and gas campaign funds,[599] is chair of the House Committee on Energy and Commerce.

The U.S. Senate, by a vote of 98–1 in January 2015, decided that climate change was real. That amendment was sponsored by Sen. Sheldon Whitehouse (D-R.I.) The lone senator to vote against that bill was Roger Wicker (R-Miss.).

By a 50–49 vote, on a non-binding resolution sponsored by Sen. Brian Schatz (D-Hawaii), the Senate also decided climate change was "significantly" caused by humans. However, under threat of a filibuster by the Republican leaders, the non-binding legislation needed 60 votes to pass. The rejection of the majority will was a vote that pleased the shale gas industry.

Sen. James Inhofe (R-Okla.), a "climate change denier," and chair of the Environment and Public Works committee, had surprisingly lobbied his fellow Republicans to vote for the first amendment, which was added to a bill supporting the completion of the Keystone XL pipeline. He argued there was "Biblical evidence" that the climate is always changing[600] and, thus, he muted the Democrats' attempt to force the conservatives, several of whom were up for re-election or planned to run for the presidency, to formally declare themselves to be climate deniers. However, Inhofe also believes climate change caused by humans is a liberal hoax. "The hoax is there are some people so arrogant to think they are so powerful they can change the climate," Inhofe said, arguing, "Man can't change the climate."[601]

Also believing climate change is a hoax, and voting to support the Keystone XL and natural gas as a primary fuel was Sen. Ted Cruz (R-Texas), who later stated, "The federal government has no business attempting to massively reorder the global economy resulting in policies that kill jobs and keep people from rising out of poverty, all in the name of a theory that can't be proven or disproven."[602]

"Only in the halls of Congress is this a controversial piece of legislation," said Sen. Schatz.[603]

About three-fourths of Americans believe climate change is real, according to independent polls conducted in September and October 2015 by the University of Michigan[604] and the University of Texas.[605] Both polls found that about 80–90 percent of Democrats and about 56–59 percent of Republicans believe climate change is real.

Sen. Inhofe had raised the issue about climate change in a two-hour speech in the Senate in 2003 when he called climate change "the greatest hoax ever perpetuated on the American people."[606] It was not just his own view, but those of a majority of conservatives, among them talk show pundit Rush Limbaugh who helped frame the discussion for religious and political conservatives to follow. Inhofe would later write *The Greatest Hoax: How the Global Warming Conspiracy Threatens Your Future* (2012). That book, which called for diminishing the role of the EPA, rose to best-sellers lists, fueled by fundamental conservatives heeding the siren song of conservative talk show pundits, all of whom saw the book as the antidote to Al Gore's *An Inconvenient Truth* (2006).

It was probably Gore who had pushed President Bill Clinton to actively support a stronger environmental policy and to support a meeting of leaders of more than 100 nations in Kyota, Japan, in 1997. Those nations signed the Kyota Protocol that declared global warming was a fact, that it was largely caused by the rise in carbon dioxide emissions, and that those emissions were primarily caused by humans. The Protocol required mandatory caps on greenhouse gasses. However, the U.S. Senate never ratified that agreement, the result of intense lobbying by the fossil fuel industry. A previous meeting in Rio de Janeiro in 1992 had made caps on emissions only voluntary, after heavy lobbying by President George H.W. Bush who considered all sides of the issue before finally accepting the views of senior staff, led by John Sununu, his chief of staff, that mandatory requirements were not politically feasible.

The United States in the first decade of the 21st century had become more conservative, as evidenced by the election of George W. Bush as president, with the more conservative Dick Cheney as vice-president. During his campaign, Bush had said in 2000 that if elected he would require industry to cap carbon dioxide emissions. However, as president, influenced by the conservative Competitive Enterprise Institute, an increasingly conservative Congress, and probably by Cheney, Bush backed off from his campaign pledge. Within months of his election, President Bush pulled the United States out of the Kyota Protocol, which more than 190 nations had signed. During 2002 and 2003, the chief of staff of the president's Council on Environmental Quality modified or deleted sections of reports by government officials that showed not only was global warming a fact, but that much of it was caused by humans. The politician who placed policy over scientific reality was Philip A. Cooney, who had been a lobbyist for the American Petroleum Institute.[607]

PRESIDENT OBAMA ISSUES FORMAL POLICY

In his March 2014 address to the nation, President Obama finally issued a major policy statement on global warming and climate change that many environmentalists believed was overdue. That speech, which Al Gore called the "best presidential address on climate change ever,"[608] outlined a plan not only to strictly enforce clean air regulations, but also to establish stricter emission rules on power plants; significantly increase wind and solar energy, especially in federal buildings and military bases; and increase energy efficiency standards in everything from auto emissions to appliances.

The President emphasized climate change is a reality that must be addressed vigorously:

> "The overwhelming judgment of science—of chemistry and physics and millions of measurements—has put all that to rest. Ninety-seven percent of scientists, including, by the way, some who originally disputed the data, have now put that to rest. They've acknowledged the planet is warming and human activity is contributing to it. . . .
>
> "So the question now is whether we will have the courage to act before it's too late. And how we answer will have a profound impact on the world that we leave behind not just to you, but to your children and to your grandchildren. . . .
>
> "We don't have time for a meeting of the Flat Earth Society."[609]

In public speeches that month, the President again attacked climate deniers and those who did little to reduce the problems of global warming. At the League of Conservation Voters, President Obama placed the blame on Congress for failure to accept changes that needed to be made to improve the climate:

> "It's pretty rare that you encounter people who say that the problem of carbon pollution is not a problem. . . . When you talk to folks, they may not know how big a problem, they may not know exactly how it works, they may doubt that we can do something about it, but generally they don't just say, no, I don't believe anything scientists say. Except . . . in Congress. In Congress. Folks will tell you climate change is a hoax or a fad or a plot. It's a liberal plot."[610]

In a commencement speech at the University of California at Irvine in June 2014, the president further alienated many in the fossil fuel industry:

> "[T]he question is not whether we need to act. The overwhelming judgment of science, accumulated and measured

and reviewed over decades, has put that question to rest. The question is whether we have the will to act before it's too late. For if we fail to protect the world we leave not just to my children, but to your children and your children's children, we will fail one of our primary reasons for being on this world in the first place. And that is to leave the world a little bit better for the next generation."[611]

However, about a year later, the President pleased the fossil fuel industry by issuing a conditional permit for Shell to drill for oil in the Arctic about 70 miles off the coast of Alaska,[612] beginning in Summer 2015 if local, state, and federal agencies agreed. Since 2008, Shell has paid about $7 billion for permit fees and development costs for the right to drill exporatory wells in the Chukchi Sea.[613] In September 2015, Shell abandoned drilling in the sea, citing low possibility of economic benefits on oil/gas recovery.

In February 2014, the U.S. National Academy of Sciences (NAS) and the Royal Society of Great Britain (RS) had issued a 36-page publication, warning that climate change "will threaten food production, freshwater supplies, coastal infrastructure, and especially the welfare of the huge population living in low-lying areas."[614]

One month after the NAS and RS joint report, the American Association for the Advancement of Science (AAAS) issued *What We Know*, a publication targeted to non-scientists. In that report, AAAS explained global warming was an established fact, which 97 percent of climate scientists concluded is caused by humans, and that there are significant consequences for the Earth's existence if measures are not taken to reduce the problem.

The publication was necessary because, "The public has been misinformed by a colossal disinformation campaign," said Dr. James McCarthy, former AAAS president.[615] It was a campaign that was partially fueled by a $510,000 media campaign by the Western Fuels Association,[616] continued by the oil/gas industry, spread by conservative talk radio and television, and promoted by dozens of members of Congress, who had been recipients of large campaign donations from oil/gas industry corporations and front groups during the 21st century.

PHOTO: Doug Carlton Jr.

Rig adjacent to golf course
in Arlington, Texas

PHOTO: Greers Ferry Lake Gas Watch

Entrance to a drilling site in
Faulkner County, Ark.

PHOTO: Diane Siegmund

Rail cars carrying sand and chemicals to a natural gas site
pass close to a child care center near Wyalusing, Pa.

PHOTOS: Michael Bagdes-Canning/Marcellus Outreach Butler

Summit Elementary School; Butler, Pa.

137

	Pulmonary	Neurologic	Reproductive	Dermal	Hematologic
Paticulate Matter	X			X	
Hydrogen Sulfide	X	X		X	
Ozone	X				
Carbon Monoxide		X	X		
Nitrogen Oxides (NOx)	X				
Sulfur Dioxide	X				
Volatile Organic Compounds (VOCs)	X	X	X	X	X
Benzene, Toluene, Ethylene and Xylenes (BTEX)	X	X	X	X	X
Naturally Occurring Radioactive Materials (NORMs)			X	X	X

GRAPH: Environment America Research and Policy Institute

CHAPTER 7
Public Health Issues

In the years after fracking came to the Marcellus Shale, the Pennsylvania Department of Health forbid staff from speaking to the public about health issues related to fracking. A State-Impact Pennsylvania investigative report in 2014 revealed employees were given a "buzzword list"—"If anybody from the public called in and that was part of the conversation, we were not allowed to talk to them," Tammi Stuck, a public health nurse for 36 years, told reporter Katie Colaneri.[617] She said if anyone called about health concerns, staff were ordered to pass the information to a supervisor for response. There were seldom any responses.

The Department also ordered staff to get permission at least a month in advance before they gave public speeches or attended public meetings, even if fracking was not the topic, according to Marshall P. Deasy III, a program specialist/investigator. He said in his 20 years with the Bureau of Epidemiology, "Community Health wasn't told to be silent on any other topic that I can think of."[618]

The policy was initiated in 2011, possibly on "advice" from the governor's office, after a Department consultant attended a community health meeting, and a resident asked about health issues related to fracking. The Department claimed there was no correlation between health issues and fracking, and also denied the "buzzword" list was ever distributed or that staff were told not to respond to citizen concerns.[619]

The Department of Health also rejected requests by Food & Water Watch in July 2014 to provide public documents about fracking-related health complaints. Only a threat of an injunction by the state's Office of Open Records yielded more than 100 pages of documents. Those documents revealed:

"Between March 30, 2011 and April 6, 2015, the DOH logged 87 complaint records filed by concerned residents, health professionals, state legislators and agencies on behalf of Pennsylvania residents. Respiratory issues, asthma, and throat and nose irritation were the most common health problems reported by residents, followed by noxious odors, skin problems, abdominal issues and noise pollution. Residents also complained of cancer, and extreme hair loss. Doctors even phoned in from "seeing unusual numbers of skin lesions/rashes in residents."[620]

"We fought for almost a year and with multiple administrations for these documents," according to Food & Water Watch, "[and] now we know why. DOH's gross irresponsibility in its failure to respond to the serious health concerns of the people it is charged to protect must be documented and challenged."[621]

Pennsylvania HB 1950, which had preceded Act 13 of 2012, initially included a provision to provide up to $2 million a year to fund the Department of Health for "collecting and disseminating information, preparing and conducting health care provider outreach and education and investigating health related complaints and other uses associated with unconventional natural gas production activity."[622] That provision, supported by numerous public health and environmental groups, was deleted in the final bill.

"What are you so afraid that we're going to uncover?" Dr. Eli Avila asked two years after he resigned in 2012 as the state's secretary of health. He said the "lack of any action speaks volumes" about the fear that such a study might reveal health problems.[623]

Gov. Tom Corbett and his well-oiled legislature shut down 15 of 60 public health clinics, planned to shut down nine more, and laid off 73 nurses and support staff. In July 2014, the state Supreme Court issued an emergency injunction to prevent the state from shutting down more health clinics. Under the Corbett administration, Pennsylvania ranked 44th of 50 states in *per capita* public health spending.[624] None of the $632 million in impact fees collected in 2011–2014[625] were allocated to health agencies.

The Scranton *Times–Tribune*, near the end of Gov. Corbett's four-year term, editorialized:

> "Pennsylvania's government has not only failed, but refused, to assess the public health impact of the natural gas industry. The administration and Legislature have ignored a government commission's recommendation to establish a health registry to keep track of public health issues near drilling, processing and transmission sites. . . .
>
> "Public health should be a nonpartisan issue. But so far, that has been expressed primarily in a bipartisan consensus not to worry about it. Mr. Corbett and his Republican majorities in the Legislature have not established a program or provided the modest funding it would need, and Democrats have not forced the issue." [July 29, 2014][626]

Dr. Tom Wolf, who defeated Corbett's bid for re-election, in March 2015 restored the jobs of 26 public health nurses and reopened 16 county-based clinics that Corbett had shut down.

Colorado, North Dakota, and Pennsylvania record health-related complaints from the gas and oil fields, but don't make them public, citing privacy concerns. Several states—among them Ohio, Oklahoma, Texas, Wyoming, and West Virginia—don't track health-related complaints, according to research conducted by StateImpact.[627]

However, "No state is requiring enough upfront collection of baseline data and ongoing monitoring to adequately protect local water supplies and public health," according to an investigation by OMB Watch.[628]

Part of the reason for a lack of research is explained by Dr. John Hughes of Aspen [Colorado] Integrative Health, which is conducting preliminary studies. "The state has no incentives to do this type of research," Dr. Hughes told the *Aspen Daily News*, because it gets "tons of tax revenue from the industry."[629]

The willful and deliberate failure to fund public health studies and clinics has generally pleased the the oil and gas industry, but puts public health at risk.

Both the process itself and human error are "known to produce a variety of physical and chemical hazards that may cause negative health effects," according to a study conducted by a team of public health scientists at the University of Colorado at Denver.[630]

A team of 13 physicians and scientists from Columbia University and the University of Pennsylvania, analyzing data of 198,000 patient records, concluded there was an increase in dermatological and neonatal illness, and a 27 percent increase in hospitalization for cardiovascular disease for persons living in high-density drilling areas than for those in a "frack free" area.[631]

Researchers at the University of Missouri determined that 11 chemicals used in fracking blocked estrogen and androgen hormones. The findings were published in the peer-reviewed journal *Endocrinology*.[632]

"Associated pollution has reached the stage where it is contaminating essential life support systems," according to The Endocrine Disruption Exchange (TEDX), a national clearinghouse that says it "focuses primarily on the human health and environmental problems caused by low-dose and/or ambient exposure to chemicals that interfere with development and function."[633] The chemicals in the fracking mixture can lead to compromising the neurological, immune, kidney, and cardiovascular systems, according to a team of TDEX researchers led by Dr. Theo Colborn who chased down wastewater trucks to get samples she later analyzed to identify chemicals in the fracking mixture. Dr. Colborn concluded that about one-third of all chemicals in the fracking mixture may cause cancer, while almost 90 percent of the toxins in a fracking mixture could cause damage to the skin, eyes, ears, nose, and throat.[634]

Physicians, Scientists, & Engineers for Healthy Energy (PSE), in a letter to President Barack Obama, argued:

> "There is a growing body of evidence that unconventional natural gas extraction from shale (also known as 'fracking') may be associated with adverse health risks through exposure to polluted air, water, and soil. Public health researchers and medical professionals question the continuation of current levels of fracking without a full scientific understanding of the health implications. . . .
>
> "There is a need for much more scientific and epidemiologic information about the potential for harm from fracking. To

facilitate a rapid increase in fracking in the United States without credible science is irresponsible and could potentially cause undue harm to many Americans."[635]

Christopher Portier, director of the National Center for Environmental Health, calls for more research studies that "include all the ways people can be exposed [to health hazards], such as through air, water, soil, plants and animals."[636]
Some of that support comes from the National Science Foundation (NSF), which awarded a team of scientists a $12 million five-year grant, beginning October 2013, to study the technology of fracking in the Rocky Mountain states. The team is led by Dr. Joseph Ryan, professor of civil, environmental, and architectural engineering at the University of Colorado at Boulder. Separate teams are investigating fracking's effects upon groundwater and aquifer systems and possible methane migration into wells; and fracking's risks to air quality. Other scientists on the project come from California State Polytechnic University at Pomona, Colorado School of Mines, Colorado State University, the National Oceanic and Atmospheric Administration, the National Renewable Energy Laboratory, the University Corporation for Atmospheric Research, and the University of Michigan.
The team's goal, said Dr. Ryan, is "to provide a framework for society to evaluate the trade-offs associated with the benefits and costs of natural gas development."[637]

Three Pennsylvania health systems are collaborating on a long-term health analysis of persons living in central and northeastern Pennsylvania. With an initial grant of $1 million from the Degenstein Foundation, Geisinger Health System, Guthrie Health, and Susquehanna Health are establishing an electronic database of as many as three million patients living within in Marcellus Shale to determine if there are correlations between certain health problems before and after the industry began fracking. Full results aren't expected for more than two decades. Part of the reason is that it takes time to get funding, establish protocols and parameters, and train staff. Another reason why the study may take two decades is because although causes of certain health problems are well-known early in the treatment—for example, chemical burns—many illnesses, such as cancers, do not manifest themselves for several years.

The American Nurses Association's House of Delegates passed a health care policy statement in June 2012 to use evidence-based information to educate health professionals about the relationships between energy development and health issues. The issue was first raised by the Pennsylvania State Nurses Association (PSNA):

> "Human and ecological health risks are directly related to the use of coal-fired power plants, mountaintop removal of coal, offshore and onshore oil and natural gas drilling, and hydraulic fracturing or 'fracking.' Research demonstrates that increased rates of asthma attacks, cardiovascular diseases and lung cancer are all associated with our current reliance on fossil fuels. As our population ages, our vulnerability to these fossil fuel-related exposures will continue to increase. Children are already at higher risk because of their increased susceptibility to respiratory illness."[638]

More than 250 of New York's leading health professionals, scientists, and environmentalists, and 66 health and environmental groups, in a 10-page letter with several attachments, sent in October 2011 to Gov. Andrew Cuomo, pointed out:

> "In Texas, Wyoming, Louisiana, North Dakota, Pennsylvania, and other states, cases have been documented of worsening health among residents living in proximity to gas wells and infrastructure such as compressor stations and waste pits. Symptoms are wide-ranging, but are typical for exposure to the toxic chemicals and air and water pollutants used in oil and gas development and can often be traced to the onset of such operations.[639] . . .
> "The totality of the science—which now encompasses hundreds of peer–reviewed studies . . . and hundreds of additional reports and case examples—shows that permitting fracking in New York would pose significant threats to the air, water, health and safety of New Yorkers. At the same time, new assessments from expert panels also make clear that fundamental data gaps remain and that the best imagin-

able regulatory frameworks fall far short of protecting our health and our environment."[640]

In calling for the health impact assessment, the group stated that a "comprehensive assessment of health impacts is likely to include information—such as mounting costs for health care and air and water pollution mitigation—that could inform how DEC and other agencies, such as the Department of Health (DOH), evaluate and assess cumulative impacts and how DEC reviews any proposed gas development permit applications."[641] The letter reviewed the scientific studies that showed links between fracking and cancer, and urged the governor not to allow "a carcinogen-dependent industry into our state."[642]

One month after the health care professionals and environmentalists sent the letter to Gov. Cuomo, 18 agencies that assist cancer victims also pleaded with the governor to continue the moratorium.

The Independent Oil and Gas Association (IOGA) dismissed the scientific evidence and compared those who signed the letter as "actors and celebrities [who] don't understand the safety of the science behind [fracking]."[643]

Despite the IOGA babble, the call for a health impact statement helped lead to an analysis of fracking's effects by the state's Department of Health and the eventual banning of fracking within New York State.

The Advisory Board of the U.S. Department of Energy concluded, "The public deserves assurance that the full economic, environmental and energy security benefits of shale gas development will be realized without sacrificing public health, environmental protection and safety."[644] The report suggested the EPA act unilaterally to study the effects of fracking because coordination with state environmental agencies is "not working smoothly."

Dr. Helen Podgainy Bitaxis, a pediatrician in Coraopolis, Pa., who is active in public health issues, says she doesn't want her patients "to be guinea pigs who provide the next generation the statistical proof of health problems as in what happened with those exposed to asbestos or to cigarette smoke."

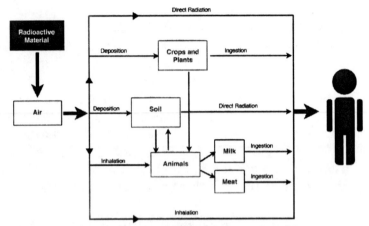

Pathways for Radiation Migration Through Air

GRAPHICS: Ivan White © 2012 Grassroots Environmental Education, Inc.

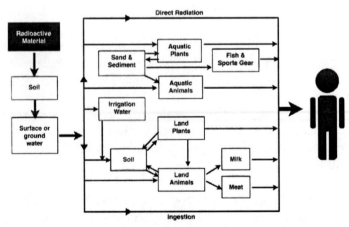

Pathways for Radiation Migration Through Soil and Water

146

CHAPTER 8
Air Pollution and Health Effects

High-volume horizontal fracking impacts the air, increases greenhouse gas levels, and has direct effect upon the health of those living near wells and compressors.

"Air concentrations of potentially dangerous compounds and chemical mixtures are frequently present near oil and gas production sites," according to research conducted by a seven-person team headed by Dr. Gregg P. Macey of the Center for Health, Science, and Public Policy at the Brooklyn Law School. Benzene, formaldehyde, hexane, and hydrogen sulfide were "the most common compounds to exceed acute and other health-based risk levels" at test sites in Pennsylvania and Wyoming, according to the team's findings.[645]

The oil and natural gas industry, according to the EPA, "is the largest source of emissions of volatile organic compounds (VOCs), a group that contribute to the formation of ground-level ozone (smog)."[646] The shale gas industry accounts for about 450,000 tons of air pollution a year, according to an investigation conducted by the Environment and Policy Center of Environment America.[647] VOC emissions are about 200 times greater in nonconventional wells than from conventional wells, according to the EPA.[648]

High concentrations of ozone can lead to shortness of breath, throat irritation, coughing, chest pain when inhaling, wheezing and asthma attacks, increased susceptibility to pulmonary and respiratory inflammation, and heart attacks, according to the American Lung Association.[649]

Researchers from the University of Colorado School of Public Health concluded that fracking may contribute to "acute and chronic health problems" from air pollution.[650] The research team, headed by Dr. Lisa McKenzie, had found "potentially toxic petroleum hydrocarbons in the air near the wells," citing benzene, ethylbenzene, toluene, trimethylbenzenes, and xylene. These toxins, the four-person team pointed out:

> "can adversely affect the nervous system with effects ranging from dizziness, headaches, fatigue at the lower exposures to numbness in the limbs, to lack of muscle coordination, tremors, temporary limb paralysis, and unconsciousness at higher exposures. Air-borne chemicals from fracking could irritate the respiratory system and mucous membranes with effects ranging from eye, nose, and throat irritation to difficulty in breathing and impaired lung function."[651]

A separate group of University of Colorado researchers monitoring the air in the rural Uintah Basin of Utah found significantly higher volatile organic compound levels than EPA standards. The levels in winter 2013 matched that of urban Los Angeles in summer. Uintah County has a population of about 30,000—and, at the time of the research, 11,200 oil and gas wells. The research team noted:

> "Levels above this [EPA] threshold are considered to be harmful to human health, and high levels of ozone are known to cause respiratory distress and be responsible for an estimated 5000 premature deaths in the U.S. per year. Because of the photochemical nature of ozone production, tropospheric ozone pollution has traditionally been considered an urban, summertime phenomenon. . . .
> "[The levels from the Uintah Basin are] to the best of our knowledge, among the highest-ever reported mole fractions of alkane non-methane hydrocarbon in ambient air. Mole fractions for the aromatic compounds reach or exceed those reported from the most heavily polluted inner cities."[652]

Biochemist Dr. Ronald Bishop suggests fracking not only "is highly likely to degrade air, surface water and groundwater quality, to harm humans, and to negatively impact aquatic and forest ecosystems," but that "potential exposure effects for humans will include poisoning of susceptible tissues, endocrine disruption syndromes, and elevated risk for certain cancers."[653]

About half the air samples monitored at homes and farms near well sites in five states (Arkansas, Colorado, Ohio, Pennsylvania, and Wyoming) showed higher-than-acceptable levels of benzene and formaldehyde. Benzene levels, according to the data, ranged from 35 to 777,000 times normal levels; formaldehyde levels were 30–240 times normal. Both chemicals can cause cancers, but not be detected for several years, according to Dr. David Carpenter, director of the Institute for Health and the Environment at the State University of New York at Albany, and principal investigator.[654] Volunteers from Global Community Monitor (GCM) assisted on the data collection, leading Energy in Depth to denounce the study. Katie Brown of EID told The Center for Public Integrity that GCM is "a group that dubiously claims no amount of regulation will ever make fracking safe [or] could make a constructive contribution within the scientific community."[655]

Persons living near compressor stations are also at risk. Research conducted by chemist-microbiologist Wilma Subra, one of the nation's leading environmental scientists, reveals nasal and throat irritations, nausea, bronchitis, and frequent nosebleeds are just some of the health problems.[656]

EFFECTS UPON CHILDREN

The EPA reports, "Children are inherently more vulnerable to environmental hazards because their physiology is still developing."[657]

Dr. Sheila Bushkin, a physician who has a master's degree in public health, agrees:

> "Fetuses and children are disproportionately vulnerable to the deleterious effects of exposure to environmental toxicants. Although health impacts from industrial chemicals already exist in our population, the magnitude of risk would be greatly increased if High-volume Hydraulic Fracturing (HVHF) is permitted . . . Exposure to industrial chemicals and to ionizing radiation cause greater injury during development and early life. This may result in greater likelihood of birth defects, cognitive and behavioral development and life-long disabilities. Likewise, environmental exposures to these substances place pregnant women at greater risk from compli-

cations of gestation, resulting in increased maternal illnesses and mortality. From an ethical point of view, it is the responsibility of the medical community and legislative leaders to protect the health of the people. . . "[658]

Exposure to nonconventional gas drilling "before birth increases the overall prevalence of low birth weight by 25 percent, increases overall prevalence of small for gestational age by 17 percent and reduces 5 minute APGAR scores," according to research by Elaine L. Hill at Cornell University. (An APGAR score is a measurement of health of a new-born baby.) "[E]xposure within at least 1.5 miles is very detrimental to fetal development," Hill concluded.[659]

Building upon Hill's study, and with a different methodology, a team of four researchers from Princeton, Columbia, and MIT studied births in Pennsylvania and confirmed that proximity to natural gas drilling could be harmful to the fetus.[660] Both studies suggest that air pollution from fracking was the primary reason for fetal development problems.

A research team from the University of Pittsburgh, led by Dr. Shaina L. Stacey of the Graduate School of Public Health, analyzed birth records of more than 15,000 babies born in southwestern Pennsylvania in the years after fracking came to the state. Their analysis suggests there may be a direct link between mothers living near active natural gas wells and lower-than-expected birth weights.[661]

Some of the chemicals in the fracking fluids "are neurological poisons with suspected links to learning deficits in children [while others] are asthma triggers," Dr. Sandra Steingraber told members of the Environmental Conservation and Health committee of the New York State Assembly.[662] Dr. Steingraber, one of the leading authorities and public speakers about the effects of fracking, also pointed out, "Some, especially the radioactive ones, are known to bioaccumulate in milk. Others are reproductive toxicants that can contribute to pregnancy loss."[663]

HEALTH PROBLEMS AND DISTANCE FROM FRACKING

Persons living within one kilometer of well sites experience significant higher rates of skin and respiratory problems than do persons whose primary residence is more than two kilometers from the site, based upon a two-year study led by Dr. Peter M. Rabinowitz, associate professor of occupational and health science at the University of Washington. Among causes of medical problems, noted Rabinowitz, could have been "flaring of gas wells and exhaust from diesel equipment."[664] The study was published in September 2014 by the National Institute of Environmental Health Sciences. The 11-person research team also concluded that the initial information about health problems warranted significant additional investigation and analysis. However, because the Southwest Pennsylvania Environmental Health Project assisted in the research, the Marcellus Shale Coalition dismissed the research by calling the health agency "a local activist group," and charged that the study "was designed to put selective and unproven data behind a pre-determined and biased narrative."[665]

A *ProPublica* investigation by Abrahm Lustgarten revealed cases where individuals living near wells developed health problems far greater than expected if they had lived away from the wells. In one case, a Colorado woman living less than a half-mile from a well was forced to wear an oxygen mask every time she left her house. According to Lustgarten, who talked with residents in Colorado, Pennsylvania, and Texas, "most common complaints are respiratory infections, headaches, neurological impairment, nausea, and skin rashes. More rarely, they have reported more serious effects, from miscarriages and tumors to benzene poisoning and cancer."[666]

A team of researchers from the University of Maryland's School of Public Health, using both original research and a systematic review of available peer-reviewed scientific articles, analyzed the potential for health and environmental problems in the parts of western Maryland where fracking could occur if a state-imposed moratorium was lifted. Their conclusions, published in July 2014, were:

> "[There is] increased risk of sub-chronic health effects, adverse birth outcomes including congenital heart defects and neural tube defects, as well as higher prevalence of symptoms such as throat & nasal irritation, sinus problems, eye

burning, severe headaches, persistent cough, skin rashes, and frequent nose bleeds among respondents living within 1500 feet of UNGDP [Unconventional Natural Gas Development and Production] facilities compared to those who lived >1500 feet. Major determinants of these relationships include the concentration of the pollutants in the environment, frequency and duration of exposures encountered by individuals. . . .

"[T]here is a **High Likelihood** UNGDP related changes in air quality will have a negative impact on public health in Garrett and Allegany Counties. The extent of the impact will be based on population vulnerability, proximity to the sites, and the success of public health prevention strategies implemented by the State and local communities and control measures taken by the industry to minimize exposures."[667]

A survey conducted by Wilma Subra, Nadia Steinzor, and Lisa Sumi determined the most prevalent medical problems of those living near gas wells in the Marcellus Shale are sinus and respiratory problems (88 percent), behavioral changes (80 percent), neurological impacts (74 percent), muscle and joint pain (70 percent), ear, nose, and throat problems (70 percent), digestive and stomach problems (66 percent), skin problems (64 percent), and visual problems (63 percent.)[668] The researchers concluded, "The status quo—in which science and policy changes proceed slowly while gas development accelerates rapidly—is likely to worsen air and water quality, resulting in negative health impacts and possibly a public health crisis."[669]

'We Can No Longer Spend Time Outdoors'

In July 2011, Pam Judy told the Murrysville, Pa., city council her experiences with living within 800 feet of a compressor station. Prior to 2009, when the station was built, the Judy family experienced no significant health problems; however, in the two years since, Judy told the council:

"Due to the noise and the fumes from the engines and dehydration unit that settle in our yard we can no longer spend time outdoors. Shortly after operations began, we started to experience extreme headaches, runny noses, sore/scratchy throats, muscle aches and a constant feeling of fatigue. Both of our children are experiencing nose bleeds and I've had dizziness, vomiting and vertigo to the point that I couldn't stand and was taken to an emergency room. Our daughter has commented that she feels as though she has cement in her bones. "[O]ur son . . . developed blisters in his mouth and throat, had extreme difficulty swallowing, and [had to go] to the emergency room of a nearby hospital."[670]

Based upon medical tests that revealed measurable levels of benzene and phenol in her body, Judy got the DEP to do an air quality study. The tests identified 16 chemicals, most of them carcinogens, including acetone, benzene, carbon tetrachloride, chloromethane, styrene, toluene, and xylene. Judy told the Council she compares what is happening in the Marcellus Shale to the problems in the asbestos industry:

> "Both our government, and the asbestos industry, through very elaborate public relations schemes, led us to believe there was no harm in being exposed to asbestos. Only to find out years later the true cancer risks. I truly believe we could be facing a similar situation as a result of the Marcellus industry. And for those of who have been exposed it could be too late."[671]

Janet McIntyre, who lives about 1,500 feet from two well pads in Butler County, Pa., told Matt Walker of the Clean Air Council:

> "There's this blue haze that emits over this whole area here. It's just this blue haze—it's everywhere. You can't be outside more than 10 minutes. You get massive headaches, your eyes are red, your lips become tinny-tasting. I lived here 20 years and never had an issue until they came to town. If it's hot and you're outside, it feels like your skin is burning and is going to crawl off—it's bad."[672]

Kathryn King, who lives in Carroll County, Ohio, about 90 miles south of Cleveland, told the Associated Press she not only objected to the noise from nearby drilling operations, but that odors from the process often burn her nose.[673] Al Butz, a soybean farmer who also lives in Carroll County, told the AP the rural county was "a pristine place" before the fracking industry arrived, "then all hell breaks loose."[674]

AIR POLLUTION AND HEALTH EFFECTS IN TEXAS

Data from the Texas Cancer Registry in 2009 revealed that six counties in the Dallas–Fort Worth area "had the highest incidence of invasive breast cancer in the state," according to the CDC.[675] The report didn't show a direct link to chemicals used in fracking, but the data suggested it may have been more than a coincidence between nearby concentrations of unconventional wells and the high breast cancer rates.

The release of hydrogen sulfide (also known as "sour gas"—it smells like rotten eggs) is a problem in all drilling operations. But in Karnes County, Texas, it is far more than just annoying. Data filed with the Railroad Commission of Texas in 2014 revealed the Person Field in the Eagle Ford Shale had an average concentration of hydrogen sulfide of 16,399 parts-per-million (ppm), about 16 times the lethal dose. Data reported from the Panna Maria field revealed an average concentration of 24,408 ppm. Hydrogen Sulfide attacks the neurological and respiratory systems.

A joint investigation by the Center for Public Integrity and *InsideClimate News* pointed out:

> "H_2S and some other chemicals emitted during oil and gas production are so dangerous that the federal government has developed safety standards for workers who encounter them on a regular basis.
> "But there are no clear federal standards to protect people living near drilling sites—including children, the sick and the elderly—who intermittently breathe varying amounts of toxic emissions for years on end."[676]

Vapor clouds released during fracking operations by Chesapeake Energy possibly led to illness among residents of Arlington, Texas, according to an investigation by Peter Gorman for the *Fort Worth Weekly*.[677] One resident, Jane Lynn, told Gorman that in December 2011:

> "I was literally in the vapor cloud, and the odor was overwhelmingly strong. Within minutes there was tightness in my chest. Later that day I began getting heart palpitations—

which I've only gotten previously when exposed to gas escaping from a well site."[678]

Ranjana Bhandari said she was in traffic when fumes from a well began to affect her sinuses. "The symptoms lasted for hours and then returned when I was stuck at a railway crossing in the vicinity a few days later," Bhandari said.[679] Jean Stephens said after walking out of her store she was overcome by nausea and breathing difficulties from a vapor cloud. "It was just a severely nauseating foul odor [that] I'd never smelled . . . before," she said.[680] The response from Chesapeake, according to the *Weekly*, was the vapor clouds were only clean steam.

Mayor Calvin Tillman of the small Texas town of DISH, surrounded by wells and compressor stations, told the *Fort Worth Weekly*, "Some days you can hardly breathe anywhere in DISH. . . . We knew something was terribly wrong." However, when the residents complained to state agencies, the agencies "basically asked the companies to investigate themselves, and they came up clean every time."[681]

And so the residents voted to commit $10,000 of their annual $70,000 budget to fund a study of possible air pollution. According to the report, Wilma Subra studied "fugitive emission sources of hazardous air pollutants emanating from the oil and gas sector, [including] emissions from pumps, compressors, engine exhaust and oil/condensate tanks, pressure relief devices, sampling connections systems, well drilling (hydraulic fracturing), engines, well completions, gas processing and transmissions as well . . . mobile vehicle transportation emissions."[682] The testing, says Subra:

> "confirmed the presence in high concentrations of carcinogenic and neurotoxin compounds in ambient air near and/or on residential properties. The compounds in the air indicate quantities in excess of what would normally be anticipated in ambient air in an urban residential or rural residential area. Many of these compounds verified by laboratory analysis were metabolites of known human carcinogens and exceeded both Short-term and Long-term effective screening levels . . ."[683]

Among the known carcinogens released into the air, presumably from wells and compressor stations, were benzene, carbon disulfide, naphthalene, trimethylbenzene, and xylene toxins. Dr. Alisa Rich, an environmental scientist and president of the research firm that conducted the study, told the *Fort Worth Weekly*, "it's a toxic soup out there."[684]

The response by the Texas Pipeline Association was to deny the scientific conclusions and question the study's methodology. In March 2012, Calvin Tillman resigned as mayor and moved out of DISH, citing health concerns for his family. Tillman said his two sons, ages 4 and 7, were having respiratory problems and nosebleeds, which he attributed to the air pollution from the gas industry.[685]

Lynn Buehring, a bookkeeper and retired teacher, closed her office in Karnes City, Texas, in March 2015 because noxious fumes from nearby gas facilities affected her and her clients, according to reporting by David Hasemyer for *InsideClimate News*. "The smell and how it would make you sick. . . . I just couldn't take it anymore," she told Hasemyer. "The foul air," wrote Hasemyer, "has caused Buehring blinding headaches, aggravated her asthma, burned her eyes and brought on nausea so bad that she'd have to lie down." Several complaints filed by Buehring with the Texas Commission on Environmental Quality (TCEQ) "have gone nowhere," Buehring said.[686] The TCEQ received 70 complaints about air quality in 2014; the previous year, there were only 10 complaints.

A 20-month investigation by the Center for Public Integrity, *InsideClimate News*, and the Weather Channel revealed not only are thousands of oil and gas facilities "allowed to self-audit their emissions without reporting them to the state," but that the TCEQ "doesn't even know some of these facilities exist." The reporters concluded that the Texas air monitoring system "is so flawed that the state knows almost nothing about the extent of the pollution in the Eagle Ford [shale]."[687]

Part of the reason the TCEQ didn't know about some of the places in Texas that were emitting high levels of toxins and why the system was so flawed can be traced to a government mind-set that fracking was good and important—the environment and health issues were of lesser importance. Between 2000 through January 2015, Texas issued 24,995 permits just in the Barnett Shale.[688] However, the state legislature had cut the TCEQ budget to $372 million in 2014, down from $555 million in 2008; the budget for air monitoring dropped almost in half during that time.[689] Even if the budget wasn't cut, it's possible the TCEQ would still be reluctant to be vigorous in enforcement of air and water pollution in the Eagle Ford and other Texas shales—all three commissioners were appointed by Gov. Rick Perry, a conservative Republican who served 15 years, beginning in 2000, and whose philosophy about government and the environment was that the EPA should be eliminated.

Pollution from Wells and Compressor Explosions

In June 2010, a "blowout" at a well owned by EOG Resources spewed about 35,000 gallons of toxic chemicals and about one million gallons of hydrofracking fluid into the air for 16 hours in the Moshannon State Forest in central Pennsylvania. The state DEP fined EOG about $400,000 and shut down operations for 40 days.[690]

More than 900 persons were evacuated, and thousands were ordered to remain in their houses with doors and windows closed, when three explosions followed by a black plume rose to about 1,000 feet from the Chevron Oil Refinery plant in Richmond, Calif., in the late evening of Aug. 6, 2012.[691] That cloud of toxic hydrocarbons began drifting over cities in the San Francisco Bay area.[692] A leaking eight-inch diameter pipe, first placed in the plant in the 1970s, was believed to be the source of the plume and fire. In addition to five employees who were injured, more than 5,700 residents sought medical evaluation and treatment; about 160,000 persons live in the affected area.[693]

"We have been fighting Chevron since I was elected," Mayor Gayle McLaughlin told NPR's *Democracy Now!* With a new Council membership, McLaughlin says "Chevron has no longer dominated City Hall,"[694] part of the reason why Richmond had previously accepted most of what Chevron was dealing.

Antonia Juhasz, author of several books about the oil industry, reported in the *Los Angeles Times* that Chevron, the state's largest employer, has a record that leads to incidents that threaten the public health and welfare. "Rather than use its $27 billion in 2011 profits to run the cleanest, safest and most transparent refinery possible," said Juhasz, "Chevron operates a refinery that is in constant violation of federal and state law and a daily threat to the health and safety of its workers and neighbors." According to Juhasz:

> "Since at least April 2009, the refinery has been in non-compliance of the Clean Water Act and the National Pollutant Discharge Elimination System in every quarter but one. Until July 2010, the refinery had been in 'high-priority violation' of Clean Air Act compliance standards, the most serious level of violation noted by the EPA, since at least 2006. . . . Chevron has been assessed hundreds of thousands of dollars in penalties for repeated Clean Air Act violations—nearly 100 citations in just the last five years, including 23 in 2011 alone.
>
> "A 2008 study by UC Berkeley and Brown University researchers concluded that the air inside some Richmond homes was more toxic than that outside because of harmful pollutants from the refinery being trapped indoors.
>
> "The Contra Costa County Health Services Department lists the residents of Richmond as one of the 'most at-risk groups' in the county: They are hospitalized for chronic diseases at significantly higher rates than the county average, including for female reproductive cancers, which are more than double the county rate. Chevron is one of four refineries in Contra Costa County where nearby incidence of breast, ovarian and prostate cancers are the second highest in California, and where nearby residents suffer higher rates of asthma, child-hood asthma and asthma-related deaths."[695]

Cal/OSHA fined Chevron $963,200, the maximum allowed under state law, for 25 safety incidents that occurred before, during, and after the explosion in Richmond.[696] Chevron is appealing those citations. However, Chevron has paid about $10 million to settle claims filed by individuals.[697]

Trying to regain its influence on city politics, Chevron spent more than $3 million in 2014[698] on a negative campaign against the new city council, while backing a slate of Chevron-supported council candidates. Progressive candidates won the four open council seats. Council member Tom Butt, who spent about $50,000, was elected mayor, defeating his Chevron-backed opponent, 51–35 percent.

Diesel Fuel Pollution

For decades, the oil and gas industry wasn't concerned about diesel fumes. "They're temporary sources, and they are not causing accumulated effects," said Nicholas (Corky) DeMarco, executive director of the West Virginia Oil and Natural Gas Association.[699] DeMarco was wrong.

Diesel fuel is soluble in water and can pollute ground, water, and air. It contains a number of toxic chemicals and carcinogens, including benzene, toluene, ethylbenzene, and xylene, commonly known as BTEX, which has significant negative effects upon health.[700]

To put the mixture of water, proppants, and chemicals into the earth requires 40,000 horsepower diesel engines, spewing exhaust into the atmosphere and running at near-jet levels 24 hours a day.

Barb Harris, a specialist in environmental toxins, points out, "Pre-natal exposure to diesel fuel combustion products is a known risk for low birth weight [which is] a major indicator of child health, and is associated with multiple health and learning issues throughout life."[701]

DIESEL FUEL IN TRANSPORTATION

As many as 200 trucks, each one with two 100–200 gallon fuel tanks, bring water to every well every day, and then remove the wastewater; dozens of trucks bring supplies; trains bring sand and other chemicals. The problems are magnified by trucks idling at sites, and by leaks and spills while on the road. Diesel fuel, if spilled onto highways, is also more difficult to wash away than gasoline and will often leave a greasy slick that becomes a hazard to driving conditions.

The New York State Department of Environmental Conservation estimates almost 4,000 truck trips per well during the first 50 days of development is needed for each gas well that uses the fracking procedures;[702] this is slightly more than twice the trips necessary for conventional wells.

Citing research conducted by Dr. Michael McCawley of West Virginia University, the New York State Department of Health concluded, "heavy vehicle traffic and trucks idling at well pads he likely sources of intermittently high dust and benzene concentrations, sometimes observed at distances of at least 625 feet from the center of the well pad . . . These emissions have the potential to contribute to community odor problems, respiratory health impacts such as asthma exacerbations, and longer-term climate change impacts from methane accumulation in the atmosphere."[703]

A Pennsylvania law[704] implemented in 2009 forbids trucks to idle more than five minutes per hour. "Idling of these heavy-duty engines produces large quantities of dangerous air pollutants that can be particularly harmful to young children, the elderly and people with respiratory problems, such as asthma, emphysema and bronchitis," John Hanger, the state's DEP secretary, told *The Trucker* magazine in 2009.[705]

A federal law enacted in 2010 requires the use of ultra-low sulfur diesel fuel (ULSF) in transportation,[706] which has resulted in a "cleaner" fuel, and more efficient engines. This has resulted in a significant reduction in particulate emissions and nitrogen oxide. However, the Diesel Technology Forum acknowledges only about "one-third of all the heavy-duty commercial trucks on the road in the U.S. are 2007 or later model year and have the most sophisticated emissions control technology."[707]

DIESEL FUEL AT THE WELL SITE

Although technology has eliminated some of the pollution from diesel-fuel transportation, another problem exists. "Diesel engines have long been associated with refinery and drilling rig disasters because they can overheat and rev to the point of explosion when their intake valves suck in hydrocarbon vapor," according to the *San Francisco Chronicle*.[708]

Dr. Ronald Bishop, professor of biochemistry at the State University of New York at Oneonta, believes "intensive use of diesel-fuel equipment will degrade air quality [that could affect] humans, livestock, and crops."[709]

Dominick Mas, a former operating engineer in the Marcellus Shale, says he saw "a lot of fluids coming off the equipment and going into the creeks." He says the pollution was obvious— "you could see it in the puddles you're walking through."

Diane Siegmund, a clinical psychologist from Towanda, Pa., wants to know how diesel fuel got into some people's water supply. "It wasn't there before the companies drilled wells; it's here now," she says.

The industry is beginning to take advantage of a natural gas glut from over drilling, and the resulting lower costs, to retrofit pumps and rigs to run on natural gas. The largest manufacturing companies—Baker Hughes, Halliburton, and Schlumberger—began retrofitting their equipment in 2012.[710] Their decisions had little to do with the desire to protect public health, and everything to do with the desire to reduce costs and maximize profits.

Although the use of natural gas to power equipment at the drilling sites is a significant improvement, use of diesel fuel in the chemical mix is not.

Although the Safe Water Drinking Act and the Energy Policy Act of 2005 gave blanket exemptions to the oil and gas industry, they did prohibit the use of diesel fuel in the fracking mixture. In 2003, three major natural gas corporations had signed an agreement with the EPA not to use diesel fuel in the fracking process. However, between 2005 and 2009, 12 companies in 20 states used about 32.7 million gallons of diesel fuel, according to an investigation by the Democratic members of the House Committee on Energy and Commerce.[711] Companies which used the most diesel fuel, according to the Committee, were BJ Services (11.6 million gallons), Halliburton (7.2 million gallons), RPC (4.3 million gallons), Sanjel (3.6 million gallons), and Frac Tech (2.6 million gallons). States in which the companies used the most diesel fuel were Texas (16.7 million gallons), North Dakota (3.1 million gallons), Wyoming (2.9 million gallons), Louisiana (2.9 million gallons), and Colorado (1.3 million gallons).[712]

An EcoWatch investigation revealed that between January 2011 and August 2012 the natural gas industry used kerosene and diesel fuels 1 and 2 as part of the fracking fluid mix on 448 separate occasions.[713] EcoWatch determined that wells in four states had more than 90 percent of all violations—Arkansas (171), Texas (142), New Mexico (57), and Pennsylvania (45). The oil and gas industry injected more than 32 million gallons of diesel fuel between 2005 and 2009.[714] A subsequent investigation by the Environmental Integrity Project (EIP) revealed that between 2010 and July 2014, drillers from 33 companies illegally used diesel fuel to frack at least 351 wells in 12 states. About 39 percent of the volume of diesel fuel used for fracking was in Texas, according to the EIP; drillers in Colorado (28 percent) and North Dakota (13.5 percent) used the second and third highest volume of diesel fuel. However, the numbers may be significantly higher, according to EIP, because of a refusal by companies to provide accurate information to FracFocus about all chemicals used in the fracking process.[715]

"The list of toxic chemicals exempted by the Halliburton loophole is staggering, but to find that the one item still restricted is nonetheless being used without regulation or consequence is unacceptable," EcoWatch observed, and argued, "Not only do these health hazards raise concern about injection through groundwater supplies to shale layers deep beneath the earth, but also [by] air transmission through flaring and fugitive emissions."[716]

Energy in Depth (EID), massaging the data, claimed the EIP study was flawed and "grossly inflated" because most of the instances involved kerosene; EID claimed that although the EPA had first identified kerosene as a diesel fuel in May 2012, it didn't finalize its rules until February 2014.[717] Another of the rebuttals was that although the EPA may have had certain regulations against use of diesel fuel, it had given some states primacy in permitting and using diesel fuel in fracking, but EID didn't acknowledge that in its study.

Diesel fuel storage tanks at a fracking site in the Permian Basin.

PHOTO: Hannah Hamilton/USGS

CHAPTER 9
Water Pollution

To frack the earth, energy companies siphon massive amounts of water from public rivers and lakes. To do so, they must first get permission from regional or state agencies. In some cases, it isn't difficult to convince public officials to allow energy companies to take what is necessary to drill for oil and natural gas.

Even if the water taken from public waterways is clear, by the time it has been used in fracking it is polluted. Between 2009 and the end of 2014, more than 175 million gallons of wastewater spilled in more than 21,000 incidents from "ruptured pipes, overflowing storage tanks and other mishaps or even deliberate dumping," according to an AP analysis.[718]

An extensive study of fracking commissioned by the European Union revealed, "Risks of surface and ground water contamination, water resource depletion, air and noise emissions, land take, disturbance to biodiversity and impacts related to traffic are deemed to be high in the case of cumulative projects."[719] The 292-page study recommended that no fracking be allowed near areas where water is used for drinking.

"To naively believe that you can put 4.5 millions of gallons of contaminated water into the ground and that it will never seep into our aquifers or riverbeds is not realistic," Dr. Walter Tsou stated in September 2011 at a public hearing organized by the Citizens Marcellus Shale Commission.[720] Dr. Tsou is past president of the American Public Health Association and former Philadelphia health commissioner.

Dr. Tom Myers, a hydrologist, agrees. In a research study published in the April 2012 issue of *Groundwater*, Dr. Myers pointed out that fracking "could allow the transport of contaminants from the fractured shale to aquifers."[721] His research suggests that normal movement takes "tens of thousands of years to move contaminants to the surface, but fracking the shale could reduce that transport time to tens or hundreds of years" to contaminate the water supply.

"Even in the best case scenario, an individual well would potentially release at least 200 [cubic meters] of contaminated fluids," according to Daniel Rozell and Dr. Sheldon Reaven of the State University of New York at Stony Brook. Their study, published in August 2012, concludes, "This potential substantial risk suggests that additional steps be taken to reduce the potential for contaminated fluid release from hydraulic fracturing of shale gas."[722]

TOTAL DISSOLVED SOLIDS AND CHEMICALS IN WATER

Research from a team of 16 scientists from four Texas universities and Inform Environmental revealed the presence of groundwater pollution in the Barnett Shale. Working independent of the gas and oil industry, the team determined the presence of the chemicals may be related to fracking operations. The team "detected multiple volatile organic carbon compounds throughout the region, including various alcohols, the BTEX family of compounds [benzene, toluene, ethylbenzene, and xylene] and several chlorinated compounds," none of them naturally occurring in the area, in 381 of 550 public and private water wells.[723] The study, published in the June 20-15 issue of the peer-reviewed *Environmental Science and Technology*, cautioned that the "data do not necessarily identify UOG [unconventional oil and gas extraction] activities as the source of contamination; however, . . . many of the compounds we detected are known to be associated with UOG techniques."[724]

The American Academy of Pediatrics recommends "families with private drinking water wells in NGE/HF [natural gas extraction/ hydraulic fracturing] areas should consider testing the wells before drilling begins and on a regular basis thereafter for chloride, sodium, barium, strontium, and VOCs."[725]

Total Dissolved Solids (TDS) and several toxic chemicals were responsible for machinery corrosion at several industrial plants, including U.S. Steel and Allegheny Energy, along the Monongahela River near Pittsburgh. Tests by the Pennsylvania

Department of Environmental Protection in October 2008 measured TDS contamination at twice the maximum acceptable levels.[726] Pittsburgh officials advised its residents to drink only bottled water. The EPA called the pollution "one of the largest failures in U.S. history to supply clean drinking water to the public."[727] Although the DEP later determined the problem was probably caused by untreated wastewater from fracking operations, Stephen W. Rhoads, president of the Pennsylvania Oil and Gas Association in 2009, claimed most of the TDS came not from wastewater but from abandoned mines.[728]

Near Hennessey, Okla., a tank leaked about 480 barrels (about 20,000 gallons) of hydrochloric acid used in fracking jobs in July 2014.[729] Research by a team of 11 scientists found "arsenic, selenium, strontium and total dissolved solids (TDS) exceeded the Environmental Protection Agency's Drinking Water Maximum Contaminant Limit (MCL) in some samples from private water wells located within 3 km. [1.87 miles] of active natural gas wells." While some wells did not have the high levels, the team also noticed, "Lower levels of arsenic, selenium, strontium, and barium were detected at reference sites outside the Barnett Shale region as well as sites within the Barnett Shale region located more than 3 km from active natural gas wells." The study, published in the Summer 2013 issue of the peer-reviewed journal *Environmental Science & Technology*, also found that methanol and ethanol were found in 29 percent of the wells closer to drilling site wells." [730]

Nathaniel R. Warner, principal researcher of a Duke University study, concluded that increased salinity of water in the Marcellus region suggests that "homeowners living in these areas are at higher risk of contamination from metals such as barium and strontium."[731]

A *Pittsburgh Post–Gazette* story about the health problems of families living near natural gas wells in Washington County noted that nitrate levels were significantly above acceptable health levels.[732] According to the *Post–Gazette*, "Small children can be sickened by drinking water with [high] nitrate levels . . . and infants drinking water or formula with high nitrate levels can die from 'blue baby syndrome.' "[733]

Dr. Amelia Paré, a plastic surgeon trying to determine the

cause of skin lesions for several of her patients living near natural gas wells and wastewater ponds in southwestern Pennsylvania, tested the urine from two families. In one family, she found phenol, hippuric acid, and mandelic acid. In the other family, she found high levels of arsenic.

She says she had asked the Pennsylvania Department of Public Health and the health departments of Allegheny and Washington counties to test the wastewater and the families' well water, "but they declined." Local anti-fracking activists had provided water buffalos for the families. The lesions decreased in severity and health improved for each of the family members when they didn't use their well water, says Dr. Paré, who says she is not opposed to gas drilling but to the health problems from how it is drilled.

Tammy Manning, of Franklin Forks, Pa., told anti-fracking activist Iris Marie Bloom:

> "All of a sudden our water turned dark grey and then we noticed that it was actually erupting from the well head with a lot of force. You would hear it begin to hiss and then the water would spray out three to four feet in a circle around the well. . .
> "[An official who measured methane levels in our home] told us, as his methane detector was sounding off, that the levels were so high that we should not use the kitchen stove, as it could start a flash fire, and we should leave the bathroom window and door open and fan going during showers, as methane could build up and cause an explosion risk. He also told us the utility companies and fire department would have to be notified of our levels.
> "I asked him if we could continue living in our home. He said it was not for him to decide.
> "We were concerned that our well might explode and it is very close to the house. So to keep the pressure from building up, we ran the water constantly. Our granddaughter's bedroom is above the kitchen and she began vomiting in the morning when she first woke up. She wasn't running a fever and after vomiting she was fine. We thought she was just waking up hungry so we left crackers on her night stand.
> "By March [2012] our methane levels had nearly doubled. The DEP asked the gas company to vent our well and give us a water buffalo and disconnect our well entirely. Once the well was disconnected, our granddaughter was fine.

"The Friday before our well was vented the DEP tested the free gas coming out of our well and said it was 82% methane coming out. I was quite concerned.

"Also, besides the methane, we had carbon monoxide coming out of [our] faucet. Our water tests also showed very high unnatural levels of some dangerous heavy metals. We bought camp showers for bathing our grandchildren as we were advised that the metals can pose serious health problems and can be absorbed through the skin and inhaled, not just ingested."[734]

Although the problems didn't begin until after WPX Energy began drilling near her property, the DEP later claimed there was no indication that natural gas drilling was the cause of the family's problems.

In Monroeton, Pa., after a well was drilled near her home in 2009, Jodie Simons said water coming from her well had a "milky grey haze." The water, says Simons, "stinks awfully; it has a scummy, rotten and nasty smell." She told psychologist Diane Siegmund that she and her son soon got rashes with oozing blisters. Simons' daughter, says Siegmund, developed nosebleeds, nausea, and severe headaches. The DEP found higher-than-expected levels of chloride, magnesium, calcium, potassium, and sodium in the water, says Siegmund, but told the family the water was safe to drink.

Three families living near Ligonier, Pa., in the southwestern part of the state, were victims of leaks from a three million gallon WPX Appalachia wastewater fracking pit in 2013. Dolly Coffman, the daughter of Ken and Mildred Geary, told the Pittsburgh Post–Gazette that her parents' water has a "foul, chemical odor to it and it's been smelly and bad like that for a year, but they didn't know what to do."[735] WPX provided fresh drinking water to the families, but the families were still using the polluted water for other activities.

In November 2007, the Hallowich family—Stephanie, an accountant; Chris, a high school history teacher; Alyson, 5; and Nathan, 3—whose case had led to a right-to-know issue— moved into a newly-built home about 25 miles southwest of

Pittsburgh. The family had a two-story house constructed upon 10 acres they had previously bought. It was to be their dream home; it soon became their nightmare.

"It's ruined our lives," Chris told the *National Geographic Daily News*. "It's ruined our plans that we had for the kids. It's ruined what we thought was our perfect ten acres," he said.[736] What ruined their lives was a toxic soup from nearby natural gas drilling. Around them, according to the *Daily News*, was "an industrial panorama [of] four natural gas wells, a gas processing plant, a compressor station, buried pipelines, a three-acre plastic-lined holding pond, and a gravel road with heavy truck traffic." For about a year and a half, the family unknowingly drank contaminated well water.[737] Stephanie Hallowich said the family "had a lot of air issues, where it smells really bad; we get burning eyes, burning throats, headaches, ringing ears; we don't know what's coming out."[738]

Range Resources and the Pennsylvania DEP denied the problems the Hollowichs experienced were from fracking. And so the family paid for independent tests. What was coming out, polluting the air and water, were high levels of manganese; from the leaking plastic-lined impoundment dam came acrylonitrile, ethyl benzyl, styrene, tetrachlorethylene, and toluene.[739]

The family, which drank and bathed in polluted water, rented a 1,500 gallon water tank, which they filled every three weeks. The cost exceeded the $300–$400 a month they received from Range Resources for mineral rights. Unable to get Range Resources or the DEP to acknowledge what the family now knew, Stephanie took her case to the media—to radio, newspapers, and television; local, regional, and national. But it was her neighbors, the ones who were also exposed to the pollution and health hazards, who complained—not about the companies but about Stephanie. "Our whole community revolves around the church [and] people yell at me when I go there now," she said.[740] Hannah Abelbeck and Elizabeth Berkowitz, producers of *Faces of Frackland*, suggest that the neighbors saw Stephanie "drawing attention to her problems as a threat to their interests, since many people expect great royalty checks when their properties are drilled."[741] It is a common problem that divides communities in the Marcellus Shale region.

Unable to sell their house or get compensation for having

polluted water, the Hallowichs sued Range Resources, three subcontractors and suppliers, and the DEP.

Matt Pitzarella of Range Resources acknowledged that his company may have made "some mistakes—poor communication with a landowner, choosing a bad location for an access road, things like that." But, said Pitzarella, "if we make a mistake we own up to it and make it right."[742] The Hollowichs, said Pitzarella, "are in an absolutely unique situation . . . . Not only will you not see it [lack of communication] in the future, you won't see it now, and you won't see it since then."[743]

Perhaps Pitzarella's comments were inspired by advice he received from some of Range Resources staff who had been military psychological operations officers. At a conference of energy company staff and executives in November 2011, Pitzarella explained:

> "Really all they do is spend most of their time helping folks develop local ordinances and things like that. But very much having that understanding of psy ops in the Army and in the Middle East has applied very helpfully here for us in Pennsylvania."[744]

Others may dispute that Range Resources improved its communications and significantly reduced human error, or that it "owns up" for every error, especially since the company sues or threatens to sue those who challenge it, and has walked out of hearings.[745] About the time Pitzarella was spinning how "responsible" his company was, the DEP issued it a citation for a faulty cement casing on a well in Lycoming County. Under the Corbett administration's "educate don't penalize" philosophy, Range Resources didn't remediate the problem or deal with a documented methane leak that contaminated a stream and water supply for several water wells. Under a new governor, the DEP cited Range for violating the Clean Streams Law and the Oil and Gas Act, and issued a notice in June 2015 it was assessing an $8.9 million civil penalty.[746] Nevertheless, for the Hallowichs—who suffered from pollution and neighbor greed, while a private corporation and a state agency failed to protect their health—the problem finally ended in August 2011 when they agreed to a settlement that, although they reluctantly agreed to allow it to be sealed, led to the purchase of their house and property.

The settlement didn't end their problems. Three months after the family agreed to the settlement, they again sued Range Resources, this time for violating the non-disclosure agreement and the gag order imposed by Judge Paul Pozonsky. The Hallowich family claimed the company falsely stated it paid $550,000 in the settlement. According to reporting in the *Pittsburgh Post-Gazette,* the Hallowichs' petition "states that Range Resources intentionally and fraudulently filed a Realty Transfer Tax Statement of Value with the state Department of Tax Revenue tax bureau to publicly embarrass Stephanie and Chris Hallowich, inflate the family's tax obligations on the sale of their home, and 'garner a public relations windfall' because the company had paid more than the full market, appraised value of the property."[747] Range Resources denied the allegations.

Equipment failure may have led to high levels of chemicals in the well water of at least a dozen families in Butler County, Pa. in late 2010. Township officials and Rex Energy, although acknowledging that two drilling wells had problems with the casings, claimed there were pollutants in the drinking water before Rex moved into the area. John Fair disagrees. "Everybody had good water a year ago," Fair told Iris Marie Bloom in February 2012. Bloom says residents told her the color of water changed (to red, orange, and gray) after Rex began drilling. Among chemicals detected in the well water, in addition to methane, were ammonia, arsenic, chloromethane, iron, manganese, t-butyl alcohol, and toluene.[748] Not acknowledging that its actions could have caused the pollution, Rex did provide fresh water to the residents but then stopped doing so on Feb. 29, 2012, after the Pennsylvania DEP said the well water was safe. The residents disagreed and staged protests against Rex.[749] Environmentalists and other residents trucked in water jugs to help the affected families. Joseph P. McMurry of Marcellus Outreach Butler (MOB) stated that residents' "lives were severely disrupted and their health has been severely impacted. To unceremoniously 'close the book' on investigations into their troubles when so many indicators point to the culpability of the gas industry for the disruption of their lives is unconscionable."[750]

Karen Harbert of the U.S. Chamber of Commerce said she didn't want to see states enact restrictive financial and environmental laws.[751]

FISH KILLS

A spill of 4,700 gallons of hydrochloric fluid from an unsecured storage tank at a Chief Oil & Gas drilling site in Bradford County, Pa., led to a fish kill in a tributary of Towanda Creek.[752]

In May 2012, Pennsylvania DEP Secretary Michael Krancer, upset that the DRBC extended the moratorium on removing water for fracking operations, had said Delaware "smells like the tail of a dog."[753] Krancer later said, "I'm not battling with people . . . I'm dialoguing with people."[754] Among the people he didn't "dialogue" with were the Pennsylvania Fish and Boat Commission, and two dozen former DEP scientists and officials.

The Pennsylvania Fish and Boat Commission several times asked the DEP to place the Susquehanna River on the "impaired waterways" list, which would mean it would receive a priority to be cleaned up. Krancer refused the request.

Stuart Gansell, former director of the Bureau of Watershed Management, on behalf of 22 retired DEP scientists, engineers, and senior administrators, wrote a letter to Krancer, telling him they didn't understand why the DEP refused to declare the river was polluted. Krancer's response to DEP professionals was as dismissive as his rejection of the Fish and Boat Commission's request. Declaring 90 miles of the Susquehanna River as polluted would be nothing more than a "publicity stunt," Krancer replied, and suggested that many of the concerned former DEP staff didn't have the professional qualifications to fully understand reasons why he refused to declare the Susquehanna was polluted.

A week after the DEP staff had sent their letter, the Fish and Boat Commission filed a formal request to the DEP to reconsider its position to not declare portions of the Susquehanna to be polluted and to reclassify a 90 mile portion as "a high-priority impaired and threatened river."[755] Krancer also rejected that request.

The problems the Hallowichs, Tammy Manning, the Simons

family, and what others faced aren't isolated cases, but are only one part of the story of water pollution from fracking operations. There are thousands of other problems. Residents living near gas wells "have filed over 1,000 complaints of tainted water, severe illnesses, livestock deaths, and fish kills," according to the Environmental News Service.[756]

Radiation in the Drilling Process

The Pennsylvania DEP claims radiation from fracking had at most only a minimal effect upon public health. In a comprehensive report published in January 2015, the last month of the Tom Corbett administration, the DEP concluded:

"There is little potential for additional radon exposure to the public due to the use of natural gas extracted from geologic formations located in Pennsylvania.

"There is little or limited potential for radiation exposure to the public and workers from the development, completion, production, transmission, processing, storage, and end use of natural gas. There are, however, potential radiological environmental impacts from fluids if spilled. Radium should be added to the Pennsylvania spill protocol to ensure cleanups are adequately characterized. There are also site-specific circumstances and situations where the use of personal protective equipment by workers or other controls should be evaluated.

"There is little potential for radiation exposure to workers and the public at facilities that treat oil and gas wastes. However, there are potential radiological environmental impacts that should be studied at all facilities in Pennsylvania that treat wastes to determine if any areas require remediation. If elevated radiological impacts are found, the development of radiological discharge limitations and spill policies should be considered.

"There is little potential for radiation exposure to the public and workers from landfills receiving waste from the oil and gas industry. However, filter cake from facilities treating wastes could have a radiological environmental impact if spilled, and there is also a potential long-term disposal issue. TENORM [Technologically Enhanced Naturally Occurring Radioactive Material] disposal protocols should be reviewed to ensure the safety of long-term disposal of waste containing TENORM.

"While limited potential was found for radiation exposure to recreationists using roads treated with brine from conventional natural gas wells, further study of radiological environmental impacts from the use of brine from the oil and gas industry for dust suppression and road stabilization should be conducted."[757]

However, with landfills rejecting drill cuttings that exceed accepted radioactive limits, the companies are storing them at their own well sites, posing additional risk to public health,[758] or are trucking it to Ohio, New York, New Jersey, and Idaho.[759]

Radium-226, 200 times higher than acceptable background levels, was detected in Blacklick Creek, a 30-mile long tributary of the Conemaugh River near Johnstown, Pa. The radium, which had been embedded deep in the earth but was brought up in wastewater, was part of a discharge from the Josephine Brine Treatment Facility, according to research conducted by Drs. Nathaniel Warner, Cidney A. Christie, Robert B. Jackson, and Avner Vengosh, and published in the October 2013 issue of *Environmental Science & Technology*.[760] The EPA fined Fluid Recovery Systems, which bought the Josephine Brine company, $83,000.[761]

Pennsylvania found a way to keep information about radioactive waste disposal and rejection from public knowledge. The DEP's Bureau of Radiation Protection refused a request by the Delaware Riverkeeper Network in 2014 to disclose amounts and locations of radioactive levels in gas development. The DEP claimed releasing the information "would be reasonably likely to result in a substantial and demonstrable risk of physical harm or harm to the personal security of an individual, or to the public at large, due to the health risks associated with exposure to radioactive materials should the security of those materials be compromised as a result of public knowledge as to their location and quality."[762] However, in denying the validity of DEP's claims, the state's independent Office of Open Records clearly noted the DEP failed to establish "how the release of this information . . . would result in harm."[763]

Although the Pennsylvania DEP claims minimal effects from radiation, other research definitively reveals significant public health concerns from horizontal fracking.

In addition to bringing up methane and elements from as deep as two miles below the earth's surface, wastewater contains radioactive elements; among them are Uranium-238, Thorium-232, and radium. Radium in wastewater brine is three times higher from fracking than from conventional wells, according to the U.S. Geological Survey.[764]

Radium-226, which emits alpha and gamma radiation, is derived from Uranium-238, and has a half-life of 1,601 years; it is absorbed into the skeletal system, and can lead to anemia, bone cancer, leukemia, and lymphoma.[765] Radium decays into radon, one of the most radioactive and toxic gases. Radon is the second highest cause of lung cancer, after cigarettes, according to the EPA; for non-smokers who develop lung cancer, radon exposure is the leading cause of death.[766] Like carbon monoxide, it is odorless, colorless, and detectable only with sophisticated measuring instruments.

Radon concentration in areas near fracking sites have 39 percent higher readings than areas where no fracking occurs, according to research conducted by Johns Hopkins scientists and published in the April 2015 issue of *Environmental Health Perspectives*.[767] That study looked at almost two million readings in 866,735 buildings, most of them houses in every Pennsylvania county, between 1987 and the end of 2013. The researchers determined the increase in radon levels in houses and commercial buildings parallels the increase in fracking operations. Their conclusion was, "The development of unconventional natural gas in the Marcellus shale in Pennsylvania has the potential to exacerbate several pathways for entry of radon into buildings."[768]

Dr. Marvin Resnikoff, a physicist, suggests that up to 30,000 cancer deaths in New York State from radon-contaminated gas mined in the Marcellus Shale, could be attributed to fracking operations. Dr. Resnikoff concluded:

> "The long-term environmental risks and public health concerns of radon in Marcellus Shale natural gas formations are far too serious to be ignored. The potential impacts of radon must not be swept under the rug. Nor should these impacts be sacrificed to short-term economic policies or to unrealistic and/ or inaccurate assessments of the benefits of natural gas development in New York State."[769]

Research conducted at Penn State revealed that both vertical and horizontal fracking were responsible for bringing to the surface, within 90 days of drilling, radioactive materials similar to those deposited during the Paleozoic era, 250 to 540 million years ago. Professors Lara O. Haluszczak, Arthur W. Rose, and Lee R. Kump, whose research was published in the academic journal *Applied Geochemistry*, noted that the wastewater not only included elements not put into the earth, but that the wastewater had extraordinary high levels of salinity.[770]

Analysis of wastewater from 13 sites in New York state revealed the presence of Radium-226. Some of the radium was more than 250 times the acceptable range for discharge into the environment, according to research conducted by Dr. Resnikoff.[771] A USGS analysis of well samples collected in Pennsylvania and New York between 2009 and 2011 revealed that 37 of the 52 samples had Radium-226 and Radium-228 levels that were 242 times higher than the standard for drinking water. One sample, from Tioga County, Pa., was 3,609 times the federal standard for safe drinking water, and 300 times the federal industrial standard.[772]

An investigation by *New York Times* reporter Ian Urbina, based upon thousands of unreported EPA documents and a confidential study by the natural gas industry, concluded, "Radioactivity in drilling waste cannot be fully diluted in rivers and other waterways." Urbina learned that wastewater from fracking operations was about 100 times more toxic than federal drinking water standards; 15 wells had readings about 1,000 times higher than standards.[773] Urbina's article, part of the "Drilling Down" series, was attacked by conservative politicians and talk show pundits, and by the natural gas industry. "Drilling Down [is] a series of articles with ominous headlines and dubious facts," Energy for America argued.[774] However, because Urbina documented his sources and verified the facts, the *Times* didn't have to print corrections. "Because of that series," says David Braun of Californians Against Fracking, "we could take the information to the legislators to help convince them about the effects of fracking."

The *Columbus* (Ohio) *Dispatch* reported that in 37 of the 52 samples the USGS analyzed, "radioactivity from Radium-226 and Radium-228 was at least 242 times higher than the

175

drinking water standard and at least 20 times higher than the industrial standard."[775]

"The radium will be bio-accumulating," said Dr. Vengosh, "You eventually could get it in the fish."[776] The Marcellus Shale Coalition (MSC), in an email to *Bloomberg News*, dismissed the research, claiming that since the plant did not discharge waste from Marcellus Shale drilling after May 2011, the study reveals "the outdated nature of the data used for the report."[777] However, what the MSC claimed probably wasn't accurate. Dr. Vengosh accurately fired back: "Based on the isotopes that we measured we can see that the effluent that's coming from Josephine in the last three years, including two months ago, still has the fingerprint of the Marcellus."

Accidents and Intentional Pollution

One of the worst floods in U.S. history hit the north-central region of Colorado the second week of September 2013. In its wake, it left at least eight deaths,[778] destroyed more than 1,500 homes, damaged at least 18,000 more, washed out bridges and miles of roads and railroad tracks, and damaged or destroyed at least 2,300 separate agricultural fields.[779]

The flood also exposed the problems of safety in the natural gas industry, even with an extraordinarily quick response by the oil and gas industry.

Anadarko Petroleum reported about 13,500 gallons of oil and gas polluted the St. Vrain River near Platteville; about 5,250 gallons of oil polluted the South Platte River near Milliken.[780] The cause, according to Anadarko, was the force of debris-laden rushing water that struck holding tanks. More than 27,000 gallons from all sources polluted the rivers.[781] The flood damaged as many as 13,000 of the 20,000 wells in Weld County,[782] and also washed away up to five feet of top soil, according to Shane Davis, a biologist who had worked for the Colorado State Parks system.[783]

However, Gov. John Hickenlooper, a strong proponent of fracking, claimed, "The several small spills we've had have been very small relative to the huge flow of water coming through."[784] It was an unusual claim, especially since Hickenlooper didn't mention the problems oil and gas pollution was causing aquatic life, and Mike King, executive director of the Colorado Department of Natural Resources, had said not only was "the scale "unprecedented," but, "We will have to deal with environmental contamination from whatever source."[785]

Statements on behalf of the oil and gas industry claimed anti-fracking activists exaggerated the damage. Environmentalists were "taking away from the real issues to grind a political ax at a time when people have real problems," said Dan Kish, senior vice-president of the pro-industry Institute for Energy Research.[786] The *Colorado Independent* reported the Colorado Oil and Gas Association (COGA) "played down reports of damaged equipment."[787] What wasn't exaggerated or political was the amount of destruction the flood caused, and the pollution of agricultural fields and waterways.

About 11:45 p.m., April 19, 2011, a blowout at a Chesapeake Energy well (Atgas 2H) about 13 miles west of Towanda, Pa., spilled several thousand gallons of frack fluid and other emissions into Towanda Creek, which flows into the Susquehanna River.[788] Chesapeake's Brian Grove, two days after the blowout, told the Scranton *Times–Tribune*, "Initial testing from Towanda Creek indicates little, if any, significant effect to local waterways as a result of an apparent surface equipment failure [and] fluids from the well are fully contained."[789] However, containment would not be for another three days; Grove's statement of how thousands of gallons of toxic chemicals spilled onto agricultural fields and into a creek could be negligible strains the level of reason.

Nevertheless, there was another problem with the gas well's blowout that magnified the problem. It took about 12 hours for an emergency response team to arrive on site; the team had to fly from Texas to Pennsylvania.[790] John Hanger, former DEP secretary, told *ProPublica* the state had a contract with a private company, CUDD Well Control, to provide emergency response to any place in the state in less than five hours.[791] CUDD had an emergency operation based in Bradford County, site of the well blowout. A CUDD executive, according to *ProPublica*, had offered its assistance, but the DEP didn't request its help; Chesapeake also rejected CUDD's assistance, and called another Texas company, Boots and Coots, owned by

Halliburton. The DEP and Secretary Michael Krancer "didn't respond to calls and emails from *ProPublica*."[792]

Rory Sweeney of Chesapeake later told *ProPublica* that an in-house control specialist was at the site within 30 minutes and three more arrived within eight hours, and reduced the flow by 70 percent before the Texas team arrived. However, the DEP, in issuing a violation, asked "why Chesapeake took 12 hours to have a well control service company at the site when there are other well control service companies located closer to the Atlas 2H Well."[793]

About two weeks after the spill, Maryland Attorney General Douglas F. Gansler, with support from Gov. Martin O'Malley and Maryland's Department of the Environment, notified Chesapeake it was filing suit against the energy company[794] for violating the Resource Conservation and Recovery Act (RCRA) and the Clean Water Act (CWA). The Susquehanna, according to the suit, provides about 45 percent of the freshwater of the Chesapeake Bay, giving Maryland legal standing to sue. Although the contaminated water was eventually contained and did not enter the Chesapeake Bay, Chesapeake Energy agreed in June 2012 to contribute $500,000 to the Susquehanna River Basin Commission to assist in water quality monitoring, and to change some of its practices that impact water quality and the environment.[795]

Chesapeake was cited 576 times for violations in Pennsylvania between 2008 and the end of 2014, but fined only 20 times.[796]

PIPELINE SAFETY

Equipment failure at a drill site in Susquehanna County, Pa., led to a spill of several thousand gallons of fluid for almost a half-hour in January 2012, causing "potential pollution," according to the DEP. In its citation to Carrizo Oil and Gas, the DEP "strongly" recommended that the company cease drilling at all 67 wells "until the cause of this problem and a solution are identified."[797]

A 4,000 gallon spill of hydrochloric acid near Canton, Pa., July 4, 2012, resulted in pollution to a tributary of Towanda Creek.[798]

In March 2013, three major spills left Americans wondering about pipeline safety.

Near Mayflower, Ark., 4,000–7,000 barrels (168,000–294,000 gallons) of heavy Canadian crude oil spilled from a leak in ExxonMobil's 60-year-old 20-inch Pegasus Pipeline, according to EPA estimates.[799] The spill, which wasn't stopped for 12 hours, killed ducks and turtles, but did not kill other aquatic life, according to ExxonMobil.[800] Twenty-two homes were evacuated. ExxonMobil tested the strength of the pipe in 2006, learned there was "susceptibility to seam failures," but, apparently did nothing in the seven years between the test and the rupture, according to the Pipeline and Hazardous Materials Safety Administration.[801]

An underground spill in northwestern Colorado flowed into Parachute Creek, which empties into the Colorado River. Within two weeks of the spill in March 2013, Williams contractors had removed 60,648 gallons of hydrocarbon material and 5,418 gallons of oil.[802] Eventually, more than 100,000 gallons of contaminated water were recovered; there is no estimate of how much was not recovered. By June, the pollution had spread over 10.6 acres.[803] At one point, the Colorado Department of Public Health and Environment (CDPHE) recorded benzene concentration at 18,000 parts per billion (ppb); the safety level for aquatic life is less than 5,300 ppb; the level for drinkable water is 5 ppb.[804] It wasn't until five months after the spill that the CDPHE determined that benzene was finally below toxic levels in the water. Although other toxic hydrocarbons were released, they were not measured; CDPHE scientists had decided that the level of benzene in the water would be the key to determining if the water was safe. Between 2004 and 2014, about 720,000 gallons of oil spills were unrecovered in Colorado, according to the Colorado Oil and Gas Conservation Commission and an analysis by the *Denver Post*.[805]

The same month of the pollution of Parachute Creek, Carrizo Oil and Gas was responsible for an accidental spill of 227,000 gallons of wastewater, leading to the evacuation of four homes in Wyoming County, Pa.[806] Two months later, another malfunction at a well, also in Wyoming County, sent 9,000 gallons of wastewater onto the farm and into the basement of a nearby resident.[807] The DEP didn't fine Carrizo for those violations; of

108 violations the DEP issued against Carrizo, as of the end of June 2015, only four resulted in fines.[808]

In July 2013, XTO Energy, a division of ExxonMobil, agreed to a settlement of a $100,000 fine and to spend $20 million to upgrade its wastewater management system. Three years earlier, failure to close a valve to a series of wastewater tanks had led to a spill of about 57,000 gallons of brine, heavy metals, several fluids and elements in fracking fluid, and radioactive strontium into a tributary of the Sugar Creek, which flows into Susquehanna River in Pennsylvania.[809] According to the federal lawsuit that preceded the settlement, the spill also "drained through the surface soils and into the groundwater, which was then released in seeps to a spring and in an unnamed tributary."[810] The EPA and Department of Justice pursued the case after a DEP field inspector found the spill. Although XTO acknowledged its responsibility, a company executive claimed, there was "no lasting environmental impact."[811] Within two months of the settlement, based upon a Grand Jury indictment, Pennsylvania Attorney General Kathleen Kane filed criminal charges against XTO, charging it with willful violations of the Clean Streams Law and the Solid Waste Management Act. The case had been referred by the DEP during the Rendell Administration to the attorney general's office for criminal prosecution more than two years earlier;[812] however, under a Republican-appointed attorney general the investigation had languished. Kane, a Democrat, took office in January 2013. XTO's response to the criminal charges was that they were an "unprecedented and an abuse of prosecutorial discretion."[813]

Between 2006 and October 2014, about 18.4 million gallons of oil and chemicals polluted the air, soil, and waters of North Dakota, according to an investigation by Deborah Sontag and Robert Gebeloff of *The New York Times*.[814]

Two spills within four years polluted the Yellowstone River that flows into Montana and North Dakota. The first spill, in 2011 from the ExxonMobil Silvertip pipeline, released about 63,000 gallons of oil;[815] cleanup costs were about $135 million. The second spill was about 50,000 gallons of oil from the Bridger Pipeline's Poplar line in January 2015.[816] After residents complained about foul odors in their drinking water, Glendive,

Mont., city officials ordered the water plant to stop taking water from the river,[817] and advised residents to use bottled water or to boil faucet water for cooking and drinking.[818]

The spills into the Yellowstone River were significant, but less deadly than a wastewater/brine spill of about three million gallons from a four-inch saltwater collection pipe near Williston, N.D., in January 2015. That spill, which wasn't detected for 12 days, first affected Blacktail Creek and the Little Muddy River, and then the Missouri River almost two weeks later.[819] The North Dakota Department of Health confirmed "high readings" of pollution where the Little Muddy enters the Missouri.[820]

Methane in the Nation's Water Supply

Methane, a greenhouse gas, is the primary chemical in natural gas. As a compressed natural gas, it can be used for heating, cooking, and as a fuel for transportation.

The gas industry claims there is no migration from fracking into aquifers and homeowners' wells—and if there is migration, it occurred naturally before there was fracking. Scientists agree there was methane that migrated into wells before fracking; they disagree with the industry's claim that fracking isn't responsible. Barry Russell, president of the Independent Petroleum Association of America, argues, "No evidence directly connects injection of fracking fluid into shale with aquifer contamination." Elizabeth Ames Jones, former chair of the Railroad Commission of Texas, claims it is "geologically impossible for fracturing fluid or natural gas to migrate upward through thousands of feet of rock."[821] Fracking "has never been found to contaminate a water well," Christine Cronkright, communications director for the Pennsylvania Department of Health, said in 2013. One year later, Energy in Depth tried to add credence to Cronkright's claim, boldly stating, "There has never been a single case of water contamination from hydraulic fracturing."[822]

Independent research studies and numerous incidents of water contamination prove otherwise. In Pennsylvania alone, the DEP determined there were at least 243 credible cases between 2008 and 2014 where fracking operations affected or destroyed private drinking wells. A *ProPublica* investigation revealed methane contamination was widespread in drinking water in areas around fracking operations in Texas, Colorado, Wyoming, and Pennsylvania.[823]

Scientists at Duke University concluded, "Methane contamination of shallow drinking water systems [is] associated with shale-gas extraction." The data and conclusions, published in the May 2011 issue of the *Proceedings of the National Academy of Sciences (PNAS)*, revealed not only did most drinking wells near drilling sites have methane, but those closest to the drilling wells, about a half-mile, had an average of 17 times the methane of those of other wells.[824]

In response to that article, Aubrey McClendon, Chesapeake Energy CEO, in a strong letter to the university, of which he is a graduate and benefactor, said the research was "more political science than physical science." He claimed, with no evidence, the scientists' "goal was to attack all forms of natural-gas drilling, presumably so that the supply of natural gas would decline and therefore the price of natural gas would rise and their beloved 'green fuels' could become somewhat less uneconomic than they are today."[825]

A study published in the July 2012 issue of *PNAS*, revealed that mineral-rich fluids deep in the Marcellus Shale are migrating to the surface, which disputes numerous industry claims that layers of impervious rock protect the escape of injected fluids in fracking. Although the scientists couldn't establish direct links between fluids used in fracking and the presence of methane and other hazardous materials in aquifers and drinking water, they did find:

> "[T]he coincidence of elevated salinity in shallow groundwater with a geochemical signature similar to produced water from the Marcellus Formation suggests that these areas could be at greater risk of contamination from shale gas development because of a preexisting network of cross-formational pathways that has enhanced hydraulic connectivity to deeper geological formations."[826]

One year after that second study from Duke, a third study, again led by Dr. Robert B. Jackson, revealed that in northeastern Pennsylvania, "Methane was detected in 82% of drinking water samples, with average concentrations six times higher for homes < 1 km from natural gas wells [and ethane] 23 times higher in homes < 1 km from gas wells."[827]

In response, Kathryn Z. Klaber of the Marcellus Shale Coalition recited an industry talking point: "Private water well quality and construction, as well as methane migration, is a longstanding public health issue in Pennsylvania, dating back decades."[828] She didn't address the critical issue that the migration was higher in wells closer to natural gas drilling.

A three-year study in Colorado, based upon methane samples from about 300 locations, revealed there was a broad level of water pollution caused by natural gas drilling.[829] A *ProPublica* summary of the September 2008 report noted:

> "The researchers did not conclude that gas and fluids were migrating directly from the deep pockets of gas the industry was extracting. In fact, they said it was more likely that the gas originated from a weakness somewhere along the well's structure. But the discovery of so much natural fracturing, combined with fractures made by the drilling process, raises questions about how all those cracks interact with the well bore and whether they could be exacerbating the groundwater contamination."[830]

METHANE EXPLOSIONS

Methane itself isn't toxic, but is highly flammable; in large concentrations in water it will replace oxygen and can cause asphyxiation. Numerous explosions of freshwater wells near natural gas rigs attest to the problem that develops when fracking is used to bring methane from oxygen-deprived shale to the surface.

A natural gas explosion damaged a house in Bainbridge Twp., Ohio, Dec. 15, 2007. There were no injuries. "Early in the investigation, responders recognized that natural gas was entering homes via water wells," leading to the evacuation of 19 homes,[831] according to the Department of Natural Resources (DNR). During the subsequent week, Dominion East Ohio, the contractor, said it "disconnected 26 water wells, purged gas from domestic plumbing/heating systems, installed vents on six water wells, plugged abandoned in-house water wells, plumbed 26 houses to temporary water supplies, provided 49 in-house ethane monitoring systems for home-owner installation, and [provided] bottled drinking water to 48 residences."[832] The DNR determined the primary contributing factors to the explosion were inadequate cementing on the casing, a decision to proceed with the fracking process "without addressing the issue of minimal cement behind the production casing," and a 31 day period following fracking that "confined the deep, high pressure gas . . . within [a] restricted space."[833]

Gas explosions destroyed two homes in McKean County, Pa., in December 2010 and February 2011. In March, several public officials—including township supervisors, county commissioners, and first responders—met in secret at the Bradford Twp. Volunteer Firehall with representatives from an insurance company, a drilling company, and two gas and oil companies to discuss what was becoming a rash of incidents. In a blistering editorial, Marty Robacker Wilder, editor of the *Bradford Era*, skewered the public officials:

> "How dare they meet in private on an issue of such great public concern?
> "We honestly don't care how they justify this secret meeting — we've heard every excuse in the book over the years—because nothing can forgive their behavior not just as a violation of the Sunshine law but as a violation of the public trust.
> "We would like to tell you, in detail, about what was discussed inside the meeting hall but all we received was a secondhand report about these 'mysterious' explosions which occurred, one on Dec. 12 in which two people were injured, and a second Feb. 28 in which a lucky homeowner was out shoveling snow when his house exploded.
> "Sheriff Brad Mason was kind enough to speak to our reporters after the meeting but, in all honesty, he merely reiterated what's already been said: Gas migration is being explored as a possible factor in the house explosions and similar incidents that have reportedly occurred in the Bradford area.
> "Since nobody seems to be able to pinpoint the exact source of these explosions, there has been broad speculation that ongoing oil drilling activity in the general vicinity may some-

how be triggering gas to migrate into people's homes.

"As we all know, holes have been poked in this historic Bradford oilfield for more than 100 years and many abandoned and uncapped wells continue to make what's been called "Swiss cheese" of our underground terrain.

"Could this provide an explanation for the 'mysterious' explosions? And could that, in turn, also be part of the reason for a closed-door meeting?

"After all, nobody wants to point the finger at anyone or anything that revolves around the oil and gas industry. Some believe it's the panacea for all the ills of our economically depressed region—regardless of any consequences.

"Already, there is a nationwide debate simmering on the environmental concerns over how a new source of energy, the Marcellus Shale, is being extracted from the ground. A couple explosions here and there in a little town known as Bradford, Pa.—one of the birthplaces of the oil industry—could really stir the pot."[834]

Almost two weeks after Wilder's editorial, U.S. Sen. Bob Casey (D-Pa.) sent an equally strong letter to the secretary of the U.S. Department of Energy:

"According to the Environmental Protection (PA DEP) Emergency Response Program, there have been dozens of gas migration incidents in northwestern Pennsylvania recently.

"Some of those have led to explosions, leading to injury and the destruction of at least two homes. The belief that the source of the explosions is some type of thermogenic gas migration caused by extensive drilling appears to be widespread. . . .

"[It] appears that the gas may have migrated from deep underground during periods of high barometric pressure coupled with seismic activity and extensive new deep drilling activities. The lack of reliable data on old oil and gas wells, which number in tens of thousands, and the deterioration of old well casings may also have contributed to the gas migration.

"The McKean County homes were located about two and half miles from each other in neighborhoods . . . where oil and gas drilling activities had caused methane gas infiltration into drinking water wells, leading to taste and smell impacts. Schreiner Oil, the company involved, was ordered by the PA DEP to restore the water and has been providing bottled water to the impacted neighborhood. The explosion of the two houses in close proximity to this troubled area certainly appears to be more than coincidence, yet the phenomenon is poorly understood and there is currently no way of preventing or even predicting when such incidents may occur. . . ."[835]

Sen. Casey urged the Secretary "to coordinate with local, state, and other federal entities to ensure that appropriate actions to protect public health are implemented."

Three persons in Pearsall, Texas, were injured after an explosion and fire at a wastewater disposal well in January 2012. The Railroad Commission said the cause of the explosion was welders who were working near a truck that was unloading the water into a collection tank. "Sparks from the welding may have ignited vapors around the storage tank, causing the explosion," according to an official statement.[836] Human error may have caused the explosion, but the wastewater had enough volatile fumes that could have impacted health and safety even if welding sparks hadn't caused the problem.

"Today's methods make gas drilling a filthy business. You know it's bad when nearby residents can light the water coming out of their tap on fire," says Larry Schweiger, president of the National Wildlife Federation. One-fifth of all Pennsylvania's residences,[837] about three million, the second highest number in the nation,[838] draw their drinking water from shallow wells. There are no Pennsylvania regulations about construction or maintenance of private drinking wells, although well drillers are required to submit well completion data to the Department of Conservation and Natural Resources. Pennsylvania's water wells also tend to have higher concentrations of numerous minerals and chemicals than the standard suggested by the Department of Environmental Protection.

In May 2011, the Pennsylvania DEP fined Chesapeake Energy $900,000 for allowing methane gas to pollute the drinking water of 16 families in rural Bradford County the previous year.[839] The DEP noted there may have been methane emissions from as many as six wells in five towns. The DEP also fined Chesapeake $188,000 for a fire at a well in Avela,

Pa., that injured three workers.

One year after Chesapeake consented to paying almost $1.1 million, high levels of methane seeped into streams and private water wells in homes in Bradford County from a Chesapeake well. Human error contributed to the escape of methane gas, according to the DEP, which responded after a complaint by the Clean Air Council. The CAC had commissioned a study that showed twice the normal airborne methane levels.[840] Dr. Bryce Payne, an environmental scientist who monitored air and water pollution in a two square mile area, determined:

> "Two methane plumes were detected. One larger plume substantially increased in size over a few hours, which suggests large amounts of methane were being emitted into the air. A smaller plume was also detected about 2 miles west of the larger one. The data and observations suggest natural gas has spread through an extensive underground area beyond where the plumes were found."[841]

However, DEP Secretary Michael Krancer, in a letter to the CAC in July, claimed, "The situation is, and at all times was, under control by DEP. Indeed, at this point in time the situation is for the most part over."[842] But the situation was not "for the most part over."

Dr. Payne told the AP, "[It] is clear that at this point the event and the damage to groundwater and the domestic wells it supplies is certainly not over, and there is no foreseeable end in sight."[843] His team found "Methane concentrations as high as 94 percent just below the soil surface; an airborne methane plume covering about 1.6 square miles; and bubbling in Towanda Creek."[844]

Chesapeake and the DEP, Dr. Payne told the Scranton *Times–Tribune*, "are not adequately addressing that issue [of impact beyond 2,500 feet]. And they are not willing to provide any information to indicate why it is that they are concluding that everything is getting better and better."[845] The response, provided by a deputy press secretary, was "As Secretary Krancer pointed out in his letter, a letter that is still entirely accurate and that we stand by, we have an active investigation under way to monitor the situation as it unfolds."[846]

Bypassing the Halliburton Loophole

Civil penalties and restoration fees were imposed against companies in West Virginia that willfully violated the Clean Water Act by dumping sand, dirt, rocks, and other fill materials into swamplands and streams. While that Act (part of the "Halliburton Loophole") excludes charges and penalties for companies that pollute the nation's waters while using fracking, it does apply to pollution that may not be directly attributed to the actual process.

Between 2009 and 2011, Trans Energy dumped fill material into about 2.5 miles of streams and 1.3 acres of wetlands. The company and the EPA settled in September 2014 on a $3 million civil penalty, and agreed to pay about $13 million to resore the areas.[847] Chesapeake Appalachia settled on a $3.2 million civil penalty and a $6.5 million payment to restore 27 sites it polluted.[848] XTO Energy, a subsidiary of ExxonMobil, settled in December 2014 on a $2.3 million fine and $3 million in restoration costs for polluting about three acres of wetlands.[849]

Following a two-year vigorous investigation by the EPA, Chesapeake Appalachia, a division of Chesapeake Energy, paid a $600,000 fine for water pollution in West Virginia.[850] The U.S. District Court accepted Chesapeake's guilty plea on three criminal violations of the Clean Water Act. Court documents revealed the corporation hired contractors who polluted the wetlands, and dumped 60 tons of crushed stone and gravel in December 2008 that buried a natural waterfall. Chesapeake's reason for the violation was that it needed a path for its trucks carrying water and other supplies to a natural gas fracking site. The fine, one of the largest handed down in a criminal case prosecuted by the EPA and Department of Justice, represented 0.00571 percent of the corporation's $10.5 billion profit in 2011.

The Pennsylvania DEP fined Vantage Energy $999,900 in December 2014 for a series of spills, including a deliberate dump of two truckloads of wastewater, which polluted small streams.[851] In April 2015, Arlington, Texas, fined Vantage $84,000 for a spill of 42,800 gallons of fracking fluid that spilled onto city streets and into the storm sewers.[852]

Cleaning up spills can lead to additional groundwater pollution. Research conducted by the U.S. Geological Survey reveals that chemicals used to clean up those spills can release naturally-occurring arsenic.[853] "Where you're trying to clean solvents, dry cleaning fluids or petroleum, what's often done is you add organic carbon to the ground," says USGS researcher Barbara Bekins, noting that the use of the cleaners can mobilize arsenic.[854]

"Once the genie's out of the bottle, it could take years if not decades to clean up contamination if we don't get this right," said Collin O'Mara, secretary of the Delaware Department of Environment and Energy, explaining why his state cancelled a meeting of the Delaware River Basin Commission (DRBC), and why his state opposed attempts to take water from the Dela-ware River until further studies are done.[855] "[W]e keep saying in every way possible that it's much more important to be right than to try to move fast," said O'Mara.

Saving Money to Maximize Profits

Almost all instances of willful spills involve owners trying to save money and "maximize profits" by not properly disposing of the fracking fluids. About three billion gallons of wastewater were illegally dumped into nine injection wells and then migrated to aquifers in drought-ravaged central California.[856] The water included higher than normal levels of nitrates, thallium, and arsenic. The state shut down eleven injection wells in July 2014.[857]

For a three month period, beginning in November 2012, Michael Guesman and Mark Goff opened a valve on storage tanks and allowed thousands of gallons of wastewater to spill into a tributary of the Mahoning River near Youngstown, Ohio. Guesman and Goff told a federal court they were acting under specific orders of Benedict Lupo, owner of Hardrock Excavating.[858] They were each sentenced to three years probation. In August 2014, Lupo was sentenced to 28 months in prison and a $25,000 fine.[859]

Between 2003 and 2009, Robert Allan Shipman and his

company, Allan's Waste Water Service, illegally dumped several million gallons of wastewater, sludge, and grease onto the ground and streams in six southwestern Pennsylvania counties. "He was pouring the stuff in any hole he could find," said Nils Frederiksen, press secretary for the state's Office of the Attorney General.[860] "God only knows where he put it all. What we found was incredible. But who knows where else . . . all the other places he dumped it," said Greene County Sheriff Richard Ketchum.[861]

According to the Grand Jury presentment, "This activity would typically occur after dark or during heavy rain so that no one would observe the illegal discharge."[862] The state charged Shipman with 98 criminal counts, including criminal conspiracy, theft by deception, receiving stolen property, forgery, tampering with public records; the state charged Shipman's company with 77 criminal counts. Information in the Grand Jury presentment charged:

> "The drivers stated that every action they took while working for Allan's Waste Water was done at Shipman's direction. Some drivers testified they believed they were fired for their unwillingness to engage in illegal activities. Other drivers voluntarily quit for the same reason. Most drivers complied with Shipman's requests because they did not want to lose their jobs."[863]

In February 2012, Shipman plea bargained to being charged with only 13 criminal charges, and accepted 13 charges against his company. Judge Farley Toothman sentenced him to seven years probation, fines, restitution, and a requirement to work five hours a week for seven years with a water conservation group. Deputy Attorney General Amy Carnicella called the sentence too lenient. The Office of the Attorney General formally appealed to Superior Court, claiming the sentence "did not fit the crime" and should have carried at least 16 months imprisonment, according to state sentencing guidelines.[864]

While it's understandable that some employees were willing to overlook illegal activities in order to cling to their jobs in a depressed economy, there were others in the community who knew about the pollution, yet did nothing to stop it. Aaron Skirboll of *AlterNet* noted, "When the arrest finally came in

March 2011, area residents were elated that the man behind the worst kept secret in Greene County was going to pay for his crimes."[865] These unindicted co-conspirators had little to be elated about; they were the ones who knew, yet did nothing.

The EPA Investigations

THE EPICENTER OF THE MARCELLUS SHALE

Most of the residents of Dimock Twp., a rural area with about 1,500 residents in northeast Pennsylvania, are farmers and blue collar workers struggling in a depressed economy. The largest permanent employers in the county are the schools, government, and a hospital. About 8.7 percent of the population was below the poverty line, according to the 2000 census.[866] And then Cabot Oil & Gas showed up, offering thousands of dollars for leases and royalties for mineral rights.

The natural gas boom helped revitalize Dimock and brought a temporary prosperity to many of its landowners, who quickly became believers in natural gas exploration. But all was not peaceful.

Fifteen residents sued Cabot in November 2009, claiming the company contaminated their drinking water.[867] The cause was failure of well casings. Tests conducted by the DEP during the last years of the Ed Rendell administration had revealed there was higher than expected methane gas in 18 wells that provided drinking water to 13 homes near the gas wells. The build-up of methane gas had also led to well explosions and DEP warnings to citizens to keep their windows open.[868] Among the provisions of a consent order, the state fined Cabot $120,000, and required Cabot to provide fresh water to those families whose water had been affected by the excess methane gas; it also intended to sue Cabot to recover $11.8 million in costs for the state to run a 12.5 mile water pipeline from Montrose to bring fresh water to the residents.[869] Cabot denied its fracking operation was responsible for the elevated levels.[870]

Within a year, DEP secretary John Hanger settled with Cabot; in exchange for the state not running a pipeline and suing the company, Cabot would make a $4.1 million payment to the affected residents of Dimock. Hanger told the Scranton

Times–Tribune that because the Republicans in the November 2010 election won control over the governorship and both houses of the legislature, he doubted they would allow funding of the pipeline.[871] Among those who would have been influential in killing the pipeline was State Sen. Donald White who had received $94,150 from the oil and gas industry between 2000 and 2010.[872]

On Nov. 30, 2011, after the DEP, now under the Tom Corbett administration, declared the water to be safe to drink, Cabot stopped delivering water.

And then something strange happened. The town of Binghamton, N.Y., about 35 miles north, said it would provide a tanker of fresh water to the residents who were affected by the drilling. However, at a township meeting in December, the supervisors of Dimock Twp., enthusiastically supported by most of the 140 residents, most of them with economic ties to the natural gas industry, refused the offer, although there would be no legal obligation upon the township. When Binghamton mayor Matthew T. Ryan asked, "Why not let people help?" he was rebuffed by one of the township's three supervisors who snapped, "Why should we haul them water? They got themselves into this. You keep your nose in Binghamton."[873]

In January 2012, after declaring the water "contains levels of contaminants that pose a health concern"—including arsenic barium, manganese, and other toxins—the EPA decided it would bring water to residents in Dimock. Cabot's response was that the EPA was wasting taxpayer money in its investigation of Cabot's environmental and health practices.[874] The response by Pennsylvania's DEP was almost as inflammatory as the water in the taps. DEP head Michael Krancer disagreed with the EPA findings, and called the agency's knowledge of fracking to be "rudimentary."[875]

In his second meeting with a Congressional panel, this time on May 31, 2012, Krancer increased his insistence that the federal government should defer to the states. The federal government should not have a pre-emptive role, said Krancer, who argued, "The question is a fundamental one: Are you in a better place in Washington to tell us what to do?"[876] Krancer later declared the EPA was "rogue and out-of-control."[877]

By now, a community group, Enough Already, supported by

Energy in Depth, had become active in opposing the rights of the minority who still had problems with their water, and were not silent in their views of Cabot and of the effects of fracking. Energy in Depth declared:

> "Folks in Dimock have had enough. They are, in fact, declaring 'Enough Is Enough' and starting to speak out against the ongoing effort aimed at maligning their community, aided and abetted by a fawning media inclined to believe anything that is said about natural gas producers as long as it happens to be negative. . . .
>
> "Dimock residents have put up with a continuous barrage of insults and hyperbole over the last three years as Josh Fox, Bobbie Kennedy, Jr. and like-minded charlatans have abused them in the name of natural gas obstructionism. They have watched and listened as professional activists use the power of the press to perpetuate the myth that Dimock water is polluted. They have tolerated the incessant meddling of others in the affairs of their community."[878]

Shortly after Energy in Depth's attack upon those it claimed were "charlatans," Enough Already changed its name to Enough is Enough, calling into question just how independent the organization of Dimock citizens truly was. The newly-named organization soon launched an online petition to the EPA, claiming that continued water testing by the Philadelphia regional office "threatens our livelihoods and is destroying our community reputation." It demanded the EPA "reign in this rogue Regional Office."[879]

In mid-March, following preliminary tests on several of the wells serving Dimock residents, the EPA acknowledged arsenic, barium, manganese, glycol, and potentially explosive methane gas remained in the water, but claimed the water "did not show levels of contamination that could present a health concern."[880] A *ProPublica* investigation revealed that four of the five water samples it obtained showed methane levels exceeding Pennsylvania standards.[881]

"We are deeply troubled by Region 3's rush to judge the science before testing is even complete, and by their apparent disregard for established standards of drinking water safety," said Claire Sandberg, executive director of Water Defense.[882]

She questioned why EPA Region 3's handling of the Dimock case differed from how other EPA regional offices handled similar cases in Texas and Wyoming when it didn't release the information until all testing was completed. Dr. Ronald Bishop, professor of chemistry at the State University of New York at Oneonta, told *ProPublica*, "Any suggestion that water from these wells is safe for domestic use would be preliminary or inappropriate."[883]

In July 2012, 32 families and Cabot agreed on a financial settlement that included a non-disclosure statement. In a public statement, Victoria Switzer, who had been outspoken in her condemnation of how Cabot dismissed concerns about health and environment, issued a brief public statement that said her family "was relieved to put this behind us and hopeful that we will be able to live out our lives in the home we have invested so much of our time and resources in."[884] In what may have been seen solely as a public relations statement, she now suggested she would "advise anyone living in a gas field with concerns or disputes involving a gas company to try to work with them." However, four families in the suit chose not to "work with" Cabot. Ray Kemble said he didn't settle because terms of the agreement would restrict what he could say publically.[885] Kemble buys fresh water, afraid to use the local water that had turned brown; he posted anti-Cabot and anti-fracking signs on his property.[886]

The same month that lawyers for Cabot and the residents announced their tentative agreement, the EPA declared that water in Dimock was safe to drink, and allowed Cabot to stop delivering fresh water to four families. The following month, the Pennsylvania DEP allowed Cabot to resume fracking in the seven wells that were already drilled, after almost a two and one-half year suspension.[887]

A year later, Neela Banerjee of the *Los Angeles Times* revealed that the Philadelphia regional office had concerns about closing the investigation into Dimock's water quality. Based upon an internal EPA PowerPoint presentation, *Banerjee* reported:

> "[S]taff members warned their superiors that several wells had been contaminated with methane and substances such as manganese and arsenic, most likely because of local natural gas production.

"The presentation, based on data collected over 4 1/2 years at 11 wells around Dimock, concluded that 'methane and other gases released during drilling . . . apparently cause significant damage to the water quality.' The presentation also concluded that 'methane is at significantly higher concentrations in the aquifers after gas drilling and perhaps as a result of fracking and other gas well work.'"[888]

The problems for Dimock residents may not have ended just because one state and one federal agency said the water was safe to drink. The Agency for Toxic Substances and Disease Registry (ATSDR), an agency of the U.S. Department of Health and Human Services, based upon extensive analysis of data of metals in the water—the DEP was primarily interested in methane contamination—concluded not only are there "important data gaps for evaluating water quality in private wells," but that:

"[T]here is a possible chronic public health threat based on prolonged use of the water from at least some of these wells—assuming future exposure to these contaminants at these concentrations is not reduced. Based on the potential quality control issues, a potential health threat for the remaining wells cannot be disregarded. Additional characterization of the ground-water quality and a thorough review of any changes in concentration over time are indicated."[889]

It recommended there be no further use of the well water "sampled to date at this site," and that "distribution of alternative residential water supplies should be considered until potential exposures are further understood and mitigated as needed."[890]

In May 2014, with Cabot banned by the EPA from drilling in a nine square mile region of Dimock, Gov. Tom Corbett honored the company with a Community ImPAct Award for charitable contributions.[891]

PAVILLION, WYO.

In December 2011, following a three-year scientific investigation, the EPA issued a 121-page preliminary report that cited groundwater pollution as a direct result of fracking operations near Pavillion, Wyo.[892] The EPA determined there was contamination in 11 drinking wells, and that 169 gas wells had been drilled within the area that provides drinking water. Rejecting alternate possibilities for the pollution, EPA scientists noted that at least 10 compounds—including the carcinogens benzene and 2-butoxyethanol—found in the water supply were specific to nearby fracking operations. The EPA also disclosed that potassium and chloride levels were higher than expected background levels. However, the EPA "was hamstrung by a lack of disclosure about exactly what chemicals had been used to frack the wells near Pavillion," according to *ProPublica,*[893] which noted that Encana, owner of the nearby wells, refused to give federal officials a list of chemicals it used in fracking operations but did claim there was no correlation to its mining and the pollution of the water supply.

Dr. Tom Myers ran his own tests at Pavillion, and determined the EPA's "conclusion is sound." According to his report, published in April 2012, "Three factors combine to make Pavillion-area aquifers especially vulnerable to vertical containment transport from the gas production zone or the gas wells— the geology, the well design, and the well construction."[894] He noted, as had the EPA, that the area was unique because "the vertical distance between the water wells and fracking wells is much less at Pavillion than in other areas."[895]

Tests run by the U.S. Geological Survey several months after the EPA tests confirmed EPA findings.[896] The USGS tests showed the presence of diesel compounds, and high levels of ethane, methane, and phenol in drinking water. Encana continued to deny any relationship between fracking and contaminated water.

Possibly influenced by political realities, the EPA abandoned the research and ceded authority to Wyoming.

PARKER COUNTY, TEXAS

When Steven and Shyla Lipsy moved into a new house on a 14 acre lot in Weatherford, Texas, in the northeastern part of the state, they didn't expect that within a year their house would have lost about 95 percent of its value,[897] and they would be fighting a major gas extraction company and the state's oil and gas regulatory agency.

A 5,000-gallon water tank, drilled in 2005, provided excellent water to a cabin and boathouse until 2009. That's when Range Resources Corp. and Range Production Co. drilled two wells (Teal Unit 1H and Butler Unit 1H) about a half-mile from the Lipskys' house.[898] Not long afterward, according to a complaint filed by the Lipskys, the water was polluted by flammable methane; Steven Lipsky videotaped himself holding a garden hose spouting flames not water. An independent analysis of the water showed methane.[899] Range Resources denied any connection between their wells and water contamination. The Lipskys, according to a court filing, were forced to move from their house into their cabin when the fumes became toxic; they moved back to the house a month later after disconnecting their well and water tank, and bringing fresh water by the truckload.[900]

When the Lipskys believed the Railroad Commission of Texas wasn't moving fast enough to identify and protect the water sources, they contacted the EPA.[901]

In December 2010, the EPA ordered Range Production to protect the Lipsky and Steve Hurst families and their water supplies after determining there were high levels of methane and benzene in their water wells, that the methane and benzene in the water was nearly identical to that of the fractured wells, and there was an "imminent and substantial endangerment" threat.[902] Al Armendariz, EPA regional administrator, says the EPA acted when Texas regulators "acknowledge[d] that there is natural gas in the drinking water wells," claimed it was there before the fracking, and then delayed action. While Range Production continued to deny it caused the problem, and the Railroad Commission of Texas eventually agreed with Range, [See: *Hearing Before Texas Railroad Commission*[903]], Armendariz was blunt in his assessment: "We know they've polluted the aquifer."[904] While not making a direct conection between fracking and methane migration, Armendariz told the Dallas *Morning News*, "[W]e are confident that the natural gas that is now in the drinking water for these two homes is the natural gas that they [Range] are producing from a production well nearby which they hydro-fracked during the summer of [2009]."[905]

In March 2011, the Railroad Commission cleared Range of any connection to the methane and benzene in the water supply of the houses owned by the Lipskys and Hurstes.

U.S. Sen. James Inhofe (R-Okla.) and other political leaders with ties to the fossil fuel industry demanded the EPA conduct an investigation to determine if the EPA went after Range Resources because of political influence by the Obama administration.

A year later, the EPA withdrew its Imminent and Substantial Endangerment Administrative Order, thus ending EPA actions. [See: U.S. v. Range Production.[906]] Texas Railroad Commissioner David Porter claimed the EPA actions against Range were a case of "fear mongering, gross negligence and severe mishandling,"[907] and demanded the EPA fire Armendariz. Commissioner Michael Williams in a statement to the Texas Tribune, declared:

> "I have maintained from the very beginning of this case that the EPA was once again overreaching and engaged in a purposeful witch hunt against Texas natural gas producers.
> "There was not then nor now, as EPA's dismissal demonstrates, evidence that shows that Range Resources in any way harmed drinking water in Parker County."[908]

However, the Inspector General's investigation, completed in December 2013, revealed the EPA was justified in protecting drinking water, and raised questions as to why the EPA withdrew its actions after Sen. Inhofe had questioned the original order.[909] That report also revealed that Range refused to cooperate in any EPA investigation as long as there was a disciplinary order against it. However, the Inspector General's report also noted that even after the order was lifted, Range refused to cooperate.

Partially based upon the Railroad Commission's conclusions, and after the EPA withdrew actions against Range, Range Resources sued the Lipskys and consultant Alisa Rich for $4 million for Conspiracy to Commit Defamation. The Supreme Court of Texas dismissed the Range lawsuit in December 2014.[910]

THE LONG-AWAITED EPA STUDY

Kate Sinding, senior attorney for the Natural Resources Defense Council, told the *Los Angeles Times*, "We don't know what's going on, but certainly the fact that there's been such a distinct withdrawal from three [recent] high-profile cases [Dimock, Pa.; Pavillion, Wyo.; Parker County, Texas] raises questions about whether the EPA is caving to pressure from industry or antagonistic members of Congress."[911] Those studies were not included in a major study commissioned in 2010, and due in 2012, by Congress to determine if fracking contaminates water resources. The draft was finally published in June 2015, with the warning at the bottom of every page, 'DRAFT—DO NOT CITE OR QUOTE." The final study was scheduled to be published a year later.

Numerous scientists were skeptical of the EPA's study.

"We won't know anything more in terms of real data than we did five years ago," said Dr. Geoffrey Thyne, a former research professor at both the Colorado School of Mines and the University of Wyoming, and a member of the EPA's 2011 Science Advisory Board. He was clear that the study "was supposed to be the gold standard. But they went through a long bureaucratic process of trying to develop a study that is not going to produce a meaningful result."[912]

Among the many reasons why there was doubt about the study's credibility was because politics and the industry's self-preservation polluted the credibility of the study. The EPA had wanted to do a lifecycle study, from acquiring chemicals, proppants, and water to final disposition of wastewater. The American Petroleum Institute, America's Natural Gas Alliance, and numerous trade groups and politicians objected, demanding the study limit itself solely to the actual fracking process at the well site. The EPA was restricted to getting information about chemicals used in the process by relying upon FracFocus, the database that is flawed because all data provided is voluntary, and there are no outside checks on omission of data. Further, to get accurate baseline measurements, the EPA needed access to the wellpads. However, only Cheaspeake Energy allowed any access, and even then placed severe restrictions.

"The process of trying to finalize a prospective study with Chesapeake was very, very difficult," Dr. Robert Puls, a former EPA scientist, said. The industry, "did want to collaborate. . . . [T]here was a tug and pull on it, with us saying, 'We need to do this,' and them saying, 'We don't want to do that,'" Dr. Puls told Neela Banerjee of *InsideClimate News.*

The preliminary conclusion of the EPA study were, "[T]here are above and below ground mechanisms by which hydraulic fracturing activities have the potential to impact drinking water resources." Howevever, this was mitigated by the EPA conclusion:

> "We did not find evidence that these mechanisms have led to widespread, systemic impacts on drinking water resources in the United States. Of the potential mechanisms identified in this report, we found specific instances where one or more mechanisms led to impacts on drinking water resources, including contamination of drinking water wells. The number of identified cases, however, was small compared to the number of hydraulically fractured wells."[913]

The refusal by most of the industry to cooperate, the political pressures from a conservative-dominated Congress, and the back-and-forth negotiation between Chesapeake and the EPA were documented in a series of emails obtained by Greenpeace through the Freedom of Information Act.[914]

Most of the mainstream media that published stories about the report summarized the executive summary, with headlines that reflected fracking did not contaminate water supplies.[915] Few pointed out the deficiencies in data collection, limitations on the scope of the research, and political pressures to adhere to what the fossil fuel industry wanted.

Mario Salazar, a former EPA engineer, in 2012 told *ProPublica,* "In 10 to 100 years we are going to find out that most of our groundwater is polluted. A lot of people are going to get sick, and a lot of people may die."[916]

CHAPTER 10:
Wastewater Removal

Horizontal hydraulic fracturing requires not only the injection of millions of gallons of water, sand, and chemicals, many of them toxic, into the earth, but also their removal. To dispose of wastewater requires tanker trucks, each with a capacity of 80 to 120 barrels (3,360–5,040 gallons) of the toxic water, to travel to an area that has storage tanks, and then transport the wastewater for disposal.

Among toxic chemicals in wastewater are numerous salts (including chlorides, bromides, calcium, magnesium, and sodium), metals (including barium, iron, and strontium), and dissolved organic chemicals (benzene, toluene, and arsenic). Although nitrates in groundwater can come from several sources, they are also in the wastewater from fracking.

No matter what public relations practitioners of the natural gas industry believe, or want citizens to believe, there are five methods to remove wastewater, all of which "present significant risks of harm to public health or the environment," says Rebecca Hammer, attorney with the Natural Resources Defense Council, and Dr. Jeanne VanBriesen, professor of civil and environmental engineering at Carnegie-Mellon University, authors of the NRDC's 113-page study, *In Fracking's Wake*.[917]

Use of Wastewater as a Deicer and Dust Suppressant on Roads

The salinity of wastewater makes it an ideal deicer. In some cases, wastewater is used to reduce dust on dirt roads.

Because of the toxicity, wastewater will be absorbed into dirt roads; when it dries, the dust can become toxic. On paved roads, the wastewater will be washed into nearby streams, creeks, and rivers, further poisoning the environment. Salts spread on roads and agriculture fields could contain ammonium, arsenic, bromide, diesel hydrocarbons, iodide, lead, mercury, and various volatile organic compounds.[918] Certain forest animals will also be tempted to lick the salt-laden wastewater that includes not just the toxins but also radioactive elements. A USGS study in 2012 of the roads of Vernon Twp., Pa., which uses brine as a deicer, revealed significantly increased levels of Radium-226 and strontium.[919]

In 2009, Ultra Resources sent about 155,000 gallons of wastewater to nine Pennsylvania towns to be used as dust suppressant. DEP records reveal the wastewater contained about 700 times the level of radium allowed in drinking water.[920] The EPA does not recommend that wastewater be used on roads.[921]

Until October 2012, the practice was forbidden in Pennsylvania. However, the DEP granted Integrated Water Technologies a 10-year general permit to spread chemical salts from the Marcellus Shale wastewater on all public roads and fields in the state.[922]

In an appeal filed with the Environmental Hearing Board, Citizens for Pennsylvania's Future argued that the original permit allowed the company only to process wastewater, and charged that DEP "misled the public about the nature and scope of the permit" to now allow "beneficial use." The DEP action, said PennFuture, was "arbitrary, unreasonable, and contrary to law" by deliberately broadening that permit without a required public hearing. The DEP called the appeal "baseless [and] an attempt to manufacture a controversy."[923]

The appeal was definitely not baseless. "We need to know what the company is claiming it can remove from the fracking flowback wastewater," said Dr. John Stolz, director of the Center for Environmental Research and Education at Duquesne University.[924] The problem, said Dr. Stolz, "is that flowback chemical composition can vary widely depending on how long the fluids have been in the [well]. It becomes saltier the longer it's been underground and the quality of the water itself can be very different."[925]

The use of wastewater on roads has become more prevalent. Brine is spread on roads in the western part of New York state, between the Pennsylvania border and Lake Ontario.[926] Just in the northwestern part of Pennsylvania, more than three million gallons were spread in 2014, according to the state's DEP.[927] However, Pennsylvania does have a few rules—the wastewater can't be put onto roads within 150 feet of a water body or when it's raining, and can not be more than a half-gallons per square yard.[928]

Storage in Open Pits

Wastewater stored in open pits is subject to floods, heavy winds, and hurricanes that can cause ground pollution. The wastewater, even without problems brought about by natural disasters, also evaporates, causing air pollution. A report by the Citizens Marcellus Shale Commission in October 2011 pointed out that Pennsylvania has "no requirement to store [fluids and wastewater] in a closed loop system," and some companies will mix the wastewater with freshwater. "Drilling wastes frequently are stored in pits with synthetic liners, which often have leaked and caused pollution," according to the Commission, which conducted public hearings throughout Pennsylvania.[929]

Additional problems occur when gas companies deliberately disregard health and environmental issues.

"In the beginning we would dig a hole and then we'd just throw plastic in it. That was more or less to make the home-owners feel comfortable about us drilling on their property," Scott Ely, a former employee of GasSearch Drilling, a sub-sidiary of Cabot Oil and Gas, told Laura Legere of the Scranton (Pa.) *Times–Tribune.* Ely said he once watched Cabot employees push a "big, goopy concoction" of sand, gels, and fracking fluid acids over a bank. The oil and gas industry "had no care for what spilled anywhere. It was the most reckless industry I've ever seen in my life," said Ely.[930]

That recklessness became more apparent in February 2015 when inspectors from the California's Central Valley Regional Water Quality Control Board found that more than one-third of active wastewater disposal sites in Kern County, which produces more than 80 percent of all oil in the state, did not have required permits. None of the 933 pits (578 of which were active) had linings to prevent the wastewater from seeping into the land and groundwater, or covers to protect residents from the effects of evaporation. The *Los Angeles Times* determined, based upon official data the oil companies supplied to the state, that about 80 billion gallons of wastewater were dumped into the pits in 2013, and that benzene levels were about 700 times higher than federal standards.[931] The *Times* correctly observed that the presence of the injection wells "represents a tremendous failure by the state, which wrongly issued injection permits for wells in areas with protected groundwater." [Feb. 12, 2015].[932]

Rules issued by the U.S. Bureau of Land Management now require companies to store wastewater in closed tanks or above ground pits that have impervious plastic liners.[933] However, in keeping with the industry-wide mind-set, Kathleen Sgamma of the Western Energy Alliance said that regulations "add more delays and cost onto an already excessively bureaucratic federal process."[934]

Recycling

Many natural gas drillers use wastewater as a primary source of water to fracture the shale. "On-site recycling can have significant cost and environmental benefits as operators reduce their freshwater consumption and decrease the amount of wastewater destined for disposal," say the NRDC's Rebecca Hammer and Dr. Jeanne VanBriesen. However, they conclude, recycling "can generate concentrated residual by-products (which must be properly managed) and can be energy-intensive."[935] When used as a primary fluid in hydraulic fracturing, recycled wastewater is not subject to the Safe Water Drinking Act.

Sewage Treatment

During the first years of fracking in Pennsylvania, "the gas industry treated our rivers as a convenient place to dispose of their waste," Myron Arnowitt, Pennsylvania state director for Clean Water Action, told *EcoWatch*.[936] Between February 2007 and August 2011, two treatment plants in western Pennsylvania—Hart Resources Technology (HRT) and Pennsylvania Brine Treatment (PBT)—dumped thousands of gallons of frackwaste into the Allegheny River. In April 2011, the DEP finally ordered the industry to cease deliveries to the 15 facilities that had currently taken waste-water.[937] The next month, the EPA urged the DEP to take stricter measures in handing wastewater.[938]

Two years after the companies stopped accepting the waste, HRT and PBT, now merged into Fluid Recovery Services, accepted a settlement of an $83,000 fine and a demand to spend up to $30 million to bring their plants into compliance if they wished to continue to receive wastewater. [939]

However, Myron Arnowitt told the *Pittsburgh Post-Gazette*, "The companies say they're only accepting wastewater from conventional well drilling operations, but we haven't seen any confirmation of that from anyone, including the EPA. It's still extremely salty, and regardless of where it's coming from, it's having a negative impact on water quality."

Because of the nature of the Marcellus Shale deposit in Pennsylvania, as opposed to neighboring states, natural gas companies transport most of the wastewater to other states for reuse or disposal.

The bromides of wastewater, if mixed with the chlorine of sewage treatment plants, produce trihalomethanes (THMs), which can cause several forms of cancer. Jeff Kovach, general manager of Tri-County Joint Municipal Water Authority in Washington County, Pa., told Natasha Khan of *Public Source* he had no problem with THM levels prior to the development of fracking.[940]

Stricter rules have reduced the amount of untreated wastewater dumped into the rivers. However, present methods still can't remove all the salt and some other chemicals and radioactive elements. "You are asking sewage plants to do something that they were not designed to do," says Robert F. Kennedy Jr.[941]

The Pennsylvania DEP reports that between July and December 2011, 30,786 gallons of frack waste from Pennsylvania were sent to a plant in Kearny, N.J.; and 478.9 tons of radioactive drill cuttings (also known as "sludge") were sent to a plant in Carteret, N.J.; there is dispute about 162,000 gallons of "liquid waste" sent to Elizabeth; Cabot Oil self-reported it delivered the waste to the Elizabeth plant; New Jersey DEP says it didn't receive it.

"These plants aren't designed to safely process this waste before dumping it into our rivers and landfills," says Tracy Carluccio, deputy director of the Delaware Riverkeeper Network.

In June 2012, the New Jersey senate voted 30–5, and the assembly voted 56–19, to ban frackwaste. The vote appeared strong enough to be veto proof, but Gov. Chris Christie vetoed it. The veto was not justified, says Carluccio, because "the main responsibility of the State is to protect residents' health and safety and a ban on toxic frack waste would do exactly that. The Governor's veto is an inexcusable cop-out without legal foundation, exposing New Jersey's communities and drinking water to just what we don't need—more pollution." The legislature didn't override the veto, probably because some Republicans believed such an action could have been politically embarrassing for themselves and the popular governor. That lack of action left New Jersey open to being Pennsylvania's dumping ground—and the continued butt of jokes from New York comics.

In addition to waste treatment plants legally accepting toxic fracking wastewater, several plants have illegally accepted and discharged the wastewater.

In New York, the Cayuga Heights wastewater treatment plant thought it could handle the wastewater from Pennsylvania. During 2009, the plant accepted more than three million gallons—and then discharged it into the southern tip of Cayuga Lake, according to research by Walter Hang, president of Toxics Targeting.[942] The practice of accepting Pennsylvania waste stopped after Hang notified authorities the plant didn't have the required permits, and local citizens protested.

Wastewater has also been dumped into abandoned coal mines. The discharge into streams, even long after the mines have been abandoned, is an environmental problem. But, the levels of bromides in the waterways are significantly higher after fracking came to the western part of Pennsylvania. An analysis of bromides in streams near abandoned coal mines in 1999, according to analysis by the U. S. Geological Survey, revealed a sampling of less than 0.6 mg/liter.[943] By the end of 2012, water analyzed by the West Virginia University Water Research Institute, determined that bromides were as much as 10 times that level. "My guess is, if bromides are turning up in mine water, it's probably because someone has dumped it there," said Paul Ziemkiewicz, WRI director. [944]

Underground Injection

Flowback and wastewater are usually deposited into the 151,000 Class II injection wells,[945] which are exempt from the Resource Conservation and Recovery Act, and have fewer restrictions than Class I wells. Injection wells go as deep as two miles into the earth. Layers of steel tubing and concrete protect the waste from leaking, but concrete breaks down and heavy pressure from a number of sources can cause leaks.

About two billion gallons of oil and gas wastewater are injected into underground injection wells every day, according to the EPA.[946] More than 30 trillion gallons of toxic waste from all sources, including fracked wells, have been injected into almost 700,000 underground waste and injection wells in 32 states over the past few decades, according to a data analysis by *ProPublica*.[947]

Between 2007 and 2010 more than 17,000 violations were issued against companies for problems associated with injection wells; more than 220,000 well inspections revealed "structural failures inside wells are routine," Abrahm Lustgarten reported for ProPublica.[948]

A *ProPublica* investigation by Abrahm Lustgarten revealed:

> "Records from disparate corners of the United States show that wells drilled to bury this waste deep beneath the ground have repeatedly leaked, sending dangerous chemicals and waste gurgling to the surface or, on occasion, seeping into shallow aquifers that store a significant portion of the nation's drinking water."[949]

The EPA reports:

> "Underground injection has been and continues to be a viable technique for subsurface storage and disposal of fluids when properly done. ...EPA recognizes that more can be done

to enhance drinking water safeguards and, along with states and tribes, will work to improve the efficiency of the underground injection control program."[950]

The editorial board of the Lincoln (Neb.) *Journal–Star* disagreed about injection wells being a "viable technique" for wastewater storage: "We're sure that Nebraskans would much prefer that their state be known for its clean air, pure water and fertile soil than as a dump for waste that no one else will take." [March 4, 2015].[951]

On Jan. 17, 2001, gas collected in an underground injection well near Hutchinson, Kansas, ignited, destroyed two buildings, damaged 25 others, killed two residents, and forced hundreds to evacuate. ONEOK Inc., which identifies itself as "among the largest natural gas distributors" in the U.S., later acknowledged there was a leak of 143 billion cubic feet of gas.[952] Lustgarten observed:

"Among a small community of geologists and regulators . . . the explosions in Hutchinson—which ranked among the worst injection-related accidents in history—exposed fundamental risks of underground leakage and prompted fresh doubts about the geological science of injection itself.

"Geologists in Hutchinson determined that the eruptions had sprung from an underground gas storage field seven miles away. For years, a local utility had injected natural gas between 600 and 900 feet down into old salt caverns, storing it in a rock layer believed to be airtight so that it could later be pumped back out and sold. The gas had leaked out and migrated miles into abandoned injection wells once used to mine salt, then shot to the surface.

"'It was an unusual event,' said Bill Bryson, a member of the Kansas Geological Survey and a former head of the Kansas Corporation Commission's oil and gas conservation division. 'Nobody really had a feeling that if there was a leak, it would travel seven miles and hit wells that were unknown.'

"Though regulated under different laws than waste injection wells, gas storage wells operate under similar principles and assumptions: that deeply buried layers of rock will prevent injected substances from leaking into water supplies or back to the surface.

"In this case the injected material had done everything that scientists usually describe as impossible: It migrated over a large distance, travelled upward through rock, reached the open air and then blew up.

"The case, described as 'a continuing series of geologic surprises and unexpected complexities' by the Kansas Geological Survey, flummoxed some of the leading injection experts in the world.

"Perhaps more troubling was that some of the officials assumed to be most knowledgeable about injection wells and the risks of underground storage seemed oblivious to the conditions that led to the accident."[953]

Lustgarten also identified later problems with injection wells:

"In 2010, contaminants from such a well bubbled up in a west Los Angeles dog park. Within the past three years, similar fountains of oil and gas drilling waste have appeared in Oklahoma and Louisiana. In South Florida, 20 of the nation's most stringently regulated disposal wells failed in the early 1990s, releasing partly treated sewage into aquifers that may one day be needed to supply Miami's drinking water."[954]

Two years after publication of Lustgarten's in-depth investigation of injection well failures, the Government Accountability Office (GAO), in a separate investigation, confirmed there were significant problems. Acknowledging both staffing and budget constraints, the GAO, nevertheless, issued a scathing report against EPA oversight. The GAO concluded the EPA failed to adequately collect data, regulate, and inspect Class II wells, and even if the EPA followed all recommendations, "it will be several years before EPA can provide updated information at a national level to Congress, the public, and others on the UIC [underground injection control] program."[955]

Because of the nature of the Marcellus Shale, underground injection in most areas of the shale isn't possible, so companies operating in the state transport wastewater to Ohio. In Ohio, which doesn't require disclosure of the chemicals going into an injection well, about 12.2 million barrels of wastewater, 53 percent of it trucked in from Pennsylvania and West Virginia— including 90 percent of all of Pennsylvania's wastewater—were injected into underground wells in 2011, according to an investigation by Rick Reitzel of WCMH-TV, Columbus.[956] Two years later, Ohio's 201 injection wells accepted 16.3 million barrels of wastewater, about half of it from out-of-state drillers.[957]

The Cincinnati City Council, in August 2012, banned injection wells within the city.[958] In support of proposed legislation to ban injection wells throughout Ohio, Cincinnati Vice-Mayor Rozanne Qualls observed:

> "While the risks of injecting these wastes are still largely unassessed, what we do know is that fracking wastewater is known to contain toxic levels of chemicals that are hazardous to human health and the environment.
>
> "The short-term windfall from fracking must not outweigh our obligation to protect public health and our land and water resources." [959]

EARTHQUAKES

Explosions are only one problem associated with injection wells. Fracking may dislodge rock and lead to seismic activity.

More than 200 earthquakes were caused by the operations of one 10,800 foot deep injection well near Greeley, Colo., according to research conducted by Dr. Anne Sheehan, professor of geophysics at the University of Colorado.[960]

Between 1967 and 2009, the central part of the United States recorded an average of 21 earthquakes a year; between 2010 and 2012, with the significant increase in fracking and injection wells, there were 300 earthquakes, according to a study conducted by Dr. William L. Ellsworth, and published in the academic journal *Science*.[961]

An investigation by U.S. Geological Survey (USGS) scientists suggests that earthquakes near oil and gas drilling sites are "almost certainly man-made." Research data presented to the April 2012 meeting of the Seismological Society of America suggests that high-pressure forcing of wastewater into injection wells can cause low-level earthquakes. According to the six-member team, headed by Dr. William L. Ellsworth:

> "A remarkable increase in the rate of [magnitude-3.0] and

greater earthquakes is currently in progress. . . . A naturally-occurring rate change of this magnitude is unprecedented outside of volcanic settings or in the absence of a main shock."[962]

Three USGS scientists who were part of the study presented a companion report in December 2012 to the American Geophysical Union. An earthquake near Trinidad, Colo., that registered a magnitude level of 5.3 in August 2011, "renewed interest in the possibility that an earthquake sequence in this region that began in August 2001 is the result of industrial activities," according to the team. Focusing upon earthquakes in the Raton Basin of northern New Mexico and southern Colorado, they concluded:

"[T]he majority, if not all of the earthquakes since August 2001 have been triggered by the deep injection of wastewater related to the production of natural gas from the coal-bed methane field here. The evidence that this earthquake sequence was triggered by wastewater injection is threefold. First, there was a marked increase in seismicity shortly after major fluid injection began in the Raton Basin. From 1970 through July of 2001, there were five earthquakes of magnitude 3 and larger located in the Raton Basin. In the subsequent 10 years from August of 2001 through the end of 2011, there were 95 earthquakes of magnitude 3 and larger. . . . Second, the vast majority of the seismicity is located close (within 5km) to active disposal wells in this region. . . . Finally, these wells have injected exceptionally high volumes of wastewater. The 23 August 2011 [magnitude 5.3] earthquake, located adjacent to two high-volume disposal wells, is the largest earthquake to date for which there is compelling evidence of triggering by fluid injection activities; indeed, these two [wells injected] more than 7 times as much as the disposal well at the Rocky Mountain Arsenal that caused damaging earthquakes in the Denver, CO, region in the 1960s."[963]

"If we have more wells, we have more chance of events [and] if we have more events, there's more probability of higher magnitude events," Dr. Murray Hitzman told CNN.[964] Dr. Hitzman is professor of geology at the Colorado School of Mines and chair of the committee formed by the National Research Council to study earthquake probabilities from fracking.

The Railroad Commission of Texas claimed not only is fracking "an environmentally safe process," but that "the Commission has no data that links hydraulic fracturing activities to earthquakes."[965]

Research by Dr. Cliff Frohlich, senior research scientist at the University of Texas, identified 62 earthquakes in 14 clusters in south-central Texas that occurred between November 2009 and September 2011. Dr. Frohlich concluded that of the 14 clusters, four were not near either wells or injection wells, two were near injection wells, but eight were near wells, suggesting that the pressure of forcing water, proppants, and chemicals into the earth may have led to the earthquakes.

"You can't prove that any one earthquake was caused by an injection well, but it's obvious that wells are enhancing the probability that earthquakes will occur," says Dr. Frohlich.[966]

Within a week in March 2010 five small earthquakes, the largest measuring 2.8 on the Richter Scale, were detected near Cleburne, Texas, a Dallas suburb. In 2011, four earthquakes hit the area; fourteen earthquakes in June and July 2012, each one between 2.3 and 3.5 on the Richter scale, were recorded.[967]

Two earthquakes in May 2012 struck Timpson, Texas, sitting in the middle of the Haynesville Shale formation about 50 miles east of Cleburne. The first one registered 3.9 on the Richter Scale; the second one, a week later, registered 4.3. "I've had people who've lived here all their lives who told me they've never experienced anything like this," said Larry Burns, Timson's emergency response coordinator.[968] Within a 10 mile radius of Timpson are seven to ten waste injection wells; within Shelby County are more than 540 natural gas wells. Burns said he suspected pressure in injection wells was the cause for the earthquakes, and gas field workers in the area had told him of the possible connection.[969]

Sixteen earthquakes, the largest measuring 3.6 on the Richter scale, shook northern Texas during a three-week period in November 2013. Of the 16 earthquakes, five measured at least 3.0, enough to shake objects in a house but not cause extensive damage.

"If you look at the history of earthquakes in Texas," says geo-physicist Paul Caruso, "it's unusual to have any down there."[970]

In November 2014, the Railroad Commission adopted new

rules to require all companies to conduct seismic research studies within a 100 square mile radius prior to drilling new injection wells, and to provide that information to the Commission before it will issue a permit.[971] However, the rules don't require the companies to provide geologic analysis that the wells could affect underground faults.

Arkansas, in March 2011, issued a moratorium on the use of injection wells following a series of 700 small earthquakes within a six month period. The Arkansas Oil and Gas Commission issued the emergency temporary moratorium less than a week after the largest quake, measuring 4.7 on the Richter Scale, hit north-central Arkansas; that quake was the largest in the state in 35 years.[972] Although there was no direct evidence that injection well fracking caused the earthquakes, during the next three months after the moratorium there was a significant decrease in earthquakes, said Matt DeCample, assistant to Gov. Mike Beebe.[973] Arkansas in August 2011 banned wastewater wells within a 1,150 square mile region in the Fayetteville Shale.[974]

Most of California's earthquakes are caused by natural geological phenomenon, including tectonic plate shifting; Oklahoma's earthquakes are caused by what humans do to the earth. Between 1972 and 2009, two to six earthquakes were recorded per year in Oklahoma.[975] In 2010, after natural gas companies had increased their operations, the state recorded 1,047 earthquakes; 103 of them could be felt by residents, according to the Oklahoma Geological Survey (OGS).[976] There were about 800 earthquakes of at least a 3.0 magnitude in Oklahoma in 2015, significantly higher than 2009 when only 20 earthquakes with magnitudes of 3.0 or higher were recorded.[977]

There are more than 3,000 active injection wells in Oklahoma.[978] The significant increase in earthquakes "do not seem to be due to typical, random fluctutations in natural seismicity rates," according to a USGS study.[979] The USGS pinpointed, "Deep injection of wastewater is the primary cause of the dramatic rise in detected earthquakes and the corresponding increase in seismic hazard in the central U.S."[980] However, the oil/gas industry denied the causes of the earthquakes, the

Oklahoma Geological Survey hedged on public announcements of the cause, and Gov. Mary Fallin told the *Tulsa World* in February 2015, long after USGS scientists made specific links between wastewater disposal, fracking, and earthquakes, "We know a lot of it's just natural earthquakes that have occurred since the beginning of the earth, but there has been some question about disposal wells."[981]

Two weeks after wastewater injection began in Marietta, Okla., a series of earthquakes, the largest one registering 3.4 on the Richter scale, caused minor damage in homes and businesses in September 2013. Love County Disposal, owner of the injection well, shut down the well after the Oklahoma Corporation Commission ordered it to reduce both volume and pressure to avoid possible future earthquakes.[982]

A cluster of seven earthquakes hit Oklahoma the second weekend in July 2014; the largest registered 4.3 on the Richter scale.[983] By the middle of the year, Oklahoma had surpassed California for the number of earthquakes. A 4.1 earthquake in January 2015 led the Oklahoma Corporation Commission to order an injection well operated by SandRidge Energy to be closed.[984]

The most destructive earthquake in Oklahoma was a 5.7 magnitude quake recorded near Prague, Nov. 6, 2010. That quake, felt more than 800 miles away, destroyed 14 houses, damaged 200 buildings, and killed two persons. Geologists affiliated with the U.S. Geological Survey, the University of Oklahoma, and Columbia University independently concluded the earthquake was probably triggered by an earthquake of 8.8 magnitude in Chile in February 2010. Dr. Katie Keranen, a professor at the University of Oklahoma and a former scientist with the USGS, linked the earthquake with injection wells.[985]

"When you overpressure the fault [by wastewater injection sites], you reduce the stress that's pinning the fault into place and that's when earthquakes happen," said Dr. Heather Savage a geophysicist at Columbia. A research team headed by Dr. Nicholas van der Elst of Columbia University confirmed that large-magnitude earthquakes a half-world away can trigger swarms of minor earthquakes near wastewater injection sites. "The fluids [in wastewater injection wells] are driving the faults to the tipping point," and leaves the faults "critically loaded" that can lead to "dynamic triggering," said Dr. Van der Elst in the July 2013 issue of the academic journal *Science*.[986] The Oklahoma Geological Survey rejected the connection.[987]

The series of fracking-related earthquakes caused residents to purchase insurance. When the Prague earthquake hit, fewer than four percent of state residents had earthquake insurance, according to Kelly Collins of the state's Insurance Commission. Within three years, and continuing earthquakes, about 12–18 percent of residents purchased insurance, specifically to cover earthquakes.[988]

"As long as you keep injecting wastewater along that fault zone . . . you're going to continue to have earthquakes," said Dr. Arthur F. McGarr, at the federal Earthquake Science Center.[989] Dr. McGarr developed a formula based upon field research that reveals the volume of fluid and solids put into the earth is proportional to the size of any subsequent earthquakes. He says there is no way to determine when or if earthquakes will occur, only that when they do occur, the formula of volume-to-earthquake size applies.[990]

The Oklahoma City *Oklahoman* editorial board wasn't convinced. In continuing to promote the half-truth that fracking was decades old, the Board stated it would "like a little more certainty in the link between earthquake frequency and fracking activity before concluding that a decades-old technique is suddenly rocking the universe and needs to be stopped."[991] This is the newspaper that proudly declares it embraces fracking and is published "in the town that fossil fuel built."[992]

There were no earthquakes in Kansas in 2012; there were also no injection wells that were used for the fluids from horizontal fracking; all wells had been used solely from fluids from vertical fracking. In 2014, there were 120 earthquakes. In January 2015, several earthquakes, ranging from 2.7 to 4.1 on the Richter scale, were measured in Kansas.[993] In the first week of March 2015, several earthquakes struck south central Kansas; the most severe was one that registered a 3.8 magnitude on the Richter scale, with the epicenter 1,400 feet from an injection well.[994] By the end of 2015, southern Kansas and northern Oklahoma experienced more than 700 earthquakes, including 42 in one week in October.[995] "[T]here is a strong correlation between the disposal of saltwater [in injection wells] and the earthquakes," Rick Miller, geophysicist with the Kansas Geological Survey, told the Lawrence (Kansas) *Journal–World*.[996]

There were no earthquakes recorded in the Youngstown, Ohio, region prior to 2011. Between January 2011 and February 2012, 109 earthquakes were recorded, according to research by Dr. Won-Young Kim of Columbia University, and published in the *Journal of Geophysical Research: Solid Earth*.[997] The quakes, all within one mile of Northstar-1, a 9,000 foot deep injection well owned by D&L Energy, registered between 2.1 and 4.0 magnitude on the Richter scale. The Ohio Department of Natural Resources (DNR),[998] had initially claimed there was no correlation, and then revised its opinion.[999] It is "virtually certain" that the fracking process caused a tremor on Dec. 31, 2011, measuring 2.7 on the Richter scale, Dr. John Armendariz, a seismologist at Columbia University, told CNN.[1000] An investigation by Mike Ludwig of *Truthout* suggests that the earthquakes may have been caused because the state twice allowed D&L Energy to increase the maximum pressure to 2,250 pounds per square inch.[1001] Gov. John Kasich (R), who strongly favors gas drilling, placed a moratorium in January 2012 on injection well drilling in the area around Northstar 1.[1002] Two months later, the Ohio DNR issued a new set of regulations for natural gas drillers that use fracking. Among the tougher rules are those which prohibit brine injection into Precambrian rock formations and plugging existing injection sites, better monitoring of all injection fluids and wastewater, and the requirement that all drillers planning to use fracking must submit more comprehensive geological data when requesting permits.
Gov. Kasich has since placed a moratorium on injection wells within five miles of a well site.

With a wink and a half-truth, the industry for years has consistently argued that fracking doesn't cause earthquakes; it has used a narrow definition to refer only to the actual fracking process, but knew its critics would use a wider definition, arguing fracking is the complete process, including the use of injection wells and everything before and after the actual frack. And, for years, it was possible to believe that fracking itself didn't cause earthquakes. Recent research disproves even this attempt to distort the truth. Recent scientific studies suggest not only can earthquakes be triggered by deep injection wells but the actual process of fracking can also trigger earthquakes.

Researchers at the University of Alberta concluded there was a high correlation between wastewater injection wells and earthquakes in western Alberta. "Certainly that region is not immune to earthquake faulting, but I would say having actual earthquakes in that area is relatively recent, relatively new," said Dr. Jeff Gu, one of the study's authors.[1003]

The first sentence of a 240-page research report by the National Research Council (NRC), working with the U.S. Department of Energy, is: "Since the 1920s we have recognized that pumping fluids into or out of the Earth has the potential to cause seismic events that can be felt."[1004]

The report, published in June 2012, pointed out:

> "Although only a very small fraction of injection and extraction activities at hundreds of thousands of energy development sites in the United States have induced seismicity at levels that are noticeable to the public, seismic events caused by or likely related to energy development have been measured and felt in Alabama, Arkansas, California, Colorado, Illinois, Louisiana, Mississippi, Nebraska, Nevada, New Mexico, Ohio, Oklahoma, and Texas."[1005]

"We've known for decades that injection of fluids can and does trigger earthquakes in some cases," Dr. John Cassidy, a seismologist with Natural Resources Canada, told the *Toronto Star*.[1006]

"Pumping water into the ground is one of the most mechanically destructive things that we can do," says Dr. Richard Ketcham, associate professor of geological sciences at the University of Texas.[1007]

A series of 77 earthquakes, each one between 1.0 and 3.0 on the Richter Scale, hit Poland Twp., Ohio, in the northeastern part of the state, during a one-week period in March 2014.

Research conducted by Robert Skoumal, Michael Brudzinski, and Brian Currier of Miami University of Ohio, and published by the *Bulletin of the Seismological Society of America* in January 2015, concluded that the earthquakes were triggered by fracking operations. The earthquakes "occurred in the Precambrian basement, a very old layer of rock where there are likely to be many pre-existing faults," said Dr. Skoumal.[1008]

Fifty earthquakes within two miles of a drilling site in Garvin County, Okla., may have been caused by fracking operations. The earthquakes, all within a 24-hour period, beginning Jan. 18, 2011, may have been "induced by hydraulic fracturing," according to Dr. Austin Holland of the Oklahoma Geological Survey. Each earthquake measured between 1.0 and 2.8 on the Richter scale.[1009] A series of five earthquakes, the largest measuring 4.3 on the Richter Scale, hit central Oklahoma in April 2013.[1010]

Cuadrilla Resources acknowledged its use of hydraulic fracturing was the cause of two small earthquakes near Lancashire, England, in 2011.[1011] A 2.3 magnitude quake in April, and a 1.4 quake the following month led Cuadrilla to suspend operations and initiate an independent report.

An investigation conducted by the British Columbia Gas and Oil Commission revealed that 272 small earthquakes in the Horn River Basin may have been caused not from injection wells but by fracking between April 2009 and December 2011. Of the 272, the Commission determined that 38 measured between 2.2 and 3.8 magnitude on the Richter scale; the others were less than 2.2 magnitude. No earthquakes were detected prior to the beginning of natural gas fracking, according to the Commission. The Commission called for "pre-emptive steps to ensure future events are detected and the regulatory framework adequately provides for the monitoring, reporting and mitigation of all seismicity related to hydraulic fracturing, thereby ensuring the continued safe and environmentally responsible development of shale gas in British Columbia."[1012]

Conclusion

Rebecca Hammer and Dr. Jeanne VanBriesen of the NRDC conclude "[T]here are not sufficient rules in place to ensure any of [the wastewater disposal methods] will not harm the people or ecosystems."[1013] They believe only recycling and underground injection methods, with adequate safety standards, should be used to deal with wastewater. The NRDC position is that the use of waste-water as deicer on roads, storage in open pits, and sewage treatment with discharge into rivers "present such great threats that they should be banned immediately."[1014]

PHOTO: Robert Donnan

Frack pit near Buffalo, Pa.

Engine noise, diesel fumes, and traffic congestion are some of the stresses that lead to mental and physical problems in the Fracking Zone.

PHOTOS: Diane Siegmund

CHAPTER 11
Psychological Effects From Fracking

Persons living within a mile of an active well "are more at risk" of increased stress and mental health issues, says Raina Rippel, director of the Southwest Pennsylvania Environmental Health Project. Among the causes are constant noise and artificial light from the rigs and well pads, flaring, the rotten-egg smell of hydrogen sulfide from the drilling, diesel fumes from trucks and generators, traffic congestion, and a ceaseless roar of trucks on roads not designed for heavy traffic. "All can impart an aberrant stress response on the body that could make people more susceptible" to physical illness, says Dr. Reynold Panettieri, professor of pulmonary medicine at the University of Pennsylvania.[1015]

Research by Dr. Michael McCawley of West Virginia University points out, "Light pollution, which can be generally defined as excessive, misdirected, or obtrusive artificial light, has been shown to affect the mating, predation and migration behaviors of many nocturnal wildlife species, in turn affecting entire eco-systems. In addition, there is evidence . . . that light pollution may, in humans, influence melatonin suppression, circadian rhythms and health."[1016]

Lies by landmen and those working in the industry are the largest source of "stressors" of residents in the Marcellus Shale, according to an ethnographical study conducted by researchers from the Graduate School of Public Health at the University of Pittsburgh. About 79 percent of respondents reported the greatest stress was being "denied or provided false information," followed by industry and politician corruption (61 percent), the failure of the industry to respond to complaints (58 percent), a feeling of "being taken advantage of" (52 percent), and the effects of noise pollution and financial damages (both 45 percent.)[1017]

Anxiety in the absence of information can produce mental and physical problems, says Kathryn Vennie of Hawley, Pa., a clinical psychologist who says she and other psychologists in areas where there is a high level of fracking are seeing patients "who are seeking support because of the disruption to their environment." However, because of a lack of information about the fracking process and the chemicals used, "We won't know the extent of patients becoming anxious or depressed," she says.

NOISE POLLUTION

Excavation and "earth moving, plant and vehicle transport during site preparation has a potential impact on both residents and local wildlife, particularly in sensitive areas," according to a 292-page report for the European Commission. Based upon extensive field analysis, the report states:

> "Well drilling and the hydraulic fracturing process itself are the most significant sources of noise. Flaring of gas can also be noisy. For an individual well the time span of the drilling phase will be quite short (around four weeks in duration) but will be continuous 24 hours a day. The effect of noise on local residents and wildlife will be significantly higher where multiple wells are drilled in a single pad, which typically lasts over a five-month period. Noise during hydraulic fracturing also has the potential to temporarily disrupt and disturb local residents and wildlife. Effective noise abatement measures will reduce the impact in most cases, although the risk is considered moderate in locations where proximity to residential areas or wildlife habitats is a consideration.[1018]

Noise pollution "is not only an environmental nuisance but also a threat to public health," says Zsuzsanna Jakob, regional director for the World Health Organization. According to Drs. NAA Castelo Branco, an occupation medical researcher; and Mariana Alves-Pereira, a physicist and environmental scientist:

> "Vibroacoustic disease (VAD) is a whole-body, systemic pathology, characterized by the abnormal proliferation of extra-cellular matrices, and caused by excessive exposure to low frequency noise (LFN) . . .

"In both human and animal models, LFN exposure causes thickening of cardiovascular structures. Indeed, pericardial thickening with no inflammatory process, and in the absence of diastolic dysfunction, is the hallmark of VAD. Depressions, increased irritability and aggressiveness, a tendency for isolation, and decreased cognitive skills are all part of the clinical picture of VAD."[1019]

Pennsylvania law gives natural gas companies authority to operate compressor stations continuously at up to 60 decibels,[1020] the equivalent of continuous conversation in restaurants. However, most other states don't have limits. Compressor stations don't have to be enclosed, which would reduce noise levels.

Calvin Tillman, the former mayor of DISH, Texas, says the natural gas industry "seem[s] to think it's okay to put a compressor in that constantly emits 100 decibels of noise," louder than a lawnmower, and the industry thinks, "That's okay, and that's well within their rights to do so, and they don't give us any relief." The industry, says Tillman, "can make a compressor station that sounds very similar to the air conditioner on your home. . . . But they won't unless you make them. And we've had to fight and fight and fight, and we've had to spend money."[1021]

"There is noise twenty-four [hours], seven [days a week]," Jennifer Palazzolo, a resident of Erie, Colo., told the *Colorado Independent*.[1022] She said the noise from the trucks going to and from the Canyon Creek well site "is waking people up at night and this is only the beginning." The Denver–Julesburg Basin, northeast of Denver, is the site of hundreds of well pads, many within sight of schools. One well pad, reported the *Independent*, is within a hundred yards of the Aspen Ridge Charter School; Erie High School "is also surrounded by drill pads."[1023]

Matt Pitzarella of Range Resources acknowledges "There are lights, some noises, some road dust." However, he claimed, "[W]ithin a year it's all gone and everything is put back together."[1024]

Calvin Tillman, Jennifer Palazzolo, and thousands of others do not agree.

Workers at a
well site near
Naples, Fla.

PHOTO: Preserve Our Paradise

PHOTO: Vera Scroggins

Workers take soil samples at a
Cabot Oil and Gas site in Susquehanna County, Pa.

CHAPTER 12
Worker Issues

Every method of extracting energy from the earth yields death and injury to the workers and residents. More than 100,000 coal miners were killed, often from structural failures within the mines, gas poisonings, explosions, and roof collapse. Long-term catastrophic effects from mining include pneumoconiosis, also known as Black Lung Disease, the result of the inhalation of coal dust within the mines. Worker and resident health problems often don't become known until decades after a new energy source is mined. For coal mining, although there were several protections brought about by the United Mine Workers, it wasn't until 1969 when the Federal Coal Mine Health and Safety Act became law that health and environmental protection advanced. Congress improved the Act in 1977 and 2006.

In the Great Recession, beginning about 2007, people became desperate for any kind of job. The fossil fuel industry, just entering the age of high volume horizontal fracking, responded with high-stress, high-paying jobs.

The industry, for the most part, is non-union or independent contractors without benefits. The billion dollar corporations like it that way. It means there are no worker safety committees, no workplace regulations monitored by the workers, and no bargaining or grievance rights. Health and workplace benefits for workers who aren't executives or professionals are often minimal or non-existent. Hire them fast. Lay them off even faster.

Some corporations, to keep profits high and expenses low, hire migrant or undocumented workers. GPX/GXP of Texas is one of those corporations. In March 2013, it admitted hiring at least 19 undocumented workers two years earlier to work in gas fields near Williamsport, Pa.; it agreed to pay a $25,000 fine and forfeiture of a $250,000 cash bond. Its operations manager was fined $5,000.[1025]

The labor movement is mixed about fossil fuel development.

The United Steelworkers (USW) initiated nationwide strikes against several refineries in February 2015.[1026] The work stoppage, said USW International Vice President Gary Beevers "is about onerous overtime; unsafe staffing levels; dangerous conditions the industry continues to ignore; [and] the daily occurrences of fires, emissions, leaks and explosions that threaten local communities without the industry doing much about it."[1027]

Many unions—including the Amalgamated Transit Union, the Communications Workers of America, and the United Auto Workers—have strongly supported environmental awareness, the necessity to reduce global warming, and the necessity to develop alternative renewable energy. The Solidarity Committee of the Capital District & New York Solidarity opposes fracking because, says John Funiciello, editor of the 1,000-circulation *Solidarity Notes*, "Labor has to be much more than just wages and working conditions." However, many unions have largely supported the fossil fuel industry because of the lure of jobs. Between 2009 and mid-2014, employment in all areas of the oil and gas exploration industry rose from about 100,000 to 133,000, according to the U.S. Bureau of Labor Statistics.[1028]

"At a time when the U.S. construction industry was in the midst of what was arguably a depression, . . . one of the few, if not only, bright spots, were the jobs that were created by virtue of domestic oil and gas development," said Sean McGarvey, president of the AFL–CIO Building and Construction Trade Department.[1029]

"The shale became a lifesaver and a lifeline for a lot of working families," Dennis Martire told the Associated Press in April 2014. Martire is the mid-Atlantic regional manager for the Laborers International Union of North America (LIUNA), which represents about 630,000 members.[1030] LIUNA has also supported the development of the Keystone XL pipeline, and actively lobbied members of Congress not only to support the pipeline's construction but also to encourage President Obama to approve its completion. The pipeline, according to LIUNA, "would be a secure energy lifeline for America [which would] unlock good, family-supporting jobs for America at a time when families are losing their homes and desperately need good jobs."[1031] The U.S. Chamber of Commerce claimed the pipeline would create up to 250,000 jobs;[1032] the American Petroleum Institute doubled that projection, claiming the pipeline, by 2035, will have created almost a half-million jobs;[1033] TransCanada, the pipeline's creator, believed there would be 20,000 new jobs in construction and manufacturing, and 465,000 jobs overall.[1034] However, a 40-page economic analysis by Cornell University's Global Labor Institute (GLI) concludes that the TransCanada figures are "so opaque as to make meaningful review impossible." The GLI pointed out, "The project will create no more than 2,500–4,650 temporary direct construction jobs for two years, according to TransCanada's own data supplied to the State Department. Even that job creation, according to the GLI, "could be completely outweighed by the project's potential to destroy jobs through rising fuel costs, spill damage and clean up operations, air pollution and increased GHG [greenhouse gas] emissions,"[1035] all of which will affect the workers.

Death and Injuries at
Well Pads, Compressors, and Refineries

Increased employment in the oil and gas fields "have come with more fatalities, and that is unacceptable," said John E. Perez, secretary of labor.[1036] The oil and gas industry has a worker fatality rate about 6.5 times that of all other industries, according to the Bureau of Labor Statistics;[1037] during the drilling phase, the rate is almost 12 times the national average for all industry.[1038] More than 660 workplace deaths in the oil and gas fields were reported between 2007 and the end of 2014, according to an investigation by the *Houston Chronicle.*[1039]

In North Dakota, in 2007, a year before the economic boom from fracking the Bakken Shale, worker fatalities were about 7.0 per 100,000 employees. Five years later, the death rate was 17.7 fatalities per 100,000 employees.[1040] Thomas R. Nehring, director of emergency medical services for the North Dakota Health Department, told the *New York Times* that ambulance calls in the Bakken Shale increased about 59 percent from 2006 to 2011.[1041]

Greg Bish was killed in Elderton, Pa., in 2010, when he used a propane blowtorch to thaw a valve on a tank filled with toxic waste.[1042] Matthew Smith, 36, died at a frack site in Weld County, Colo., after a two-inch metal pipe fitting hit him in the head in November 2014, according to the Occupational Safety and Health Administration (OSHA).[1043] Smith and two other men, both taken to the hospital in critical condition but who survived, tried to thaw a frozen high-pressure water line.[1044] OSHA fined Halliburton $7,000 for safety violations. An investigation by Mike Soraghan for *EnergyWire* revealed that using blowtorches to de-ice valves is a common practice in the industry, one that led not only to the deaths of Bish and Smith, but numerous other injuries.

Three Chesapeake Energy workers were burned from an explosion and subsequent fire at a compressor station near Avella, Pa., in February 2012. The Pennsylvania DEP attributed the cause to ignition of escaping vapors from five steel tanks that held wet gas, also known as condensate, a product in natural gas. "If you've ever been to one of those well sites, the fumes are very bad," Avella fire chief Eric Temple told the *Pittsburgh Tribune–Review*.[1045]

An oil rig exploded near Victoria, Texas, in the Eagle Ford Shale in August 2013. The rig was operated by Enron Oil and Gas (EOG) for Nabors Drilling USA.

"We heard the blast; it kind of rumbled things at the house," Walter Scott, a nearby resident, told the *Victoria* (Texas) *Advocate*. He says he looked up, "and it was enormous—500 feet of fire in the air."[1046]

For two days, the fire burned before it could be contained, but the Texas Commission on Environmental Quality claimed there were no health problems. Even if there were no subsequent health problems, which strains the levels of credibility, there may have been problems at the rig.

Scott Marshall, a rig worker, told the *Advocate* the protective concrete "did not have adequate time to dry [when] a superior commanded him to go into the oil rig's cellar to tighten the bolts while it was flooding and filling with gas from the blowout that occurred earlier." The concrete had dried for only about an hour; normal drying time, said Marshall, was four to six hours. "If I would have gone down there," said Marshall, "I would have drowned or been hit by the gas."[1047]

A worker was killed in southwestern Pennsylvania when an explosion and fire in February 2014 destroyed a well pad owned by Chevron, but worked primarily by subcontractors. During the four days the fire burned and emitted toxic smoke, Chevron PR decided it would be a good idea to mitigate the discomfort of the nearby residents; its solution was to provide coupons for free pizza and soda.[1048] While developing its PR mitigation campaign, the multi-national corporation had denied the DEP access to the site for almost two days, citing safety concerns, although the regulatory agency has unrestricted access to all sites at all times. The DEP later cited Chevron with nine violations, including hazardous venting of gas, discharge of toxic fluids into the ground, emission of fugitive air contaminants, and failure to properly secure a lock pin at the well head that allowed methane to escape.[1049] In its summary report, the DEP cited working conditions as a significant problem. According to the Bureau of Investigations:

> "Some of the WSMs [well site managers] had decades of experience in the oil fields. However, others had virtually no background in the oil and gas industry. They worked, for example, in information technology, food service, or as a construction laborer. Having limited oil field experience reduces effectiveness of the oversight a WSM can provide.
> "Several well site managers expressed some frustration about the demands on their time. Several stated that documentation and paperwork took an inordinate amount of time. Another WSM said that he was preoccupied making calls to obtain equipment or trying to determine the whereabouts of delayed equipment and personnel. One WSM stated that observing all of the contractors working on site to be daunting task, though one former WSM found the challenge manageable. Workload and distractions may explain why a con-

tractor's employee with no well-site experience was allowed to work on a pressurized well even though he was not approved for any work, as required by Chevron policy."[1050]

Fifteen months after the tragedy, Chevron agreed to a Consent of Civil Penalty settlement for $939,552.[1051] The parents of Ian McKee, 27, who died in the fire, settled for $5 million.[1052]

Roberto Andrade Magdaleno and Amos B. Ortega were killed, and nine others injured, at a gas well explosion near Orla, Texas, in April 2014.[1053] The men were employed by Ameriflow Energy Services, which was subcontracted to RKI Exploration & Explosion. The explosion may have been caused by a pressure build-up when a valve at the well head was not properly closed, according to Loving County Sheriff Billy Hopper.[1054] Contributing factors could have been worker fatigue and inexperience.

In October 2014, 28-year-old Dustin Payne, a combat Marine veteran, was killed in Williston, N.D., when supervisors at Nabors Completion and Production Services, ordered him to weld a tank with oil residue.[1055] Prior to completing that job, Payne had texted his fiancé: "I'm literally going to be welding something that's full of oil. Don't [feel] comfortable welding this at all. Dangerous as [expletive]."[1056] OSHA investigators called the death "willful," placed Nabors on a list of "severe violators," but fined the company only $97,200 in April 2015.[1057]

An investigation by the *Houston Chronicle* revealed that subsidiaries of Nabors Industries, with headquarters in Bermuda, had reported 18 fatal accidents between 2007 and 2014.

Dominick Mas, an operating engineer who worked in the Marcellus Shale, says he saw "a kid who almost got himself killed when he walked into a line of excavators." Part of the problems, says Mas, was company safety inspectors "really didn't know their job. They followed a protocol."

Two storage tanks from a well owned by Antero Resources exploded in New Milton, W.Va., injuring five persons. An executive with Antero told Reuters, "We do not know the ignition source, but we suspect it was a methane explosion."[1058] One of

230

the injured, Charlie Arbogast, a rigger and trucker, suffered third degree burns on his hands and face. "You come to the rigs, you do what you do and you don't ask questions," Diana Arbogast, his wife, told the *Pittsburgh Post-Gazette*.[1059]

On Wednesday, Feb. 18, 2015, four men were hospitalized following an explosion at the 750-acre ExxonMobil refinery in Torrance, Calif., which processes about 155,000 barrels (6.5 million gallons) of crude oil daily.[1060] The explosion, which destroyed cars, could be felt for at least five miles, according to KNBC-TV. Cal Tech scientists said the explosion was the equivalent of a magnitude 1.7 earthquake. The cause of the explosion was high pressure in an electrostatic precipitator; ironically, that machinery regulates air pollution. Gray ash, fiberglass, aluminum oxide, and glass wool covered much of the refinery and spread for several city blocks,[1061] leading the South Coast Air Quality Management District to issue a smoke advisory. The ash itself was from a fluid catalytic cracking unit that converts the crude oil into gasoline.[1062] ExxonMobil says the ash was non-toxic, but "may cause irritation to the skin, eyes, and throat," according to an email sent to *Climate Progress*. The United Steel Workers, which represents workers at the plant, told *Climate Progress* the ash and subsequent methane flaring could have a greater health impact than what ExxonMobil stated.[1063]

The same month of the fire at the ExxonMobil refinery, three workers—Arturo Martinez Sr., Arturo Martinez Jr., and Rojelio Salgado[1064]—were killed in a drilling rig explosion near Rankin, Texas. The three men were employed by Mason Well Services, under contract to Parsley Energy. The fourth member of the crew was not injured. OSHA cited Mason for five serious violations. A year earlier, it had cited Mason for similar workplace violations.[1065]

Toxic Exposure

Workers at drilling sites are exposed to dozens of toxic chemicals. At six sites in Colorado and Wyoming, researchers from the National Institute of Occupational Safety and Health

(NIOSH) found that workers who routinely inspected chemicals and wastewater in tanks were exposed to higher levels of airborne benzene, a carcinogen, than acceptable safe levels.[1066]

It may be months or years before most workers learn the extent of possible injury or diseases caused by industry neglect. For José Lara of Rifle, Colo, the effects no longer matter. Lara, an employee of Rain for Rent, died at the age of 42 in August 2011 from pancreatic and liver cancer.

Lara's job was to power-wash wastewater tanks owned by natural gas companies.

In a six-hour deposition three months before he died, Lara said he was never provided a respirator or protective clothing. "The chemicals, the smell was so bad. Once I got out, I couldn't stop throwing up. I couldn't even talk," Lara said in his deposition, translated from Spanish.[1067]

John Dzenitis of KREX-TV reported:

> "Both the industry and the Colorado Oil and Gas Conservation Commission, the state's agency meant to protect public health and regulate oil and gas, have denied the existence of high levels of hydrogen sulfide in Colorado. In 1997, the Colorado Department of Public Health and Environment wanted to monitor for hydrogen sulfide at oil and gas facilities after they were designated as confirmed sources of the deadly gas by the EPA.
> "The COGCC stepped in and told them not to, claiming there were no elevated levels in the state. The public health department listened, and tells us they haven't pursued any monitoring of hydrogen sulfide at oil and gas facilities since."[1068]

No matter what Colorado officials did or didn't do, it didn't keep the Occupational Safety and Health Administration (OSHA) from issuing Rain for Rent nine violations for exposing Lara to hydrogen sulfide and not adequately protecting him from the effects of the cyanide-like gas.[1069] Nevertheless, hydrogen sulfide monitoring is exempt from regulations of the Clean Air Act.[1070] That could have changed if a bill had survived Congressional inaction. Reps. Jared Polis (D-Colo.), Maurice Hinchey (D-N.Y.), and Rush Holt (D-N.J.) introduced a resolution (HR 1204[1071]) that would have closed that loophole. The bill was introduced in March 2011 but under Republican leadership was buried in the Committee on Energy and Power.

OSHA fined Williams Energy and two subcontractors $27,000 in July 2013 for failing to advise workers they were exposed to hazardous chemicals. According to the *Denver Post*:

> "As workers began digging for super-concentrated hydro-carbons, the companies 'did not inform (them) of the nature, level and degree of exposure likely as a result of participation in such hazardous waste operations,' OSHA documents said.
> "Workers dug trenches . . . to find and remove toxic material, documents said. 'This condition potentially exposed employees to benzene and other volatile organic compounds,' [according to OSHA.]"[1072]

Tom Bean, a former gas field worker from Williamsport, Pa., says he doesn't know what he and his co-workers were exposed to. He does know it affected his health:

> "You'd constantly have cracked hands, red hands, sore throat, sneezing. All kinds of stuff. Headaches. My biggest one was a nauseating dizzy headache ... People were sick all the time ... and then they'd get into trouble for calling off sick. You're in muck and dirt and mud [a chemical-based brown sludge used as a coolant for the high-speed drills and for for lubrication] and oil and grease and diesel and chemicals. And you have no idea [what they are] . . . It can be anything. You have no idea, but they [Management] don't care . . . It's like, 'Get the job done.' . . . You'd be asked to work 15, 18 hour days and you could be so tired that you couldn't keep your eyes open anymore, but it was 'Keep working. Keep working. Keep working.'"[1073]

A truck driver who was afraid to identify himself told a community meeting in Heber Springs, Ark., more truths:

> "Now, these drivers will not get up and stand up . . . because their wages are better, they get $15 or $20 an hour; they think they are in high cotton. But just the same, anything I had to say, when they told me go clean a frac truck out, I said under OSHA regulations I've got to have a hazmat suit on. They laughed me out of it.

"It's a culture of fear that's in the oil industry right now, similar to the coal miners in West Virginia [that] told ya', 'Well we really did not want to tell you the truth about how bad the coal mine is.' It's bad out there, too. A majority of people at a rally last weekend [said] . . . 'We're for the Industry! Yeah, yeah, yeah! Because we're making money.' But I guarantee you what's going on, they're dumping illegally . . .

"This industry is a bunch of liars. And until we find out if this stuff is safe, we need to have a moratorium."[1074]

HEALTH PROBLEMS WITH SILICA SAND

Crystalline silica sand (SiO_2), also known as fine-grained silica sand or quartz, has a variety of uses. It is the prime compound for glass; as a part of the process to manufacture mortar, concrete, and bricks; in some paints; as a base for salt water aquariums; and to enhance soil for certain agricultural plants. The best sand, mined near the Great Lakes, is known as "Northern White" or "Ottawa." A lesser quality sand, known as "Brown sand," is mined in Texas. Sand has been mined for centuries, but with the development of horizontal fracking, and the necessity of silica sand as a proppant, mining operations increased significantly, primarily in Wisconsin, Minnesota, and Iowa. In Wisconsin, the number of active mines went from five in 2010 to 63 by 2015.[1075] Marcellus Shale wells used about 13 billion pounds of silica sand in 2014, up from 9.6 billion in 2013; projections for 2015 are companies in the Marcellus will use about 16 billion pounds.[1076] In all shale plays, about 95 billion pounds of sand were used in 2014.[1077]

CSX, the largest freight railroad on the east coast, in 2011 purchased 900 cube hopper cars designed specifically to move frack sand.[1078] The railroad reported it carried more than 12,000 carloads of sand into the Marcellus Shale region in 2011, an increase of 40 percent from the previous year.[1079] Norfolk Southern, with track in the Marcellus Shale, in 2012 spent about 14 percent of its entire capital expenditures for cube hoppers.[1080] About three-fourths of all freight Norfolk Southern moves into the Marcellus Shale region is frac sand.[1081]

Because of the boom in the oil industry in North Dakota and the gas extraction industry in the Marcellus Shale, both the result of the development of horizontal fracking, ACF Industries reopened its rail-car manufacturing plant in Milton, Pa., in June 2013. The facility, which had closed in 2009, by 2016 will hire about 330 employees to make tank cars and stationary propane tanks.[1082] The state gave ACF a $483,000 grant to purchase equipment and to assist in employee training. ACF is owned by Carl Icahn, whose net worth is estimated at $30 billion.[1083]

The increase in silica sand mining has brought with it the same issues as those that are part of the fracking problem. The impact, according to a report from the Wisconsin Department of Natural Resources (WDNR), includes "noise, lights, hours of operation, damage and excessive wear to roads from trucking traffic, public safety concerns from the volume of truck traffic, possible damage and annoyance resulting from blasting, as well as concerns regarding aesthetics and land use changes." In many mines, blasting is necessary to loosen geological formations, some of them 400 million years old; this blasting "can result in noise, vibration, and fugitive dust emissions," according to the WDNR.[1084]

The sand is loaded onto hopper cars or tractor-trailers by conveyer belts. The trucks and trains bring the sand to the shale fields, where it is unloaded by conveyer belts onto trucks that bring the sand to the well. At each step, silica sand dust can be released into the air. It's common to see 100-car trains, each of them uncovered hopper cars, bringing sand to the fracking zones.

The Occupational Safety and Health Administration estimates about 25,000 oil and gas workers are exposed to silica dust, with almost two-thirds of them exposed to levels that are unsafe.[1085] Persons working around silica sand are advised to use respiratory protection. However, not all companies require their workers to wear protective suits and breathing masks when working around silica sand.

Inhalation of excessive amounts of silica can lead to silicosis, a respiratory disease marked by an inflammation and scarring in the upper lobes of the lungs. The more common form, chronic silicosis, is caused by inhalation of low or moderate levels of the silica dust over a long period, and is usually not observed until 10 to 30 years after first exposure. However, higher levels of exposure can be observed within weeks of exposure. Inhalation of the silica dust can also cause kidney and autoimmune diseases, lung cancer, tuberculosis, and Chronic Obstructive Pulmonary Disease (COPD).

A research study conducted by the National Institute for Occupational Safety and Health revealed about 79 percent of all samples it took in five states (Arkansas, Colorado, North Dakota, Pennsylvania, and Texas), exceeded acceptable health levels, with 31 percent of all samples exceeding acceptable health levels by 10 times.[1086] "When sand was handled—that is, when it was transported by machines on site, or whenever these machines that move sand were refilled—dust, visible dust was created," noted Eric Esswein, the principal researcher of the NIOSH study.[1087]

The increased sand mining, similar to coal field strip mining, has destroyed agricultural and scenic land, and has led to the land "being ground up and shipped away, only to become toxic, radioactive waste somewhere else," journalist Pilar Gerasimo reported in the *Dunn* (Wisc.) *County News*.[1088]

The EPA and most states do not regulate crystalline silica sand. Those that have some regulation are understaffed. Mike Ludwig, writing in *Truthout*, reports that although the Wisconsin Department of Natural Resources "requires mine operators to monitor silica dust emissions and report them to the state . . . DNR officials rarely visit the mines in person."[1089]

Because of health and environmental concerns raised by Gerasimo and others, several communities established temporary moratoriums on sand mining; Dunn County, Wisc., had placed a six month moratorium, beginning January 2012, on non-metallic mining, and then extended it in July for three more months.[1090] The moratorium ended at the end of October, but the county says it has not received any requests to drill unconventional wells or for sand mining.

The supervisors of Allamakee County, in the northeast part of Iowa, near Wisconsin, established an 18-month moratorium on mining of silica sand, beginning February 2013. That decision was partly influenced by a strong campaign by the newly-formed Allamakee County Protectors. By the time of the vote, more than 98 percent of residents had talked to or had written the supervisors to support a moratorium, according to Supervisor Dennis Koenig.[1091] The moratorium was to allow the Protectors to study the effects of mining silica sand and, if the supervisors determined the moratorium should end, to write regulations to protect the environment and health of the residents and workers, including the transportation of silica sand in open railroad hopper cars. The county's board of supervisors in June 2014 struck down the existing ordinance and established the strictest regulations for sand mining in the nation. That ordinance includes setback distances, requires sand mining be done only in daylight hours, and protects homes, businesses, and numerous bluffs from sand mining.

One of the major concerns is that frack sand mining could pollute the Jordan Aquifer, which provides water to about 300,000 residents.[1092] But, sand is also the key to methane migration into aquifers and well water at fracking sites because its function is to keep the fractures open to allow methane to flow from the shale.

The effects of chemicals and the sand used in the fracking process so concerned the AFL–CIO that three senior health and safety officials in May 2012 wrote to divisions within the Department of Labor to express the concern of the 11 million member federation.[1093] Referring to the NIOSH study, the AFL–CIO observed:

> "Many of these exposures were well in excess of permissible and recommended levels, putting workers at risk of silicosis, lung cancer and other diseases. These findings coupled with concerns about health risks posed by chemical additives used in the fracking process and the well-documented safety hazards in this industry warrant immediate attention and action."[1094]

The AFL–CIO urged the federal government to consider that "the development of new energy sources, and exploration of existing energy sources, must be done safely without putting workers in danger," and suggested "effective regulation and oversight."[1095]

NIOSH and OSHA issued a joint Hazard Alert in June 2012 about the effects of crystalline silica, noting there were seven primary sources of exposure during the fracking process.[1096] In the Alert, NIOSH and OSHA issued several suggestions of how the industry could monitor air samples and worker health, control dust, and provide respiratory protection. However, the Hazard Alert was only advisory; it carried no legal or regulatory authority. About 14 months after issuing the hazard alert, OSHA issued a set of proposed rules to limit worker exposure to crystalline silica sand. The rules included permissible levels of exposure, mandated training for workers, and required records be maintained of workers' medical exams and exposure to silica sand. OSHA determined that if the proposed rules became permanent, they would "save nearly 700 lives and prevent 1,600 new cases of silicosis per year."[1097]

As expected, the oil and gas industry objected. And, as expected, the objections had nothing to do with worker health issues. The new rules would "create profound detrimental economic consequences," according to a statement filed by almost 80 organizations, led by the American Petroleum Institute and the Independent Petroleum Association of America. The U.S. Chamber of Commerce claimed new rules to protect worker health were built on "a chain of assumptions."[1098] In January 2015, OSHA issued a 42-page pamphlet, *Hydraulic Fracturing and Flowback Hazards Other Than Respirable Silica*,[1099] that discussed non-silica hazards in fracking and a section about worker rights.

Pay, Fatigue, and Traffic Accidents

Drivers for oil and gas operations are exempt from several sections of Federal Motor Carrier Safety Administration law. Because of work shifts that often exceed the 14 hours limit to commercial truckers, gas industry truckers are putting in work shifts of as many as 20 hours, leading not only to fatigue but accidents as well.

Analysis by Food and Water Watch of data from the Pennsylvania Department of Transportation revealed, "After fracking began, the average annual change in truck accidents trended upward in the counties with fracking wells (after trending down before fracking started) and continued to decline in unfracked counties after fracking began."[1100] Heavy-truck crashes, according to Food and Water Watch, "rose 7.2 percent in heavily fracked rural Pennsylvania counties (with at least one well for every 15 square miles) but fell 12.4[%] in unfracked rural counties after fracking began in 2005."[1101]

Vehicle deaths—often the result of driver fatigue—are about eight times the national average, according to data compiled by NIOSH.[1102] Traffic fatalities are "one of the key risk areas of the business," said Marvin Odum,[1103] president of Shell Oil, the U.S. subsidiary of Royal Dutch Shell.

In addition to problems for drivers, there are problems of truck integrity and traffic issues.

About 40 percent of all gas and oil trucks inspected by the Pennsylvania State Police between January 2009 and February 2012 had enough mechanical problems that they were ordered off the road, according to reporting by Ian Urbina in the *New York Times*.[1104]

Fire chiefs in Bradford County, Pa., which has the state's largest number of wells, said that response times of fire trucks and ambulances are delayed because of the increased traffic. "It's just about impossible [for firefighters] to get through [U.S. Route 6] to get to local fire stations," North Towanda Fire Chief Terry Sheets told a special meeting of emergency responders, politicians, and gas company representatives. "Ninety percent of the time it takes 20 minutes to [drive] those two miles," through the main part of town, said Wysox Fire Chief Brett Keeney. The fire departments, said Sheets, spend up to four hours per day "doing traffic control for disabled, broken-down vehicles in the road," according to reporting by James Loewenstein of the *Daily Review*.[1105]

Health- and Crime-Related Issues

MARCELLUS SHALE

In addition to job-related health problems, there has been a significant increase of sexually transmitted diseases in counties that have fracking. Analyzing data from the Pennsylvania Department of Public Health, Food and Water Watch reported, "The average annual number of gonorrhea and chlamydia cases increased by nearly a third (32.4 percent) in the most heavily

239

fracked rural Pennsylvania counties once fracking began—62 percent more than the 20.1 percent increase in rural unfracked counties."[1106]

Alcoholism and drug addiction are also prevalent in the gas fields. "One of the workers in Lycoming County [in northeast Pennsylvania] was a coke addict," says operating engineer Dominick Mas, "and many were alcoholics." He says the workers were tested for illegal substances, "but faked it by bringing in fake urine." Pennsylvania communities "have experienced steep upticks in drunken driving, traffic violations and bar fights," according to a Food and Water Watch data analysis of Pennsylvania State Police records."[1107] In the six heaviest-drilled Pennsylvania counties, arrests for drunken driving were up about 65 percent over counties with fewer wells, according to data analyzed by the Multi-State Shale Research Collaborative (MSRC)[1108] Drug abuse was up 48 percent in Pennsylvania counties that had more than 400 wells than in counties with fewer wells.[1109]

Counties with the most wells also had the highest crime rates. The MSRC noted a 17.7 percent increase in violent crime in the six Pennsylvania counties that had at least 400 wells, compared to counties with fewer wells. In contrast, both urban and rural areas with no wells had declining violent crime rates during the same time.[1110]

Property crimes in Pennsylvania, Ohio, and West Virginia also showed a high correlation to the number of wells drilled, according to MSRC data.[1111]

BAKKEN SHALE

The oil and gas boom in North Dakota has led to the lowest unemployment rate in the nation.[1112] While the nation's rate hovered about 5.3 percent,[1113] North Dakota averaged 3.0 percent unemployment in July 2015.[1114]

Energy company workers are earning annual incomes in the high five figure range; even entry-level jobs in local businesses are paying a minimum wage of $12 an hour in some parts of the state.[1115] However, North Dakota also has one of the nation's highest dropout rates,[1116] as high school boys leave

240

before graduation to take oil field jobs, and then are stuck in dead-end jobs or are laid off because the industry has begun to downsize to try to recover from overdrilling and falling prices.

Businesses in the mining region that may have been marginally profitable at one time began showing double-digit profits. As business took advantage of the housing boom, the cost of living increased significantly.[1117] The boom led to a housing shortage, with hundreds of workers living in tents, RVs, campers, or their cars, often paying as much as $1,200 a month for lot rentals and substandard housing.[1118] Motels and hotels, with inflated rates, quickly filled up. The rent for a one bedroom 700 square foot apartment in Williston, N.D., which advertises itself as "Boomtown U.S.A.," was almost $2,400 a month, the highest in the nation.[1119] Two bedroom apartments, if available, were renting for at least 50 percent more. The boom also led to significant reduction of the water supply, and strained the capacity of local government to provide adequate services.[1120] However, by 2016, with the price of crude oil plunging, the cost-of-living diminished when companies began leaving the state.

Not decreasing was crime, which increased 7.9 percent in 2012 over 2011 in the Bakken Shale, and has continued to rise, according to the the state's Uniform Crime Report.[1121] Attorney General Wayne Stenehjem said at the time most of the increase in violent crime was in counties where there was horizontal fracking.[1122] A year-long campaign by local and state officials led the federal government to declare the region to be a High Intensity Drug Trafficking Area area,[1123] and to open a Drug Enforcement Administration permanent office in Bismark, N.D., in 2014. By the end of the year, the FBI established a permanent field office in Williston, N.D. "Towns across the Bakken region have been hit hard by a spike in crime in the wake of our state's massive population boom, [which is why] we need more robust federal support to fight lawlessness right in the eye of the storm," said State Sen. Heidi Heitkamp,[1124] who had led the fight for increased federal law enforcement. The area has also seen significant increases in violence to Native Americans and women. "Combined with a lack of housing and law enforce-ment, the dramatic changes have brought crime such as human trafficking, drug activity and violence, especially against women, that are overwhelming local resources," the *Indian County Today* media network reported.[1125]

A coalition of Native American organizations in May 2015 requested the United Nations to intervene to protect women from sexual violence in the Bakken Shale. The area, populated by "man-camps," extensive temporary housing for oilfield workers, has also been a hub for sex trafficking, according to an investigation conducted by the Forum News Service.[1126] Only 13 percent of sexual violence attacks result in arrests, according to the Lakota Law Project.[1127]

Lisa Brunner, writing for *Indian Country Today*, said that many of the region's temporary workers:

> "treat Mother Earth like they treat women. . . . They think they can own us, buy us, sell us, trade us, rent us, poison us, rape us, destroy us, use us as entertainment and kill us. . . . [T]he level of violence that is occurring against Mother Earth . . . equates to us [women]. What happens to her happens to us. We are the creators of life. We carry that water that creates life just as Mother Earth carries the water that maintains our life."[1128]

Although the region embraced the fracking boom, Williams County Commissioner Dan Kalil says the people "look forward to the day when we no longer need [the FBI], when the problems are gone and we can go back to the community that we were."[1129]

Looking Out for Profits Not Worker Health

Contractors and subcontractors often choose not to report many health problems and injuries to OSHA, and prefer to pay worker salaries rather than allow them to file workmen's compensation claims. Injuries often include chemical and radiation burns, and bruises and contusions from direct contact with machinery. Workers have reported contact with chemicals without appropriate protective equipment, inhalation of sand without masks, and repeated emergency visits for heat stroke [and] heat exhaustion," says Dr. Pouné Saberi, a public health physician affiliated with Protecting Our Waters.[1130]

Underreporting of injuries is common in the oil and gas industry and distorts the actual safety rate of problems, which is higher than the national average, Peg Seminario, AFL-CIO Director of Safety and Health, told Tim McDonnell and James West of *Mother Jones* magazine.[1131]

Hospitals are experiencing increased debt because gas and oil field workers often don't have health insurance and are deliberately falsifying their identities, usually because they don't want to let their employers know they have been injured and possibly lose their jobs.

Much of that problem is that gas industry subcontractors usually don't give their employees health care insurance, but the hospital has a responsibility to treat all persons who come to its emergency room.

Multiple pressures weigh on the people who work in this high-risk, high-reward industry, including the need to produce on schedule and keep the costs down," reports Gayathri Vaidyanathan of *EnergyWire*.[1132]

"Almost every one of the injuries and deaths you will happen upon, it will have something to do with cutting a corner, to save time, to save money," attorney Tim Bailey told *EnergyWire*.[1133]

PHOTO: Frank Finan

Workers, without HazMat protection,
clean up toxic fluids near Lathrop Twp., Pa.

A drilling rig dominates a farm and agricultural land near Troy, Pa.

PHOTO: Bob Nilsson

CHAPTER 13
Effects Upon Agriculture, Livestock, and Wildlife

Building roads, pipelines, compressors, and drilling pads lead not only to decreased agriculture production and the destruction of the environment, but also to numerous accidents that impact the health of people and animals. Since 2005, fracking operations have directly changed or destroyed about 360,000 acres of land, according to data compiled and analyzed by the Research and Policy Center of Environment America.[1134]

Destroying the Food Supply

About 360 million years ago, the Bakken shale began forming in the area now known as South Dakota, Montana, and Saskatchewan.[1135] It is now about 200,000 square miles, lying between 4,500 and 7,500 feet below the earth's surface.[1136]

Oil in the shale was discovered in 1953.[1137] However, because the shale is only 13 to 140 feet thick, using conventional drilling methods were only marginally profitable. And then came the process of high-volume hydraulic horizontal fracturing, which allowed energy companies to drill into the earth, and then snake their pipes and tubes horizontally for as much as a mile. Because oil in the shale has a lower viscosity than oil in most shales, it is preferred by the industry.

Energy company landmen, buying land and negotiating mineral rights leases, became as pesky as aphids in the wheat fields. However, the landmen didn't have to do much sweet talking with the farmers, many of whom were hugging bankruptcy during the Great Recession. The farmers yielded parts of their land to the energy companies in exchange for immediate income and the promise of future royalties. Even if the farmers didn't want to lease part of their land, many didn't have a choice—others, not them, owned the subsurface mineral rights.

The first leases went for as little as $10 an acre plus royalties for five years. But, all leases had a termination clause; if the company didn't begin active drilling within those five years, the farmer or landowner would again hold mineral rights. To avoid having to renegotiate the lease at a substantially higher price, as landowners became more sophisticated in the value of their land, the energy companies drilled shallow wells, taking just enough oil and gas to show activity. The Energy Information Administration reported:

> "Although total U.S. crude oil production generally decreased each year from a peak in 1970, the trend reversed in 2010, and in 2012 production was the highest since 1995. These increases were led by escalating horizontal drilling and hydraulic fracturing, notably in the North Dakota section of the Bakken formation."[1138]

By the end of 2013, there were 189 rigs, up from 55 rigs four years earlier,[1139] and 10,800 oil wells, with an additional 7,200 expected to be drilled by the end of 2018, according to Lynn Helms, the state's mineral resources director.[1140]

In 2006, oil production in the North Dakota fields was about 2.2 million barrels;[1141] the next year, with EOG Resources as the primary developer of oil extraction, it was about 7.4 million barrels. Recoverable oil and gas continued to increase. Energy companies mined about 290 million barrels of oil in 2012,[1142] almost 360 million barrels in 2013,[1143] and 401 million barrels in 2014.[1144] There may be 3.65 billion barrels of recoverable oil, according to the *Oil and Gas Journal*.[1145]

Drilling for oil also yields natural gas; there are about two trillion barrels of natural gas in the shale, of which about 530 million barrels may be economically recoverable.[1146] North Dakota has 16 natural gas processing plants, with a capability to produce about 800 million cubic feet of gas a day, according to Justin J. Kringstad, director of the state's Pipeline Authority.[1147] "That represents a tripling of gas processing capacity in just the last five years. By the end of 2012, gas processing and transport capability [rose] to just over 1.1

billion cubic feet per day," Kringstad told David Fessler, an energy investment analyst.[1148] However, because the companies are primarily drilling for oil, natural gas becomes a secondary product. The industry flared about 375 million cubic feet of natural gas a day, about 28 percent of all oil drilled, into the atmosphere during 2014.[1149] That began to change in October 2014 when mandatory reductions established by North Dakota went into effect. The state imposed incremental reduction targets that would continue to 2020 when no more than 10 percent of all drilling could be flared.[1150]

Another problem caused by drilling is the inevitable spill. In September 2013, a quarter inch hole in a 20-year-old six-inch diameter pipeline owned by Tesoro Logistics leaked. Almost 21,000 barrels (about 865,000 gallons) of crude oil spilled onto about seven acres of agricultural fields in northwest North Dakota. The AP reported the field onto which the oil spilled had already been harvested; only a 40-foot thick layer of clay beneath the surface kept the oil from contaminating water sources. The people and Tesoro were "very lucky," said a state environmental geologist. The state and Tesoro both claimed there were no environmental impacts from the spill, something environmentalists didn't believe was accurate. "If you have an oil spill, some species of wildlife are going to be impacted, no matter where you have a spill," said Wayde Schafer of the Sierra Club.[1151] Steve Jensen, who first reported the oil on his field, told James MacPherson of the Associated Press the clean-up was like "an excavation war zone," and that it would be years before he could grow anything on the land.[1152] Tesora initially estimated the clean-up costs at $4 million,[1153] but later revised the costs at $20 million.[1154] Almost two years after the spill, Patty Jensen said clean-up crews "are working 24 hours a day, seven days a week, but it's so big and it's not as easy to clean up as they thought it would be."[1155] The crews are expected to continue working until 2018, five years after the spill.

Following the spill, the AP investigated state records and learned there were more than 300 pipeline spills between 2011 and 2013, none of which were reported to the public; like most states, North Dakota doesn't require public disclosure.[1156]

Even if the impossible occurred, and there were no workplace injuries and deaths, no spills and no damage to agriculture, and the environment and to public health, there is one problem that can't be solved. The Bakken Shale lies directly below one of the most fertile wheat fields in the United States. To get to the shale, companies must destroy the fields.

Those fields are primarily amber durum wheat. High in protein and one of the strongest of all wheats, amber durum, when mixed with semolina flour, is a base for most of the world's food production. It is used for all pastas, pizza crusts, couscous, and numerous kinds of breads. Red durum, a variety, is used to feed cattle. North Dakota farmers in late Summer harvest about 50 million bushels (about 1.4 million tons) of amber durum,[1157] almost three-fourths of all amber durum harvested in the United States.[1158] About one-third of the production is exported, primarily to Europe, Africa, and the Middle East.[1159] The destruction of the wheat fields, from a combination of global warming and fracking, will cause production to decline, prices to rise, and famine to increase. However, an AP feature didn't mention anything about the impact of fracking upon the wheat, and attributed a decrease in production in 2014 and a resulting price increase for breads and pastas solely to increased rains during Spring planting and Fall harvesting.[1160]

A further problem is the delay in transportation of grain from North Dakota to the middle of the country and the West Coast. BNSF, the primary rail carrier for both grain and Bakken crude oil, has delivered the oil on time to the refineries in the Gulf and mid-Atlantic ports, while giving grain a lesser priority. Angel Gonzales of the *Seattle Times* reported that "train trips from the Midwest's grain belt to the Pacific Northwest took 22 days, nearly twice as long as normal."[1161] To maintain their schedules, some ships had to leave before the assigned loads arrived; in other cases, those shipping the grain had to pay an additional $30,000–$50,000 a day to wait for the grain.[1162]

The destruction of farms isn't unique to North Dakota. Wherever energy companies drill is a loss of agriculture.

Canada's National Farmers Union (NFU) called for a moratorium on fracking. Jan Slomp, coordinator of the NFU in Alberta, said the farmers "are in the heart of Alberta's oil and gas industry where our ability to produce good, wholesome food is at risk of being compromised by the widespread, virtually unregulated use of this dangerous process."[1163]

The NFU, which represents about 47,000 farmers in England and Wales, called for compensation for its members if land values fall because of fracking. Dr. Jonathan Surlock, the NFU's chief advisor on renewable energy, told the *London Telegraph* the farmers were concerned that supermarkets might refuse to buy products grown above fracking sites, significantly reducing income even if there was no pollution caused by the fracking.[1164]

EFFECTS UPON CALIFORNIA AGRICULTURE

Beneath some of the nation's richest agricultural land in drought-ravaged central California lies the Monterey Shale, a 1,750 square mile formation. The Energy Information Administration (EIA) had originally believed there was as much as 15.4 billion barrels (647 trillion gallons) of recoverable oil, about four times that of the Bakken Shale.[1165] This created widespread economic speculation by oil and gas companies, as well as by various Chambers of Commerce. With funding from the Western States Petroleum Association, a USC study glowingly claimed extracting the available oil and gas would result in as many as 2.8 million jobs and almost $25 billion in additional revenue for the state of California.[1166] However, in 2014, two years after the initial estimates, the Energy Information Administration revised its figures, noting that under current technology only about 600 million barrels (2.5 trillion gallons) are economically recoverable under current technology.[1167] "Our oil production estimates combined with a dearth of knowledge about geological differences among the oil fields led to erroneous predictions and estimates," said John Staub, EIA petroleum exploration and production analyst.[1168] Nevertheless, just as it took about five decades for scientists to develop horizontal fracking after they developed vertical fracking, it may be only a decade or two until oil and gas company scientists develop a modification of horizontal fracking, with increased environmental and health risks, to be able to capture and recover most of the 15.4 billion barrels of oil and gas that still lie within the shale. When that happens, the impact upon agriculture and the nation's food supply will be be not just significant but catastrophic.

More than 200 different crops are grown in the central valley, including about 70 percent of the world's supply of almonds,[1169] most of the grape production, and 90 percent of all domestic wine sold in the United States.[1170] The Sun-Maid farm cooperative, headquartered in the Central Valley, is one of the world's largest producers of raisins and dried fruits. Overall, the valley produces one-fourth of all food consumed in the United States, with a value of about $17 billion a year, according to the U.S. Geological Survey.[1171]

Earthquake faults and massive tectonic plates of irregular shaped rock can make drilling risky and expensive. Another problem is that the shale, thicker than other major shales, lies between 6,000 and 15,000 feet beneath the earth. The deeper the drillers have to go, the more water, proppants, and chemicals they have to use, all of which raises the cost of exploration and drilling. Nevertheless, the landmen are buying leases and setting up what could be the biggest oil and gas boom in the country. If that occurs, even if miraculously there would be no groundwater contamination from fracking, and there won't be massive flaring because the drillers want oil not gas as a primary product, the oil and gas industry will outbid farmers for water, putting the nation's food crops at risk.

In July 2014, California ordered seven companies to stop injecting flowback water into 11 injection wells near Bakersfield in the central valley, and began to review data from about 100 other wells, citing concern that the waste could be affecting the aquifers in the state that was suffering from an extended drought.[1172]

DESTROYING PENNSYLVANIA AGRICULTURE

In Pennsylvania, 17,000 acres have been lost to the development of natural gas fracking.[1173] That land is not likely to be productive for several years because of "compaction and landscape reshaping," according to a study by the Penn State Extension Office.[1174] U.S. Geological Survey scientists conclude there is a "low probability that the disturbed land will revert back to a natural state in the near future."[1175]

The presence of natural gas drilling companies has led to decreased milk and cheese production. Penn State researchers Riley Adams and Dr. Timothy Kelsey concluded:

"Changes in dairy cow numbers also seem to be associated with the level of Marcellus shale drilling activity. Counties with 150 or more Marcellus shale wells on average experienced an 18.7 percent decrease in dairy cows, compared to only a 1.2 percent average decrease in counties with no Marcellus wells. In contrast, the average county experienced a 6.4 percent decline in cow numbers. . . .

"Higher drilling activity in all counties was associated with larger average declines in cow numbers. For example, counties with fewer than 5,000 cows in 2007 and no Marcellus wells averaged a loss of 2.2 percent, compared to an average 19 percent decline in such counties with 150 or more wells. Counties with 10,000 or more cows in 2007 and no Marcellus wells experienced an average 2.7 percent increase in cow numbers between 2007 and 2010, compared to an average loss of 16.3 percent in such counties with 150 or more Marcellus wells."[1176]

Among other direct effects of fracking, according to Penn State agriculture specialists Dr. Patrick Drohan, Gary Sheppard, and Mark Madden, are problems attracting and keeping farm help because of the higher pay from the fracking companies, and difficulty in obtaining mulch:

[M]ulch is in high demand for erosion and sedimentation control on gas sites. As local supplies are exhausted, farmers who purchase mulch for animal bedding might expect their costs to increase. Gas companies in northeast Pennsylvania are importing mulch from Delaware, Maryland, Ohio, and Virginia.[1177]

The potential effects of fracking have caused restaurants and co-ops to refuse to buy food from farms near rigs and wells. "A restaurant doesn't want to visit [a farm] and see a drill pad on the horizon," said Ken Jaffe, a New York state farmer.[1178]

"If hydrofracking is allowed in New York State, the co-op will have to stop buying from farms anywhere near the drilling because of fears of contamination," Joe Holtz, general manager of Brooklyn's Park Slope Food Co-Operative, told *The Nation* magazine. The co-op purchases about $4 million of food a year for its 16,200 members.[1179]

Destroying the Wildlife and Land

Strangling logic, Energy in Depth claimed if the Marcellus Shale was not being fracked, struggling dairy farmers would not be able to reap the financial rewards of having gas wells on their fields.

No matter what twisted logic EID believes, fracking results in significant impact upon wildlife. "In addition to loss of habitat, other potential direct impacts on wildlife from drilling . . . include increased mortality . . . altered microclimates, and increased traffic, noise, lighting, and well flares," according to a 900-page Environmental Impact Statement (EIS) published by the New York Department of Environmental Conservation in September 2011.[1180] The impact, according to the report, "may include a loss of genetic diversity, species isolation, population declines . . . increased predation, and an increase of invasive species,"[1181] and critical changes of migration routes. The report concludes that because of fracking, there is "little to no place in the study areas where wildlife would not be impacted, [leading to] serious cascading ecological consequences."

A research study by the U.S. Geological Survey a year later, focusing upon two areas of Pennsylvania, reached the same conclusion. According to that report:

> "Landscape disturbance associated with shale-gas development infrastructure directly alters habitat through loss, fragmentation, and edge effects, which in turn alters the flora and fauna dependent on that habitat. The fragmentation of habitat is expected to amplify the problem of total habitat area reduction for wildlife species, as well as contribute towards habitat degradation. . . .
> "Changes in land use and land cover affect the ability of ecosystems to provide essential ecological goods and services, which, in turn, affect the economic, public health, and social benefits that these ecosystems provide."[1182]

Also affected are humans who may eat fish that once lived in polluted streams and rivers, and cattle and other livestock that ate grass from polluted fields, drank polluted water, or absorbed airborne chemicals that were not detected during USDA inspections. Even vegetarians fear the effects of fracking. Toxic compounds "accumulate in the fat and are excreted into milk," says Dr. Motoko Mukai, environmental toxicologist at Cornell University.[1183]

Every well "will generate a sediment discharge of approximately eight tons per year into local waterways, threatening federally endangered mollusks and other aquatic organisms," says Dr. Ronald Bishop, professor of biochemistry at the State University of New York at Oneonta.[1184]

Katy Dunlap, Eastern Water Project Director for Trout Unlimited, told columnist Morgan Lee of the *Daily Gazette* (Schenectady, N.Y.):

> "[T]he most significant impact our members are seeing on the ground is erosion and sedimentation resulting from drilling-related activities, such as construction of well pads, new roads and pipelines.
> "This is of particular concern to TU because science has demonstrated at least 15 different direct negative effects from sedimentation on trout and salmon, ranging from stress, altered behavior, reductions in growth and direct mortality."[1185]

Concerned about what appeared to be black skin lesions and diseases affecting small mouth bass in the Susquehanna River, the Pennsylvania Fish and Boat Commission asked DEP secretary Michael Krancer to investigate causes of pollution, possibly from nearby fracking operations, and to declare the river to be polluted. In a letter made public by the *Sunbury Daily Item*, Krancer not only refused to declare the river was polluted, but claimed. "[T]he lesions and sores on the fish . . . are a complex problem and the reasons are not fully understood."[1186] Dr. William Yingling, a physician and fisherman, disagreed:

> "The evidence points strongly to the fact that the problems . . . are being caused by chemical endocrine disruption pollution in the watershed. If these black skin lesions on the smallmouth bass are spindle cell tumors or show cellular changes of melanoma, this presents serious questions about human

cancer risks from the chemical pollution in the watershed. We have been told [by DEP] that the black spots will not harm us and the fish are safe to eat. That may be correct. But if the spots represent the influence of chemicals in the water and they show evidence of malignant change in the fish, what would be the effect on human health?"[1187]

Dr. Yingling suggested that histopathologists and scientists from the U.S. Geological Survey and the National Institute of Environmental Health Sciences analyze the problem. Krancer told the *Daily Item* the DEP added water quality gauges that "will continue to operate and staff will assist when they can,"[1188] but refused to even speculate that water pollution caused by fracking could have contributed to the health problems of fish. Nevertheless, data strongly suggests the problem was not present before fracking operations began on a large scale.

Research "strongly implicates exposure to gas drilling operations in serious health effects on humans, companion animals, livestock, horses, and wildlife," according to Dr. Michelle Bamberger, a veterinarian, and Dr. Robert E. Oswald, a biochemist and professor of molecular medicine at Cornell University. Their study, published in *New Solutions*, a peer-reviewed academic journal in environmental health, documents evidence of milk contamination, breeding problems, and cow mortality in areas near fracking operations as higher than in areas where no fracking occurred. Drs. Bamberger and Oswald noted that some of the symptoms present in humans, from what may be polluted water from fracking operations, include rashes, headaches, dizziness, vomiting, and severe irritation of the eyes, nose, and throat. For animals, the symptoms often led to reproductive problems and death. At one farm, they documented 17 cows that died within an hour after being exposed to fracking fluids. Of the seven farms they studied in detail, "50 percent of the herd, on average, was affected by death and failure of survivors to breed."[1189]

Energy in Depth attacked the study as unscientific and "laughable at best, and dangerous for public debate at worst."[1190] EID claimed the use of anonymous sources negated the study. Drs. Bamberger and Oswald followed acceptable scientific protocol. However, there are many reasons why there have not been additional studies or why sources aren't specifically named. Elizabeth Royte, a distinguished science/environment writer, explains the problem:

> "Rural vets won't speak up for fear of retaliation. And farmers aren't talking for myriad reasons: some receive royalty checks from the energy companies (either by choice or because the previous landowner leased their farm's mineral rights); some have signed nondisclosure agreements after receiving a financial settlement; and some are in active litigation. Some farmers fear retribution from community members with leases; others don't want to fall afoul of "food disparagement" laws or get sued by an oil company for defamation (as happened with one Texan after video of his flame-spouting garden hose was posted on the Internet. The oil company won; the homeowner is appealing).
>
> "And many would simply rather not know what's going on. 'It takes a long time to build up a herd's reputation,' says rancher Dennis Bauste, of Trenton Lake, North Dakota. 'I'm gonna sell my calves, and I don't want them to be labeled as tainted. Besides, I wouldn't know what to test for. Until there's a big wipeout, a major problem, we're not gonna hear much about this.' Ceylon Feiring, an area vet, concurs. 'We're just waiting for a wreck to happen with someone's cattle,' she says. 'Otherwise, it's just one-offs'—a sick cow here and a dead goat there, easy for regulators, vets and even farmers to shrug off."[1191]

An explosion at a gas compressor station in October 2006 led to the end of a successful horse breeding and boarding farm near DISH, Texas. According to an in-depth investigation by Peter Gorman for the *Fort Worth Weekly*:

> "[Lloyd] Burgess, who had been out of town, returned to discover that one of his mares had aborted her foal. Two weeks later, the same thing happened to a second mare. . . .
>
> "Several months later one of his stallions got sick and finally had to be put down. Then a mare went blind. Then another stallion, a valuable quarter horse, got sick and was saved only when a friend offered to take if off Burgess' property, away from the compressor stations on Burgess' back fence line, to nurse it back to health. . . .

"'After the explosion and what happened to my horses, all my boarders took their horses out of there,' said Burgess. . . . 'Who could blame them? This was going to be my retirement, but now it's valueless.'"[1192]

It probably also wasn't just a coincidence that vegetation died in the areas where fracking occurred. Gorman reported:

"[Burgess's] fence used to be lined by huckleberry trees, planted as a windbreak back in the 1930s and '40s. The wind blows through pretty freely now, however, since most of the trees have recently died. . . .

"[Jim] Caplinger said that when he moved into DISH, his street 'was lined with willows, and now they're almost all gone. They just died. And I've lost three elm and hackberry trees as well.' . . .

"[City Commissioner Bill Cisco] agreed that something very troubling is happening to the trees. 'We've lost a lot of trees in DISH. And a number of them were cedar trees, which are almost impossible to kill. Those trees breathe just like we do, so when they start dying, you've got to pay attention,' he said. 'They're the canary in the coal mine for our air.' Thus far, dozens of trees in the little town and right outside it have died."[1193]

DISH, in the middle of the gas-rich Barnett Shale and 15 miles from where the nation's first natural gas well was fracked, by the middle of 2012 had 60 wells and 11 compressor stations with 36-inch diameter pipes to transport the gas.[1194]

FISH KILLS AND ANIMAL DEATHS FROM FRACKING

Fracking fluids released by Nami Resources into Kentucky's Acorn Fork Creek in 2007 probably led to large fish kills, according to research conducted by the U.S. Geological Survey and the U.S. Fish and Wildlife Service (FWS). Among those killed was a minnow-like fish, the Blackside Dace, a federally-threatened species. Nami, which claims the release was by an "independent contractor," paid a $50,000 fine in 2009 for violation of the Clean Water and Endangered Species Act. "This is an example of how the smallest creatures can act as a canary in a coal mine," said ecologist Tony Velasco of the FWS[1195]

In September 2009, about 65,000 fish, mussels and mud-puppies died in a 30 mile stretch of Dunkard Creek that flows in southwestern Pennsylvania and northern West Virginia.[1196] It was one of the nation's largest fish kills. The cause was determined to be the presence of massive amounts of golden algae, which produce toxins that are lethal to certain aquatic life. The golden algae are found only in coastal areas, primarily the Gulf and southern states.[1197] Extensive research for an environmental mystery determined that the algae could flourish in areas where there was fracking because of the extremely high salt content brought up in flowback water. The Pennsylvania Fish and Boat Commission charged Consol Energy with "illegal, toxic discharges [that were] willful, wanton and malicious."[1198] Consol, which the Commission said was dumping flowback into Morris Run that flows into Dunkard Creek, denied the charges but agreed to pay $5.5 million to the U.S. Department of Justice and the West Virginia Department of Environmental Protection to settle civil claims and create a new water treatment plant for its operations.[1199]

A fire at a well in Monroe County, Ohio, probably led to a fish kill of about 70,000 in June 2014, according to the Ohio Department of Natural Resources.[1200] The fire started in tubing to a fracked well owned by Statoil North America, and quickly spread to 20 trucks at the site, including four blender trucks that store the fluids for fracking.[1201] The fire fouled the air with thick black smoke and polluted at least five miles of Opossum Creek, which flows into the Ohio River. The fire led to the evacuation of 25 families.[1202] Compromising the public health and safety was the action of Halliburton, the primary contractor on the actual fracking of the site. The EPA had immediately requested a list of all chemicals used at the site; Halliburton, citing state-protected trade secret guarantees, provided only some of the full list. The company provided a full list five days after the initial request, according to the *Columbus Dispatch*.[1203]

Wyoming's mule deer population has declined by half, possibly because of the addition of natural gas wells, according to wildlife biologist Dr. Hall Sawyer and biometrician Ryan Nielson. The researchers noted: "Following the 2008 record of decision [on drilling], the level of winter drilling activity increased on the Mesa. It is possible that this increased winter disturbance affected fawn survival or adult reproduction."[1204]

Fracking probably caused 82 percent of pronghorn sheep to leave their traditional wintering grounds in Wyoming's PAPA and Jonah gas fields, according to research by the Wildlife Conservation Society.[1205] More than half of all prong-horns live in Wyoming.

Wastewater accumulating in puddles was probably the cause of the death of a dog and horse in southwestern Pennsylvania, according to a *ProPublica* investigation. A veterinarian concluded, "The dog's organs began to crystallize, and ultimately failed" because of ethylene glycol present in the wastewater.[1206]

Seven families were evacuated near Towanda, Pa., in April 2011, after about 10,000 gallons of wastewater contaminated an agricultural field and a stream that flows into the Susquehanna River, the result of an equipment failure.[1207]

Near Floresville, Texas, Fred and Amber Lyssy raise pigs, goats, and cattle on a 564-acre farm they have refused to allow oil and gas companies to lease mineral rights. In February 2013, five of their dogs died "mysterious, agonizing deaths," according to an in-depth investigation of the of the impact of fracking in the Eagle Ford Shale, conducted by by the Center for Public Integrity, *InsideClimate News*, and the Weather Channel and reported in *Bad Oil, Bad Air*.[1208] The cause of the deaths may have been leaks of hydrogen sulfide from a nearby Hunt Oil complex, documented by the Texas Commission on Environmental Quality (TCEQ) but which didn't fine Hunt for the air pollution violation. Three weeks after Hunt promised to fix the problem, there were other hydrogen sulfide leaks, yet the TCEQ still didn't fine the company. In November 2013, a sixth dog died, the same agonizing death as the first five dogs.[1209]

Seventeen cows died after drinking contaminated water in a pasture near a natural gas rig in Caddo Parish, La., in April 2010. Neither the owner nor the subcontractors reported the spill. Only after residents called the sheriff's department to report that cows were foaming at the mouth and bleeding were HazMat teams dispatched, according to the *Shreveport Times*.

Chesapeake Energy, which owned the well, acknowledged, "During a routine well stimulation/formation fracturing operation by Schlumberger for Chesapeake, it was observed that a portion of mixed 'frac' fluids, composed of over 99 percent freshwater, leaked from vessels and/or piping onto the well pad."[1210] Chesapeake, of course, didn't state that it wasn't the 99 percent freshwater that caused the deaths but the one percent of whatever was in the toxic mix. Energy in Depth continually pushes the fiction about how pure the mixture is:

> "Did you know that 99.51% of hydraulic fracturing fluids are made up of sand and water? The last 0.49% is made up of household items like sodium chloride (table salt), guar gum (used in ice cream and baked goods) and citric acid (lemon juice)."[1211]

Despite the fiction of the purity of the fracking solutions, Chesapeake and Schlumberger were each fined a token $2,000 by the Louisiana Department of Environmental Quality.[1212]

One month after the deaths in Louisiana, the Pennsylvania Department of Agriculture quarantined 28 beef cattle on a farm in Tioga County following a spill from an impoundment pit that had stored wastewater. The spill affected about 1,200 square feet of pasture in an area where the cattle normally grazed. East Resources, which was mining the land, told farmers Carol and Don Johnson not to drink the well water, but didn't provide fresh water. East Resources later denied any correlation between the eventual death of several of the cattle and the fracking operation.[1213] Even if there was no direct correlation, the mining operation caused problems for the Johnsons.

The installation of transmission lines across their property "spoiled almost every hay field I have," Don Johnson told Chris Torres of *Lancaster Farming*.[1214] In addition to having to deal with a loss of hayfields and toxic water spills, the Johnsons also had to deal with the noise from hundreds of trucks a week that use the couple's driveway to reach the well.

East Resources later told the Johnsons it wanted to put a second well and a compressor station onto property the Johnsons own a few miles away. But, Carol Johnson told the company she "had enough," and wanted to be left alone. "You sign a lot of stuff before you learn what's going on," she told *Lancaster*

Farming.[1215] The money they would earn from having leased mineral rights could help them survive the recession, but she says it would never be enough to compensate for the problems that the natural gas company created.

'All Farmers Are Stewards of the Land'

Joe Bezjak, of Smithfield, Pa., learned the same lesson as the Torres family. In his case, not only did a natural gas gathering company ruin part of his field, but a county judge put him in jail four days for protesting the pollution of his land.

Bezjak says he "hadn't even had a parking ticket" in his life and now, at the age of 73 in December 2012, he was in jail.

His story begins in 2005, a year after he retired after four decades as a teacher and principal, when he allowed Atlas Pipeline Partners to lease almost 700 acres of his land that he, his wife, Mickey, and relatives owned in southwestern Pennsylvania. On that land, Bezjak and his family raised about 200 head of black Angus cattle. Atlas paid him $10 an acre. Years later, as farmers became more sophisticated in lease management options, the companies paid as much as $1,700 an acre. But this was still before the fracking boom, and Atlas paid only what it had to pay to get the rights to drill and put in pipes. And so, it put in a six-inch diameter transmission pipe to transport natural gas from a nearby well to a compressor station. Five years later, before the lease expired, the company created several shallow wells to assure continued mineral rights ownership. For seven years, the Bezjak family, cattle, and the natural gas company co-existed.

"There weren't any problems at first," Bezjak says, but Laurel Mountain Midstream, a joint venture of Williams Companies and Atlas Pipeline Partners, seven years later wanted to install a 16-inch diameter line. Bezjak objected, but learned the company had the right to put in as many lines as it wanted in its right-of-way.

Laurel Mountain Midstream came onto his property in April 2012, without his permission, and tore down parts of his fence that divided the property. Bezjak had been careful to separate his herd by age and sex to maintain the quality of his herds. Without the fence, the cattle had wandered loose and interbred. "That ruined the herd," he says, lamenting. "More than 40 years of careful work to separate the lines was destroyed."

The company was "all friendly at first and said they would put in the fence and make it right," says Bezjak. But it never was right.

In May, he confronted the workers about not properly restoring his fence. Someone from Laurel Mountain Midstream called the police, complaining that Bezjak threatened them with a .22 single-shot rifle. "I rode up to them on my Quad, and I had a rifle mounted on it; it's always been there, but I didn't use it or threaten anyone with it," he says, quickly pointing out he never used it—"I'd have trouble shooting anything." On advice of his State Police, he took the rifle off the Quad.

He'd walk onto his land, sometimes to see what the company was doing, sometimes to talk with the workers. One State Trooper told him, "Every time they see you, they call us." But still there were no heated arguments. During the Summer, one judge of the Common Pleas Court ordered the company to fix the fence to Bezjak's satisfaction; another judge ordered him not to talk to Laurel Mountain workers.

On Nov. 9, 2012, he saw workers pumping what he believed to be wastewater from a ditch onto his property. He says he told the workers to stop. "All farmers are stewards of the land," he says, "and I'm an environmentalist. I didn't like what they were doing." Neither did the state Department of Environmental Protection, which issued Laurel Mountain a notice of violation for discharging industrial waste into the waters of the commonwealth.[1216]

The DEP had previously cited Laurel Mountain in June for discharging pollution into public water[1217] and in August not only for discharging industrial waste but also for failing to properly store, transport, process, or dispose of residual waste.[1218] Laurel Mountain Midstream also failed to obtain a DEP precertification to lay pipelines. "Our concern is when pipelines cross a stream or waterway, we need to know what they are going to do with the pipe" to avoid polluting the waters, says John Poister, a DEP community relations coordinator.

On Nov. 28, Laurel Mountain workers used a backhoe to pile mud and dirt on contaminated water. Bezjak called the DEP, whose inspectors in the region quickly responded to problems, but was told that there were not enough personnel available to investigate that day. But, the DEP did tell him that that covering up a spill was not acceptable, and it would investigate. So, Bezjak "decided to confront them myself to stop it." He says he "was concerned they were destroying the environment, and told them so." At the time, he believed the polluted water had come from a terra cotta pipe that had carried sulfur water from surface mining decades earlier. However, the water could also have come from a pipe laid down by Laurel Mountain. "We just don't know," says Poister.

Whatever its origins, several hundred gallons of acidic water drained onto parts of the land. "I had been farming that land 40 years," Bezjak says, "and had developed a good pH value and good grass." And now, not only was contaminated water lying in sulfur-red pools on his land, but heavy rains were washing it into the wetlands and into a small stream. The DEP had previously measured the water to have a pH reading of 3.36, acidic and dangerous. A private company Bezjak hired confirmed the water as toxic.

Bezjak readily acknowledges, "I got upset and was arguing with them because I saw what I'd worked for my whole life being damaged by these people who lied to me. So I stopped them, and by 10 a.m. there were three State Police troopers in my driveway and I was handcuffed." He was not arrested. "I wasn't looking for trouble," says Bezjak, "but they were damaging the land." A supervisor, says Bezjak, "even apologized to me and said the company would 'make it right.'"

Two weeks later, Bezjak got a summons. In Court four days after he received the summons, Common Pleas Court Judge Nancy D. Vernon "was livid that I had defied a court order not to talk to the workers." She didn't want to hear about the supervisor's apology, nor did she allow Bezjak to present evidence about environmental pollution. What upset her, he says, "is that I violated her order not to talk with the workers, and then in court told her she had to do whatever she had to do." He had planned to face the charge, pay whatever fine the judge imposed, continue Christmas shopping, and then pick up a friend he had taken for chemotherapy. "I never expected to go to jail over this," he says.

Citing Bezjak for contempt, Vernon ordered him handcuffed

and jailed without bail in the Fayette County Prison for four days, beginning Dec. 14.

On the Friday afternoon that Vernon threw Bezjak into jail, Laurel Mountain Midstream, by DEP direction, came onto his property to backfill the trench to temporarily stabilize the site during Winter. "We were concerned about acid water seepage, and wanted to minimize the potential of accelerated erosion," says John Poister. DEP, says Poister, planned to require Laurel Mountain Midstream to fully restore the site during the Spring thaw.

In a prepared statement, Julie Gentz said her company's goal is "to have good working relationships with our landowners and other stakeholders. This situation is a rare occurrence, but there have been a number of disagreements with this particular landowner that we have tried to resolve amicably."[1219]

Like many farmers, Bezjak doesn't oppose drilling for natural gas. What he does oppose is polluting the public health and environment. And, like most farmers, Bezjak's experience is that the landmen for companies involved in fracking "start out trying to be your friend, but they just lie and lie and lie to you."

He now keeps his cattle out of that area, "because I don't know what's in that water, and I don't want to contaminate the herd or cause health problems for them or the public."

Horses in a pasture near Dimock, Pa., are dominated by a nearby gas rig.

PHOTO: Vera Scroggins

263

PHOTO: Doug Duncan/USGS

Storage tanks for produced water in the Marcellus Shale

PHOTO: Robert M. Donnan

Jefferson Compressor Station (Saxonburg, Pa.)

PHOTO: U.S. Department of Transportation

Derailment of a Canadian Northern train, with DOT-111 tanker cars, near Cherry Valley, Illinois; June 19, 2009.

PHOTO: John Trallo

A 36-inch diameter gas pipeline, built by Williams Co. in 2010, cuts through an agricultural area and near houses in Wyoming County, Pa. There are 2.6 million miles of pipelines in the United States. There have been about 11,000 incidents and $6.5 billion in damage in the United States in the two decades beginning 1995.

265

MAP: Natural Resources Defense Council

CHAPTER 14
Trucks, Trains, Pipelines, and Compressors: Transporting Fossil Fuel

Because methane is explosive and flammable, problems can occur anywhere from the first exploratory hole to delivery in pipelines to homes and businesses. There is at least one major natural gas explosion, fire, or leak every week, according to documentation compiled by Natural Gas Watch. Dozens more each week are less hazardous.[1220] Many of the problems are fracking-related.

Trucks

HEALTH AND ENVIRONMENTAL HAZARDS

About 40 percent of all freight is shipped by truck[1221] on America's four million miles of roads.[1222] Each day, interstate carriers transport about five million gallons of hazardous materials.[1223] Not included among the 800,000 daily shipments are those of intrastate carriers, which don't have to report their cargo deliveries to the Department of Transportation. "Millions of gallons of wastewater produced a day, buzzing down the road, and still nobody's really keeping track," Myron Arnowitt, the Pennsylvania state director for Clean Water Action, told Aaron Skirboll of *AlterNet*.[1224]

On the evening of Aug. 9, 2012, a Halliburton truck carrying 4,000 gallons of hydrochloric acid pulled off Interstate 80, and into a convenience store market/gas station near New Columbia, Pa., after the driver noticed a leak. "There was a huge plume in the air and it was just getting bigger and bigger," resident Amanda Friend told the Sunbury (Pa.) *Daily Item*.[1225] Customers immediately fled when they saw the plume, which eventually engulfed the store, according to the *Daily Item*. HazMat teams detected a hole that had leaked at least 250 gallons of the acid. Larry Maynard, White Deer Twp. EMA director, told the *Daily Item* that because of the increased truck traffic due to fracking operations, EMA officials trained "many times" to deal with the probability of hazardous materials leaks. "If you want a shock, just park along I-80 and watch what goes by," Maynard said.[1226] The problem of a toxic fume that would affect several body systems was diminished only because a westerly wind blew the plume away from the village.

The gas leak could have been detected and the problem resolved much quicker had the truck been equipped with a leak-sensor camera, which detects hazardous materials leaks within 15 microseconds. In a letter-to-the-editor to the *Daily Item*, Flora Eyster, who lives near New Columbia, suggested:

> "Not only should the DEP and others require and monitor compliance, but every truck hauling these materials should be required to 1. Install and use this device 2. Be trained in a legally-required response and 3. Be REQUIRED to follow certain communication paths and immediate call-in response from drivers."

The toxic air cloud is only one of dozens of problems a year. Several thousand gallons of wastewater spilled into a storm drain that empties into Pine Creek, near Jersey Shore, Pa., when a truck hauling 4,600 gallons of wastewater crashed on the way to a gas well.[1227] A chain-reaction accident involving three tanker trucks in Canton, Pa., led to the spill of 1,300 gallons of diesel fuel and 400 gallons of wastewater into Chartiers Creek.[1228] In Karnes County, Texas, an open valve on a truck allowed about 1,000 gallons of drilling fluids to spill along eight miles of road.[1229]

ROAD DAMAGE IN THE FRACKING ZONE

Testifying before the Pennsylvania House Transportation Committee in June 2010, Scott Christie, the state's deputy secretary for the Department of Transportation, stated:

"Marcellus shale drilling is having a significant impact on the Commonwealth's roadway system; especially the secondary roads. Most of the drilling is taking place in rural areas with access via low-volume secondary roads. Most of these roadways do not have sufficient strength to withstand the large amount of trucks and other vehicles that are a part of Marcellus shale drilling. . . . The damage caused by this additional truck traffic rapidly deteriorates from minor surface damage to completely undermining the roadway base. Additionally, we have also found the sudden increase in heavy truck traffic has caused deterioration of several of our weaker bridge structures . . ."[1230]

Damage on state-maintained roads for each unconventional well in Pennsylvania is between $13,000–$23,000 per road, according to the Rand Corp.[1231] The state's Department of Transportation estimated the cost to repair roads caused by heavy truck damage is about $265 million.

To cover road and bridge damage, as well as to fund a better mass transit system for the urban areas, Tom Corbett, in his last year as governor, and the legislature authorized $2.3 billion a year for five years, beginning in 2014. The $11.5 billion package is being paid for by motorists, not the oil/gas industry. The increased funding comes from a 28-cent tax increase on each gallon of gas purchased at pumps, and increases in fees for several Penn-DOT services, including registrations, license renewals, and certificates of title.

Karnes County (Texas) Judge Barbara Shaw—whose husband works in the oil Industry and who believes oil "is a natural resource that's given by God to allow us to function"[1232]—told the *Houston Chronicle* in February 2013 that although her county had about $18 million in the bank, "I have at least $100 million in road damage" caused by the gas industry's heavy equipment.[1233] Seven months later, the Texas Transportation Commission finally approved $40 million to repair roads in the Barnett and Eagle Ford shales.

In November 2013, Nine months after Karnes told the public about road damage in her county, the Texas legislature authorized the transfer of $2.9 billion from the economic stabilization fund, funded by the taxpayers, to the Department of Transportation to cover costs of bonds previously purchased to fund improvements to the state's roads and bridges.[1234]

In North Dakota, road damage to county and township roads could cost the state as much as $7 billion by 2023, according to the Upper Great Plains Transportation Institute at North Dakota State University, which conducted the study for the state legislature.[1235] Between 2012 and 2014, the state had to pay about $800 million for road repair, much of it from fracking operations in the western part of the state.[1236]

The New York State Department of Transportation in 2011 estimated that if gas fracking came to the state, most likely only in several counties that border Pennsylvania, the cost of road damage would be about $90–156 million on state roads, and $121–222 million for local roads. "There is no mechanism in place allowing State and local governments to absorb these additional transportation costs without major impacts to other programs and other municipalities in the State," according to the Department.[1237]

Fracking's 'Bomb Trains'

Disregarding potential health and environmental effects, the railroads have become the principal means to move fracked oil to the refineries.

Canada's railroads carried only about 500 carloads of crude oil in 2009. Four years later, Canada's railroads had carried almost 140,000 carloads.[1238] U.S. railroads in 2014 carried more than 415,000 tanker cars of crude oil, more than 40 times what was hauled in 2008, according to the Federal Railroad Administration.[1239]

For short line railroads, horizontal fracking has been a boost in productivity and profits and, according to *Marcellus Drilling News*, "the revival of short line railroads.[1240] The expansion of gas drilling "has resulted in a spike in demand primarily for 'frac' sand, but also pipe, chemicals and other materials," reported *Progressive Railroading*.[1241] The monthly trade magazine also forecast that a possible increase in fracking operations "of about 43 percent over the next two years" [2012 and 2013] will have significant impact on rail traffic. The North Shore Railroad System, which owns six short lines in northeastern Pennsylvania, hauled its first carload of silica sand in 2008; four years later, it handled about 10,000 carloads.[1242] Lycoming Valley Railroad, one of the short lines owned by North Shore, reported hauling 130 car loads of silica sand in 2008; in 2013, it hauled 7,510 carloads into one of the fastest developing areas of the Marcellus Shale boom, according to the joint Railroad Authority (JRA). North Shore income, primarily because of increased traffic to fracking sites, was about $2.2 million in 2014, a 64 percent increase from six years earlier. However, it hauled only about 5,000 carloads in 2015 as the boom had begun to fade.

The increase in fossil fuel production and the use of railroads to move fracked oil has led to tragedy.

Almost three-fourths of the oil produced in the Bakken Shale is transported by trains through middle America to Gulf Coast and mid-Atlantic refineries;[1243] the remainder is transported to terminals in the Pacific Northwest and southern California. By January 2015, railroads were transporting about one million barrels per day,[1244] up from about 55,000 barrels per day in 2014.[1245]

The U.S. Pipeline and Hazardous Materials Safety Administration (PHMSA) had issued a safety alert in January 2014, pointing out that "crude oil being transported from the Bakken region may be more flammable than the traditional heavy crude oil."[1246] That alert followed previous advisories to notify railroads and shippers of "the importance of proper characterization, classification, and selection of a hazardous materials packing group as required by the Federal hazardous materials law . . . and Hazardous Materials Regulations."[1247] It was an acknowledgment, Steve Horn of *DeSmogBlog* reported, that Bakken shale oil "may be more chemically explosive than the agency or industry previously admitted publicly."[1248] An investigation by Horn had previously disclosed that fires and explosions from Bakken shale oil could cause evaporative losses of explosive volatiles benzene, toluene, hexane, and xylene.

About 92,000 of the 106,000 tanker cars currently in service were built before 2011 when stricter regulations mandated new design. The older cars (DOT-111), which have a 30,000 gallon capacity of crude oil, have an "inadequate design" and are susceptible to leaks and explosions in derailments, according to the National Transportation Safety Board.[1249] Railroad accidents in 2013 in the United States accounted for about 1.15 million gallons of spilled crude oil, more than all spills in the 40 years since the federal government began collecting data, according to PHMSA.[1250] The U.S. Department of Transportation predicts an average of 10 derailments a year in a two decade period beginning 2015, with estimated damage of about $4.5 billion.[1251]

Forty-seven persons were killed, and more than 30 buildings destroyed from fire, explosions, and smoke on a 73-car unmanned train that rolled down a seven mile incline and derailed in Lac-Megantic, Quebec, July 6, 2013. Seventy-two tanker cars of the Montreal, Maine & Atlantic railroad were hauling crude oil from the Bakken Shale to a New Brunswick refinery owned by Irving Oil.[1252] The accident released about 1.5 million gallons of crude oil; it was the worst rail disaster in North America since 1989.[1253] Firefighters had earlier disabled the brakes in order to put out a fire from a broken oil line on one of the train's five locomotives.[1254] However, the hand brakes may not have been reset after the initial fire was extinguished. The environmental cleanup is expected to cost about $200 million, all of it paid by the government; the railroad declared bankruptcy.[1255]

Less than a week after the disaster in Lac-Megantic, three tanker cars on a Norfolk Southern train carrying 90,000 gallons of ethanol exploded near Columbus, Ohio. The explosion led to the evacuation of about 100 residents within a mile of the accident.[1256]

Three months later, a Canadian National train carrying oil and gas derailed in Gainford, Alberta; three of the tanker cars carrying liquefied natural gas had leaks and were on fire as a result of the derailment.[1257]

In November 2013, a 90-car Genesee & Wyoming train, carrying about 2.7 million gallons of crude oil from the Bakken Shale,[1258] derailed near Aliceville, Ala., spilling about 750,000 gallons into surrounding wetlands;[1259] fire and toxic smoke burned for more than a day.[1260]

The following month, a 106-car BNSF train hauling Bakken

Shale crude oil slammed into a 112-car train that had derailed near Casselton, N.D. Explosions, fire, and toxic smoke led Cass County officials to urge evacuation of all residents within five miles of the accident.[1261] About 475,000 gallons of crude oil were spilled, according to estimates by the National Transportation Safety Board. BNSF carries about three-fourths of all oil from the Bakken Shale.[1262]

A week later, 45 homes were evacuated in Plaster Rock, New Brunswick, after a Canadian National train carrying propane and crude oil from the Bakken shale derailed and caught fire.[1263] In April 2014, 17 DOT-111 cars of a CSX train carrying fracked-oil from the Bakken Shale derailed near Lynchburg, Va.[1264] Four cars exploded, spilling about 50,000 gallons of crude oil into the James River.[1265] However, the Department of Environmental Quality (DEQ) said most of the oil was burned off from the explosion, and only 390 gallons remained in the river. Firefighters had to wait two hours before the railroad released information about the chemical contents of the cars.[1266] A year later, the DEQ fined CSX $379,574, including $18,574 for the cost to investigate the spill.[1267]

Two CSX trains carrying crude oil from the Bakken Shale derailed one year apart in Philadelphia. No injuries or spills were reported in each incident. In January 2014, seven cars of a 101-car train derailed on the Schuylkill Arsenal Bridge in a residential section near I-76. Six of the cars carried crude oil; one carried silica sand.[1268] It took more than a week to remove the railcars from the bridge. A year later, 11 tanker cars of a 111-car train derailed near a sports complex and the Philadelphia Naval Yard.[1269]

Within three weeks in February and March 2015, five trains carrying crude oil derailed in Alberta, Ontario, West Virginia, and Illinois. Four of the derailments involved explosions, fires, and evacuations.

There were no injuries or explosions when 12 cars of a Canadian Pacific train derailed near Bellevue, Alberta, Feb. 14, 2015.[1270] The next day, 29 cars of a 100-car Canadian National Railway train derailed near Timmons, Ontario.[1271] That train was carrying crude oil from Alberta. One day later, 14 tanker cars of a 109-car CSX train carrying three million gallons of crude oil[1272] exploded near Mount Carbon, W.Va., shooting a fireball of thick, black flames as high as 300 feet, according to witnesses.[1273] One car, leaking crude oil, fell into the Kanawha River, causing two water treatment plants to shut down their intake valves;[1274] another derailed car destroyed a house. Emergency Management officials ordered evacuation of about 1,000 people from two small towns along the train's path.[1275] The CSX and CN engines were hauling the newer model Casualty Prevention System Circular 1232 (CPC-1232) models tank cars, designed to replace the DOT-111 cars.[1276] Flames shot several hundred feet high when eight cars of a 105-car BNSF train derailed near the confluence of the Galena and Mississippi rivers, forcing a one-mile evacuation. Two of the cars were hauling sand; the others were hauling crude oil from the Bakken Shale. The fire from two of the cars was so intense that firefighters were forced to abandon the site, and allowed the flames to burn out.[1277]

The fifth derailment was near Gogama, Ontario, March 8, 2015. Five of the 10 tanker cars that derailed fell into the Mattagami River system. The cars were the newer CPC-1232 models[1278] that met requirements for better containment after Canada imposed stricter rules on tanker cars following the Lac-Megantic disaster. Although there were no immediate injuries reported in the five derailments, the effects of the fires and toxic smoke on train crew, residents, and emergency responders may not be known for years.

It isn't always tanker cars that derail. Sixteen hopper cars carrying silica sand on a Norfolk Southern train derailed in western Lucas County, Ohio, in February 2014. [1279] Ten hopper cars of a Progressive Railroad train hauling silica sand derailed near Chetek, Wisc., in January 2013.[1280] Seven hopper cars of a Southwestern Pennsylvania Railroad train carrying silica sand derailed near Uniontown, Pa., in January 2015.[1281] Three months later, seven cars of a Lehigh Railway train hauling fracking sand derailed near North Bradford, Pa. Two of the cars left the tracks and overturned in a field.[1282] The hopper cars have a capacity of 130 tons of silica sand. Although there were no reported injuries, several tons of sand spilled, possibly leading to airborne pollution and subsequent health issues.

Another side effect of increased oil transportation from the Bakken Shale has been significant delays for Amtrak passenger service, especially the Empire Builder, which runs from Chicago to Portland and Seattle. The delays became so severe that Amtrak issued an alert, warning passengers that the delays, because of increased freight shipments, "averaged between three and five hours,"[1283] causing passengers to miss connections to Eastern and Midwest trains.[1284] The problem, noted by journalist Mark Karlin, is:

> "BNSF Railway Company owns the tracks that Amtrak uses in North Dakota. Given that the tracks are the property of BNSF, it decides which trains have de facto priority passage (even though a federal law is supposed to give priority to Amtrak), and it has allegedly given fracking and oil container car shipments passage scheduling times that impede the passage of Amtrak passenger trains (and, apparently, trains carrying farm goods that are perishable)."[1285]

While significant delays are annoying, more annoying would be if a passenger train was on a side rail waiting for an oil-laden freight train to pass—and a 100-tanker freight train derailed.

SUGGESTIONS AND REGULATIONS

The derailment and explosions of what fracktivists described as "bomb trains" became so severe that in May 2014 the Department of Transportation (DOT) declared the movement by trains of fracked crude oil posed an "imminent hazard" and ordered all railroads carrying at least one million gallons of fracked crude oil from the Bakken Shale to provide details to state officials.[1286] The DOT established a $175,000 fine per day for non-compliance. BNSF, CSX, and Union Pacific tried to block that order, but failed.[1287] The railroads further decided that proposed federal rules to limit speed were not appropriate. A DeSmogBlog investigation showed that the major railroads in a dozen meetings at the White House not only lobbied against reductions in speed,[1288] but claimed that such actions would affect their economic integrity and, as the president of CSX claimed, would "severely limit our ability to provide reliable freight service to our customers."[1289]

The railroads also claimed that although they were required to provide information to state authorities, that information was not public record because what was carried, how much, and the routes were "sensitive security information."[1290] North Dakota, Wisconsin, and several other states denied those claims.[1291] Oregon required public disclosure following a series of articles by Rob Davis in *The Oregonian*.[1292] Maryland and Pennsylvania, however, agreed with the drillers and railroads, and refused to disclose information about train routes and hazardous materials on those cars, claiming without justification those records were "confidential" and "proprietary."[1293] However, in October 2014, Pennsylvania's Office of Open Records denied the policy of the state's Emergency Management Agency, and ruled that the railroads were required to provide information about trains carrying at least one million gallons of crude oil from the Bakken shale.[1294]

The DOT recommended that railroads not use the DOT-111 tank cars to ship oil from the Bakken Shale unless the cars were reinforced.[1295] Because it was only a safety recommendation, railroads were not forced to comply.[1296] Canada, however, ordered railroads to stop using the DOT-111 tank cars by May 2017.[1297]

In May 2015, the U.S. Department of Transportation announced significant new standards for tank cars built after Oct. 1, 2015, to meet more rigorous standards. The rules requir railroads to replace all DOT-111 cars within three years, and the newer non-jacketed CPC-1232 cars within five years. The DOT also require an improved braking system on all cars, and reduced maximum speed to 50 miles-per-hour in all areas, and 40 miles-per-hour in urban areas for trains with the older cars.[1298]

Within a week of the announcement of new regulations, the eighth derailment of the year occurred. A 109-car BNSF train, with 107 of the cars carrying Bakken crude oil and two cars carrying silica sand, derailed near Heimdal, N.D. The fire and toxic smoke from six cars forced evacuation of about 40 residents. "Today's incident is . . . why we issued a significant, comprehensive rule aimed at improving the safe transport of high hazard flammable liquids," said Sarah Feinberg, acting head of the Federal Railroad Administration.[1299]

BNSF, beginning January 2015, put a $1,000 per car surcharge on its DOT-111 cars, and a lesser surcharge on CPC-1232 cars that were not retrofitted to meet higher standards. The surcharge, according to BNSF, was to force shippers to use the newer DOT-117 cars and help pay for newer cars. BNSF said it was planning to eliminate DOT-111 cars by January 2016 and CPC-1232 cars by 2018. While being proactive in reducing possible tanker car explosions and, perhaps, reducing insurance industry concerns, BNSF faced a backlash from shippers. The older cars, said Jamie Heller of Hellerworx, "are federally compliant cars. From a shipper's perspective, when they bought these other cars, they were compliant, and they still are. As long as the shipper's compliant, why should there be a surcharge?"[1300]

Beginning April 2015, all oil produced in North Dakota was required to be "conditioned" to reduce volatility. The cost to North Dakota oil producers to meet state regulations was likely to be about $20 million, and to increase sales costs by about 10 cents per barrel, Lynn Helms, director of the state's Department of Mineral Resources, told the Minneapolis *Star Tribune*.[1301] Fines for non-compliance are $12,500 a day.

Pipelines

Individuals, corporations, and politicians who believe in fracking, when confronted about the problems of the "bomb trains," claim that pipelines are preferred and much safer than railroads in moving oil and gas. They're also wrong about that.

About half of the nation's 2.6 million miles of pipelines are at least 50 years old; corrosion, according to *ProPublica*, is responsible for between 15 and 20 percent of deaths, injuries, or property damage.[1302] Between 1995 and the end of 2014, there were more than 10,800 incidents of spills, contamination, injuries, and deaths, according to the Pipeline and Hazardous Materials Safety Administration. During that time, there were 371 deaths, and about $6.35 billion in damage. Spilled oil accounted for about half of the four million barrels of fluids; about 20 percent was natural gas. Other spills were jet fuel, liquefied gas, propane, diesel, gasoline, and other fuels.[1303]

However, because there are only 460 state and federal inspectors,[1304] and most spills are reported by the companies, "There is a high probability there are more pipeline 'events' than are being reported," according to oil/gas industry analyst Dory Hippauf.[1305]

An explosion in a 30-inch underground pipeline in southeast New Mexico, near the Pecos River, killed seven adults and five children in August 2000. They, and one other, were camping about 250 yards from the explosion.[1306] The explosion left a crater 86 feet long, 46 feet wide and 20 feet deep.[1307] The pipeline, owned by El Paso Natural Gas Co. (EPNG), was installed in 1950. The section of the pipe that exploded had not been inspected since the mid-1950s, according to the National Transportation Safety Board, and had "severe corrosion damage" prior to the explosion.[1308] Seven years later, the U.S. Department of Justice determined EPNG "failed to employ personnel qualified in corrosion control methods; failed to investigate and mitigate internal corrosion in two of its pipelines transporting corrosive gas; and failed to suitably monitor those two pipelines to determine the effectiveness of steps taken to minimize internal corrosion." The Department of Justice required the company to pay a $15.5 million civil penalty and to "implement modifications and comprehensive reform on the entire 10,000 miles of EPNG pipeline system."[1309] Separate payments to the families of those killed were not disclosed.

Eight persons died, 50 were injured, and 38 houses destroyed from an explosion in a 30-inch diameter natural gas pipe in San Bruno, Calif., Sept. 9, 2010.[1310] The explosion, which led to flames more than 1,000 feet high,[1311] also left a 40-foot deep crater that measured 167 feet long by 26 feet wide at the surface.[1312] Because there were no automatic shut off valves on the pipe, which was laid in 1956, it took workers almost 90 minutes to manually shut off the fire.[1313] An independent investigation ordered by the California Public Utilities Commission (CPUC) later revealed that Pacific Gas & Electric (PG&E) had diverted more than $100 million for safety improvements to executive bonuses and increased stockholder dividends. The report noted that PG&E gave a "low priority" to safety, and that its "focus on financial performance [was] well outside industry practice—even during times of corporate austerity programs."[1314] Three years after the explosion, PG&E revealed it was expecting to pay about $565 million in legal claims and settlements.[1315] Four years after the explosion, the City of San Bruno, with about 7,000 pages of recently-discovered e-mails and documents, revealed what Mayor Jim Ruane called an "ongoing, illicit and illegal relationship" between CPUC and PG&E, and called for the resignation of CPUC President Michael Peevey who, prior to becoming CPUC president was president of Southern California Edison.[1316] The CPUC, in April 2015, fined PG&E $1.6 billion; PG&E decided not to appeal. The fine, the highest in the state's history,[1317] is in addition to the $635 million the CPUC previously required for pipeline improvement.[1318] The fine and penalties are being absorbed by shareholders, and not consumers.

An explosion near Sissonville, W. Va., in December 2012 that destroyed four houses and about 800 feet of Interstate 77[1319] was probably the result of sudden pressure drop in a six-foot section of pipe that had worn to 0.078 inches of thickness, according to the National Transportation Safety Board (NTSB).[1320]

The NTSB also determined it took more than an hour for Columbia Gas Transmission employees to manually shut off the gas. For almost two decades, the federal agency had urged, but could not require, companies to install automatic shutoff valves. Jim Hall, NTSB chair from 1994 through 2000, told the Associated Press, "The companies' attitude is, in many cases, unless it's required, they're not going to do it."[1321]

A law signed by President Obama in 2012 to require automatic shutoff valves has not been implemented. The AP reports the Congressional mandate, still being fine-tuned, would require such valves only when it's "economically, technically and operationally feasible." Existing pipes will not be retro-fitted for automatic shut-off valves, but new pipes might be.

PIPELINE REGULATION

An inspector from the Pennsylvania Public Utilities Commission was on scene within two hours of a pipeline explosion in Springville Twp., Pa., but did not have jurisdiction. Pennsylvania's Gas and Hazardous Liquids Pipeline Act,[1322] like regulations in most states, includes oversight of classes 2–4, but excludes Class 1 pipelines. A Class 1 location is any area with "10 or fewer buildings intended for human occupancy within 220 yards of the center-line of the pipeline," according to PHMSA. A four-part series by the *Philadelphia Inquirer* in December 2011 had revealed no state or federal agency has jurisdiction over pipelines in Class 1 rural areas, nor are operators required to report any incidents, including property damage, injuries, or deaths associated with those pipelines.[1323]

One of the major problems in excluding Class 1 pipelines is "The vast majorities of Class 1 pipelines are and will be located in rural areas where first responders will most likely be untrained on how to handle an ignited well," according to *Jurist*, a web-based newsletter about legal issues.[1324]

Regulating Class I pipelines is "at the bottom of the state's priority list," Patrick Henderson, Pennsylvania's energy executive, said in February 2012.[1325]

"For decades, the gas industry has fought hard to protect federal and state exemptions, defeating repeated attempts by Congress and safety advocates to change it," the Philadelphia *Inquirer* reported.[1326]

A one year investigation by the Government Accountability Office (GAO), initiated by the Senate Committee on Commerce, Science, and Transportation, revealed that PHMSA regulates only about 20,000 of 200,000 miles of natural gas gathering pipelines (such as those at condensers) and only about 4,000 of the estimated 30,000–40,000 miles of hazardous liquid gathering pipelines. The report, published in March 2012, noted:

> "In response to GAO's survey, state pipeline safety agencies cited construction quality, maintenance practices, unknown or uncertain locations, and limited or no information on current pipeline integrity as safety risks for federally unregulated gathering pipelines.
>
> "Our survey revealed that only 3 of the 39 state agencies reported that they collect and analyze comprehensive pipeline spill and release data on federally unregulated pipelines."[1327]

The GAO recommended the Department of Transportation should "Collect data from operators of federally unregulated onshore hazardous liquid and gas gathering pipelines [and] establish an online clearinghouse or other resource for states to share information on practices that can help ensure the safety of federally unregulated onshore hazardous liquid and gas gathering pipelines."[1328]

"[H]undreds of miles of high-pressure pipelines already have been installed in the shale fields with no government safety checks—no construction standards, no inspections, and no monitoring," according to reporting by Joseph Tanfani and Craig R. McCoyin of the *Philadelphia Inquirer*.[1329] Only about one-fourth of all oil, natural gas, and propane pipelines have been inspected since 2006, according to Public Employees for Environmental Response (PEER),[1330] which had to file a Freedom of Information Act suit to get the public records. "PHMSA is a sleepy, industry-dominated agency that tries to remain obscure by doing as little as possible," said Kathryn Douglas, PEER legal counsel. "[T]he safety and reliability of much of this key but volatile transport grid remains unknown," PEER concluded.

In a scathing report, distributed in May 2014, the Department of Transportation's inspector general concluded PHMSA must "strengthen its management and oversight of state pipeline safety programs."[1331]

PHMSA has "very few tools to work with," Jeffrey Wiese, associate administrator of pipeline safety, told a convention of pipeline compliance officers in July 2013. The problem, he said, is that the Republican-controlled Congress has refused to increase the agency's budget since 2010, although President Obama had requested a $276 million budget for the 2013–2014 fiscal year, a 35 percent increase, to allow a stronger public education program, better co-ordination with states, and the hiring of more inspectors and staff. Although there has been a significant increase in pipelines, there are only 135 federal inspectors.[1332] Only about $1 million of the budget is taxpayer funds; the rest is from an oil spill liability fund and industry user fees. The regulatory process, said Wiese, "is kind of dying."

However, both the inspector general's report and an investigation by Elana Schor and Andrew Restuccia of *Politico* reveals

PHMSA was slow to implement Congressional direction to revise rules to tighten pipeline security and safety.[1333] Schor and Restuccia also revealed that between 2003 and 2015, PHMSA "levied just $44.2 million in fines against pipeline operators that caused more than $5.5 billion in damage."[1334]

Compressors

A natural gas compressor in Springville Twp., about 30 miles northwest of Scranton, Pa., exploded March 29, 2012, destroying the sheet metal roof of the compressor station and sending flames and thick clouds of black smoke into the air. The compressor station, owned and operated by Williams Partners of Tulsa, Okla., can process about 365 million cubic feet of gas per day, and then send it along two 24-inch diameter pipelines.[1335] The DEP initially reported the cause was possibly a leak in a gas line; it later reported the cause was a valve that was left open during maintenance work.[1336] One worker suffered minor injuries. One day after the explosion, and against orders by the DEP, Williams again was sending natural gas into the pipelines.[1337] One year after the explosion and fire, the DEP decided the release of about one ton of methane into the atmosphere didn't violate the allowable limits set by the state; it didn't fine the company for air or water pollution. Unlike Nuclear Regulatory Commission regulations, the Pennsylvania DEP doesn't have the power to penalize a company for operator error, according to a DEP spokesperson.[1338] In May 2013, a Williams compressor in Brooklyn Twp., about nine miles northeast of the Springville Twp. station, caught fire. Two weeks later, an explosion and fire hit a Williams compressor station in Branchburg, N.J. Two workers were hospitalized; others refused medical treatment.[1339] In March 2014, an explosion at a Williams LNG plant in Plymouth, Wash., injured five employees and led to the evacuation of 400 residents within a two mile radius of the explosion.[1340]

An investigation of federal records by Natural Gas Watch revealed Williams had several violations, and may have been responsible for a natural gas explosion that destroyed two homes and injured five persons near Appomattox, Va., Sept. 14, 2008.[1341] The company paid a $952,000 fine and replaced about 2,500 feet of the 53-year-old pipeline.[1342] Between Dec. 3, 2011 and April 23, 2014, there were nine fires and explosions at Williams facilities,[1343] including one at a gas processing plant in Opal, Wyo., that sends gas through five major pipelines. The plant accounts for about two percent of all gas transmission in the United States.[1344] The explosion and subsequent fires burned for more than two days, sending toxic fumes into the air, and leading to a temporary evacuation of about 100 families.[1345]

There have been more than two dozen incidents at compressor stations, about half at Williams stations, since 2008. Explosions and subsequent fires destroyed two separate compressor stations in Wyoming in December 2011. Two workers were injured near Pinedale, Wyo.[1346]; there were no reported injuries at a fire and explosion that destroyed the Falcon compressor station, owned by Enterprise Products, in Sublette County, Wyo.[1347] Two workers were injured when an explosion and fire damaged a compressor and destroyed several buildings near Nine Mile Canyon, Utah, in November 2012.[1348] One worker was killed and two others injured in an explosion and fire at a BP compressor station in Bayfield, Colo., in June 2012.[1349] Two workers died and one was seriously burned from an explosion at a compressor station in Tyler County, W. Va., in April 2013.[1350] A month later, an explosion at a facility near Crockett, Texas, caused about $7.5 million property damage[1351] Four workers were killed and four injured in October 2015 at a Williams compressor station in Gibson, La.[1352]

Fires and explosions, deaths and injuries, aren't limited to the United States. A gas tank exploded in Qatar in February 2014, killing 12.[1353] A gas leak explosion in Kaohsiung, Taiwan, in August 2014, led to the deaths of 28 and injuries of at least 267.[1354] A truck carrying LNG and butane exploded near a maternity hospital in Mexico City in January 2015, killing a nurse and two babies, and injuring about 70.[1355]

Even *if* mining fossil fuel could be safe, with no health or environmental effects—an impossibility in itself—moving the oil and gas would still be a hazard.

PRIOR PRINTER'S NOS. 2689, 2765, 2777, 2837

PRINTER'S NO. **3048**

THE GENERAL ASSEMBLY OF PENNSYLVANIA

HOUSE BILL

No. **1950**

Session of 201?

Report of the Committee

To the Members of the

We, the unde
the Senate a†
consideri†
"An ac†
and

(b.2) Trade secret or confidential proprietary information.--When an operator submits its stimulation record under subsection (b.1), the operator may designate specific portions of the stimulation record as containing a trade secret or confidential proprietary information. The department shall prevent disclosure of a designated trade secret or confidential proprietary information to the extent permitted by the act of February 14, 2008 (P.L.6, No.3), known as the Right-to-Know law, or other applicable State law.

3
14
15
16
17
18
19

...se of Representatives.)

. SCARNATI

…RY JO WHITE

(Committee on the part of the Senate.)

CHAPTER 15
Gagging the Public and the Health Care Industry

In Durango, Colo., Cathy Behr, an emergency room nurse, almost died in August 2008 because of exposure to fracking fluids—and a refusal by a gas driller to release what was in that fluid. According to reporting by Abrahm Lustgarten:

> "Behr [was] treating a wildcatter who had been splashed in a fracking fluid spill at a BP natural gas rig. Behr stripped the man and stuffed his clothes into plastic bags while the hospital sounded alarms and locked down the ER. The worker was released. But a few days later Behr lay in critical condition facing multiple organ failure.
>
> "Her doctors searched for details that could save their patient. The substance was a drill stimulation fluid called ZetaFlow, but the only information the rig workers provided was a vague Material Safety Data Sheet, a form required by OSHA. Doctors wanted to know precisely what chemicals make up ZetaFlow and in what concentration. But the MSDS listed that information as proprietary. Behr's doctor learned, weeks later, after Behr had begun to recuperate, what ZetaFlow was made of, but he was sworn to secrecy by the chemical's manufacturer and couldn't even share the information with his patient.
>
> "News of Behr's case spread to New York and Pennsylvania, amplifying the cry for disclosure of drilling fluids. The energy industry braced for a fight.
>
> "'A disclosure to members of the public of detailed information . . . would result in an unconstitutional taking of [Halliburton's] property,' the company told Colorado's Oil and Gas Conservation Commission. . . .
>
> "Then Halliburton fired a major salvo: If lawmakers forced the company to disclose its recipes, the letter stated, it 'will

have little choice but to pull its proprietary products out of Colorado.' The company's attorneys warned that if the three big fracking companies left, they would take some $29 billion in future gas-related tax and royalty revenue with them over the next decade.

"In August the industry struck a compromise by agreeing to reveal the chemicals in fracturing fluids to health officials and regulators—but the agreement applies only to chemicals stored in 50 gallon drums or larger. As a practical matter, drilling workers in Colorado and Wyoming said in interviews that the fluids are often kept in smaller quantities. That means at least some of the ingredients won't be disclosed. . . .

"Asked for comment, Halliburton would only say that its business depended on protecting such information."[1356]

The section of the law in Colorado is known as Form 35.[1357] "If there's a chemical in the environment and there are potentially other people being exposed, it does not make sense, from a health standpoint, to not be sharing that information," Dr. Michael Pramenko, former president of the Colorado Medical Society, told the *Denver Post*.[1358] "This is just about transparency so that nobody is harmed and the environment is not harmed," said Dr. Mitchell Gershten, who told the *Post* he had treated a rig operator for liver problems after exposure to one of the fracking chemicals.[1359]

In most states, health care professionals may request specific information, but the company doesn't have to provide that information if it claims it is a trade secret or proprietary information, nor does it have to reveal how the chemicals and gases used in fracking interact with natural compounds. If a company does release information about what is used, health care professionals in most states are bound by a non-disclosure agreement. That agreement not only forbids them from warning the community about water and air pollution that may be caused by fracking, but also forbids them from telling their own patients what the physician believes may have led to their health problems.

Ohio's legislature, with a strong assist from the oil and gas industry, protected the industry from disclosing chemicals that the industry determined were "trade secrets," and to wait at least 60 days after drilling begins to disclose other chemicals.

That legislation (SB315), passed in 2012, also limited access of health care workers to know what chemicals were used in drilling operations except for "an incident related to the production operations of a well," and then only for a health issue and only if obtained from the company itself.[1360]

Attorney General Mike DeWine, a conservative Republican, called for stronger regulation of the fossil fuel industry and for the industry to dislose all chemicals used to extract natural gas from the Marcellus and Utica shales. "We need to do it right, we need to do it with safeguards, but we need to do it," DeWine told the AP in March 2012." [1361] Heavy lobbying by the fossil fuel industry and reluctance by Ohio's Republican-controlled legislature kept DeWine's proposals from becoming law.

Dr. Mike Robichaux, in a letter to the editor of the *Arkansas Times*, wrote about the health effects caused by the Deepwater Horizon/BP–Halliburton oil spill, and the ability of Big Energy to keep information hidden:

> "I have seen and treated over 100 individuals [between 2010 and 2013] whose health was severely affected by this disaster and many of them will likely remain ill for the rest of their lives. You probably have not heard or read much about these illnesses because the responsible parties have the ability to suppress this type of information and keep it out of the mainstream press."[1362]

With the discovery of the New Albany Basin as a possible source of economically-feasible natural gas extraction, Illinois enacted a series of rules in 2014, not unlike those of most other states. The new rules require a list of all chemicals used in the fracking process to be submitted to the Department of Public Health, with a redacted list available to the public. The oil/gas industry objected to providing a full list because it claimed it would expose "trade secrets."[1363]

The Sanford sub-basin is a 150 mile shale in a northeast to southwest formation in central North Carolina. Although extracting natural gas, ethane, and butane may not be economically beneficial, the state government enthusiastically encouraged drillers to come into the state, and expects 55 wells by 2017.[1364] To regulate those wells, the Republican-controlled legislature pushed a set of regulations that appear to be industry-friendly. One of those regulations requires companies to inform the Department of Environment and Natural Resources of the content of the drilling fluids, even if they are trade secrets. In case of a spill or explosion, the state would release, to emergency responders, the contents of the fluids. However, private citizens, even those directly exposed to the effects from drilling operations, are forbidden to know what's in the fluids. Emergency responders who tell citizens, or citizens who find out the contents of the chemical cocktails and reveal them, can be convicted of a misdemeanor.[1365] The initial bill had made disclosure a class I felony until Democratic legislators vigorously objected.

In June 2015, the American Medical Association finally spoke out for full disclosure of chemicals used in fracking. The AMA policy calls for "government agencies [to] record and monitor the chemicals placed into the environment for extracting petroleum, oil and natural gas. Monitoring for fracking chemicals should focus on human exposure in well water and surface water and government agencies should share this information with physicians and the public."[1366]

PENNSYLVANIA'S GAG

In Pennsylvania, a strict interpretation of Act 13 forbids physicians who sign the non-disclosure agreement and learn the contents of the "trade secrets" from notifying others, including medical specialists about the chemicals or compounds, thus delaying medical treatment. The clauses that established that restriction were buried on pages 98 and 99 of the bill.

Like Colorado's gag order, Pennsylvania's clause was primarily written by the oil and gas industry, with assistance from the right-wing American Legislative Exchange Council (ALEC).[1367]

"I have never seen anything like this in my 37 years of practice," says Dr. Helen Podgainy Bitaxis, a pediatrician. She says it's common for physicians, epidemiologists, and others in the health care field to discuss and consult with each other about the possible problems that can affect various populations. Her first priority, she says, "is to diagnose and treat, and to be proactive in preventing harm to others." The new law, she says, not only "hinders preventative measures for our patients, it slows the treatment process by gagging free discussion."

The law is not only unprecedented, but "complicate[s] the ability of health departments to collect information that would reveal trends that could help us to protect the public health," says Dr. Jerome Paulson, director of the Mid-Atlantic Center for Children's Health and the Environment at the Children's National Medical Center in Washington, D.C. Dr. Paulson, who is also professor of pediatrics at George Washington University, calls the law "detrimental to the delivery of personal health care and contradictory to the ethical principles of medicine and public health." Physicians, he says, "have a moral and ethical responsibility to protect the health of the public, and this law precludes us from doing all we can to protect the public." His strongest statement about why there needs to be a moratorium on horizontal fracking came in June 2014 when he concluded, "Neither the industry, nor government agencies, nor other researchers have ever documented that [unconventional natural gas extraction] can be performed in a manner that minimizes risks to human health. There is now some evidence that these risks that many have been concerned about for a number of years are real risks."[1368]

Pennsylvania requires physicians to report to the state instances of 73 specific diseases, most of which are infectious diseases. However, the list also includes cancer, which may have origins not only from chemicals used to create the fissures that yield natural gas but also in the blowback of elements, including arsenic and radium present within the fissures. Thus, physicians are faced by conflicting legal and professional considerations.

DEP secretary Michael Krancer said "nothing could be further from the truth or more nonsensical" than persons claiming there was a gag order in Act 13.[1369] Patrick Henderson, Gov. Tom Corbett's energy executive, not only claimed there was no gag order, but that Act 13 "seeks to foster health professional access to the information, and implicit in that is the free exchange of information with their patient so they can, together, make informed decisions."[1370] Dr. Eli Avila, Pennsylvania's health secretary, also dismissed arguments that the law restricts physicians. Dr. Avila, who resigned in September 2012, had claimed that physicians could share information with patients and other medical providers, as well as to report health concerns to the state.[1371]

House Speaker Sam Smith, responding to criticism by physicians, health professionals, and others about the impact of Act 13, called their complaints "outrageous" and "irresponsible."[1372] He also claimed, "Pennsylvania has the most progressive hydraulic fracturing disclosure law in the nation. It is designed for transparency and access, and it provides unfettered access to physicians or other medical professionals who need information to treat their patients."[1373]

No matter what Krancer, Henderson, Dr. Avila, and Smith claimed, they were wrong. Act 13, like regulatory acts in other states, specifically restricts physicians from disclosing corporate "trade secrets" to patients and other medical providers.

Dr. Alfonso Rodriguez, a nephrologist from Dallas, Pa., sued the Pennsylvania DEP secretary, attorney general, and the chair of the Public Utilities Commission in July 2012 after treating a trucker who was splashed by fracking fluid. That trucker now suffers from kidney failure and is on dialysis. Dr. Rodriguez says he had tried to find out the chemical ingredients of that fluid:

> "That's when we ran up against the firewall. They keep telling me the stuff's benign, just soap and emulsifiers. And I say, 'Just send me the sheet.' They say, 'We can't do that. It's not allowed.'"[1374]

The complaint charged: "the Medical Gag Rule is an unconstitutionally overbroad content-based regulation of speech [that requires] a confidentiality agreement as a condition precedent to receive information needed for the ethical and competent treatment of a patient in an emergency situation to prohibit communications which are not narrowly tailored to advance a compelling governmental interest."[1375] He argued the gag rule, which he believes violates his 1st and 14th Amendment rights, requires him either to violate a non-disclosure agreement or his oath as a physician. As a physician, says Dr. Rodriguez, he is "ethically required to secure any information necessary to provide competent treatment to his patients. In order to secure needed information from gas drillers and/or their agents and/or vendors in emergency situations, plaintiff is required by the Medical Gag Rule, upon request, to waive rights secured to him."[1376]

A federal judge dismissed Dr. Rodriguez's suit in October 2013. The ruling by U.S. District Judge A. Richard Caputo had nothing to do with the truth of Rodriguez's allegations; it had to do with a legal technicality on the physician's standing:

> "Plaintiff does not allege that he has been in a position where he was required to agree to any sort of confidentiality agreement under the act. Therefore, to the extent that plaintiff's alleged injury-in-fact is an inability to exercise his First Amendment rights, he has not yet indicated that he has been prevented from engaging in any sort of communication as a result of the act. Similarly, plaintiff has failed to indicate that he has been forced to waive any of his fundamental constitutional rights."[1377]

A federal appeals court in March 2015 upheld the decision of the district court.[1378]

"The confidentiality agreements are worrisome," says Peter Scheer, a journalist/lawyer who is executive director of the First Amendment Coalition. Physicians who sign the non-disclosure agreements and then disclose the possible risks to protect the community can be sued for breach of contract, and the companies can seek both injunctions and damages, says Scheer.

In pre-trial discovery, a company might be required to reveal what it claims are trade secrets and proprietary information; the court would then determine if the chemical and gas combinations really are trade secrets or not. The court could also rule the contract is unenforceable because it is contrary to public policy, which places the health of the public over the rights of an individual company to protect its trade secrets, says Scheer. However, the legal and financial resources of the natural gas corporations are greater than those of individuals, and they can stall and outspend most legal challenges.

Leaders of the five largest clinical medical societies, in a joint statement published in the *New England Journal of Medicine*, attacked politicians for encroaching upon physician–patient rights and confidentiality in four broad areas, one of which is encompassed in the gag orders related to fracking operations in Colorado, Ohio, Pennsylvania, and Texas. The executives of the American Academy of Family Physicians, the American Academy of Pediatrics, the American College of Obstetricians and Gynecologists, the American College of Physicians, and the American College of Surgeons, declared:

> "[L]egislators in the United States have been overstepping the proper limits of their role in the health care of Americans to dictate the nature and content of patients' interactions with their physicians. Some recent laws and proposed legislation inappropriately infringe on clinical practice and patient–physician relationships, crossing traditional boundaries and intruding into the realm of medical professionalism. . . .
>
> "By reducing health care decisions to a series of mandates, lawmakers devalue the patient–physician relationship. Legislators, regrettably, often propose new laws or regulations for political or other reasons unrelated to the scientific evidence and counter to the health care needs of patients."[1379]

Although Pennsylvania and some other states are determined to protect the natural gas industry, not everyone in the industry agrees with the need for secrecy. Dave McCurdy, president of the American Gas Association, says he supports disclosing the contents included in fracturing fluids. In an opinion column published in the *Denver Post*, McCurdy argued, "We need to do more as an industry to engage in a transparent and fact-based public dialogue on shale gas development."[1380]

The Natural Gas Committee of the U.S. Department of Energy also agrees. "Our most important recommendations were for more transparency and dissemination of information about shale gas operations, including full disclosure of chemicals and additives that are being used," said Dr. Mark Zoback, professor of geophysics at Stanford University and a Board member.[1381]

Also in agreement is the general population. About 91 percent of Pennsylvanians and 90 percent of Michigan residents believe fracking companies should disclose all chemicals in fracking and the health risks, according to a University of Michigan/Muhlenberg College poll released in May 2013.[1382]

In his State of the Union address in 2012, President Obama called for full disclosure of all chemicals and compounds used in fracking on federal and Native American lands.[1383] The Obama Administration and the Department of the Interior had originally wanted all chemicals and compounds to be disclosed at least 30 days prior to drilling. However, in a concession to the natural gas industry, the rules issued in May 2012 allow the industry to disclose the chemicals only after drilling at a site had been completed.[1384]

Both Dave McCurdy's statement and the Department of Energy's strong recommendation about full disclosure were known to the Pennsylvania General Assembly when it created the law that restricted health care professionals from disseminating certain information that could help reduce significant health and environmental problems from fracking operations.

"Companies have to realize that they need to be transparent about what they are doing and they need to take the people's concerns seriously," Maria van der Hoeven, director of the International Energy Agency, said. If the companies don't, Van der Hoeven said in August 2012, "there's a very real possibility that public opposition to drilling for shale gas will halt the unconventional gas revolution and fracking in its tracks."[1385]

Drs. Michelle Bamberger and Robert E. Oswald, like most of those in the health and environmental professions, not only call for "full disclosure and testing of air, water, soil, animals, and humans," but also point out that with lax oversight, "the gas drilling boom . . . will remain an uncontrolled health experiment on an enormous scale."[1386]

In June 2013, ConocoPhillips and ExxonMobil finally disclosed the content of the toxic soup in six of its wells—in Poland.[1387] Full disclosure in the United States is still voluntary.

Disclosing Some of the 'Fracking Soup'

Like almost every industry, the natural gas industry, even if it takes public funds and accepts tax benefits from local, state, and federal governments, believes that transparency is something that applies to someone else. It even has federal support to avoid being responsive to the people. The natural gas industry is exempt from the Emergency Planning and Community Right to Know Act (EPCRA),[1388] which allows citizens to learn what chemicals and toxins are used at facilities.

Although most states require disclosure of chemicals used in fracking, except for what the industry claims are "trade secrets," most of the states which allow fracking do not require full disclosure of chemicals prior to drilling.[1389]

Fourteen of the 29 states in which fracking occurs have no requirements to disclose the chemicals that are in the fracking mixture or that which is brought to the surface by the process itself, according to data compiled by the Natural Resources Defense Council (NDRC).[1390] The chemical disclosure rules that are in place, says the NRDC, may be "woefully inadequate to provide sufficient public health protection—underscoring the need for federal rules that require all oil and gas companies fracking anywhere in the country to fully reveal the chemicals they're using."[1391]

In July 2010, Range Resources announced it became the first company to voluntarily disclose "the composition of each of the hydraulic fracturing components for all the wells operated by Range Resources with the Pennsylvania Department of Environmental Protection (DEP) completed in the Marcellus Shale."[1392] Trumpeting its new practice, Range announced, "Transparency and open dialogue are vital to the continued progress of energy development." But the reality was little more than a public relations attempt to placate the public. A report issued by four investment groups ranked Range Resources second from the bottom of two dozen oil and gas producers in the Marcellus Shale on issues of transparency. (ExxonMobil was at the bottom; Encana, with a rating of 44 percent was at the top.) The report gave Range Resources a score of 3 on 32 items.[1393] But Range had more to worry about than a failing grade.

In a lawsuit filed before the Environmental Hearing Board by Loren Kiskadden and other residents of Amwell Twp., Pa., against the DEP and Range Resources *[Docket 2011-149-R*[1394]*]*, Range admitted it didn't know all the chemicals in the fracking fluids that may have polluted Kiskadden's drinking water.[1395]

About a year after Judge Thomas Renwand demanded Range to provide a list of all chemicals, Range claimed it was unable to get all the information from its sub-contractors. The failure to adequately respond to a court order infuriated Washington County President Judge Debbie O'Dell–Seneca who gave Range and 40 contractors a 30 day notice in November 2013 to provide a detailed list of what Range used in its drilling formula. Range continued to claim it was following the court's order but that getting the information was a "lengthy process,"[1396] and the company was not responsible for any health problems. It filed an appeal, which the Pennsylvania Superior Court a year later ordered quashed. In its decision, the Court ruled, "Range Resources does not have a recognizable interest in the proprietary information it seeks to protect [and] the right to assert such protection is held by the manufacturers of those products."[1397] The ruling set a statewide precedent.

In April 2014, ExxonMobil, which had received condemnation for its policies against transparency, announced it would finally disclose fracking risks. The decision came after 30 percent of its investors demanded the company change its policies. Among those who demanded transparency was Scott M. Stringer, custodian of the New York City Pension Fund, which held about $1 billion in ExxonMobil stock. "We have seen the significant risks that come from hydraulic fracturing activities," said Stringer, "[and] corporate transparency in this arena is truly necessary for assessing risk and ensuring that all stakeholders have the information they need to make informed decisions."[1398] However, ExxonMobil has no plans to release data about methane leaks. Sharon Kelly, writing for DeSmogBlog, believes, "The jury is out on whether Exxon's recently announced report is a publicity stunt or an actual reckoning with the need for public disclosure of risks."[1399]

Baker Hughes, the third largest fossil fuel service provider, didn't think that releasing the composition of the chemicals it provides to the industry should be that lengthy of a process. In October 2014, it began disclosing the names of all chemicals "without compromising our formulations—a balance that increases public trust while encouraging commercial innovation."[1400] However, a month later, Halliburton, the second largest fossil fuels service provider, which vigorously defended its right to keep chemicals secret, announced it was buying Baker Hughes for $34.6 billion, effective during the latter part of 2015.[1401]

Within two months of the purchase announcement, Halliburton, Wyoming, and environmental groups announced an agreement that would require all companies drilling in the state to submit applications to the Wyoming Oil and Gas Conservation Commission (WOGCC) on specific chemicals it wished to classify as trade secrets. The Commission, not the companies, would determine which chemicals would be classified as trade secrets. The agreement came after more than three years of legal debate, begun when several environmental groups demanded, under right-to-know legislation, the composition of chemicals used in the oil and gas fields. A decision issued by a district court against the environmental groups was overturned by the Wyoming Supreme Court.[1402]

FRACFOCUS

Databases that give solid information about fracking include "The List of the Harmed," produced by the Pennsylvania Alliance for Clean Water and Air; Marcellusgas.org, which gives full information about well location, production, and enforcement; SkyTruth (developed by geologist John Amos) and FracTracker, which provide maps and satellite images, as well as extensive data about all aspects of shale gas drilling, including health and environmental issues; and WellWiki, developed by Dr. Joel Gehman of the University of Alberta, which will include information about more than four million wells drilled in North America since 1859. Federal information sources include those from the Department of Interior/Bureau of Land Management, Energy Information Administration, Environmental Protection Agency, the National Energy Technology Lab, and the United States Geological Survey, all of which give overviews and extensive data about fracking.

However, the one database that is critical, but flawed, is FracFocus, which includes information from about 100,000 wells from 23 states.[1403] With public criticism increasing, the natural gas industry created a website where citizens can learn about fracking issues and most of the chemicals in the fracking "soup." It doesn't collect data on natural occurring radioactive material (NORM), much of which is included in drill cuttings that must be disposed.

FracFocus was initially funded by oil and gas trade associations and a $1.5 million grant from the U.S. Department of Energy.[1404] The website is maintained by two public organizations, the Groundwater Protection Council, an association of state water officials; and the Interstate Oil and Gas Compact Commission.

An investigation by *Bloomberg News* revealed that in the first eight months the website was active (April 11, 2011 through the end of the year), "Energy companies failed to list more than two out of every five fracked wells in [the] eight U.S. states" in its analysis.[1405] Those states were Arkansas, Colorado, Louisiana, Montana, Oklahoma, Texas, Utah, and Wyoming, which accounted for 64 percent of all natural gas production in 2010. Since then, Pennsylvania, the 15th largest state in natural gas production in 2007,[1406] became the leading producer of natural gas.[1407]

Because participation in FracFocus is voluntary, corporations decide not just what to report but also when to begin reporting. Thus, some wells that were fracked before April 11, 2011, were not listed on the website. Because FracFocus is a voluntary self-reporting database, accuracy can be questioned; companies can modify or delete data. FracFocus also doesn't tell state regulators when companies file disclosure reports, thus evading penalties for prompt reporting.

For the first two years, the site was difficult to navigate, and 29 percent of the chemicals reported by the industry were impossible to track through the Chemical Abstract Service (CAS).[1408] A revision, beginning June 2013, placed data into an XML database, which allowed a search of wells by several criteria, including geography and operator, and allows individuals to use GIS technology to identify chemicals used at specific wells and search the data-base by CAS numbers. The 3.0 version, released in February 2015, reduced human error, expanded search capability, allowed users to take advantage of a "machine readable" format, and updated information about chemicals and their environmental effect.[1409]

Although the natural gas industry finally released a list of most of the chemicals used in fracking,[1410] much of the composition in the fracking toxic soup is not available to the public, protected as "proprietary information" or "trade secrets." Wastewater chemicals are also not included. The Bloomberg investigation noted:

> "Gaps remain on the website even when wells are disclosed. Companies skip naming certain chemicals when they decide that revealing them would give away what they consider trade secrets. Many of the wells that are listed on FracFocus have at least one or two chemicals marked confidential. Others have far more."[1411]

A Harvard University Environmental Law study, authored by Kate Konschnik, notes:

> "[T]hree characteristics of a robust trade secret regime prevent overly broad demands for this protection: substantiation by the company, verification by a government agency, and opportunity for public challenge. FracFocus has none of these characteristics; operators have sole discretion to determine when to assert trade secrets. As a result, inconsistent trade secret assertions are made throughout the registry."[1412]

"In its current form, FracFocus is not an acceptable regulatory compliance method for chemical disclosures"[1413] was the conclusion of the Harvard study. Even after the 2.0 version was released, the study concluded, "FracFocus still fails as an acceptable regulatory compliance tool."[1414]

"FracFocus is just a fig leaf for the industry to be able to say they're doing something in terms of disclosure," said U.S. Rep. Diana DeGette (D-Colo.).[1415] Ralph Kisberg, co-founder of The Responsible Drilling Alliance, said he believes the website is little more than "a PR effort to placate people."[1416]

FracFocus replied that the attacks upon its credibility fail to "reflect the true capabilities of the FracFocus system and misrepresents the systems *[sic]* relationship to state regulatory programs," and accuarately noted the system's major defect with self-reporting: "It is up to each operating company to know and understand individual state laws regarding disclosure. It is also up to each state to enforce compliance with its own laws."[1417]

Pennsylvania, under a new governor, Dr. Tom Wolf, in June 2015 announced it was developing a user-friendly database to record and make available to the public significant information about all wells in the state. Unlike FracFocus, the new database will allow complex searches.[1418] DEP Secretary John Quigley says the database is the result of the administration's commitment to "collaboration, transparency and integrity."[1419]

The truth about what's in fracking fluids is revealed by a question from a Canadian politician. Tom Mulcair, leader of Canada's Federal New Democratic Party, asked "If you think that your method of getting to that gas is safe, why won't you reveal the contents of the fracking fluid?"[1420] Since the Industry didn't answer the question adequately, Mulcair answered it: "Because that fracking fluid contains known carcinogens and other very dangerous substances."

Denial and Non-Disclosure

The industry's typical first response to complaints from the public that fracking could contribute to health problems is to deny any correlation. Industry spokespeople will first claim the problems existed before fracking was conducted in that area. When additional evidence comes in, they will continue to debate it or offer a settlement. That's because the industry, like most corporations, prefers to settle rather than expose itself in a public trial. It's unknown how many families settled claims with the natural gas industry. As a condition of settlement, the companies demand non-disclosure contracts and the sealing of court records,[1421] effectively shutting off public comment by the plaintiffs and hiding any public health issues that arose through the discovery portion of the lawsuit. This restriction prevents others from knowing if health problems are confined to one area or may be a problem that affects public health.

In June 2012, three families from Wyalusing, Pa., settled with Chesapeake Energy after refusing to accept a non-disclosure agreement. The settlement included paying $1.6 million to Heather and Jared McMicken, Michael and Jonna Phillips, and Scott and Cassie Spencer, which included purchasing their houses and properties. Two years earlier, the families had noticed muddy water in their water wells.

Chesapeake denied its operations caused the problem, claiming excessive methane existed prior to the drilling. A filtration system provided by Chesapeake did not function properly, according to the three families.

In Washington County, Pa., EQT Production silenced 17 families in 2013 by offering $50,000 to each of them in exchange for never objecting to EQT permit filings or applications, or to any activity, including "noise, dust, light, smoke, odors, fumes, soot or other air pollution, vibrations, adverse impacts or other conditions of nuisances."[1422] EQT initially required every one of the 30 affected homeowners to sign the agreements in order for any of them to receive the $50,000. The company later relented, but still imposed non-disclosure clauses that forbid the property owners from discussing or revealing the contract clauses without EQT permission. The agreement was binding not just upon those who signed it, but also upon all heirs or others who may later purchase the property.

However, the public's health may become greater than a company's rights of secrecy. The *Pittsburgh Post–Gazette* and Washington (Pa.) *Observer–Reporter* filed petitions in August 2011 to intervene in the case of *Stephanie and Chris Hallowich v. Range Resources, et al.*, [C-63-CV-201003954[1423]] after Judge Paul Pozonsky agreed with Range Resources and the Hallowichs, who had been reluctant to agree to the company's demand for a non-disclosure clause, to seal the court records.

In its petition to unseal the records, the *Post–Gazette* argued:

> "The gas companies' interest in secrecy must yield to the greater social good of disclosing information relevant to public health and safety. Moreover, no Pennsylvania court has ever held that court records may be sealed based on nothing more than the interest in using confidentiality to promote settlements."[1424]

In an *amicus curiae* petition, Earthjustice, pointed out:

> "When these cases, alleging serious adverse health effects from gas development, are resolved, they are not being resolved in a way that provides more information to the public about the alleged health effects of gas drilling. Instead, the

defendant companies are successful at limiting the knowledge of defendants' operations—especially as they relate to public health—gained in litigation to the plaintiffs, who are bound by protective orders and nondisclosure agreements preventing them from sharing such information with the public. Litigation secrecy, like state law limits on disclosure such as Pennsylvania's impact fee law, deprives the public of information that could be used to protect public health. . . .

"Since gas companies use confidentiality so routinely in so many contexts, it is critical to counter this trend by upholding public access to court records in cases involving the health effects of gas development. . . .

"The public interest in accessing the record in this particular case is heightened by the secrecy generally promoted by the natural gas industry. If the industry were more forthcoming generally—if it did not seek exemptions from otherwise applicable federal and state disclosure requirements, did not advocate for and use state laws to limit disclosure of information such as the identity of chemicals used in drilling and fracturing, and did not routinely silence injured parties during litigation or as a condition of settlement—then an order sealing the record here might not be significant. But the calculus changes when an effort to conceal information is part of a pattern and practice limiting dissemination of information on the health impacts of gas development. Against that background, it is all the more important to ensure that health and safety-related information in court records is accessible to the public."[1425]

The Superior Court of Pennsylvania agreed, and sent the case back to the lower court for reconsideration.[1426] In a 32-page opinion rendered in March 2013, Washington County President Judge Debbie O'Dell–Seneca not only ordered the 900 pages of documents to be unsealed, she also ruled that corporations may not use the laws of privacy created for persons. "The defendants' assertions of a right of privacy under the Constitution of the Commonwealth of Pennsylvania are meritless . . . Nothing in that jurisprudence indicates that that right [of privacy] is available to business entities," the judge ruled.[1427] That ruling was "a significant development for the growing movement to restore democracy to the people," John Bonifaz told *AlterNet*.[1428] Bonifaz is founder and executive director of Free Speech for People, which was created to oppose the Supreme Court's 5–4 decision in *Citizens United v. FEC* that gave corporations and organizations the same First Amendment rights as individuals.

Four months later, additional evidence about the Hallowich settlement was revealed. Included in a 16-page settlement agreement that was not released when the 900 page transcript was released in March was a clause that imposed a lifetime gag order not just upon Stephanie and Chris Hallowich but also upon their two children.[1429] When this was revealed by the news media, Range Resources spun its response and now claimed the gag was "not something we agree with"[1430]—even though the settlement was clear that the gag order applied to all family members—and released the children from a lifetime of silence.

PHOTO: Bill Cunningham/USGS

Storage tanks of a few of the dozens of chemicals,
most of them toxic, at a drilling site in the Fayetteville Shale.

CHAPTER 16
Theological Perspectives About the Environment

"Let each of you look not to your own interests,
but to the interests of others." —*Philippians 2:4*

Jorge Mario Bergoglio, upon his election as pope in March 2013, took the name Francis, in honor of Francis of Assisi who worked for the poor, for the rights of animals and who, said the new pope, "[taught] us profound respect for the whole of creation and the protection of our environment, which all too often, instead of using for the good, we exploit greedily, to one another's detriment." The new pope later asked Christians to become "custodians of creation," and that a threat to peace:

> "arises from the greedy exploitation of environmental resources. Even if 'nature is at our disposition,' all too often we do not 'respect it or consider it a gracious gift which we must care for and set at the service of our brothers and sisters, including future generations'. . . .
> "Here too what is crucial is responsibility on the part of all in pursuing, in a spirit of fraternity, policies respectful of this earth which is our common home." [January 2014][1431]

In June 2015, the Pope released a 184-page encyclical calling for all mankind to recognize that global warming is primarily caused by mankind, and that mankind has a responsibility to take better care of the environment. "We have grown up thinking that we were [Earth's] owners and dominators, authorized to loot her," said the Pope, who noted, "Numerous scientific studies indicate that the greater part of the global warming in recent decades is due to the great concentration of greenhouse gases . . . given off above all because of human activity."[1432] He warned, "If the current trend continues, this century could see unheard-of climate change and an unprecedented destruction of ecosystems, with grave consequences for all of us,"[1433] and blamed much of global warming and the destruction of the environment upon a frenzied rush to extract fossil fuels. "Our immense technological development," said the Pope, "has not been accompanied by a development in human responsibility, values and conscience." In keeping with the tenets of Christian theology, the Pope called for mankind not only stop to stop its destruction of the environment and seek out renewable energy, but also to actively improve the lives of the poor:

> "When we fail to acknowledge as part of reality the worth of a poor person, a human embryo, a person with disabilities—to offer just a few examples—it becomes difficult to hear the cry of nature itself; everything is connected. Once the human being declares independence from reality and behaves with absolute dominion, the very foundations of our life begin to crumble, for [as John Paul II stated in 1991], 'Instead of carrying out his role as a cooperator with God in the work of creation, man sets himself up in place of God and thus ends up provoking a rebellion on the part of nature.' . . .
> "We are all too slow in developing economic institutions and social initiatives which can give the poor regular access to basic resources. We fail to see the deepest roots of our present failures, which have to do with the direction, goals, meaning and social implications of technological and economic growth."[1434]

It was the first time a pope issued an encyclical about the environment. The encyclical called for all Catholic churches to discuss the environment, how to protect it, and how to care for the weak and the poor.

About 70 percent of American Catholics believe there is global warming; about 57 percent of all Catholics believe it is caused by humans, according to a poll conducted in October 2014 by the Yale University Project on Climate Change Communication.[1435]

Opposing the Pope's encyclical and call for action were evangelical Christians, and conservative talk show hosts and Republican politicians, many of whom tried to link the Pope's message to socialism. Greg Gutfeld, a Fox News pundit, called the Pope, "The most dangerous person on the planet," and ignorantly stated, "All he needs is dreadlocks and a dog with a bandana and he could be on Occupy Wall Street."[1436] Rush Limbaugh reiterated his belief the Pope is a Marxist and incorrectly declared, "Every other word [in the Encyclical] seems to be about how unfettered capitalism is destroying the world and how the rich countries have to give more money to the poor countries to make amends."[1437] Tom Altmeyer of Arch Coal, the nation's second largest coal mining company, said if the Pope really believed in social justice, instead of claiming there is global warming caused by humans, he would talk about the value of fossil fuel energy.[1438] Sen. James Inhofe (R-Okla.), who doesn't believe there is global warming, said the Pope "ought to stay with his job" and not comment upon environmental policy.[1439] At a town hall meeting in New Hampshire, former Florida Gov. Jeb Bush, a declared candidate for the presidency, said, "[R]eligion ought to be about making us better as people and less about things that end up getting in the political realm."[1440] Sen. Ted Cruz (R-Texas) and former Sen. Rick Santorum (R-Pa.), devout Catholics and major candidates for the Republican presidential nomination, like Inhofe, call global warming a "hoax."

The Cornwall Alliance, which defines itself as "an evangelical voice promoting environmental stewardship and economic development built on Biblical principles,"[1441] believes, "There is no convincing scientific evidence that human contribution to greenhouse gases is causing dangerous global warming."[1442] It also believes fossil fuels "provide the abundant, affordable energy necessary to sustain prosperous economies or overcome poverty."

U.N. Secretary-General Ban Ki-Moon, addressing a climate change summit, which had been called by Pope Francis in April 2015, declared:

> "Climate change is intrinsically linked to public health, food and water security, migration, peace and security. It is a moral issue. It is an issue of social justice, human rights and fundamental ethics.
> "We have a profound responsibility to the fragile web of life

305

on this Earth, and to this generation and those that will follow. . . . That is why it is so important that the world's faith groups are clear on this issue—and in harmony with science. Science and religion are not at odds on climate change. Indeed, they are fully aligned. Together, we must clearly communicate that the science of climate change is deep, sound and not in doubt.

"Climate change is occurring—now—and human activities are the principal cause. . . .

"The facts of climate change are upheld by the Intergovernmental Panel on Climate Change and the major scientific bodies of every government in the world, including the Vatican's Pontifical Academy of Sciences.

"Our response has to be global, holistic and rooted in universal values.

"Climate change affects us all, but not equally. Those who suffer first and worst are those who did least to cause it: the poor and most vulnerable members of society."[1443]

Former U.N. Secretary-General Kofi Annan also praised the Pope for his "inspired leadership," and stated, "Climate change is an all-encompassing threat: it is a threat to our security, our health, and our sources of fresh water and food. Such conditions could displace tens of millions of people, dwarfing current migration and fueling further conflicts."[1444]

Every major religion has a basic tenet to protect and preserve the environment. Many of the major Eastern religions, including Hinduism, Taoism, Shinto, and Buddhism consider all life as interdependent. The responsibility of government, according to Buddhism's *Kutadanta Sutta*, is to actively protect the environment, and all its flora and fauna.[1445] The Koran of Islam warns, "And do not corrupt in the earth after being tilled." Saudi Arabia in 1994, long before much of the world began to understand the long-term effects of uncontrolled gas emissions, cautioned, "Human activities over the last century have so affected natural processes that the very atmosphere upon which life depends has been altered."[1446]

All indigenous people, from the aborigines of Australia to the Native Americans of North America, have shown respect for the land, which most believe is not theirs to own, but only to enjoy until passed to their children.

Bolivia is the only member of the United Nations to declare that in addition to the rights of all people there are also legal rights for the protection of Mother Earth. In *Defensoría de la Madre Tierra*, passed in 2010, Bolivia defined the living Earth as, "[T]he dynamic living system formed by the indivisible community of all life systems and living beings whom are interrelated, interdependent, and complementary, which share a common destiny." Among the seven specific rights are those to protect the water and air, and to live free of contamination.

Some members of Western religions have interpreted *Genesis 1:28* as God giving mankind dominion over all life and the Earth. For many corporations and politicians, this means mankind has the right to drill and use Earth's resources however they see fit, that fracking is God's gift to humanity. However, the Rev. Dr. Leah Schade, whose doctoral work was in ecological theology, disagrees. A better interpretation of *Genesis* is not that God gave mankind "dominion," but "stewardship" of what He created, says Dr. Schade. She suggests:

> "All entities—individual, corporate, governmental, and community—must restrain against those practices and human laws that bring harm to God's Creation and the human community; we must hold a mirror up to our own economic and material conscience and the reality of the ways in which we (individually and communally) are 'curved in upon ourselves,' thinking only of our desire for cheap energy that results in harming environmental and human health, or thinking of our desire for the accumulation of wealth and comfort at the expense of others and God's Creation.
> "We must raise questions about the lures of 'easy money' that comes from the shale gas and oil boom. At the same time, we must raise concern about the ethics of individuals or corporations benefitting so handsomely while God's Creation and others in society bear the costs."

Synagogues, churches, mosques, and other houses of worship often host meetings of anti-fracking organizations; religious leaders have often written letters to the editor or OpEd commentaries to explain why fracking—and a refusal by some politicians to believe the reality and long-term effects of man-made climate change—are counter to religious principles.

The Eco-Justice Programs of the National Council of Churches, a coalition of about 100,000 congregations with 45 million members,[1447] took a forceful stand on environmental protection:

> "The ways in which we generate electricity have serious implications for the health of our neighbors and all Creation. Many forms of energy, when used to generate electricity or power machinery, also pollute our air, land, and water. Renewable sources of energy are cleaner and more enduring but are still relatively expensive. And our methods of transmitting energy from place to place waste this precious resource. . . .
>
> "We fail to acknowledge that some forms of energy, like gas and oil, are finite. And, we fail to make energy available to everyone in order for them to meet basic needs. We also have not made good choices regarding how and where we get our energy. From extraction to production to transmission, the decisions we make about energy have consequences, sometimes devastating, on God's Creation and on the health and well-being of our neighbors.
>
> "God's land and water provide rich and valuable sources of energy—particularly oil. However, in order to extract the oil from oceans or land we often put the needs of ourselves over the health and well-being of the whole of Creation and in many cases before the needs of future generations.
>
> "Our heavy reliance on fossil fuels adds millions of pounds of harmful pollutants to the air every day. . . .
>
> "In addition to the potential to pollute ground and surface water supplies, communities dealing with fracking also may experience the 'boom and bust' impact that drilling a non-renewable energy source will have on the long-term health of their local economy."[1448]

In 1998, a coalition of several Episcopal churches formed Episcopal Power & Light (EPL) to help member churches purchase solar energy. Two years later, EPL broadened its focus and membership, becoming Interfaith Power & Light. More than 15,000 congregations of all faiths in 40 states[1449] are now part of IPL, which has developed a strong educational campaign to help people better understand the issues of global warming and the problems of fossil fuel extraction. The IPL mission "is to be faithful stewards of Creation by responding to global warming through the promotion of energy conservation, energy efficiency, and renewable energy [and to] protect the earth's ecosystems, safeguard the health of all Creation, and ensure sufficient, sustainable energy for all."[1450] As part of its mission, IPL helps congregations adapt renewable energy.

The Pennsylvania IPL chapter outlines why it won't condone horizontal fracking:

> "We serve God through establishing justice—and economic gains that come at the expense of harming others are unjust. Many towns in Pennsylvania have already gone through one or more cycles of boom and bust from oil and coal production. Typically, these cycles have brought riches to few but lasting economic and social problems to many, ranging from depressed economies to scarred and infertile lands. So far, the Marcellus Shale developments, especially without taxes or impact fees in place, seem more likely to continue this destructive pattern than to break from it. In addition, illegal or ethically questionable practices by drilling companies have set neighbor against neighbor."[1451]

Among the tenets of the General Resolution on Environmental Justice (1994) of the Unitarian Universalist Association (UUA) is that the church will:

> "promote programs for social, economic, and political empowerment so that all people may join together in one struggle for peace, justice, and sustainable development; support the development of democratic and ecologically responsible community organizations, labor unions, and business cooperatives; [and] develop religious education and community action programs honoring cultural and religious diversity and connecting environmental issues to other social justice concerns."[1452]

Two decades after the UUA approved the general resolution, two of the church's leaders issued a commentary strengthening that position. "Oil and other fossil fuels are making the planet uninhabitable. We must work urgently to switch to cleaner alternatives and to convince our leaders to work toward that end as well," wrote the Rev. M. Linda Jaramillo, executive minister for Justice and Witness Ministries; and Meighan Pritchard, minister for environmental justice.[1453]

Although the 800,000-member church hasn't developed a general philosophy about fracking, several congregations "are engaging in the issue," says Jessica Halperin, UUA program associate.

"Our immoderate use of the Earth's resources violates the entire biosphere, threatening the lives of millions of people and the habitats of thousands of species" is part of the policy of environmental and climate change concern of the Society of Friends (Quakers).[1454]

The Evangelical Lutheran Church in America (ELCA) and its predecessor bodies have a long-standing and honorable history of engaging in politically-charged issues routinely proclaiming a public theology that takes seriously Jesus' call to care for "the least of these" (*Matthew 25:31-46*) and his model of engaging publicly with those who control the power and wealth of a society (*Matthew 21:12-13*).

The ELCA Social Statement, "Caring for Creation," offers the following observation:

> "In our captivity, we treat the earth as a boundless warehouse and allow the powerful to exploit its bounties to their own ends (*Amos 5:6-15*). Our sin and captivity lie at the roots of the current crisis."[1455]

Several Lutheran synods, following the principles enunciated in the ELCA Assembly Report, have taken bold stands to develop resources about the environmental and health impacts of fracking, and to challenge it when necessary. The New England Synod, which represents 130 congregations, voted for a moratorium "until such a time as it can be shown that the process is developed to the point of adhering to the standards of the Clean Air and Clean Water Acts." The Northwest Washington Synod recommended the National Council name "the fossil fuel extraction and production industry as one which damages the environment."

The Upper Susquehanna Synod in the heart of Pennsylvania's Marcellus Shale called for a repeal of all environmental and health exemptions for the oil and gas industry. The decision was made in June 2014 following a two year investigation of fracking by a committee composed of pastors, scientists, teachers, lay leaders, and several individuals who work in the oil/gas industry. "Some of us would like to see a total ban on fracking," says the Rev. Dr. Schade, "while others argued fracking could be safe if there was stronger regulation." But all congregations, she says, "believe the loopholes created for the industry that exempt it from established laws protecting our water, air, and public health are unjust and need to be repealed."

Not all ECLA congregations are in agreement. Seventeen congregations from western Pennsylvania operate Camp Agape, a 257-acre summer camp for children. In 2008, the congregations leased mineral rights under the camp to Range Resources. "It's been a good experience for us [but] not without some worry and trepidation," Charles Wingert, a member of the board of directors told the Associated Press. The reason to lease the land was financial—it helped fund the program.[1456]

JEWISH VIEWS OF THE ENVIRONMENT

Four Jewish summer camps in the extreme northeast part of Pennsylvania signed mineral leases in 2008 and 2009 with the Hess Corp. They received about $400,000 for signing away oil and gas rights. With bonus payments and royalties if full production began, it could have yielded more than $4 million, according to the *Jewish Daily Forward*.[1457] The leases called for wells to be no less than 500 feet from any structure on the camps.

"This thing is so much bigger than we are," Leonard Robinson, executive director of the New Jersey YMHA–YWHA, which operates two of the camps, told the *Forward*. However, Rabbi Daniel J. Swartz of Temple Hesed in Scranton and vice-president of Pennsylvania Interfaith Power and Light, explained, "[O]nce you sign . . . you're contributing to those businesses and their profits and all the things that they're standing for."[1458] Rabbi Arthur Waskow called the decision by the camps' owners, "A profound violation of Jewish wisdom and values [that] will poison God's and humanity's earth, air, food, and water."[1459]

In September 2012, while politicians and the oil/gas industry were holding a convention in Philadelphia to tout the benefits of fracking, Rabbis Waskow and Mordechai Liebling conducted a blessing of the waters to call for clean, sustainable energy to replace shale gas extraction and fossil fuel energy.[1460]

Almost three years later, 336 rabbis of all denominations signed a letter, co-authored by Rabbi Waskow, that first discussed religious views about the environment, and then argued, "Great Carbon Corporations not only make their enormous profits from wounding the Earth, but then use these profits to purchase elections and to fund fake science to prevent the public from acting to heal the wounds."[1461] The letter identified fracking, coal burning, tar sands extraction, and offshore oil drilling as threats to the environment, and called for "a new sense of eco-social justice."[1462]

In their 6,000 year covenant with God, the Jews have considered themselves as stewards of the Earth. In *Genesis 2:15* is the requirement to care for the Earth. In *Ecclesiastes 7:13*, the Jews are told by God, "See to it that you don't spoil or destroy my world—because if you spoil it, there is nobody after you to fix it." In the 14th century, Talmudic scholar Rabbi Isaac ben Sheshet added strength to the command to care for the Earth. Based upon the writings of the *Torah* and subsequent discussions by Jewish leaders, he observed that mankind is forbidden "from gaining a livelihood at the expense of another's health."[1463] In 2000, the Central Conference of American Rabbis declared, "For 25 years, the organized American Jewish community has unanimously advocated action to reduce our nation's reliance on fossil fuels through energy conservation and the development of environmentally sound, non-nuclear energy technologies."[1464] The Commission on Social Action of Reform Judaism notes that Jews are "increasingly aware of the potentially negative environmental impact of extracting, transporting and burning fossil fuels."[1465] The Union for Reform Judaism called for a repeal of the oil/gas industry exemptions from several federal environmental laws and the "pursuit of economic incentives to reduce the dependence on fossil fuels."[1466]

The founding principal of the Coalition on the Environment and Jewish Life (COEJL)—founded by the Jewish Council for

Public Affairs, the Religious Action Center of Reform Judaism, and the **Jewish Theological Seminary of America**—is, "It is our sacred duty as Jews to acknowledge our God-given responsibility and take action to alleviate environmental degradation and the pain and suffering that it causes [and to] reaffirm and bequeath the tradition we have inherited which calls upon us to safeguard humanity's home."[1467]

In 2012, COEJL called for the Jewish community to:

• Support adequate federal and state regulation to protect groundwater sources, surface water sources, air quality, human and animal health, infrastructure and ecosystems.
• Support federal legislation to eliminate the natural-gas industry's exemption from the Safe Drinking Water Act.
• Support legislation and regulation enabling the EPA to require full disclosure of the type and amount of hydrofracking chemicals used at each well site.
• Educate their communities about the extraction of natural gas and oil by hydrofracking and about relevant Jewish perspectives on the issue.
• Support preservation of unique and/or sensitive areas by putting them off limits to gas drilling to be determined by an appropriate science-based process.[1468]

GRAPHIC: Will Sweeney/EcoWatch

313

CHAPTER 17
Bans and Moratoriums

Because of significant questions about health and pollution issues related to fracking, several countries—among them Bulgaria, France, Germany, Ireland, Luxembourg, the Netherlands, Scotland, and Tunisia—banned or placed moratoriums on all fracking operations. The Bulgarian parliament voted for the moratorium in 2012 following massive public protests by citizens who protested the government signing a five-year $68 million deal that gave Chevron significant gas rights.[1469] The German moratorium extends to 2021. The Czech government is seriously considering a ban. Fracking in several other European countries, however, may be a moot issue. Test wells in several countries reveal that extracting gas may not be commercially feasible.

Almost two-thirds of Canadians oppose fracking, according to a poll conducted in January 2012 by Environics Research.[1470] Quebec, British Columbia, New Brunswick, Newfoundland and Labrador, and Nova Scotia have banned fracking pending full studies.

In April 2011, South Africa banned Shell Oil from using fracking to extract natural gas in the Karoo Desert.[1471] "[F]racking fluid *will* contaminate the groundwater. There is not doubt at all," said Dr. Gerrit van Tonder of South Africa's Institute for Groundwater Studies at University of the Free State.[1472] However, in September 2012, the South African government lifted that moratorium on exploration following a study that correlated safe extraction with the reality that the country has about 485 trillion cubic feet of natural gas, most of it in the Karoo Desert, and drilling would boost economic recovery and lower oil dependency.[1473]

After visiting the United States in Spring 2013, and talking with people affected by fracking, Piers Verstegen, director of the Conservation Council of Western Australia, told the *Perth Sunday Times:*

> "I met sick families who could not move off their land. I saw gas bubbling out of natural groundwater springs that could be set alight. I was told stories of sick and dying livestock. I met regulators who were turning a blind eye and I saw the gas industry's armed security guards preventing farmers access to their own land.
>
> "After seeing this I vowed I would do whatever I could to stop this happening in Western Australia. I have no doubt that gas fracking is one of the most serious environmental challenges facing Western Australia."[1474]

The New York Ban

Unlike Pennsylvania, which is rushing like a runaway train about to derail and can't slow down, New York state, after a contentious six-year series of moratoriums, finally banned fracking, effective July 2015.

The decision by Gov. Andrew Cuomo, based upon a 184-page analysis from the state's Department of Health,[1475] was preceded by an extensive public education campaign by environmentalists, health care professionals, and bans or moratoriums by 160 New York municipalities. The statewide ban will extend to 2019, when the governor's second term ends.

Five years before Gov. Cuomo's decision, New York City had asked the state to ban all natural gas drilling in the watershed. Fracking presents "unacceptable threats to the unfiltered fresh water supply of nine million New Yorkers," said Steven Lawitts, commissioner of the city's Department of Environmental Protection, in December 2009. Mayor Michael Bloomberg agreed; a representative of the mayor told Reuters, "Based on all the facts, the risks are too great and drilling simply cannot be permitted in the watershed."[1476]

The New York state moratorium had begun in 2008; it was extended until July 1, 2011, following majority votes in the state senate and assembly. Signing the legislation in December 2010, Gov. David Patterson said New York "would not risk public safety or water quality."[1477] The moratorium was continued under the administration of Gov. Cuomo, who was subjected to extensive lobbying by all sides of the issue. At almost every public appearance, Gov. Cuomo was faced by non-violent anti-fracking protestors.

The Religious Action Center of Reform Judaism, reflecting the views of those opposed to fracking, had called for a continuation of the moratorium, pointing out, "The regulation standards in place are not rigorous enough to ensure that hydraulic fracturing and its attendant contaminated water disposal will be conducted safely and pose no risk to our communities."[1478]

There was also another issue that helped unite New Yorkers to oppose fracking. In Pennsylvania, most shale drilling was being done in rural and agricultural areas; New York's Southern Tier is suburban, affluent, and has a high population density.

In late Summer 2012, Sean Lennon and Yoko Ono created Artists Against Fracking, primarily to gather public support to continue New York's moratorium. Among more than 200 artists opposed to fracking are Alec Baldwin, Jackson Brown, David Crosby, Robert DeNiro, Anthony Edwards, Jimmy Fallon, Carrie Fisher, Roberta Flack, David Geffen, Daryl Hannah, Anne Hathaway, Lady Gaga, Richard Gere, Jude Law, Paul McCartney, Julianne Moore, Graham Nash, Gwyneth Paltrow, Bonnie Raitt, Tim Robbins, Mark Ruffalo, Todd Rundgren, Susan Sarandon, Ringo Starr, Martha Stewart, Uma Thurman, Liv Tyler, Debra Winger, and Bethany Yarrow. Among the groups that are a part of the Artists group are the Beastie Boys, Black Keys, Indigo Girls, Patti Smith Group, The B-52s, and White Out.

Energy in Depth (EID), the oil and gas industry's highly-funded PR front, which uses personal attacks as a prime method of rebuttal, attacked the celebrities. EID claimed Sean Lennon was nothing more than a "struggling artist" who was "using an anti-natural gas agenda to promote himself."[1479] It attacked the group of high-profile musicians, actors, and others in the creative arts, most of whom had long histories of involvement in social issues, as "artists looking for relevance" who were "regurgitating outdated and tired information."[1480]

The creative artists "who have turned into activists on this issue are being led by those who are more comfortable twisting the facts and taking part in street theater, stunts and gimmicks," said Brad Gill, president of the industry-front group Independent Oil and Gas Association of New York.

Reciting the energy industry's dogma of prosperity through fracking, Gill said the Artists "are ignoring the prosperity and environmental protection that modern natural gas development is bringing to many other states."[1481]

On Jan. 17, 2013, Artists Against Fracking took a chartered bus into Pennsylvania's Susquehanna County to look at well sites and talk with the people who were affected by the drillers. Trailing the celebrities were the media and representatives of the natural gas industry. Rebecca Roter, of Kingsley, Pa., one of the activists, explained:

> "The celebrities had come . . . to hear the voices that have been marginalized by media, by industry spin, by the state of Pennsylvania. They did not come for a sound bite or to take a photograph of flaming tap water—they wanted to reach out to us, to the people living with fracking, and to see for themselves how our lives have been affected by shale gas extraction. . . .
>
> "Yoko Ono, Sean Lennon, Susan Sarandon, and Arun Gandhi [grandson of Mohandas Gandhi] accomplished more in one day to draw national attention to the human cost of shale gas extraction than our collective voices have in six years of confronting natural gas development in Susquehanna County. Local news reports about the celebrities' tour of Susquehanna County gasfields finally included the voices and personal stories of local residents speaking about the human cost of shale extraction as they have experienced it. The press had to cover that story because the celebrities about whom they thronged were listening to us."[1482]

After talking with residents, Susan Sarandon said, "If it's been decided that these people are expendable, and that the people in this area are expendable, there's nothing to stop [the industry] from thinking that they can sacrifice other people in other places."[1483]

Arun Gandhi observed, "We are committing violence against nature, against resources, against environment and eventually this is going to destroy us, destroy humanity."[1484]

As expected, EID had its "take" on the bus tour. Tom Shepstone said the celebrities "just come up here to pick on this area and use it as part of their trendy cause."[1485] He made sure the media also heard his opinion that "It's definitely a celebrity tour, it's a stunt, and I have to wonder if it's not connected with Sean Lennon releasing an album yesterday."[1486]

'RECKLESS AND IRRESPONSIBLE AND DISHONEST'

Of Americans with wide-spread name recognition, Robert F. Kennedy Jr. is among the most vocal opponents of fracking. A month before New York's Department of Environmental Conservation appointed him to an advisory panel on fracking, Kennedy charged:

> "[T]he natural gas industry has been reckless and irresponsible and dishonest with the American public and they've lost much of their credibility. . . .
> "[B]ecause of the lack of candor by the industry, because of their reckless behavior, it's unclear whether we can get that natural gas out of the ground without causing cataclysmic environmental damage."[1487]

EID had previously attacked Kennedy when it unsuccessfully tried to ingratiate itself with the residents of the gas fields: "[They] have put up with a continuous barrage of insults and hyperbole over the last three years as . . . Bobbie Kennedy, Jr. and like-minded charlatans have abused them in the name of natural gas obstructionism."[1488] Apparently, EID erroneously believed that using both a diminutive and feminine form of Kennedy's first name somehow defused his credibility.

Public protests, a petition with more than 260,000 signatures,[1489] and more than 80,000 public comments,[1490] along with Kennedy's strong opposition, may have convinced Gov. Cuomo, who had been leaning to allow fracking under certain circumstances, to continue the four-year moratorium and conduct yet another study, specifically directed to investigating the public health effects.[1491]

New York state attorney general Eric Schneiderman said in June 2013 he was "surprised the governor hasn't done anything yet, and I attribute it to the political organization of the anti-frackers." Schneiderman said the activists "have out-organized the oil and gas industry."[1492]

Pushing Gov. Cuomo and the state's Department of Environmental Conservation to lift the moratorium, especially in the Southern Tier counties, and not allow any local jurisdiction to forbid fracking was the Washington Legal Foundation (WLF). The WLF, a conservative non-profit organization, says its primary goal is "to defend and promote the principles of freedom and justice."[1493] However, the Center for Media and Democracy, an independent investigative journalism organization, disagreed. The Center identified the WLF as having been established to "fight activist lawyers, regulators, and intrusive government agencies at the federal and state levels, in the courts and regulatory agencies across the country."[1494] The WLF believes, "Environmental extremist groups, activist courts, and the media foster the false notion that the environment is always threatened by economic activities."[1495] Dory Hippauf's "Connect the Dots" series details significant ties between WLF and Big Energy.[1496]

Karen Moreau, executive director of the New York State Petroleum Council, a part of the American Petroleum Institute, continued the industry's philosophy of hyperbolic rants upon those who disagreed with the benefits of fracking. In a prepared statement, Moreau, former general counsel for the New York Senate Republicans, stated:

> "As the largest trade association for the oil and gas industry in the world, representing companies which stand ready to invest in New York for the long term, the NYS Petroleum Council, API and its member companies will not be deterred by the tactics or fraud perpetrated upon the public by radical environmentalists, and renegade groups which would like nothing better than to shut down NY to the benefits natural gas development is bringing over 30 other states. We will vigorously continue to educate the public on the environmental and economic benefits that safe and responsible natural gas development can offer the citizens of NY."[1497]

Former Pennsylvania Gov. Ed Rendell, who first approved fracking in Pennsylvania, turned to the news media to try to convince Gov. Cuomo to end the moratorium. In an op-ed commentary in the *New York Daily News*, Rendell argued:

> "If we choose to embrace natural gas, it will help us get past a

number of significant economic and environmental challenges. On the other hand, if we let fear carry the day, we will squander another key moment to move forward together. . . .

"New York has a healthy band of vocal critics right now who continue to push a false choice: natural gas versus the environment.

"But as the former Democratic governor of a major natural gas-producing state, I know we can enjoy the benefits of gas production while also protecting the environment."[1498]

What Rendell didn't disclose were his ties to the oil and gas industry. Josh Greenman, *Daily News* opinion editor, told ProPublica if he had known about the possible conflict-of-interest, "I certainly would have disclosed that and conceivably would have made a different judgment on the piece."[1499]

'THE POTENTIAL RISKS ARE TOO GREAT'

The Binghamton (N.Y.) *Press and Sun–Bulletin* strongly objected to opening the state to fracking, editorializing: "If New York allows horizontal hydrofracking, it faces many of the challenges that produced the astonishing failures in Pennsylvania's oversight of shale gas wells. [Aug. 9, 2014][1500]

The Albany (N.Y.) *Times–Union*, the state's capital city newspaper, agreed with anti-fracking activists to continue the moratorium. In a strong editorial, the *Times-Union* outlined the problem—and then the solution:

> "The spin from the natural gas drilling industry has been so rosy for so long that an energy executive's recent acknowledgement that things have not been as good as the ads suggest must seem like refreshing candor.
>
> "Unfortunately, it only affirms what many New Yorkers have suspected all along: that there is every reason to question the industry's claims about hydraulic fracturing, or fracking. Nor is there any reason to believe that the industry's seeming frankness is anything but the latest spin.
>
> "Let's review the record.
>
> "First the gas industry sought to dismiss concerns about the new fracking process. It distorted the truth by declaring that fracking has been around for decades. The full story is that high volume, horizontal hydrofracking is a new and dramatically different version of the low-volume, vertical

method that had long been used. It covers a vastly larger area, and pumps millions of gallons of water and chemicals into the earth to fracture rock and release gas. To suggest the new process is the same as the old is like saying there's no difference between a bamboo fishing pole and a mile-long commercial net.

"The industry initially told New Yorkers that the chemicals it wanted to pump into the ground were a secret, proprietary mix. When it became clear that answer was unacceptable, industry leaders offered that there was nothing worse in the recipe than what's in a typical kitchen cabinet. When people pointed out that there's a lot under our sinks no human ought to drink, they were told it was only a tiny percentage of the fracking fluid. Then, when the millions of gallons that tiny percentage represented became apparent, we were assured it was too far underground to worry about the stuff getting into drinking water.

"Except when it did. Then the industry blamed poor well construction, not the drilling, as if those are unrelated. And it only happened once, after all. Then twice. Well, more times, really.

"Now the gas industry seems to have a new line: Mistakes were made. . . . So much for a process New Yorkers were assured was perfectly safe.

"We would like to see clean, safe natural gas extraction in New York. But given its track record, we simply don't trust a gas industry that time and again has given us the sense that it's pulling one over on us.

"All the more reason that New York needs a moratorium on drilling, so that the health and environmental concerns about it can be studied without the constant pressure of gas industry lobbyists and other advocates who keep assuring us that all is well, if we'll just take their word for it." [June 3, 2013][1501]

In December 2014, the Department of Health issued the long-awaited 184-page analysis of the effects of fracking.

"We cannot afford to make a mistake. The potential risks are too great [and] are not even fully known," said Dr. Howard A. Zucker, the state's acting commissioner of health. Citing extensive research about the impacts of fracking upon air, climate change, drinking water, surface water contamination, the community, and of the relationships between fracking and earthquakes, Dr. Zucker concluded:

"There are numerous historical examples of the negative impact of rapid and concentrated increases in extractive resource development (e.g., energy, precious metals) resulting in indirect community impacts such as interference with quality-of-life (e.g., noise, odors), overburdened transportation and health infrastructure, and disproportionate increases in social problems, particularly in small isolated rural communities where local governments and infrastructure tend to be unprepared for rapid changes. . . .

"The dispersed nature of the activity magnifies the possibility of process and equipment failures, leading to the potential for cumulative risks for exposures and associated adverse health outcomes. Additionally, the relationships between HVHF [high volume horizontal fracturing] environmental impacts and public health are complex and not fully understood. Comprehensive, long term studies, and in particular longitudinal studies, that could contribute to the understanding of those relationships are either not yet completed or have yet to be initiated. In this instance, however, the overall weight of the evidence from the cumulative body of information contained in this Public Health Review demonstrates that there are significant uncertainties about the kinds of adverse health outcomes that may be associated with HVHF, the likelihood of the occurrence of adverse health outcomes, and the effectiveness of some of the mitigation measures in reducing or preventing environmental impacts which could adversely affect public health.

"While a guarantee of absolute safety is not possible, an assessment of the risk to public health must be supported by adequate scientific information to determine with confidence that the overall risk is sufficiently low to justify proceeding with HVHF in New York. The current scientific information is insufficient. Furthermore, it is clear from the existing literature and experience that HVHF activity has resulted in environmental impacts that are potentially adverse to public health. Until the science provides sufficient information to determine the level of risk to public health from HVHF and whether the risks can be adequately managed, HVHF should not proceed in New York State."[1502]

At a cabinet meeting to present the Public Health Impact report, Dr. Zucker asked: "Would I live in a community with [fracking] based on the facts that I have now? Would I let my child play in a school field nearby? After looking at the plethora of reports behind me . . . my answer is no."[1503]

Gov. Cuomo said Dr. Zucker's findings were "powerful," and that he agreed if "your children should not live [near fracking], then no one's child should live there."[1504] The governor had several times previously emphasized he would make his decision not upon politics or the effects of lobbying by all sides, but solely upon the independent evidence and scientific literature.

The response from environmentalists, fractivists, and those in the public health professions was ecstatic about the ban, but with mild surprise; most believed that Gov. Cuomo might lift the moratorium and allow strongly-regulated limited drilling in the Southern Tier.

"Future generations will point to this day and say, 'This is when the tide began to turn against the dirty, dangerous and destructive fossil fuel industry,'"[1505] said Deborah Goldberg, an attorney representing EarthJustice and one of the strongest advocates for banning horizontal fracking.

The *Washington Post*, backing Maryland Gov. Martin O'Malley's recent decision to allow fracking into his state, attacked Gov. Cuomo, arguing, "Fracking should be well-regulated and legal, not permanently off-limits." The *Post* claimed, "The benefits of burning the fuel are just too attractive," and stated that natural gas energy "can be an important part of a transition to more sustainable fuels.[1506] However, like many, the *Post* editorial board failed to understand that the key problem was not the use of natural gas as a transition energy to renewable energy but the method itself, which would have significant environmental and health problems even if well-regulated.

'INFLUENCED BY A BUNCH OF EXTREMISTS'

The response to the New York ban from the industry was a outrage and hyperbole, focused upon the governor's political beliefs and the believed economic benefits of drilling.

Thomas West, an attorney who represented the oil and gas industry in their battles against municipalities that had enacted fracking moratoriums, said the industry probably would not take the state's ban into the court system "to fight this bizarre decision." The industry, he said, "has already written off New York as being influenced by a bunch of extremists."[1507]

Gov. Cuomo's decision was a "politically motivated and equally misinformed ban on a proven technology used for over 60 years," said Karen Moreau, executive director of the New York State Petroleum Council,[1508] disregarding the environmental and health effects studies, while continuing the industry's favorite lie that mixed vertical fracking, which had been used for six decades, with horizontal fracking that was in use less than a decade.

"For Andrew Cuomo, this was a political exercise from the start, as he cowers in fear of the environmental left and fights [New York City Mayor] Bill de Blasio for the soul of the Democratic Party," added David Laska,[1509] communications director of the New York Republican State Committee. Ed Cox, the Republicans' chair, agreed: "This study was a political charade from the start. Andrew Cuomo has given into the radical environmental Luddites in his own party."[1510]

Cox, who owns shares in Noble Energy, which had done test drilling in New York, also tore into Gov. Cuomo for what he saw as abandoning those who could profit from drilling: "To New Yorkers across upstate struggling for economic growth: New York's governor has failed you." The people in the Southern Tier, said Karen Moreau, "had to watch their neighbors and friends across the border in Pennsylvania thriving economically. It's like they were a kid in a candy store window, looking through the window, and not able to touch that opportunity."[1511]

Dan Fitzsimmons, president of the Joint Landowners Coalition of New York, said the governor, "Once again turned his back on the hard working men and women of the Southern Tier. opting instead to appease the state's environmental extremists for his own political gain." He erroneously claimed Gov. Cuomo "rejected lower taxes, lower utility bills, job creation, business growth, clean affordable energy and domestic power generation."[1512]

State Sen. Tom O'Mara claimed Gov. Cuomo's decision, "Eviscerates the hope of so many Southern Tier farmers, landowners, businesses and potential jobs in the natural gas industry." U.S. Rep. Tom Reed claimed:

"This decision makes it even more difficult to replace the good jobs that have already left due to New York's unfriendly business climate. Once again Albany shows that it wants to enact an extreme liberal agenda rather than care about individual property rights and job opportunities."[1513]

Only one-fourth of New Yorkers believed jobs creation would be permanent.[1514] Also in rebuttal, Svante Myrick, mayor of Ithaca, N.Y., which had previously banned fracking on city-owned property,[1515] said the Southern Tier's "economy is strong because of our agriculture, tourism and education," and pointed out a reality: "Drilling pads dotting our hills, trucks pounding our roads, and the fear of soil and water contamination would threaten our success in all of those areas." The gains, said Myric, "would be short-term." [1516]

Although the industry, conservative politicians, and chambers of commerce opposed the ban, the people approved. A Siena Research Institute poll, conducted in June 2013, had revealed 47 percent of New York voters opposed fracking; 37 percent favored it. Siena pollster Steven Greenberg pointed out:

"There is a clear partisan divide as it is supported nearly two-to-one by Republicans, opposed two-to-one by Democrats and divides independents virtually evenly. A majority of upstaters and plurality of New York City voters oppose fracking, while a plurality of downstate suburbanites wants to see fracking move forward."[1517]

Shortly before Gov. Cuomo announced his decision for a permanent ban, 79 percent of New Yorkers supported a moratorium and 56 percent opposed fracking, with only 35 percent supporting it, according to an impartial poll conducted by Franklin, Maslin, Maullin, Metz & Associates (FM3).[1518] A Quinnipiac College poll, conducted the week of the decision to ban fracking, revealed only 30 percent of residents in upstate New York opposed the ban; only 19 percent of the residents of New York City opposed it; only 27 percent of the residents in New York suburbs opposed it.[1519]

Even with the ban, New Yorkers would still have to deal with the effects of increased high-pressure natural gas pipelines, compressors, the storage facilities in the Finger Lakes region, trains carrying volatile oil from North Dakota into the state, and the deposit of wastewater and other byproducts from Marcellus Shale drilling. The *Poughkeepsie Journal*, in an editorial, outlined the problem:

> "The state took years before making the far-reaching decision to ban fracking in New York. It would be remarkably dumb to make such a dramatic decision only to continue to allow the byproducts of the practice to be transported and then dumped into the state in ways that could foul the environment and ruin the water." [Feb. 18, 2015].[1520]

Banning Fracking in Maryland

Gov. Martin O'Malley (D) issued an executive order to delay drilling in Maryland until a special commission determined if fracking could be conducted safely. However, oil and gas energy lobbyists influenced enough members of the legislature that it refused to fund that study. A resolution endorsed by several Maryland environmental, political, and civic organizations urged, "This statutory moratorium should only be permitted to expire if and when detailed and transparent studies prove that fracking activities will not cause harm to our public health, rural communities, natural environment, and global climate." The moratorium was scheduled to expire at the end of Gov. O'Malley's term in January 2015.

Three months earlier, the *Baltimore Sun* had suggested:

> "The longer we wait before embracing fracking, the better informed we will be. Protection of the health and welfare of people living in this state and the preservation of the environment ought to be weighed more heavily than the exploitation of a resource that is not going away. Maryland's quest to determine the 'best practices' for regulating hydraulic fracturing may not be sufficient with so much still not known about its consequences. As others have pointed out, the burden in this debate is on advocates of hydraulic fracturing to prove that these myriad concerns can be addressed before the practice is ever allowed in Maryland. In other words, prove it's harmless and then we'll talk." [Oct. 7, 2014][1521]

Following the election in November 2014, Larry Hogan, a Republican who believes in fracking, the *Baltimore Sun* editorialized: "What the next governor ought to be examining more closely, given the relatively modest and fleeting benefits of fracking versus the potential harm to local businesses, is whether these risks ought to be tolerated at all." [Nov. 27, 2014][1522]

Within a month of Gov. Hogan's inauguration, the *Frederick News–Post* cautioned, the state "should follow in New York state's footsteps and ban fracking, then acknowledge there are some kinds of economic development for which the price paid is outweighed by the ramifications to our health and environment." [Feb. 12, 2015][1523] A large public and media campaign, including a 30-second radio ad narrated by actor Edward Norton,[1524] to temporarily block fracking while examining health and environmental effects, led to overwhelming support in the state legislature. The Maryland House, by a 103–33 vote, and the Senate, by a 45–2 vote,[1525] called for a continuation of a moratorium until Oct. 1, 2017. By choosing not to sign or veto the bill, Gov. Hogan allowed it to become law.

The New Jersey Moratorium

The New Jersey legislature passed a permanent ban in 2012. However, Gov. Chris Christie signed a conditional veto to reduce that ban to one year. That ban expired in January 2013. The Legislature—by a 30–5 vote in the Senate and a 56–19 vote in the Assembly—in June 2012 had also banned disposal of fracking wastewater, targeting the possibility that companies operating in Pennsylvania would transport much of the wastewater to New Jersey. However, Gov. Christie vetoed it three months later, stating the legislation violated the Constitution's Commerce Clause, and argued:

> "Despite the vigorous public debate that surrounded last legislative term's fracking legislation and the absence of consensus on the merits of the drilling technique, there was one fact on which there could be no debate and on which the Legislature and I fundamentally agreed . . . Hydraulic fracturing is not occurring and is unlikely to occur in New Jersey in the foreseeable future."[1526]

The veto wasn't justified, says Tracy Carluccio, deputy director of the Delaware Riverkeeper Network, because, "The main responsibility of the State is to protect residents' health and safety and a ban on toxic frack waste would do exactly that. The Governor's veto is an inexcusable cop-out without legal foundation, exposing New Jersey's communities and drinking water to just what we don't need—more pollution."

Moratoriums in Colorado

There are more than 52,000 wells in the Niobrara shale of Colorado, making it the sixth largest state in natural gas production.[1527]

Following a presentation by NOAA chemist Dr. Steven Brown, Erie, Colo.—primary site of a NOAA research study that found high atmospheric levels of toxins and carcinogens—put a moratorium on fracking within the town limits.[1528]

Longmont, Colo., an 86,000 population city 10 miles north of Erie, also put a moratorium on oil and gas permits within residential areas of the city.[1529] Then, in July 2012, the Longmont city council passed an ordinance declaring the moratorium was necessary "for the immediate preservation of the public peace, health, or safety."

Gov. John Hickenlooper responded that allowing the ordinance to stand would "stir up a hornet's nest" to encourage other towns to also take local control of their zoning.[1530] The Colorado Oil & Gas Association, the Colorado Oil and Gas Conservation Commission, and the state's attorney general sued Longmont; the attorney general charged that Longmont was usurping the power of the state.[1531] The city's residents were undeterred. Led by Michael Bellmont, a musician and insurance agent who said he drew inspiration from the ban on fracking passed in 2010 by the Pittsburgh, Pa., city council,[1532] the citizens easily got enough signatures to force a ballot initiative onto the November ballot to amend the city charter to ban fracking and waste storage. Against opposition from local and regional newspapers and a $500,000 advertising onslaught by the natural gas industry,[1533] the citizens approved that initiative.[1534] A month later, Gov. Hickenlooper announced the state would not pursue legal action against Longmont because,

329

"there is uncertainty on whether [Colorado] has legal standing to sue, because nothing has been taken from the state."[1535] However, he also said the citizen action is "a clear taking from oil and gas mineral rights owners," and that the state would "help and support" any suit from citizens or the gas industry against the city.[1536] In July 2013, the state joined with the Colorado Oil and Gas Conservation Commission in a suit to overturn the ban.[1537] One year later, Boulder District Court Judge D. D. Mallard ruled against the city for violating existing state law.[1538] By the end of 2014, Longmont had spent about $136,000 defending the ban and appealing the decision.

In February 2013, the Fort Collins City Council, by a 5–2 vote, banned fracking.[1539] "Nine months ago there was little support for banning fracking in Colorado, and there were hardly any organized groups willing to take it on," wrote Dr. Gary Wockner, one of Colorado's leading anti-fracking activists.[1540] However, three months later, by a 4–3 vote, the City Council rescinded that ban.[1541] The reason was that both Gov. Hickenlooper and Prospect Energy, aided by Energy in Depth, had threatened to separately sue the city if the ban was not overturned. Shortly after the vote, Citizens for a Healthy Fort Collins began a campaign to put a measure on the November 2013 ballot to establish a five-year moratorium to allow time to study and analyze health, safety, and environmental effects.[1542]

Fort Collins voters, in April 2015, elected a mayor and two council persons who are pro-fracking. The election was marked by a significant increase in campaign advertising from the previous elections and major donations from independent organizations, which aren't required to disclose individual donors. Supporting candidates favorable to fracking were Citizens for a Sustainable Economy ($77,595) and the Larimer Energy Action Project ($45,000). Fort Collins Common Sense, an anti-fracking group, spent $70,434.[1543]

The actions of the state's governor and Big Energy didn't deter the city council of Boulder, a city of about 100,000 in north-central Colorado, from unanimously approving a one-year moratorium in June 2013. The emergency ordinance was passed a week before a Boulder County moratorium on issuing new

330

permits would expire. The *Daily Camera* reported that all but one of the 40 persons who spoke during the public comments part of the meeting opposed fracking within the city limits[1544]

The Colorado Oil & Gas Association contributed almost $900,000 in a lobbying and media campaign to try to convince Colorado citizens to vote against moratoriums on the November 2013 ballots in Boulder, Broomfield, Fort Collins, and Lafayette. The Association claimed its six-figure campaign was to counter the "hype and misinformation" from anti-fracking activists.[1545] In contrast, anti-fracking groups only raised about $26,000, according to the *Denver Post*.[1546] Frackfree Colorado, in response to the industry's own "hype and misinformation," created two short "Frack Checked" videos, each one with specifc information that exposed the half-truths by the oil and gas industry.

Although being outspent more than 30-to-1, the anti-fracking community prevailed, partially because of superior knowledge and use of social media. About 76 percent of Boulder's voters and 55 percent of Fort Collins' voters approved five-year moratoriums on fracking; in Lafayette, 59 percent of voters approved a permanent ban.[1547] Broomfield's five-year moratorium passed by 17 votes, triggering an automatic recount.[1548] The recount confirmed the election.

However, in Loveland, Colo., a city of about 70,000, a two-year moratorium failed, 10,844–9,942 in June 2014.[1549] Having seen the residents of four other cities successfully approve the moratoriums, the oil and gas industry threw a media and money campaign against the ballot proposition. By the time of the election, the Colorado Oil and Gas Association, the Loveland Energy Action Project, and Coloradans for Responsible Energy Development, a front group of Coloradans for Responsible Oil Development, and funded primarily by Anadarko Petroleum and Noble Energy, had spent more than $1 million to defeat the measure. The campaign included extensive oversized post cards and flyers delivered to residents' homes and a constant robo-call campaign. Anti-fracking activists, primarily from Protect Our Loveland, spent less than $8,000. Dr. Wockner, writing for *EcoWatch,* believes the measure might have passed had "an industry-friendly city council [not have] cherry-picked a date . . . for the election that coincided with the hotly contested Republican

Governor's race [which ensured] a large turnout of very conservative voters."[1550]

The will of the people was blocked by a vigorous legal challenge by the Colorado Oil and Gas Association, TOP Operating, a major oil and gas company, and the Colorado Oil and Gas Conservation Commission, the state's regulator. In July 2014, a Boulder County district judge ruled that Longmont's ban and desire to protect the health and environment was not greater than the state's interest in developing oil and gas resources.[1551] In her decision, D. D. Mallard cited a 1992 state Supreme Court decisions *[Cty. Comm'rs of La Plata Cty v. Bowen/Edwards Assoc.;*[1552] and *Voss v. Lundvall Bros.*[1553]*]* that allowed municipalities to regulate oil and gas, but not to ban explo-ration and transportation of gas. However, those decisions were rendered long before the development of high volume hydraulic horizontal fracturing, and at a time when vertical fracking did not impose as many health and environmental risks. The 2014 decision, if not overturned by higher courts, and if pursued by the oil and gas industry, would also reverse bans in the other Colorado cities.

Against a litigious industry, the commissioners of the 300,000-resident Boulder County in November 2014 extended its own moratorium on fracking until July 2018. "We can't guarantee to our residents . . . that the regulations we put in place . . . are sufficient to protect them from potential air and water quality impacts and the impacts to the health and safety of the community," county commissioner Elise Jones told *InsideClimate News.*[1554]

The *Boulder Daily Camera*, reflecting the will of the people and opposing the views of Gov. Hickenlooper, numerous politicians, and the oil and gas industry, observed:

> "As [ex-gov Richard] Lamm pointed out, there is plenty of geography for oil and gas extraction in Colorado without doing it within eye shot of kitchen windows in communities where they've made it clear they don't want it. Voters or community leaders in Boulder, Broomfield, Fort Collins, Lafayette and Longmont are fighting for what they see as their quality of life. It would be nice to think the governor was in favor of that, too." [Nov. 22, 2014].[1555]

The Truth Behind Texas Fracking

Denton, Texas, is a city of about 123,000, about 35 miles northwest of Dallas. It has a large music and arts culture, and several manufacturing industries. Within the city are the University of North Texas and the Texas Women's University; the two universities and the public school district employ about 12,000 people,[1556] making education the largest employer, and giving the city a reputation as a college town.

Beneath the city lies the Barnett Shale. On the surface are 272 wells, pulling gas from the earth through unconventional wells, most of the wells within 250 feet of homes, schools, and recreational areas. West of Denton County is Wise County, where Mitchell Energy drilled the first test well to use horizontal fracturing. By 2014, the American Lung Association rated the area "F" on air quality.[1557]

As more companies moved to Denton, at the edge of the Barnett Shale, the residents began to see the effects of horizontal fracking in an area in which there were few restrictive regulations; what regulations existed were written into state law with the strong assistance of the oil and gas industry.

"We had been trying for more than five years to get moratoriums and slow-downs until we could write new rules," says Cathy McMullen, a nurse and president of the Denton Drilling Awareness Group (DDAG). But for five years, says McMullen, "We had been hearing the same thing—if we write new rules, the state would sue us." The moratorium, says McMullen, "was a last resort," caused by a unified industry "that refused to work with us."

By July 2014, less than five months after DDAG began a petition drive to call for the moratorium, it had almost 2,000 signatures, more than three times what was necessary to put a referendum on the ballot. DDAG wasn't the only group to collect signatures. Denton Taxpayers for a Strong Economy, a front group primarily funded by the oil and gas industry, also began collecting signatures. The group, as reported by the *Denton Record Chronicle*, was not registered with the state.[1558] Taylor Petition Management of Colorado Springs, Colo., paid workers $2–$4 per signature; their pitch, according to Sharon Wilson, a leading opponent of fracking who witnessed and recorded numerous approaches by those who were collecting signatures, was to tell residents these petitions would be to increase regulation.[1559] About 8,000 signed those petitions, mostly because of the verbal pitches by out-of-town workers who knew the papers they were handing the residents had no language that would increase regulation.

During a contentious eight-hour city council meeting, attended by about 500 persons,[1560] council members heard from residents about health and environmental effects of fracking; they heard about the noise from rigs, drills, and compressors; they heard about truck traffic; they heard the fears of residents worried about wells placed as close as 300 feet from parks, schools, and homes.

The council also heard from the industry and had read a four-page letter from Barry T. Smitherman, chair of the Railroad Commission, which is supposed to regulate oil and gas drilling. In that letter, Smitherman called oil and gas drilling "one of the key pillars of our Texas economy," and praised horizontal fracking and the oil/gas industry.[1561] The Perryman Group, with funds from the industry and the Fort Worth Chamber of Commerce, released a 69-page report that claimed a ban on fracking "would have substantial adverse effects on the economy and tax revenues to local entities and the state."[1562] However, Council member Kevin Roden countered the report with facts—of about $7 billion in property valuation in Denton, only about $81.5 million was from mineral/gas extraction, about 1.7 percent; that only 0.27 percent of the work force in Denton was jobs in the oil/gas industry; and that the $561,894 in property tax revenue from gas well royalties was only 1.1 percent of total tax revenue for the city.[1563]

The City Council voted 5–2 against a moratorium. That led to the DDAG ballot referendum that would bypass the council.

During the three months between the City Council election and the referendum vote, the oil and gas industry, accompanied by the Chamber of Commerce and state officials, unleashed their full resources. Those resources included a $750,000 promotion campaign fronted by the oil and gas industry; DDAG spent only about $75,000,[1564] most of it raised by individuals and through numerous fund raisers who, says McMullen, "had a passion for what we believed in." State officials threatened, "If we got a ban, the state would strip all cities of the right to use state law [*i.e.*, a referendum] for local governance." Those opposed to the ban "called us Communists, liberals, socialists, traitors, idiots," says McMullen. But, she says, "That was OK, because we knew that would be their downfall." The people, says McMullen, "knew better, so when the accusations started against us, people who knew us, who went to church with us, who went to school with us, they were very angry, and that may have turned the tide, because the people knew the opposition was willing to say anything to win."

On Nov. 4, 2014, 58.6 percent of the 25,376 voters who cast ballots[1565] approved the referendum to ban fracking. "There was never a doubt in my mind it was going to pass," says McMullen, who says because she and those who proposed the ban, "live and work in Denton, we knew the problems and the issues."

Within hours, the pro-fracking industry fought back. Many claimed the ban passed only because of the student vote; some even claimed the vote should be annulled because students—legally registered to vote in Denton—went to the polls. However, only 1,033 students between the ages of 18 and 22 voted; if none of them voted, the ban would still have passed.

State officials—most of them conservatives who publicly argued in numerous other acts of legislation that they believed in democracy and local governance—protested the vote in Denton. The Texas Oil and Gas Association,[1566] a private organization, and the Texas General Land Office,[1567] an official state agency, the morning after the vote filed petitions with the District Court for an injunction against the city, claiming the ban violated state law. George Prescott Bush, elected Land Office Commissioner in the November election, was endorsed by the Texas Oil and Gas PAC.[1568] He is the grandson of George H. W. Bush and nephew of George W. Bush. George P. Bush owns about 13 million acres of mineral rights.[1569]

DDAG and Earthworks, which had assisted in the campaign against fracking in Denton, filed intervention papers with the district court. Texas had previously granted local governments the right to regulate oil and gas operations. When state and federal officials won't stand up for the public, citizens must have the right to use local democracy to protect themselves," said Deborah Goldberg of Earthworks."[1570]

The *Dallas News*, among several Texas newspapers, supports oil and gas drilling, but also editorialized: "A homeowner has the right to not endure the sounds and smells of drilling activity, and a city has the right to determine whether drilling should be part of its future." [March 23, 2015].[1571]

Texas politicians, however, didn't agree with the vote or with the *Dallas News* editorial. Railroad Commissioner David Porter, in a statement to the media, said he was "disappointed that Denton voters fell prey to scare tactics and mischaracterizations of the truth in passing the hydraulic fracturing ban."[1572]

Christi Craddick, who succeeded Barry T. Smitherman as Railroad Commission chair in August 2014, also called the vote a "disappointment." She claimed fracking "has been plagued by a cloud of misinformation, mainly due to groups more interested in scaring people than actually understanding the complex science of minerals extraction."[1573] The reason why the ban passed was partially the fault of the Commission because, "We missed as far as an education process in explaining what fracking is [and] I think this [the vote] is the result of that, in a lot of respects, and a lot of misinformation about fracking,"[1574] she said. Denton County, said Craddick, "will lose jobs, tax revenues, business development, and the other economic benefits that come with oil and gas production, our state's most iconic and lucrative industry."[1575] Craddick, who proudly says she was "raised in a strong conservative household," said she intended to disregard the city's vote and keep issuing permits to corporations that wanted to drill in Denton. However, the ban applied only to the process of horizontal fracturing, not to drilling. Craddick's comments, says McMullen, "were ignorant."

Craddick's primary concern "is getting the permits written quickly so oil and gas companies can develop their resources,"[1576] Cyrus Reed, conservation director for the Texas chapter of the Sierra Club, told the *Houston Chronicle*. The Railroad Commission of Texas approved almost 200,000 drilling permits between January 2003 and September 2013, rejecting only 650.[1577] Reed said Texas chose "to regulate oil and gas with three elected officials, and those commissioners generally raise money from the oil and gas industry."[1578] During the 2012 and 2014 election cycles, the commissioners received almost $3 million in campaign contributions from the oil and gas industry.[1579] About one-fourth of all Texas legislators or their spouses have financial interests in companies that are active in the Eagle Ford shale, according to an investigation by *Inside-Climate News* and the Center for Public Integrity.[1580]

The Texas House of Representatives and Senate affirmed their close ties to the oil and gas industry by passing a bill (HB40) to strip municipalities of a right to impose moratoriums on fracking.[1581] The House vote was 122–18; the Senate vote was 24–7,[1582] with almost no discussion. That bill, primarily written by a former ExxonMobil lawyer,[1583] did allow local governments to regulate noise and traffic and to establish "setbacks," but those setbacks needed to be "commercially reasonable," a gift to the shale oil and gas industry. Among those voting to eliminate the Denton vote was State Rep. Myra Crownover, who represents the residents of Denton, and who has extensive ties to the oil and gas industry, according to the *Texas Tribune*.[1584] In May 2015, Gov. Greg Abbott signed the legislation that also allowed the state to retroactively block any previous ban, even if legal at the time.

Several Denton residents, protesting the state's new law, were later cited for trespassing on gas company property or for blocking public roads. The Denton City Council, by a 6–1 vote, in June 2015 voted to repeal its seven-month-old ban on fracking as unenforceable[1585] and to avoid significant costs in future litigation. At the time of the vote, Denton had spent $842,300 since 2009 writing and then defending the ban, according to the Fort Worth *Star–Telegram*.[1586] To defend itself against the two lawsuits, the Denton City Council may have to use a large portion of the $4 million it had previously set aside in its risk fund for legal challenges.[1587]

What Commissioners Smitherman, Porter, and Craddick said about the Denton referendum, combined with Reed's observation, the lack of adequate funding by the Texas legislature to the TCEQ, and the legislation to deny local governments the right to ban fracking, are all that people need to know about the truth of oil and gas regulation in Texas.

Bans and Moratoriums in Other States

Vermont banned fracking, but the action by the legislature and Gov. Peter Shumlin is symbolic since Vermont has no deep earth natural gas deposits. Nevertheless, the natural gas industry lobbied against the proposed legislation and then spoke out after the legislature passed it. Rolf Hanson of the American Petroleum Institute said the Legislature's action "follows an irresponsible path that ignores three major needs: jobs, government revenue and energy security."[1588]

There is no active fracking in Connecticut, but in May 2014, the legislature declared flowback waters to be treated as hazardous waste, and established a minimum three year moratorium on importing all waste water from other states.

Gov. Bev Perdue (D-N.C.), who supports fracking and natural gas development, vetoed a Republican-sponsored bill in July 2012 that would have removed that state's ban on fracking. "Our drinking water and the health and safety of North Carolina's families are too important [and] we can't put them in jeopardy by rushing to allow fracking without proper safeguards,"[1589] said Gov. Perdue in announcing her veto. However, in 2014, the Republican-controlled state legislature pushed through, without time for public comment, an industry-friendly bill to allow and regulate fracking. Gov. Pat McCrory, a Republican who followed Bev Perdue into office, signed that bill that allowed fracking to begin by Spring 2015. The Triassic Basin shale lies within North Carolina, but is not expected to be more than a minor producer of natural gas.

The Bureau of Land Management (BLM) had announced it would postpone awarding any leases for oil or gas drilling until at least October 2013. Its decision followed a ruling in April 2013 by the U.S. District Court for the Northern District of California that the BLM had violated the National Environmental Policy Act by awarding leases in 2011 for about 2,700 acres, primarily in the Monterey Shale. U.S. Magistrate Paul Grewal ruled the BLM failed to consider current data about fracking and that "BLM's dismissal of any development scenario involving fracking as 'outside of its jurisdiction' simply did not provide the 'hard look' at the issue that NEPA requires." [*Center for Biological Diversity, et al. v. Bureau of Land Management, et al.;* 11-06174 PSG[1590]] The ruling did not invalidate the leases.

Mora County, which lies in the Lewis Shale in the northeast part of New Mexico, is one of the nation's smallest counties. In April 2013, it became the first county to ban oil and gas drilling and to establish a residents' bill of rights. County Commissioner Alfonso Griego told *E&E News* he supported the ban because energy companies:

> "just come in and do whatever is necessary for them to make profits. There is technology for them to do it right, but it's going to cost them more money. They're not willing to do that yet. So we don't want any oil and gas extraction in the county of Mora. It's beautiful here."[1591]

In November 2013, Hawaii County, the only county on the Big Island of Hawaii, unanimously approved a ban on fracking. "We have to protect our land, we have to protect our people, we have to protect our aquifers," said Brenda Ford who introduced the bill.[1592]

Trying for a Moratorium in California

The newly-formed Californians Against Fracking, a coalition that eventually grew to more than 200 groups, rallied in Los Angeles and San Francisco on May 30, 2013, to demand a ban on fracking. The group delivered petitions with more than 100,000 signatures to Gov. Jerry Brown, an environmentalist who was leaning toward allowing fracking. Becky Bond, one of the activists and political director of CREDO, told the news media why the governor should ban fracking:

> "He has fearlessly championed clean energy, called for swift action on climate change, and fought to hold corporations accountable for polluting our state's precious natural resources. These decades of experience make him uniquely qualified to call out the cynics who are advocating that fracked oil extrac-

tion is worth risking not only California's clean water resources but the very climate that makes our planet livable. We're launching Californians Against Fracking to urge Gov. Brown to make the only choice that's consistent with his and his father's grand vision for our state—to protect our water and our planet by banning fracking in California."[1593]

Jerry Brown's father was Edmund G. (Pat) Brown, who served two terms as governor from 1959 to 1967. The protest and rallies didn't convince the legislature or the governor to ban fracking.

California's Democratic party in April 2013 passed a resolution calling for a moratorium, but three separate bills died in the Assembly. Heavily lobbied by the Western States Petroleum Association (WSPA), which spent $4.67 million for lobbying fees and expenses in 2013,[1594] the California State Assembly, by a 35–24 vote in June 2013, killed a bill that required a review of environmental and health risks before fracking could occur. The WSPA, according to Susan Frank, writing in the *Capitol Weekly*, spent more than $16 million since 2009 lobbying state legislators,[1595] including $8.9 million in 2014,[1596] the largest amount for any lobbyist in California.

The California Senate, by a 16–16 vote in May 2014, with eight senators abstaining, failed to establish a two year moratorium on fracking until additional studies could be conducted to assure public safety and the protection of the environment. Those who opposed the moratorium received about $4.1 million from oil and gas interests and several trade unions between 2009 and 2013; those who voted for the moratorium received only about $1.4 million from organizations that approved the moratorium, according to data collected by MapLight,[1597] a nonpartisan research organization that tracks donations to politicians. "Because of numerous exemptions and loopholes as well as federal regulations," says David Braun of Californians Against Fracking, "it is critical that Gov. Brown and other leaders protect the public health."

Four months after the rallies, Jerry Brown signed into law regulations to allow fracking. However, the law, far tougher than that of most states, requires oil and gas companies to provide full disclosure of its methods. "Oil companies will not be allowed to frack or acidize in California unless they test the groundwater, notify neighbors and list each and every che... on the Internet," State Sen. Fran Pavly, who wrote the new law, told Reuters. Pavley is a Democrat with a master's in environmental planning. She said the law was "a first step toward greater transparency, accountability and protection of the public and the environment."[1598]

A year after the California Assembly and Senate failed to establish a fracking moratorium, Santa Cruz County, Calif., which borders the Pacific Ocean in central California, became the state's first county to ban fracking and gas and oil development.[1599] In November 2014, 57 percent of the voters in San Benito County[1600] and 67 percent of Mendocito County voters[1601] banned fracking.

The Pennsylvania Protest

While other states and countries are either banning or suspending fracking until full health and environmental analyses can be completed, Pennsylvania has been "handing out permits almost like popcorn in a theater," says Diane Siegmund, a psychologist from Towanda, Pa. Pennsylvania issued 19,205 permits for unconventional wells from 2007 to June 2015.[1602]

By a 9–0 vote of the city council in November 2010, Pittsburgh became the first Pennsylvania city to ban natural gas drilling. The Council cited health concerns as its reason to ban drilling.[1603] City Councilman Doug Shields called the natural gas industry "arrogant," and said it placed profits ahead of citizen health and the environment. Among the leases that Huntley & Huntley signed were for 1,060 acres of land beneath 15 cemeteries under jurisdiction of the Catholic Cemeteries Association of the Diocese of Pittsburgh.[1604] Council president Darlene Harris, responding to industry claims it would bring jobs to the region, was brutally honest—"They're bringing jobs all right. There's going to be a lot of jobs for funeral homes and hospitals. That's where the jobs are. Is it worth it?"

Two months after the Pittsburgh city council voted to ban fracking, the city council of Philadelphia, the state's largest city, voted to ban all fracking in the Delaware River Basin.[1605]

In a press release dated Nov. 18, 2011, Gov. Tom Corbett explained Pennsylvania's all-out support for natural gas: "Today's delay—driven more by politics than sound science—is a decision to put off the creation of much-needed jobs, to put off securing our energy independence, and to infringe upon the property rights of thousands of Pennsylvanians."[1606] The delay Corbett referred to was a decision by the Delaware River Basin Commission to extend the moratorium on taking water from the Delaware River to be used for fracking operations. In June 2013, the Wayne County commissioners and Corbett issued scorching attacks upon the DRBC, both attacks focusing upon economics while excluding health and environmental concerns. The Commissioners of the 53,000-resident county in northeast Pennsylvania argued:

> "Wayne County residents and landowners are being unequivocally deprived of economic benefits and the freedom to make land use decisions. This taking of basic property rights and the denial of economic liberties does not bode well for Wayne County landowners, residents or their families.
>
> "The Wayne County Commissioners believe that the . . . moratorium on natural gas development has exceeded a reasonable time period and, because of such, has caused irreparable economic damage to those who entered into good faith leasing agreements with natural gas developers."[1607]

Corbett claimed that because of the moratorium, "Operators interested in developing natural gas have closed offices and laid of employees,"[1608] again focusing upon the perceived economic boom rather than environmental and health issues.

In a scorching editorial, the *Pocono Record* attacked Corbett for being "uncomfortably beholden to the drilling companies":

> "Someone should remind the governor that Pennsylvania's Constitution calls for the protection of its clean air and water. Gas companies are having diverse and, frequently, adverse impacts on both air and water as they extract natural gas from deep in the ground. So the four-state DRBC had good reason to impose the moratorium. . . .
>
> "The Delaware River warrants protection not only for its 'natural, scenic, historic and esthetic values' but as the source

of drinking water for more than 15 million people, including Philadelphia and New York City residents."[1609]

Corbett managed to partially compromise the financial integrity of the DRBC for its stand on behalf of health and environment issues. He cut $500,000 from the DRBC budget for the 2014–2015 fiscal year, an action approved by the Republican-controlled legislature. The state's previous allotment was $934,000. The governor and legislature did not cut funds from the Susquehanna River Basin Commission, Chesapeake Bay Commission, the Ohio River Valley Water Sanitation Commission, and the Interstate Commission on the Potomac River.[1610]

Pennsylvanians want a moratorium on fracking. And it's not just a few thousand, but a majority of the state's residents.

A joint University of Michigan/Muhlenberg College study published in May 2013 reveals that only 49 percent of Pennsylvanians support shale gas extraction and 58 percent of all Pennsylvanians want the state to order "time out" until the health and environmental effects of fracking can be fully analyzed. That same study revealed that 60 percent of Pennsylvanians believe fracking poses a major risk to groundwater resources, 28 percent disagree, and 12 percent have no opinion.[1611]

Petitions with more than 100,000 signatures requesting a moratorium were delivered to Gov. Corbett in April 2013.[1612] The day the petitions were delivered, State Sen. Jim Ferlo proposed legislation to call for a moratorium.[1613] In September, he formally introduced a bill [SB 1100] to put a moratorium on future permits. Part of the reason for that bill is because, says Ferlo, "I've lost all faith with the DEP at the leadership level." The bill doesn't stop current drilling "because contracts have already been established and [we] can't legally reverse that." But, says Ferlo, "We need to take a step back and see what we're doing, to do an impact study of the effects of fracking."

In June 2013, the Democratic State Committee approved a resolution to establish a moratorium.[1614] So, if almost three-fifths of all residents want fracking to stop, who's opposing the moratorium?

Just about anyone in a political leadership position in both major political parties oppose a moratorium. They tend to be the ones who from their own houses can't see drilling rigs, well pads, frack pits, and frack trucks that block public roads. They tend to be the ones who have deliberately twisted the facts and now squawk about how fracking the earth has helped create jobs and improve the economy, while ignoring the problems already proven that affect their constituents' health, environment, and food supply.

The Democrats' resolution had begun in February. Sue Lyons, an attorney, had proposed the resolution. However, the Rules Committee of the Democratic Party State Committee did not allow it to go forward, questioning its legality. To make sure the resolution was not in the best interest of Pennsylvania, the party even contacted the Department of Environmental Protection, the same DEP that had policies that block full transparency, that had a policy to "educate" rather than penalize gas companies that violate state pollution standards, and for two years had been run by a political crony of Gov. Corbett. The State Committee accepted the DEP opinion about fracking and agreed the resolution was out of order.

Enter Karen Feridun, Patti Rose, and Berks Gas Truth. With a massive grassroots campaign, in less than two months they convinced the delegates to the State Committee not only to get the resolution out of committee but also onto the floor for the vote.

The Democratic leadership, parroting the Republicans, didn't accept that democracy prevailed in the state central committee. Vice-chair Penny Gerber, who lives in Montgomery County, which is exempt from fracking, called fracking "a thriving industry."[1615] Gerber is an associate at a PR firm whose clients include large energy companies.

Former Gov. Ed Rendell, who had opened fracking in Pennsylvania, told the *Patriot–News* of Harrisburg the resolution was "ill-advised,"[1616] and then used the same arguments spewed forth by Tom Corbett and the Republican leaders to claim fracking improved the economy and "helped create wealth in the poorest areas of Pennsylvania," avoiding any references to the detrimental effects that Feridun had so eloquently brought forth, but which the mainstream news media had ignored.

What Rendell didn't say, although it isn't any secret, is that he is a special counsel to one of the nation's largest law firms that represents Big Energy. Among his chores was to intervene on behalf of Range Resources, one of the nation's largest drilling companies, to get the Environmental Protection Agency to drop a water contamination suit.[1617]

Eighteen House Democrats signed a letter to the party's state chair to say they also opposed the moratorium. The letter claimed a moratorium "would create immense burdens and significant adverse impacts on the citizens of Pennsylvania with no discernible gain." The Democrats stretched credibility with a statement that the state "passed comprehensive legislation creating among the most stringent of regulatory frameworks in the country for gas drilling," and claimed, "With adequate enforcement and proper oversight by state and federal agencies, the process of hydraulic fracturing poses negligible risk to individuals."[1618] All Democratic candidates for governor in the 2014 election, like Rendell, emphasized the economy, barely throwing a few nuggets to the environmentalist opposition and not mentioning health effects from fracking operations.

Piling on, as expected, were the state's Republicans. Rep. Tom Marino, whose district lies above one of the most active parts of the Marcellus Shale, demanded that President Obama "should show some leadership and admonish the Pennsylvania Democratic Party for their job-killing stance that would only further hurt the struggling economy."[1619]

The Pittsburgh *Tribune-Review*, one of the nation's most conservative newspapers, declared the proposed moratorium "certainly makes us wonder why the hardworking folks who've found good, solid jobs in the Marcellus shale industry would ever vote for state Democrats again."[1620]

Karen Feridun, however, on the floor of the Democratic State Committee had the best reason to establish a moratorium until full health and environmental effects can be known. The moratorium, she said, would "put people over profits."

Dr. Tom Wolf, the state's former revenue director, who defeated Tom Corbett in the November 2014 general election, opposes a moratorium. Although far more environmentally-concerned than his predecessor, he believes fracking "can create good-paying jobs, be a leader in energy production, and help decrease our dependence upon foreign oil" while also allowing for a "safe and secure environment."[1621] A major part of his campaign was to pledge to increase and enforce reulation upon fracking, and to secure a severance tax on drilling, using the revenue to improve the public school system. His opposition was not only the fossil fuel industry, but also the Chambers of Commerce and the Republican-controlled General Assembly.

PHOTO: Jacques-Jean Tiziou
(courtesy of Protecting Our Waters)

Protesters at the Shale Gas Outrage;
September 20, 2012, in Philadelphia.

PART III
Some Social Issues

How Fossil Fuel Interests Attack Clean Energy

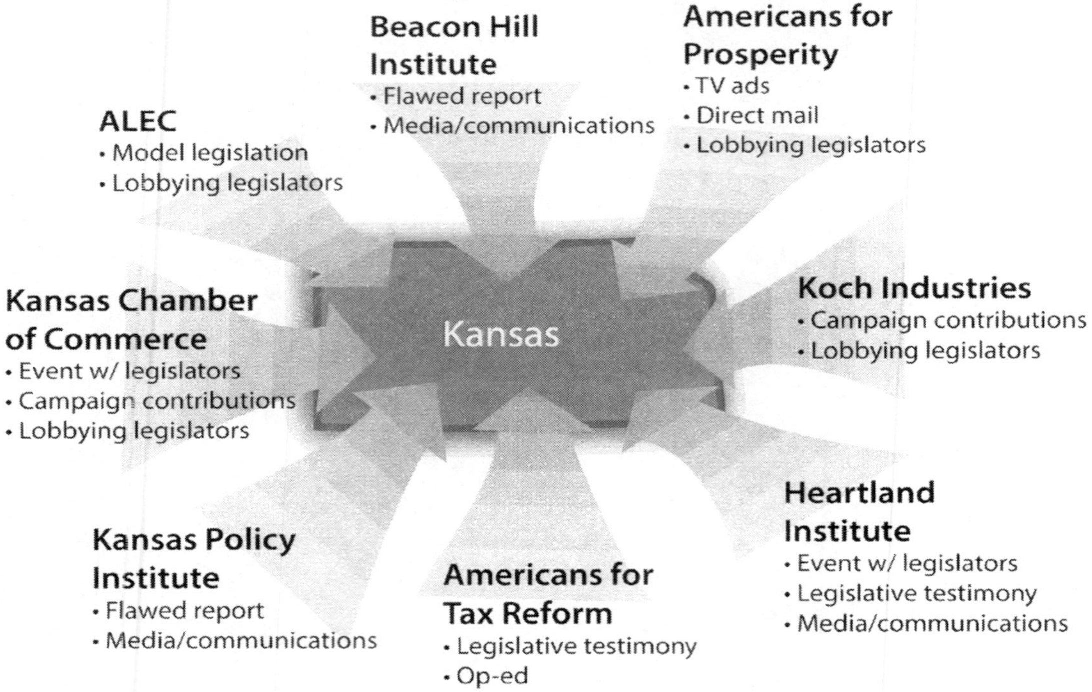

Beacon Hill Institute
- Flawed report
- Media/communications

Americans for Prosperity
- TV ads
- Direct mail
- Lobbying legislators

Koch Industries
- Campaign contributions
- Lobbying legislators

Heartland Institute
- Event w/ legislators
- Legislative testimony
- Media/communications

ALEC
- Model legislation
- Lobbying legislators

Kansas

Americans for Tax Reform
- Legislative testimony
- Op-ed

Kansas Chamber of Commerce
- Event w/ legislators
- Campaign contributions
- Lobbying legislators

Kansas Policy Institute
- Flawed report
- Media/communications

GRAPHIC: Energy & Policy Institute

CHAPTER 18
Government and Industry Response

The response by the industry and its political allies to the scientific studies of the health and environmental effects of fracking "too often has been to argue that hydraulic fracturing can't possibly cause any problems," says Fred Krupp, president of the Environmental Defense Fund.[1622]

The energy companies claim water wells were tainted before gas operations began, "but only after fracking began did these chemicals show up in people's water supplies," says Dr. Amy Paré, a physician from McMurray, Pa. Hit hard by the public response to industry denials, and faced by numerous photos and videos showing contaminated water, the industry is now testing the water before and after fracking occurs.

"In general [the fossil fuel industry tends] to be dismissive of individual complaints while expressing an understandable need for further research and concern for the health of individuals, but really shying away from any connection with their own activities," Abrahm Lustgarten told NPR's "Fresh Air Report." Lustgarten says:

> "You won't hear the drilling industry say, 'This isn't an issue and we don't have to study this.' You won't hear them say they don't care about [individuals with health problems]. But you will hear them say [certain individuals appear] to not like the industry and maybe [they have] health issues and maybe [they want] the industry to leave. . . And by the way, we need to do a whole lot more research and it's a decade-long effort and let's just get started and not talk about blame at this point.'"[1623]

The volume and intensity of response "has approached the issue in a manner similar to the tobacco industry that for many years rejected the link between smoking and cancer," say Drs. Michelle Bamberger and Robert E. Oswald.[1624]

The response to a suit claiming health problems caused by a three million gallon wastewater spill was not to address the issue but to attack the attorney representing three families. Matt Pitzarella, director of corporate communications and public affairs for Range Resources–Appalachia, accused of the spill that began in November 2010, said the suit "isn't about health and safety; it's unfortunately about a lawyer hoping to pad his pockets, while frightening a lot of people along the way."[1625]

The Haney, Kiskadden, and Voyles families of Washington County, Pa., had sued the company after developing "a multitude of health problems, including nose bleeds, headaches and dizziness, skin rashes, stomach aches, ear infections, nausea, numbness in extremities, loss of sense of smell and bone pain," according to the *Pittsburgh Post–Gazette*.[1626] However, a major part of the suit accused Range, several subcontractors and two water testing labs for not revealing full and accurate reports of the contamination, and for advising the families, as the *Post-Gazette* reported, to believe it was safe to "drink, cook, and bathe in the contaminated water." A separate lawsuit filed by Beth Voyles, in May 2011, had asked for a *writ of mandamus* to force the Pennsylvania DEP to fulfill its obligations to protect public health. That suit stated Voyles had "numerous health ailments, including but not limited to rashes, blisters, lightheadedness, nose bleeds, lethargy, and medical testing [that] revealed elevated levels of arsenic, benzene, and toluene in her body."[1627] Those chemicals are found in fracking fluids, which are alleged to have been present in a nearby impoundment that had a capacity of about 321,000 barrels of wastewater. "Even after numerous reports of holes and tears in the primary impoundment liner that needed to [be] repaired [were reported], the DEP still required no monitoring system," the *Canon-McMillan Patch* reported.[1628]

REPORTING THE TRUTH

Although the natural gas industry may lie or shade the truth

350

to the media, the people, and the governing bodies, they are required to report the truth to the Securities and Exchange Commission. In its annual SEC 10-K report, filed December 2011, Cabot Oil & Gas Corp. admitted:

"Our business involves a variety of operating risks, including:
• well site blowouts, cratering and explosions;
• equipment failures;
• pipe or cement failures and casing collapses, which can release natural gas, oil, drilling fluids or hydraulic fracturing fluids;
• uncontrolled flows of natural gas, oil or well fluids;
• fires;
• formations with abnormal pressures;
• handling and disposal of materials, including drilling fluids and hydraulic fracturing fluids;
• pollution and other environmental risks; and
• natural disasters.

"Any of these events could result in injury or loss of human life, loss of hydrocarbons, significant damage to or destruction of property, environmental pollution, regulatory investigations and penalties, suspension or impairment of our operations and substantial losses to us.

"Our operation of natural gas gathering and pipeline systems also involves various risks, including the risk of explosions and environmental hazards caused by pipeline leaks and ruptures. The location of pipelines near populated areas, including residential areas, commercial business centers and Industrial sites, could increase these risks."[1629]

The final word about the health of the public is not that from any of the scientists who have been researching the health and environmental issues surrounding the fracking process. The final word is "mitigation," and it belongs to Rex W. Tillerson, CEO of ExxonMobil, which had revenue of about $394 billion in 2014.[1630] In a speech to the Council on Foreign Relations in June 2012, the same speech that he declared not only are people illiterate in the sciences, but that opponents of the energy industries are manufacturing fear, Tillerson laid out a corporate truth:

"And so when people manufacture this fear that we can't allow this to go forward . . . our answer is 'yes we can,' because we will have a technological solution and we will have risk mitigation and risk management practices around those resources to ensure they can be developed in a way that mitigates risk—it doesn't eliminate it, but when you put it into the risk versus benefit balance, it comes back into a balance that most reasonable people in society would say, 'I can live with that.'" [1631]

Thus, the energy industry is telling the people there will be accidents. There will be deaths. There will be health and environmental consequences. But, they are acceptable because "mitigation" allows a corporation to accept errors, injuries, illnesses, environmental destruction, and even death if they believe there is a "greater [financial] good" that outweighs those risks.

It is no different than when Ford in the 1970s manufactured the Pinto, a sub-compact with a gas tank positioned at the rear of the vehicle, and didn't put in a heavyweight bumper or much reinforcement between the rear panel and the gas tank. This design flaw led to a higher-than-expected number of fires if the car was hit from the rear. To recall each Pinto and modify the fuel tank would cost $11 per vehicle,[1632] about $121 million total. But, Ford figured that cost would be greater than what it would have to pay out for injuries and deaths caused by the defect, which it figured to be about $50 million. Thus, "mitigation" and "risk management" suggested it would be cheaper to allow injuries and deaths than to fix the problem. A leaked memo published in *Mother Jones* magazine[1633] led to public outrage, additional lawsuits, a government-mandated recall, and one of the worst public relations disasters in Ford's history.

In the four decades since the Pinto was first introduced, the energy industry learned nothing from the fall-out over Ford's decision to sacrifice lives to cost-ratio benefits.

CHAPTER 19

The Chilling Effect of Business and Government Collusion

After graduating from the Shady Side Academy in Pittsburgh, Pa., Tim DeChristopher bounced around, first for two years at Arizona State, then to Missouri to work in outdoor education for almost four years, and then to Utah to become a wilderness guide with at-risk and troubled youth. It was in Utah where he would do something that exposed corruption—and lead to his imprisonment.

The Bureau of Land Management, in the last month of the George W. Bush presidency (December 2008), fearing that the incoming Obama Administration would suspend the sale of public land, planned to auction several hundred thousand acres of public land, including 149,000 acres (116 parcels) in southern Utah. Environmentalists protested, and filed suits to block the sale. Also protesting was the National Park Service.

Big Energy, which already had leases on 5.6 million acres of land in Utah,[1634] was there to scoop up what it could at bargain basement prices in order to drill for gas and oil. The decision to hold the sale of public land during the transition to a new administration may have had a philosophical base in the environmental policies of Ronald Reagan almost three decades earlier. Interior Secretary James Watt had declared environmentalism was a plot that would "weaken America," and that environmentalists were "a left-wing cult dedicated to bring down the kind of government I believe in."[1635]

DeChristopher, an economics student at the University of Utah, went to the auction to protest the sale. He says he didn't know just what he was going to do, but when one of the officials asked him if he was there to bid, he suddenly realized what he could do. On a spontaneous decision, he got a paddle and bidder number 70. After watching energy companies take about half the parcels at prices as low as $20 an acre, he bid on parcels to inflate the price, winning bids on 14 of those parcels totaling 22,500 acres. His bids, amounting to $1.8 million, would have given him prime federal land for about $80 an acre. Excluding his bids, the BLM auctioned off 126,500 acres of public land for $5.7 million, about $45 an acre.

DeChristopher's actions voided the auction. Ten days after the auction, the *Salt Lake City Tribune* editorialized:

> "President Bush will be remembered for eight years of disregard for the environment and disdain for hard science on the catastrophic effects of climate change caused by burning fossil fuels. DeChristopher will be remembered for trying to save his heritage from a government bent on taking it from him—and from all of us."[1636]

In January 2009, a federal court issued a temporary injunction, ruling that the BLM violated environmental and historic protection laws. A month later, the Obama administration revoked the sale of 77 parcels totaling more than 100,000 acres; the sale price of those parcels was about $6 million. The Obama administration allowed the sale of the other 39 parcels because they were in existing oil fields.

Newly-appointed Interior Secretary Ken Salazar said the Department had "rushed ahead to sell oil and gas leases at the doorstep of some of our greatest national icons, some of our nation's most treasured landscapes."[1637]

Although DeChristopher and hundreds of thousands of activists succeeded in reversing the BLM sale and kept the land from being carved up by drillers, they didn't succeed in obtaining justice. The federal government continued its pursuit of DeChristopher who now increased his activism, further enraging the prosecution. In April 2009, he was indicted for fraud and violation of the Federal Onshore Oil and Gas Leasing Reform Act.[1638] DeChristipher's lawyer had found out about the indictment from an Associated Press reporter, who had learned about the indictment from an oil industry lobbyist.

On the night before his trial at the end of February 2011, after the prosecution had received nine separate delays, hundreds gathered at the First Unitarian Church; among them were Peter Yarrow of Peter, Paul, and Mary, and his daughter, Bethany, who gave a concert to support DeChristopher and the rights of the people to protect their environment.[1639]

Four days later, DeChristopher was convicted. Federal District Court Judge Dee Benson had refused to allow the defense to present evidence that the auction was illegal or that other successful bidders reneged on their commitments and were not prosecuted. The trial was nothing less than selective prosecution of a citizen who the government wished to punish for disrupting an illegal action.

"The injustice in this case isn't that I am facing a trial," said DeChristopher, "It's that the jury is being denied the information to decide if my actions were justified."[1640]

During the trial, the federal prosecutor had argued that DeChristopher could have halted the auction in other peaceful ways or that he could have appealed the awarding of the land. At the sentencing hearing, four months after the trial concluded, DeChristopher accurately argued:

> "The government has made the claim that there were legal alternatives to standing in the way of this auction. Particularly, I could have filed a written protest against certain parcels. The government does not mention, however, that two months prior to this auction, in October 2008, a Congressional report was released that looked into those protests. The report, by the House committee on public lands, stated that it had become common practice for the BLM to take volunteers from the oil and gas industry to process those permits. The oil industry was paying people specifically to volunteer for the industry that was supposed to be regulating it, and it was to those industry staff that I would have been appealing. Moreover, this auction was just three months after the New York Times reported on a major scandal involving Department of the Interior regulators who were taking bribes of sex and drugs from the oil companies that they were supposed to be regulating. In 2008, this was the condition of the rule of law, for which Mr. [John] Huber [the federal prosecutor] says I lacked respect. . . .
>
> "The reality is not that I lack respect for the law; it's that I have greater respect for justice. Where there is a conflict between the law and the higher moral code that we all share,

355

my loyalty is to that higher moral code. . . .

"In these times of a morally bankrupt government that has sold out its principles, this is what patriotism looks like."[1641]

The Department of Justice's Office of Pre-Trial and Probation Services ran an extensive investigation on DeChristopher, talking with him and many of those who knew him, evaluating his intent, the probability he would repeat his actions, and any possible threat to the community. The Department recommendation was to sentence DeChristopher to probation, with no jail time. DeChristopher and his attorneys had previously rejected a plea bargain that would have given him a 30-day jail sentence and probation. However, Judge Benson disregarded the recommendation, and ordered DeChristopher to pay a $10,000 fine and serve a two year sentence.[1642]

Judge Benson openly acknowledged, "The offense itself, with all apologies to people actually in the auction itself, wasn't that bad," and said he might not have imposed a prison sentence— but that DeChristopher's "continuing trail of statements" and activism following his arrest were not acceptable.[1643] The judge may also have considered that DeChristopher had formed Peaceful Uprising, an organization devoted to protecting the environment. Possibly trying to mitigate reasonable objections that he was sentencing DeChristopher for exercising his First Amendment rights, Judge Benson also stated, "I want to make clear that Mr. DeChristopher had every right to make every pronouncement that I know of about everything, and when he was convicted of a felony, to go out in front of this courthouse and say whatever he said. He had the right to do that."[1644] Nevertheless, it was obvious that DeChristopher's exercise of free speech rights—and his statements that he would continue to protest BLM actions if he thought they were illegal—was a factor in the sentence.

Within two hours of sentencing, several dozen people in Salt Lake City protested, linking themselves together and blocking traffic. Police arrested 26, according to the *Salt Lake City Tribune*.[1645] Dozens of other demonstrations occurred at federal courthouses throughout the country, the people energized and angry that the government pursued charges against an activist who had stopped an illegal auction of public lands.

Shortly after the verdict, Robert Redford, actor/director and environmental activist, summed up the prosecution's hypocrisy:

> "He just did what he thought was his constitutional right. In the meantime we have all these guys on Wall Street sending this country into the tank. And no one's going to jail."[1646]

The *Salt Lake City Tribune* editorial board nominated DeChristopher as one of 16 persons for consideration as Utahn of the Year.[1647]

DeChristopher's defense team—led by Patrick Shea, a nationally-known civil rights lawyer and former national director of the Bureau of Land Management; and Ron Yengrich, a Salt Lake City attorney with a national reputation for taking civil rights and labor issues on behalf of the "underdog" and oppressed—appealed the court verdict. However, a three-judge panel of the U.S. Court of Appeals for the Tenth District affirmed the sentence of the district court:[1648]

> "[T]he Fifth Circuit holds that a sentencing court may consider a defendant's beliefs if they are 'sufficiently related to the issues at sentencing.' . . . Similarly, the Ninth Circuit held that a district court could impose a sentence based on the defendant's views where the court 'made it clear that it was increasing the sentence based on [the defendant's] lack of remorse and his threat to . . . the public when released.' [*United States v. Smith*] . . . Because the need to promote respect for the law and afford adequate deterrence were 'legitimate sentencing factors under 18 U.S.C. §3553(a),' *Smith* held the district court could consider the Defendant's views in imposing sentence.
>
> "[W]hen the [district] court imposed sentence, it noted it was 'focusing primarily on respect for the law and deterrence.' . . . Defendant's statements that he would 'continue to fight' and his view that it was 'fine to break the law' were highly relevant to these sentencing factors."[1649]

Shea had said he would be willing to take the case to the Supreme Court to argue on several possible constitutional violations, as well as to the role of the jury. DeChristopher, however, declined to pursue the appeal. The Appeals Court, said DeChristopher, "ignored the arguments in court by my defense attorneys, and ruled on a small technicality to uphold the decision"; he said he didn't want to extend the proceedings.

By the time of the Appeals Court ruling, DeChristopher had already served 14 months of his sentence, having been housed at a county jail, a private prison in Nevada that had a contract with the federal government to evaluate all prisoners, the Federal Correctional Institution at Herlong, Calif., a medium security prison; and the Federal Correctional Institution Englewood, in Littleton, Colo., a minimum security prison. He would soon be transferred to a half-way house in Salt Lake City.

DeChristopher says he had wanted to work as a social justice coordinator for the First Unitarian Church, "but the Department of Justice denied that because my proposed job was too close to my crime." He would also have been silenced for making any public statement about the justice system. He took a job as a clerk at Ken Sander's Rare Bookstore.

DeChristopher was released from federal custody, April 21, 2013, the day before Earth Day, three months short of his mandated sentence; he would be on probation for three years.

The attack upon public lands ended in October 2013 when the Supreme Court refused to review a Court of Appeals ruling that threw out an appeal by several energy companies and Utah counties to Ken Salazar's ruling that vacated the leases of 77 parcels of land near federal parks and monuments in Utah.

DeChristopher refuses to allow his supporters or attorneys to pursue a federal pardon. "When I thought about what that really meant," says DeChristopher, "it would be that I would not be a second class citizen anymore. But those who are felons, who have been released, are second class citizens their entire life. I would be a part of that, and accept that that class would include me."

In September 2013, on a full scholarship, DeChristopher began a three-year graduate program, leading to a Master of Divinity degree at the Harvard Divinity School, with intent to become an ordained Unitarian minister. The divinity school, established in 1816, was the first non-sectarian theological college in the United States.

DeChristopher says he wanted to go to Harvard "because of its academic rigor and broad understanding and respect for all religions, of all peoples and their cultures." He had applied for admission because he admires Emerson and Thoreau who were graduates; Thoreau is best known as a writer and social activist who believed civil disobedience was sometimes necessary to challenge immoral wars and laws.

Bidder 70, a compelling 73-minute documentary released in May 2012, is the story of environmental activism. For three years, Beth and George Gage followed DeChristopher, recording his life and the issues that put the people who wished to defend the environment in opposition to the government that wished to sell out the land to private corporations. The film has been a favorite at film festivals, and has drawn renewed attention to the power of civil disobedience.

Stephanie Penn Spears, editor of *EcoWatch*, says the film "topped my list for the most inspiring film."[1650] *AlterNet*, in its Earth Day 2012 edition, included *Bidder 70* as one of nine films "That Will Change the Way You Think About the World."[1651] The *Hollywood Reporter* said the film "vividly illustrates the personal consequences of daring to take on the government."[1652]

CARRYING THE PROTEST INTO MICHIGAN

In Michigan, protestors didn't bid on parcels they didn't plan to purchase. They had another tactic. For a May 2013 auction by Michigan of public lands, they sat silent, duct tape covering their mouths, emphasizing an official sign, "No Public Comment Allowed." But they also had other ways of being noticed. "They created a steady barrage of annoyances: cell phones rang for an hour and a duck quack brought a chuckle," Maryann Lesert reported in *EcoWatch*. The cell phone protest annoyed the two dozen law enforcement officers who, says Lesert, "roamed the seating area, leaning in to locate the offending ring tones and escorting out any noisemakers."[1653] Within an hour, most protestors were escorted out of the auction.

It wasn't the first time protestors had tried to interrupt Michigan's auction of public lands. A year earlier, activists protested the auction of mineral rights to 108,000 public acres, some of which went for as low as $12 an acre.[1654] Many of the activists publicly expressed their concern during the auction, only to be escorted out of the building; one was arrested for disturbing a public meeting. "I think we learned a lot today, and I think we made our voices heard to let them know that there's resistance out there for what they're doing and we're only going to go from there," Max Lockwood told WOOD-TV.[1655]

Six months later, with the auction of mineral rights to 193,000 acres, there were now more than 100 protestors in Lansing. The Natural Gas Subcommittee of the Republican-controlled House of Representatives, pushing for the state to be fracked, had recommended leasing mineral rights for all of its 5.3 million public acres. The protestors, now in organized groups, disrupted the auction. Forty were escorted out; police arrested seven.[1656] But those who believed public land shouldn't be fracked weren't done. They moved to the Lansing Public Library for workshops on how to effectively organize and protest.

The May 2010 auction, before there were protestors, brought $178 million in successful bids, almost all from representatives of Chesapeake Energy and the Encano Corp. However, a Reuters News investigation, based upon intercepted e-mails and published in June 2012, revealed that the two energy giants "repeatedly discussed how to avoid bidding against each other in a public land auction in Michigan two years ago and in at least nine prospective deals with private land owners."[1657] Such collusion is a violation of the Sherman Anti-Trust Act. The Reuters investigation triggered civil suits and criminal probes.[1658]

MIXED MESSAGE FROM THE OBAMA ADMINISTRATION

President Obama, continuing to send mixed messages about his "all of the above" approach to energy, approved more than 240 proposals in 2015 to frack land near the Chaco Culture National Historical Park in northwestern New Mexico. The area, site of a major pueblo development between 900 and 1150, is sacred to Native Americans. Fracking would "cause irreversible damage to ancient buildings, ceremonial roads and archeological sites," according to the Natural Resources Defense Council.[1659]

Jailed for Trying to Present the Truth

Alma Hasse, one of the founders of Idaho Residents Against Gas Extraction (IRAGE), went to a meeting of the Payette County Planning and Zoning Commission in October 2014 to listen to—and object to—the Commission's plans to issue a permit to Alta Mesa Idaho to expand a gas processing plant.

During the hearing, Hasse had discussed a number of issues, including questions about the three county commissioners and members of the Planning and Zoning Commission having conflicts of interest,[1660] because of possibly benefitting financially from owning property they had leased to oil and gas interests. After the commissioners closed the hearing, one of the commissioners charged Hasse with "telling whoppers" and "promulgating lies." Hasse called a "point of order" to respond to the personal attack, and to present additional information. The response by the Board's chair and the general counsel was to have her arrested for criminal trespass, and subsequently for resisting and obstructing arrest, although she had complied with the deputy's arrest.

What she did not comply with were continued demands by the police for her to identify herself. Invoking her constitutional rights, she remained silent. This led to her being jailed for eight days, including five days in isolation. "I kept saying, 'I want to talk to my attorney,'" Hasse told *Boise Weekly*.[1661] The state allows a person to be booked under a "Jane Doe" warrant, but insisted she identify herself, although she had previously identified herself during the hearing, and those involved in her arrest knew who she was.

"We find it concerning that she was removed from a public meeting very easily, and if any American is that easily removed from a public hearing, we should all be very concerned about the amount of power local officials have," says Leo Morales, executive director of the ACLU of Idaho. He says Hasse's constitutional rights were violated when she was imprisoned indefinitely without formal charges being filed. The judge imposed a $10,000 bail because he agreed with the prosecution that she was a "flight risk," although she was a long-time resident of the county, and known to most officials.

She "was given no access to a shower, no contact with her

husband, no access to clean clothes, and was forced to use the toilet while a male prison guard watched her," according to a subsequent civil suit filed against members of the Planning Commission, county commissioners, city and county police officers, and others.[1662]

Following a massive public protest, the judge dropped the bail requirement, and released her on her own recognizance after eight days of confinement.

Morales says it's "unclear why the county held her so long," and suggests the arrest and confinement is similar to those who are arrested and jailed as political prisoners.

In April 2015, almost six months after her arrest and two weeks before a jury trial was scheduled to begin,[1663] Payette County dropped all charges, but only after requiring Hasse to write a letter of apology for speaking out after the original zoning hearing was technically closed.

An Injunction Against the Truth

Monday morning, Oct. 21, 2013. Vera Scroggins, a retired real estate agent and nurse's aide, was in Common Pleas Court for Susquehanna County, Pa., to explain why a temporary injunction should not be issued against her. That injunction would require her to stay at least 150 feet from all properties where Cabot Oil and Gas had leased mineral rights, even if that distance was on public property. Because Cabot had leased mineral rights to 40 percent of Susquehanna County, about 300 square miles, almost any place Scroggins wanted to be was a place she was not allowed to be, even if the owner of the surface rights granted her permission.

Before her were three corporate lawyers, a lawyer from the county, and several Cabot employees who accused her of trespassing and causing irreparable harm to the company that had almost $1 billion in revenue the previous year.[1664]

Scroggins is someone who not only knows what is happening in the gas fields of northeastern Pennsylvania, but willingly devotes much of her day to helping others to see and understand the damage fracking causes. Since 2009, she has led government officials, journalists, and visitors from Pennsylvania, surrounding states, and foreign countries, on tours of the gas fields, to rigs and well pads, pipelines, compressor stations, and roads damaged by the heavy volume of truck traffic necessary to build and support the wells. As part of her tours, she introduces the visitors to those affected by fracking, to the people of northeast Pennsylvania who have had their health impacted, their air and water polluted.

And now in a court room in Montrose, she was accused of trespassing and forced to defend herself.

She asked Judge Kenneth W. Seamans for a continuance. She explained she only received the papers by mail the previous Thursday and was told she had 20 days to respond. She said that on Friday a sheriff's deputy came to her house with copies of the same papers that ordered her to court three days later. She explained she had tried to secure an attorney, but was unable to do so over the weekend.

Judge Seamans told her he wouldn't grant a continuance because she didn't give the court 24 hours notice. "He said that to grant a continuance would inconvenience three Cabot lawyers who had come from Pittsburgh [about 250 miles to the southwest], and I might have to pay their fees if the hearing was delayed," says Scroggins.

In four hours, Cabot called several witnesses—employees, security personnel, and subcontractors—to testify they saw her trespassing. They claimed her presence presented safety risks. "What we've seen is an increase in frequency and also the number of visitors she is putting in harm's way," Cabot's George Stark had told Staci Wilson, editor of the *Susquehanna County Independent*.[1665]

In her defense, Scroggins called three friends who had accompanied her to court. They testified she was always polite and never posed a safety risk. She says when she went onto a Cabot location, she always reported to the security or field office, and never received any written warnings or demands in the two years she was at the sites. When she was told to leave, she did, even if it sometimes took as much as an hour because Cabot security officers had often blocked her car. Cabot personnel on site never asked local police to arrest her for trespassing.

In court, Cabot personnel testified she was never a visitor, although she frequently had amicable chats with on-site managers since 2009. They claimed she was on company-owned access roads. She replied she primarily used public roads; the times her car or a chartered bus might have been on access roads they never blocked them—unlike gas industry vehicles that often keep drivers bottled up in traffic jams or which set times when residents can't use public roads, even leading to their own homes, because of heavy frack-truck traffic.

"I was blocked after going on sites and access roads several times since 2009, and kept up to an hour," says Sroggins, "but then allowed to leave." No police were called, she says. "If I'm trespassing, then charge me," she remembers saying. Cabot had never charged her, nor sent her written demands to cease her visits. None of the other five companies drilling in the county filed requests for injunctions.

For Cabot personnel, it had to be frustrating to have to deal with what they may have thought was a nosy pest who kept showing up at their work sites, possibly endangering herself, her own guests, and the workers. For Scroggins, she was there to explain drilling to many who had never seen a rig or well pad, and to videotape the truth about Cabot's operations and fracking in the Marcellus Shale. In court, Scroggins tried several times to explain she had documented health and safety violations at Cabot sites, many of which led to fines and citations by the DEP and OSHA. She tried to explain she put more than 500 videotapes on YouTube to show the damage Cabot and other companies are doing to the people. Every time she tried to present the evidence, a Cabot lawyer objected, and the judge struck the testimony from the record. However, when Judge Seamans asked her if she wished to take the stand to testify, stated she could be charged under criminal law and advised her she had the right not to speak and possibly incriminate herself—"I stopped talking."

That afternoon, Judge Seamans granted Cabot its preliminary injunction. The injunction, says Scott Michelman of Public Citizen, was "overbroad and violates her constitutional rights to freedom of speech and freedom of movement."

"I have a lot of friends who have leased mineral rights," says Scroggins, "this means I can't even go to their homes if invited." She also can't go to the recycling center—Susquehanna County leased 12.2 acres of mineral rights to Cabot.

If Scroggins is arrested, she won't be able to go to the Susquehanna County jail—it's also on those 12.2 leased acres.

Friday morning, March 28, 2014. Cabot Oil and Gas lawyers and Vera Scroggins were back in court for a hearing to modify Judge Seamans' original temporary injunction. This time, Cabot wanted the buffer zone extended to 500 feet, but couldn't show any reason why 500 feet was necessary. Unlike her first appearance when she didn't have legal representation, she now had Public Citizen, the Pennsylvania ACLU, and local attorney Gerald Kinchy represent her when she sought to vacate the order.

The original order didn't specify where Scroggins could or could not go, only that there was a 150-foot zone from any property owned or leased by Cabot. So, for slightly more than five months, she had to go to the courthouse in Montrose, dig through hundreds of documents, and figure it out for herself.

Seamans' revised order prohibited Scroggins from going within 100 feet of any active well pad or access roads of properties Cabot owns or has leased mineral rights. Land not being drilled, but which Cabot owns mineral rights, was no longer part of the injunction. That 100 feet separation was still far more than most injunctions call for; even abortion clinics typically have 15 feet exclusion zones to prevent violence, according to the brief filed in Scroggins' behalf. Although the judge agreed that his preliminary order may have been broad and violated Scroggins' First Amendment rights, the revised injunction still violated her First and Fourteenth Amendment rights. Nevertheless, Scroggins said she would sign the settlement agreement, but then refused. Seamans would later require Scroggins' attorneys to breech lawyer–client privilege and testify what they had discussed about the contract. What Seamans didn't do was to recuse himself for a confict-of-interest. On Nov. 9, 2007, he and Elexco Energy signed a mineral lease agreement for 79 acres; the lease was recorded on Dec. 21, 2007. On April 29, 2008, that lease was transferred to Southwestern Energy, and recorded on May 21, 2008.[1666]

Wednesday morning, Feb. 25, 2015. Vera Scroggins and her attorneys were again in court, trying to rebut claims she violated the injunction. Once again, she was facing Judge Kenneth W. Seamans, who had retired at the end of 2014, but had asked to be assigned to cases that originated in his court.

This time, Cabot Oil and Gas claimed that on Jan. 16, 2015, Scroggins, while leading another tour of the gas fields, walked on an access road to one of its operations. Scroggins argued she had parked in the private driveway of a friend 672 feet from Cabot property, and that the three persons she was hosting, including a French photojournalist, walked to the gate of an inactive Cabot operation, took pictures, and then walked back to the driveway where she waited for them. Cabot produced a worker who backed up the company's claim, and provided a photo of Scroggins not on Cabot land but on the driveway. Witnesses for Scroggins said she never violated the injunction.

Judge Seamans disagreed, ruled she was in contempt of court, and would fine her "somewhere between $300 and $1,000, maybe per day."

Thursday afternoon, April 23, 2015. Judge Seamans fined Scroggins $1,000 for violating his injunction; the money was to be paid to Cabot for a portion of its legal costs. Cabot "had a false witness, who was willing to perjure himself under oath, and the judge found him more credible. I am not willing to pay a fine for something I didn't do," said Scroggins.

Vera Scroggins never planned to be among the leaders of a social movement, but her persistence in explaining and documenting what is happening to the people and their environment put her there. Cabot's "take-no-prisoners" strategy in trying to shut her voice has led to even more people becoming aware of what fracking is—and the length that a megacorporation will go to keep the facts from the people.

No matter what Cabot says or does in its vigorous pursuit of a citizen-journalist, its purpose is not to protect the safety of its workers or visitors, but to shut down opposition. According to a brief filed by Scroggins' attorneys, "The injunction sends a chilling message to those who oppose fracking and wish to make their voices heard or to document practices that they fear will harm them and their neighbors. That message is loud and clear: criticize a gas company, and you'll pay for it."

And that's why the Cabot Oil and Gas Corp. wanted an injunction against Scroggins. It had little to do with keeping a peaceful protestor away or protecting worker safety; it had everything to do not only with shutting down her ability to tell the truth but also to put fear into others who might also wish to tell the truth about fracking and Cabot's operations.

Coda. About two weeks after Judge Seamans fined Vera Scroggins for trespassing, Susquehanna County DA Jason Legg filed six felony charges against her. The criminal complaint charged her with two counts each of violating Pennsylvania's wiretapping laws, specifically, intercepting communications, disclosing intercepted communications, and using intercepted communications.[1667]

On June 5, 2013, Scroggins and Craig Stevens, another anti-fracking activist, had gone to the office of Laurence Kelly, an attorney and head of the Montrose July 4th parade. Kelly denied the request by anti-fracking activists to participate in the parade, claiming they were too controversial.

Scroggins, with her video camera visible, taped that conversation. At the end of a brief discussion, Scroggins informed Kelly she was recording his comments; he ordered her to leave his office, saying that she was illegally recording him and his secretary. "I wanted to have notes of what he would say to me," says Scroggins. Two days after being denied a parade permit, Scroggins sent that tape to the district attorney through a private YouTube channel; she asked the DA, "Do not share for now." She says she sent the videotape and several legal questions about Kelly's refusal to allow anti-fracking activists to peacefully participate in the parade.

Two years later, the DA filed the wiretapping charges. Pennsylvania is one of a dozen states in which all persons must be aware they are being taped. In other states, only one party must be aware of the taping; thus, it would be legal for a person participating in that conversation to record the conversation, but it would be illegal for a third party not involved in that conversation to record the conversation. Recordings in a public place are major exceptions in Pennsylvania.[1668]

Scroggins might have been able to argue that Kelly and his secretary, although in a private office, were discussing a public issue. Further, Scroggins stated, and her actions indicate, she was recording the conversation for notes, not distribution, which is allowed in Pennsylvania. DA Legg refused to state why the case took two years to file or why the extenuating circumstances, permitted by law, did not influence his decision to file felony charges.

About six weeks after the charges were filed, Scroggins agreed to a plea bargain. The charges would be dropped if she accepted the terms of Accelerated Rehabilitative Disposition (ARD), which would include a fine, community service, and probation. If Scroggins did not comply with those terms or would be arrested during the one-year probation, the plea bargain would be revoked. She reluctantly agreed to the plea bargain because, she says, she "needed my resources to continue to fight the injunction" issued against her two months earlier.

Scroggins, and those in the anti-fracking movement, believe the filing of the felonies was a blatant case of intimidation and collusion between Cabot and local government.

Vera Scroggins, following the court hearing, asked the most important question of the day—"Who is the danger here? Me or an industry that is contaminating the air and water?"

Chilling a 'Terrorist' Anti-Fracking Group

When a dozen friends met in a basement in March 2010, they thought their purpose was to learn more about fracking and to try to help others learn about fracking's process and effects. As more persons joined the group, they moved to a church basement.

Within a few weeks, "we realized we had to become more active," says Dr. Tom Jiunta, the group's founder and the first president. And so they attended township meetings, successfully got a homeowner's association to overturn a member's decision to lease mineral rights to Encana Corp., and kept the company from further drilling after two test wells didn't prove as financially lucrative as it had hoped.

"We even developed a Code of Conduct," says Dr. Jiunta—"to conduct ourselves with professionalism, dignity, and kindness."

However, the Commonwealth of Pennsylvania had another idea of what the Gas Drilling Awareness Coalition (GDAC) of

Luzerne County, Pa., really was. In September, six months after GDAC was formed and became a non-profit organization, it learned it was a terrorist organization, featured on an intelligence bulletin created by the Institute of Terrorism Research and Response (ITRR), a private organization.

Those bulletins were distributed three times a week by the Pennsylvania Office of Homeland Security, which had a $125,000 a year no-bid contract with ITRR.[1669] The bulletins were sent not only to the state's law enforcement agencies but also to natural gas drillers, their associations and front groups. The ITRR was also "reporting" not only about GDAC and groups it considered to be eco-terrorists, but also about persons who wanted a moratorium on gas drilling, several gay rights activists, tax protestors, and even a meeting of the Pittsburgh City Council.[1670]

A bulletin in July 2010, according to reporting by Adam Federman, who has written extensively about government surveillance of persons and groups exercising their First Amendment rights, stated not only did ITRR believe that an "attack is likely [but it] might well be executed." That bulletin also described what it erroneously believed was a source for such an attack:

> "The escalating conflict over natural gas drilling in Pennsylvania may define local fault lines and potentially increase area environmentalist activity or eco-terrorism. GDAC communications have cited Northeastern Pennsylvania counties, specifically Wyoming, Lackawanna and Luzerne, as being in real 'need of our help' and as facing a 'drastic situation.'"[1671]

One bulletin, obtained by ProPublica, summarized an FBI advisory:

> "Environmental extremists continue to target the energy industry by committing criminal incidents, primarily to oppose the fossil fuel industry. To date, the energy industry has encountered little more than vandalism, trespassing and threats by environmental extremists. But this pattern is beginning to morph—transitioning to more criminal, extremist measures actions. Based on several recent reported criminal incidents toward energy companies and affiliated industries

or individuals, environmental extremism will become a greater threat to the energy industry owing to our historical understanding that some environmental extremists have progressed from committing low-level crimes against targets to more significant crimes over time in an effort to further the environmental extremism cause. The FBI also assesses—with medium confidence—extremists will continue to commit criminal activity against not only the energy companies, but against secondary or tertiary targets. This assessment includes the use of tactics to try to intimidate companies into making policy decisions deemed appropriate by extremists. Previous incidents also seem to support this assessment. Note: Lack of direct reporting on this issue prevents analysts from making a higher confidence level at this time. Potential Impact on Pennsylvania: Pennsylvania has gained a prominent position in the production of natural gas from drilling operations within the Marcellus Shale Formation. Analysts expect that groups of environmental activists and militants on the one hand—and property owners, mining and drilling companies on the other will be focusing their attention on one another in the future months as production increases."[1672]

One of those bulletins, which placed GDAC on a "watch list," Dr. Jiunta says, was accidentally sent to the wife of one of the persons who occasionally attended GDAC meetings.

That bulletin, says Dr. Jiunta, "had a chilling effect upon us. We became paranoid." Several members left GDAC; others took batteries out of their cell phones before attending meetings, out of fear of being tracked by state intelligence agencies. Dr. Jiunta says he even took different routes to work to avoid detection. "When our emails didn't go through or when our computers crashed," he says, "we believed it was because of a result of us being on a terrorist watch bulletin."

In mid-September 2010, Gov. Ed Rendell announced he was not renewing the contract between the state's Office of Homeland Security and ITRR, which expired a month later. At a news conference, the Governor stated:

"I am appalled that this contract was entered into without my knowledge. I am appalled that information was disseminated about groups that were exercising their constitutional right to free speech and to protest. They shouldn't be on any list [of security threats]. This is extraordinarily embarrassing."[1673]

370

Not long after the scandal broke, James Powers, director of the state's Office of Homeland Security, voluntarily resigned. The Commonwealth of Pennsylvania paid $40,000 to GDAC in January 2015 for wrongly including GDAC as a terrorist organization.

GDAC, which has never participated in eco-terrorism, continues to bring nationally-known anti-fracking speakers twice a year to the Wilkes-Barre/Scranton metropolitan area; most of the speakers pay their own expenses and stay overnight in member houses. Dr. Jiunta, a podiatrist, believes his background in biology and the other sciences "helps to translate the many scientific studies against fracking so people can better understand them."

There is still that fear. "I think we're still being watched," says Dr. Jiunta. He, and several thousand in the environmental movement, are probably justified to be paranoid about governmental surveillance.

'Ag-Gag' Laws

The Humane Society of the United States (HSUS) has been the nation's most effective organization to investigate and bring to the public's attention animal abuse, including cruelty to farm animals. HSUS methods have often included undercover surveillance, something that has infuriated corporations and self-righteous politicians.

Provisions of several bills make it a felony if persons even photograph farm operations while standing outside that farm's boundaries. Although the HSUS has been able to defeat Ag-Gag legislation in two dozen states, eight states—Idaho, Iowa, Kansas, Missouri, Montana, North Carolina, Utah, and Wyoming—have laws aimed at intimidating whistleblowers and investigators from documenting animal cruelty on factory farms and in slaughter-houses. The Wyoming law forbids anyone from taking "a sample of material, acquire, gather, photograph or otherwise preserve information in any form from open land which is submitted or intended to be submitted to any agency of the state or federal government."[1674] The broad interpretation of that law, pushed by ranchers and farmers, is that it is now illegal for anyone even to take water samples from public rivers or photograph contamination that seeps onto public lands. Other states have bills snaking through the process. Much of the legislation is written by the conservative American Legislative Exchange Council (ALEC). Persons who go undercover to record animal cruelty, according to the model legislation, would be placed on a terrorist registry.

"Animal welfare groups have exposed egregious animal cruelty through recordings and photos, and the industry's response hasn't been to clean up its act but to merely make it illegal to expose what's happening," says Matthew Dominguez of the HSUS.

"Ag-Gag" legislation can also impact getting the truth about fracking. Because most wells and rigs are on farm land, any recordings of what happens at a fracking site, including illegal activities, would also be illegal. This would prohibit workers, environmentalists, health care professionals and others from going to a site and documenting illegal or dangerous practices.

Robert E. Cooper Jr., Tennessee's attorney general, called proposed legislation in his state "constitutionally suspect," as "an impermissible prior restraint [that] could be found to constitute an unconstitutional burden on news gathering."[1675]

Cooper and numerous animal rights and environmental groups that oppose ag-gag laws could be right. Chief Judge B. Lynn Winmill of the federal district court for Idaho ruled that state's law unconstitutional. His 29-page decision in August 2015 declared the law violated the First Amendment because:

> "Audio and visual evidence is a uniquely persuasive means of conveying a message, and it can vindicate an undercover investigator or whistleblower who is otherwise disbelieved or ignored. . . . Prohibiting undercover investigators or whistle-blowers from recording an agricultural facility's operations inevitably suppresses a key type of speech because it limits the information that might later be published or broadcast."[1676]

The Reality of Dissent

Sanford, N.Y., sits on the Pennsylvania border in the Marcellus Shale region on the western banks of the West Branch of the Delaware River. That city's pro-fracking council issued a gag order in September 2012 to forbid citizens from speaking against fracking.[1677] In April 2013, faced by a First Amendment lawsuit filed by the Natural Resources Defense Council and Catskills Citizens for Safe Energy, the town board dropped the gag order.

The reality of power and money has kept many persons affected by fracking from filing against corporations. The corporations have full legal and financial resources, including the ability to buy expert testimony and to challenge any suit. For residents, most of whom are in the lower- or middle-class, the best they can hope for is that a law firm that specializes in environmental and social justice will take their case on a contingency basis.

The gas companies also know that spreading money can silence critics. It's not unusual for Big Energy to occasionally agree to settlements to silence those who have filed against their companies—the financial and legal teams understand it's just another way to do business—and to demand as part of the settlement a "gag order" to prohibit the injured parties from talking against the company or revealing details of the settlement, almost all of which require the plaintiff to give up any attempt to argue that fracking was the cause of health problems in lieu of the settlement. It's also not unusual to learn that a company might donate a few thousand dollars for "community benefit" or that corporations and the state environmental regulatory agencies trumpet the distribution of shale taxes to local counties.

The attitude of Big Government being hostile to the people is also a reason why many who have sustained damage do not speak up. "Perhaps it is against their culture to do so," says psychologist Diane Siegmund, "or maybe it is just because they see no way to get it made 'right.'"

She says there "is a lot of fear" among the residents, those whose lives are being uprooted, those whose health is being compromised, and those whose economic benefits may be compromised if fracking operations are reduced. "As long as the powers can keep the people isolated and fragmented," she says "the momentum for change can never be gained."

Academic integrity and independence could be tainted by significant donations by the oil and gas industry to colleges and universities.

PHOTO: Vera Scroggins

CHAPTER 20
Fracking Academic Integrity

Several colleges throughout the country, realizing they could make quick money from leases, have allowed the oil and gas industry onto their campuses. In West Virginia, both Bethany College and West Liberty University signed leases, claiming the money from royalties would help improve programs and provide for new buildings.[1678] The University of Texas at Arlington, sitting above the Barnett Shale, has 22 wells on a single pad site at the edge of campus.

At Indiana State University, President Dan Bradley, a petroleum engineer who touts fracking as "a freight train on steroids,"[1679] has permitted wells and pipes on campus.[1680]

Against significant student and community opposition, the University of Tennessee opened its 8,000 acre Cumberland Research Forest to the natural gas industry. Although the university had been trying to find a corporate partner since 2002 in order to get additional revenue,[1681] and the proposed 20-year lease would include a $300,000 a year payment plus at least 10 percent royalties, the university stated in March 2013 it was entering into agreements in order to "conduct unbiased, scientifically sound research." Dr. William F. Brown, dean of research and director of the Agricultural Experiment Station, explained, "We're hearing more questions about the environmental impact of natural gas extraction. We feel like we have an obligation to be able to answer those questions."[1682] However, because the research is funded by the natural gas industry, the ethical probability of a conflict of interest must be raised. If the university makes money from the industry, and a portion of that money is targeted for faculty research, how impartial can that research be?

SACRIFICING INCOME FOR THE ENVIRONMENT

During the Summer of 2012, representatives of natural gas companies approached Allegheny College, a small liberal arts college in the northwest part of Pennsylvania. Allegheny told staff, students, and faculty about the possibility of drilling on a 283-acre nature reserve and protected forest the college owns. "We believe in transparency," says Larry Lee, vice-president for administration. He also believed an open discussion would be a great educational experience." Although the college could make money from the drillers, most of those who would be affected opposed the drilling because it was "against the ethics of our college [which] made a commitment to environmental steward-ship and pledged to become climate neutral by 2020," said Brian Anderson, a senior at the time. "Leasing college land for natural gas extraction," said Anderson, "is a direct impediment to the goals of the college."[1683]

The issue became moot when Range Resources decided it was unable to profitably drill into the Utica Shale in that location. Part of the problem is that the depth of the shale requires longer drill bores, far more water and chemicals, and will bring up far more wastewater than drilling into the Marcellus Shale.

Political Infringement upon Science

"Offensive" is how Sen. James Inhofe (R-Okla.) described the EPA draft report that documented water pollution in Pavillion, Wyo., as being caused by the chemicals used in the fracking process. He claimed the EPA's study was an "apparent effort to reach a predetermined conclusion that hydraulic fracturing affects groundwater. Inhofe claimed the EPA's conclusions . . .

> "if made in an irresponsible way, could have devastating effects on natural gas development as well as our economy. . . . EPA's reckless process could put [600,000] jobs at risk.
> "EPA has gotten off to a dubious start and going forward, its investigation can have no credibility if it is not held to the highest standards."[1684]

Inhofe also claimed, "It is irresponsible for EPA to release such an explosive announcement without objective peer review."[1685] He and nine other Republican senators demanded the EPA classify the research as "highly influential scientific assessment," and subject it to a more rigorous "peer review" process. The senators' letter to the EPA followed a similar request by Encana, the wells' owner. However, the review process following a draft report, peer review, and public hearings, was already done by the EPA.

Inhofe's concern may have been more political than scientific. He had received $694,900 in campaign contributions from individuals and PACs associated with the oil and gas industry for the 2007–2012 election cycles,[1686] and $576,250 in the 2013–2014 cycle,[1687] according to data compiled by the Center for Responsive Politics. Inhofe became chair of the Senate's Environment and Public Works committee in January 2015.

Also attacking the EPA report were Wyoming politicians and newspapers, which echoed industry response. "The EPA may have poisoned the public debate by releasing its report," the *Casper Star–Tribune* editorialized.[1688] The newspaper called the report "clumsy" and claimed the samples were "improperly tested." Disregarding obvious ties between the industry and politics, the newspaper attacked the researchers. The *Star–Tribune* claimed the EPA research was "not about science and more about politics." In an OpEd rebuttal, James B. Martin, EPA regional administrator, stated the newspaper's claims "do a disservice to the rigorous scientific process EPA conducted," and pointed out the three year study was reviewed by EPA managers and "subjected to an initial peer review by independent experts."[1689]

Two months after the EPA draft report was published, the House Subcommittee on Energy held a hearing to question the EPA methods. The Committee determined the EPA "wells were drilled and installed without the State of Wyoming's knowledge or assistance. Without these records, it is difficult to eliminate the possibility that EPA's actions in drilling and installing the monitoring wells may have contributed to the contamination detected in the samples."[1690] In June 2013, the EPA announced it was terminating its study of the migration of methane into water supplies at Pavillion, and turning the investigation over to Wyoming.[1691] Abrahm Lustgarten of ProPublica explains a probable reason for the termination:

> "In private conversations . . . high-ranking agency officials acknowledge that fierce pressure from the drilling industry and its powerful allies on Capitol Hill—as well as financial constraints and a delicate policy balance sought by the White House—is squelching their ability to scrutinize not only the effects of oil and gas drilling, but other environmental protections as well.
> "Last year, the agency's budget was sliced 17 percent, to below 1998 levels. Sequestration forced further cuts, making research initiatives like the one in Pavillion harder to fund."[1692]

Michael Krancer, Pennsylvania's DEP secretary, had a broader concern than just the Pavillion study. "The myth that terrible chemicals are getting into the groundwater is completely myth,"[1693] he declared in November 2011. Later, before a Congressional hearing, Krancer claimed all studies that showed toxic methane gas in drinking water were "bogus," and specifically cited as "statistically and technically biased"[1694] a Duke University study.[1695] Two of the study's researchers fired back. In an OpEd commentary in the *Philadelphia Inquirer*, Drs. Robert Jackson and Avner Vengosh suggested, "Rather than working to discredit any science that challenges his views, the secretary and his agency should be working to get to the bottom of the science with an open mind."[1696]

The oil and gas industry figured out if it couldn't destroy public perception of scientific studies that showed possible connections between fracking and environmental and health effects, they would counter with their own studies. "Science is clearly on the side of development [of fossil fuel] and on the side of the industry" is the unsubstantiated opinion of Jack Gerard, American Petroleum Institute CEO.[1697]

Energy in Depth selected 137 research studies to showcase its public statements about how safe and beneficial horizontal fracking is. Included among the 137 studies were blog posts and PowerPoint presentations. However, as reported by the Public Accountability Initiative (PAI), only 14 percent of those 137 studies were subjected to peer review, and more than three-fourths of all the selected studies had either moderate or strong connections to the oil and gas industry. Many of the studies EID cited as evidence of the positive effects of fracking were "marred by conflicts of interest and lacking in academic rigor."[1698] Two of the studies—one at the University of Texas, the other at the State University of New York at Buffalo—were discredited and retracted by the institutions that published them; other studies did not disclose financial ties between the researchers and the industry.[1699] The research "by the oil and gas industry . . . is often closer to public relations than to actual science," warns Robert Galbraith, PAI research analyst and co-author of the analysis of the EID compendium of claims about the beneficial and safe effects from fracking.[1700]

Parsing the Truth in Oklahoma

Harold Hamm, billionaire founder and CEO of Continental Resources, wasn't happy. The reason he wasn't happy is because geologists had begun correlating earthquakes and deep injection wells.

Perhaps others might have asked the critical question, "What can we do to make these wastewater disposal wells safer?" But that's not what Hamm and the industry asked. They asked, "What can we do to make geologists stop saying these things that can hurt our industry?"

And so, one October day in 2013, Dr. Austin Holland, Oklahoma's state seismologist who was on the staff of the Oklahoma Geological Survey (OGS) at the University of Oklahoma, found himself in a meeting with Patrice Douglas, one of the members of the state's oil and gas regulatory agency, and Jack Stark, Continental Resources senior vice-president.[1701] "The basic [gist] of the meeting is that Continental does not feel induced seismicity is an issue and they are nervous about any dialogue about the subject,"[1702] Dr. Holland wrote in a subsequent memo to Dr. Larry Grillot, dean of the university's Mewbourne College of Earth and Energy, and Dr. G. Randy Keller, professor of geosciences and director of the Oklahoma Geological Survey (OGS).[1703] In that same memo, Dr. Holland pointed out Continental and much of the industry "are in the denial phase that [a correlation between earthquakes and deep well injection] is a possibility."

One month later, Dr. Holland was called to a meeting with Hamm and university president David Boren. Hamm was a major donor to the university, and had been targeted but did not contribute for a major $25 million contribution to fund an oil and gas exploration division. Boren, who had served one term as governor and three terms as a U.S. senator before becoming university president, was a member of the board of directors of Hamm's company, and had received about $1.6 million in cash and stocks from Continental between 2009 and 2015, according to *EnergyWire*.[1704] What Hamm and Boren were focused upon was that Holland be careful in his comments of any connection between earthquakes and injection wells.

That meeting was "just a little bit intimidating," Dr. Holland told *Bloomberg News*. In an e-mail to *Bloomberg News*, Kristin Thomas, Continental's spokeswoman, said the meeting was "cordial and an information exchange with Austin."[1705] Boren said the meeting was "purely informational." Boren explained, "Mr. Hamm is a very reputable producer and wanted to know if Mr. Holland had found any information which might be helpful to producers in adopting best practices that would help prevent any possible connection between drilling and seismic events. In addition, he wanted to make sure that the Survey had the benefit of research by Continental geologists."[1706]

Whether "intimidating" or "cordial" and "purely informational," three months after the meeting with Hamm and Boren, the Oklahoma Geological Survey, in a position statement, claimed the "majority, but not all, of the recent earthquakes [in Oklahoma] appear to be the result of natural stresses."[1707] Much of the language of that statement was drafted by Dean Grillot, who had been an industry geophysicist for 30 years.

In an email to colleagues, one of dozens obtained by Mike Soraghan of *EnergyWire*, Dr. Holland expressed concerns about the language of the position paper and says he "tried to make some changes that help me feel a little better about things."[1708] In an e-mail to Danny Hilliard, university vice-president of government relations, which *Bloomberg News* obtained, Dean Grillot, concerned about corporate interference in university affairs and preservation of academic freedom, noted that Hamm "is very upset at some of the earthquake reporting to the point that he would like to see select OGS staff dismissed."[1709] In that same e-mail, Dean Grillot noted that Hamm said he planned to meet with Gov. Mary Fallin "on the topic of moving the OGS out of the University of Oklahoma."[1710] Dean Grillot later said he didn't tell his staff and faculty about Hamm's memo because "I didn't want it to impact their day-to-day work."[1711] No staff were disciplined or fired.

Dr. Holland told *EnergyWire*, "None of these conversations affect the science that we are working on producing."[1712] The OGS eventually declared the majority of recent earthquakes in Oklahoma could be attributed to the injection of wastewater and other fluids at high pressure into deep wells.

Defending Truth in Colorado and Wyoming

Dr. Geoffrey Thyne, a hydrogeologist, didn't plan to be an expert witness for law firms. But, that's the way it turned out shortly before he planned to retire.

He had spent most of his career working in the oil/gas industry and then in academics. He didn't have problems when he worked for ARCO for seven years after earning an M.S. in oceanography from the University of South Florida. However, he did have problems in academics when he tried to tell the truth.

After five years as an assistant professor at California State University at Bakersfield, in the heart of the state's rich oil industry, he left to become associate professor/researcher at the Colorado School of Mines (CSM), a public university with a strong reputation in engineering and applied sciences. For 10 years, he taught and did research. But in 2006, as horizontal fracking began to be the way the industry was headed, he learned that research is compromised by politics.

That's the year he was asked by the Oil and Gas Accountability Project (OGAP) to evaluate an EPA study about horizontal fracking. The EPA study, conducted during the Bush–Cheney administration, had claimed there were no problems with horizontal fracking.

"I wasn't aware of the study or much about fracking," says Dr. Thyne, "but I looked at the document and said it appeared to be political, and I can't speak to a political issue." He did say there was no data to lead to the EPA conclusions, which would eventually be used to help justify the Halliburton Loophole, which exempted the industry from numerous environmental laws. But, it was Dr. Thyne's observation about the validity of the EPA report that upset the university's administration. That's when Dr. Murray Hitzman, his department chair, called him in to discuss the report.

Dr. Hitzman, who had worked in the fossil fuel industry for 18 years and in the White House Office of Science and Technology for two years before taking a professorship at CSM,[1713] wanted to know if Dr. Thyne had claimed there wasn't sufficient data. "I told him I wasn't against fracking," says Dr. Thyne, "but that I didn't think the EPA conclusions were justified." That's when the department chair said that not only couldn't Dr. Thyne say that, but he couldn't even identify himself as from CSM.

And so, Dr. Thyne went back to his job, careful never to use his affiliation when discussing anything about fracking, and began working for the university's Colorado Energy Research Institute, advocating that the university get involved with research about fracking. But, research about fracking upset some in the administration, one of whom was Dr. Myles W. Scoggins who had worked for Mobil and ExxonMobil for more than three decades, eventually becoming president of the International Exploration & Production and Global Exploration division and then executive vice president of ExxonMobil Production Company before becoming CSM president in 2006.[1714]

In 2014, the last year of his presidency, Dr. Scoggins received $380,000 in salary[1715] and, according to the Public Affairs Institute, about $800,000 from being on the boards of three oil and energy companies[1716] and, for four years (2007–2011), on the board of directors of the Colorado Oil and Gas Association (COGA).[1717] It would be the COGA that Dr. Thyne believes protested independent academic research about fracking.

This time, instead of a department head telling him never to use his university affiliation in his research and public statements, it was a university vice-president. Dr. Nigel T. Middleton, vice-president of academic affairs, told Dr. Thyne the university was dropping him to half-time employment and ordered him not to discuss fracking. Dr. Middleton also has a long history of work with the oil and gas industry. In addition to work for oil corporations, Dr. Middleton was "a principal in developing a major collaborative partnership between Colorado School of Mines and The Petroleum Institute in Abu Dhabi, through the sponsorship of the Abu Dhabi National Oil Company and major multinational corporate participants," according to his official biography.[1718]

In an official public relations statement, CSM denied terminating Dr. Thyne's employment:

> "No one in the Mines administration recalls having anything but cordial conversations with Dr. Thyne this spring. When Dr. Thyne was quoted during that time by the media, the school received inquiries about Dr. Thyne's association with Mines.
>
> "As a result, Mines officials phoned and e-mailed Dr. Thyne to inform him of the inquiries, and also to remind him of the university policy that people must be clear in public communications that the opinions they express are personal and do not represent institution positions—one way or another—on issues being discussed."[1719]

The university claimed Dr. Thyne left CSM solely because he had another job. However, that was a carefully-couched distortion of truth. CSM did not renew his contract after he did an interview with National Public Radio, and reiterated his position that there was insufficient data to justify EPA conclusions.

The American Association of University Professors had wanted to take up Dr. Thyne's case as a violation of academic freedom, but he declined because "by that time, it seemed to be a no-win situation."

The next year, he became a researcher at the University of Wyoming, from where he had received a Ph.D. in geology in 1991. This time, six years after he began working at the university, a comment made to a local newspaper led to his termination. The Cheyenne *Tribune–Eagle* had published a five-part series about fracking and water usage. He says he had told the reporter each well could use two to ten million gallons of water, but for certain wells the water used could be 350,000 to one million gallons per stage, and that there could be as many as 40 stages of drilling. The reporter took the maximum per stage, and the maximum number of stages, and noted there could be more than 40 million gallons of water used. The source of the highest possible number of gallons was unattributed. However, representatives of the industry demanded to know the source, which the newspaper revealed.

That eventually led to Dr. Thyne being called before the university's vice-president of government affairs and a representative from Noble Energy, who demanded he retract the highest number, a number Dr. Thyne had never given the newspaper. Like CSM, the University of Wyoming also demanded that Dr. Thyne deny that any of his comments represented the views of the University of Wyoming. Shortly after that meeting, Dr. Thyne was told, "Your services are no longer needed." He was never told why his employment was terminated. Because Wyoming is a "right-to-work" state, there was no grievance procedure. The university could easily claim, without having to prove the truth, that there were no more research funds to justify Dr Thyne's continued employment. David Mohrbacher, director of the university's Enhanced Oil Recovery Institute (EORI), told the *Boulder* (Colo.) *Weekly*:

> "The reason for not renewing his contract was not related to his participation with the press. It was not related to that. It was based on EORI achieving their technical, strategic objectives. We chose a different way to go, and really that's all I can say."[1720]

Dr. Thyne's last academic employment was in 2012. "With fracking booming, I thought there would be a lot of jobs," he says, but no one in academia had wanted him. A couple of years later, he found out why. "A friend told me to check out YouTube." On that social network, he found a one-minute video,[1721] which he had recorded in 2011, that stated human error in the fracking process can cause water pollution.

"I'm not naive, I understand politics," says Dr. Thyne, who acknowledges, "It's been a difficult transition," but one he accepts because he will not sacrifice his academic integrity for political convenience.

A QUESTION OF INDEPENDENCE

Public relations and academic independence clashed in a research study that was funded by organizations with oil/gas industry ties, and then was promoted by the industry in 2014 to try to convince Colorado voters to reject proposed bans and moratoriums against fracking.

Brian Lewandowski and Dr. Richard Wobbekind of the Business Research Division (BRD) at the University of Colorado determined there would be 68,000 fewer jobs and about $8 billion in lower gross domestic product (GDP) between 2015 and 2020 if Colorado cities and the state enacted bans or moratoriums.[1722] (A subsequent update, which considered the declining production of natural gas and lower prices, estimated job loss at 18,000–36,000 and a decrease in GDP of $2.2 billion–$4.4 billion.[1723])

Most studies from the university's Business Research Division have outside funding. Funding the research for the report on a fracking ban were the Metro Denver Economic Development Corp., the Denver South Economic Development Partnership, and the Common Sense Policy Roundtable (CSPR); the consortium also provided the economic modeling system for the research. An investigation by Joel Dyer of the *Boulder Weekly* revealed extensive ties between CSPR and the oil/gas industry.[1724] Four of the eight board members own oil/gas companies; most are affiliated with industry front groups. CSPR's founder and executive director is Kristin Strohm, co-founder of the Starboard Group, a lobbying and PR agency that represents the Western Energy Alliance PAC and which Dyer identifies as "possibly the most powerful fundraising entity for the Republican Party in the Western U.S." The other two staff members of CSPR are executives of EIS. Strohm's husband is vice-president of EIS Solutions, which creates front organizations for the oil/gas industry. However, none of the 11 staff members of the Metro Denver Economic Development Corp. have extensive ties to the fossil fuel industry;[1725] the board of the Denver South Economic Development Partnership includes politicians, and those employed by banks, real estate and property management companies.[1726]

Greenpeace, which co-funded requests for hundreds of University of Colorado memos and documents through the state's Open Records Act, argued that research studies at the BRD, including the fracking ban study, were submitted to the oil/gas industry for review and comment prior to their public release.[1727] Lewandowski acknowledges the sponsored research reports are reviewed "by the funding entities prior to release, but at no time are funders able to change the conclusions or the numbers."

Dyer agrees, stating there is no evidence the researchers "ever intentionally altered... research outcomes to suit the desires of CSPR or its funding partners" or that the researchers "have done anything inappropriate."[1728]

Nevertheless, one major issue overrides all others—how independent can the research be if outside companies provide the funding and direction, can review the study prior to publication, and then use parts of that study to further their own economic interests?

The Problems of Integrity in New York

The natural gas industry needs to "seek out academic studies and champion with universities—because that again provides tremendous credibility to the overall process," said S. Dennis Holbrook, executive vice-president and chief legal officer of Norse Energy and a member of the board of directors of the Independent Oil and Gas Association of New York (IOGA).[1729] Holbrook pointed out at an industry conference Oct. 31–Nov. 1, 2011, that IOGA has "done a variety of . . . activities where we've gotten the academics to sponsor programs and bring in people for public sessions to educate them on a variety of different topics."[1730]

One of the ways IOGA helped direct academic research is by its connection to SUNY's Shale Resources and Society Institute (SRSI), which sponsored lectures, workshops, and professional papers. Among those papers was "Environmental Impacts During Shale Gas Drilling: Causes, Impacts and Remedies."[1731] According to Steve Horn, who wrote an extensive analysis of the SUNY/Buffalo ties to the Marcellus Shale Industry, "Calling the final product a 'study' is a generous way of putting it. . . .

All four co-authors had ties to the oil and gas industry, as did four of five of its peer reviewers. The study didn't contain any acknowledgement of these ties."[1732]

An informal group of faculty, students, alumni, and citizens (University of Buffalo Coalition for Leading Ethically in Academic Research—UB CLEAR) stated that the Institute and the research emanating from it were not only "fatally compromised," but that it represented "not the independent search for knowledge proper to a university but a frantic and servile willingness to sell academic legitimacy to a public relations campaign for the gas industry."[1733]

The Public Accountability Project analyzed the SUNY/Buffalo study and "identified a number of problems that undermine its conclusion." Among the problems were:

> [D]ata in the report shows that the likelihood of major environmental events has actually gone up, contradicting the report's central claim; entire passages were lifted from an explicitly pro-fracking Manhattan Institute report; and report's authors and reviewers have extensive ties to the natural gas industry.
>
> "[T]he serious flaws in the report, industry-friendly spin, strong industry ties, and fundraising plans raise serious questions about the . . . Institute's independence and the University at Buffalo's decision to lend its independent, academic authority to the Institute's work."[1734]

In November 2012, six months after the Institute was created, SUNY/Buffalo closed it.[1735]

A CONFLICT AT SYRACUSE UNIVERSITY

A research study published in the March 12, 2015, issue of *Environmental Science and Technology,* concluded fracking was not responsible for methane contamination of drinking wells in Pennsylvania.[1736] The study reviewed 11,300 samples from Chesapeake Energy wells. According to the researchers, "Previous analyses [which showed correlations between proximity of fracking operations and water wells] used small sample sets compared to the population of domestic wells available, which may explain the difference in prior findings compared to ours."[1737]

Most academic publications require researchers to disclose possible conficts-of-interest and financial ties that could compromise the integrity of their research. In this case, the study first appeared online, with the authors declaring, "The authors declare no competing financial interests."[1738] However, the authors did have competing financial interests. Dr. Donald Siegel, the study's primary researcher, a hydrologist and professor of earth sciences at Syracuse University, had previously told both *Syracuse University Magazine* and *Inside Climate News* he had received financial compensation from Chesapeake; he didn't disclose the amount. Bert Smith, one of the authors, had worked for Chesapeake and was currently employed by a company that does consulting work for Chesapeake. Two months after the study first appeared, the journal published a correction that identified the industry ties of both Dr. Siegel and Smith.[1739]

Even if there was no possible conflict-of-interest, there was no indication if Chesapeake withheld any information from the database it provided the researchers, which would have skewed the results.

Selling Out Integrity in North Dakota

Among the mission statements of the University of North Dakota Department of Geology and Geological Engineering is that it "strives to develop in its engineering graduates keen insight and abilities to design an environmentally sound and sustainable future for humanity."

Like most college mission statements, it's a broad and vague goal, one that may not reflect reality. The Department is one of the better ones in the country, especially in training students to work in areas of gas and oil exploration and processing. However, their training—and research by the faculty—may be tainted by an industry bias, fueled by a $14 million gift.[1740]

The Department is now the Harold Hamm School of Geology and Geological Engineering. Hamm, CEO of Continental Resources, the ninth largest producer in the United States,[1741] provided $5 million to the renamed School; his company provided an additional $5 million; the other $4 million came from the Industrial Commission/Oil and Gas Research Program,

388

a merger of the state of North Dakota and several gas and oil corporations.

Continental Resources, which had revenue of $3.78 billion and a net profit of $977.3 million in 2014,[1742] had opened up the oil shale exploration in North Dakota, site of the Bakken Shale, and is currently the top producer of oil production in the country. Continental, which uses high volume horizontal fracturing to extract the oil, produced 62.5–65.5 million barrels of oil in 2014, an increase in production of 26–32 percent from the previous year.[1743]

Fracking Academic Integrity in Texas

When completed in 2017, the Engineering Education and Research Center (EERC) at the University of Texas will be housed in an eight-story 430,000 square foot $310 million facility. The state, with deep connections to the fossil fuel industry, authorized the university to borrow about two-thirds of that total; the rest is from private donations. Among those donations are $20 million from the Mulva Family Foundation[1744] and $5 million from Rex and Renda Tillerson.[1745] Jim Mulva, a university alumnus and long-time personal friend of the university's president, was CEO of ConocoPhillips; Rex W. Tillerson, also a university alumnus, is CEO and chairman of ExxonMobil.

It's possible the large amount of funding from fossil fuel sources won't affect the academic integrity of the teaching and research missions of the faculty. But, it's also possible there might be continued problems, similar to past academic scandals.

Research conducted by Drs. Charles G. Groat and Thomas W. Grimshaw and a team from the Energy Institute at the University of Texas placed the primary problem of methane in well water with the construction problems in both natural gas wells and drinking water wells rather than the process itself. The authors stated in February 2012, "[T]here is at present little or no evidence of groundwater contamination from hydraulic fracturing of shales at normal depths."[1746] The presence of methane and toxins in well water, said the research team, probably pre-dated natural gas fracking and that a major problem is well construction—Pennsylvania and Alaska are the only two states that do not have regulations for fresh water well construction:

> "[M]any of the water quality changes observed in water wells in a similar time frame as shale gas operations may be due to mobilization of constituents that were already present in the wells by energy (vibrations and pressure pulses) put into the ground during drilling and other operations rather than by hydraulic fracturing fluids or leakage from the well casing. As the vibrations and pressure changes disturb the wells, accumulated particles of iron and manganese oxides, as well as other materials on the casing wall and well bottom, may become agitated into suspension causing changes in color (red, orange or gold), increasing turbidity, and release of odors. . . .
>
> "The greatest potential for impacts from a shale gas well appears to be from failure of the well integrity, with leakage into an aquifer of fluids that flow upward in the annulus between the casing and the borehole . . . . In general, a loss of well integrity and associated leakage has been the greatest concern for natural gas–leading to home explosions as described in a subsequent section. . . .
>
> "[I]n most cases, [explosions of fresh water wells are] the result of naturally-occurring methane migration into aquifers and wells before shale gas development began."[1747]

Essentially, Dr. Groat's study supported the industry's claims that fracking doesn't cause health and pollution problems.

However, the Public Accountability Initiative revealed in July 2012 that research by Dr. Groat may have been compromised by a conflict of interest.[1748] The Initiative, a non-profit group, disclosed that Dr. Groat was a member of the board of Plains Exploration and Production Co., which conducts fracking operations. He owned 40,000 shares of stock, and received an annual fee for being a member of the Board; in 2011, it was $58,500, according to *Bloomberg News*.[1749] Between November 2007, when he became a member of the Board, and 2012, Groat received about $1.6 million in stock from the company.

The Initiative noted that the research by Dr. Groat and his team was distinguished by "bold, definitive, industry-friendly claims highlighted in the press release but not supported by the underlying report; evidence of poor scholarship and industry bias; and dubious and inaccurate claims of peer review" that had led the media to report there was no relationship between fracking and health and pollution problems. In response, Dr. Groat said his role "was to organize [the study], coordinate the activities and report their conclusions."[1750] He claimed he did not "alter their conclusions" and his presence on the Pioneer board had "no bearing on the results of the study."[1751]

An independent investigation initiated by the University of Texas, and released in December 2012, found "failures and inadequacies in several procedural areas,"[1752] and that the study "fell short of contemporary standards for scientific work."[1753]

Dr. Groat retired from the university the month before the report was released, and became director of the Water Institute of the Gulf. Dr. Raymond Orbach, the Energy Institute's director, who did not participate in the project, also resigned; he was not under investigation and kept his tenured status on the faculty.[1754]

A University of Texas study, published in the *Proceedings of the National Academy of Sciences* in September 2013, concluded there were minimal leaks of methane from fracked wells.[1755] The oil and gas industry was ecstatic. Energy in Depth triumphantly reported:

> "The activist fear-mongering about methane emissions has been exposed as fraudulent by the most comprehensive research on the subject to date, including data that incorporates the first-ever direct measurements of methane emissions from various segments of the production process. Perhaps now we can all come together, take a deep breath, and recognize the clear economic and environmental benefits of natural gas from shale."[1756]

Although there were no scientific flaws in the recording, analysis, and conclusions of the study, there were several problems with the methodology. Sharon Kelly, an attorney, journalist, and long-time environmentalist who analyzed the University of Texas study for DeSmogBlog, noted: "The vast majority of the wells studied used leak-control technology that has yet to be adopted at many, if not most, oil and gas wells, while others were wells that produced very little gas and consequently even serious leaks would produce relatively small emissions."[1757]

Physicians Scientists & Engineers for Healthy Energy (PSE) determined the study was "fatally flawed." Among the problems, the study measured emissions from only 489 wells, the wells were selected for the study by the oil and gas industry rather than by scientists on a random basis, the study did not measure the complete life cycle of the well (*i.e.*, transmission, compression, processing, storage and distribution), and the study didn't discuss or account for several independent research studies that showed higher levels of methane release at the wellhead. The PSE analysis of the research concluded:

> "Given the politically charged environment around unconventional natural gas development, we must question whether this study is simply an attempt to manipulate science and reverse the political discussions of fugitive methane emissions. A confirmation of high rates of fugitive methane losses as is concluded in all of the field-level studies to date . . . would discredit the 'clean natural gas' narrative. It is likely that a higher methane emission rate would necessitate more regulatory oversight of the oil and gas industry and this study may be an industry maneuver to counter that possibility."[1758]

Manipulating Pennsylvania's Universities

Within weeks of taking office in January 2011, Gov. Tom Corbett cut funding for K–12 education by $335 million. The 2012–2013 education budget remained the same as the previous year, having already suffered severe cuts in the 2011–2012 budget. He also proposed cutting funding by half for the 14 state-owned universities. Although the final cut was only 20 percent, the 110,000-student system was forced to terminate 5 percent of its workforce.[1759] However, Gov. Corbett had a strange idea how the 14 state-owned universities could restore some of their budget he had wanted to cut by half.[1760] In April 2011, he had suggested that the State System of Higher Education (SSHE) could allow natural gas drilling on the campuses that sit above the Marcellus Shale. Several months later, the state Senate passed a bill sponsored by Donald C. White that authorized state officials to lease mineral rights beneath state land to gas, oil, and other mining companies. White was the recipient of $94,150 in donations from PACs and individuals associated with the natural gas Industry, according to Pennsylvania Common Cause.[1761] The Senate passed the Indigenous Mineral Resource Development Act (SB 367[1762]), 46–3; the House passed the bill, 136–62. Although only six of the 14 universities are in the Marcellus Shale, the bill permits drilling for oil, coal, coal-bed methane, or limestone on all university campuses. Gov. Corbett signed the bill in October 2012.

In an attempt to placate the SSHE—but especially to compromise their academic integrity and acceptance of proposed revenue in exchange for supporting fracking—the new act allowed the university where the gas is extracted to retain one-half of all royalties; 35 percent would go to the other state universities; 15 percent would be used for tuition assistance at the 14 state universities.

The Legislative Assembly of the Association of Pennsylvania State College and University Faculties (APSCUF), the union that represents the system's 6,000 faculty, passed a resolution in September 2013 opposing drilling on campuses:

> "PASSHE campuses are not appropriate locations for hydraulic fracturing (fracking), that given the environmental and health hazards of the fracking process, including all of its infrastructure and associated enterprises, its presence on PASSHE campuses is inconsistent and potentially deleterious to the PASSHE educational mission as well as to the health and welfare of PASSHE community members."[1763]

The SSHE chancellor's office supports drilling on or near state universities, and has actively worked with the natural gas industry to create programs.

California University of Pennsylvania, a SSHE institution about 35 miles south of Pittsburgh, had ceded mineral rights to the natural gas industry. In a secret negotiation revealed by the *Pittsburgh Post-Gazette*, the Student Association, which owns recreational and dormitory space at the university, signed a lease in January 2011 with Antero Resources Appalachian for subsurface drilling rights on 67 acres.[1764] The lease included a confidentiality clause. The *Post-Gazette* reported in November 2011 that the Association initially refused to say if it had such a lease, "even though its offices are on campus and its executive director is a CalU employee."[1765] That executive director was the university's dean of students and acting vice-president for student affairs. The university's president told the *Post-Gazette* he didn't know about the lease. However, he had previously suggested to Antero Resources that since the university itself wasn't authorized to enter into any lease, it might wish to contact the student association.

At Indiana University of Pennsylvania, the energy resources track, begun in Fall 2009 in the Department of Geoscience prepares students for "direct entry into the energy industry with a focus on the discovery and development of energy resources and geophysical exploration techniques."[1766] The university added two faculty to the new program, which averages about 30 students, and graduated the first dozen in Spring 2014. IUP added a course in Fall 2015, Geology of Energy, which covers fossil fuel and alternative energy, according to Dr. Steven Hovan, department chair.

The Marcellus Institute at Mansfield University, which has more than 1,500 permitted wells,[1767] is "an academic/shale gas partnership," designed to educate the people about the issues of natural gas production. The university holds summer classes for teachers and week-long camps for high school students to allow them to "learn about the development of shale gas resources in our region and the career and educational opportunities availableto you after high school!"[1768]

The university's associate in applied sciences (A.A.S.) degree in natural gas production and services, begun in Fall semester 2012, has five separate tracks—Permitting and Inspection, Mudlogging/Geologic Technician, Environmental Technician, GIS Technician, and Safety Management. The degree was fast-tracked, submitted and approved in less than six months rather than the 12–18 months normally required for approval. "The industry is here and now, and we needed to take advantage of that," said Lindsey Sikorski, interim Marcellus Institute director. The program began when the university and SSHE noticed that 40 of its graduates, most in geology and geography, were working in the industry, she said.

The university "will take as many students as we can," said Sikorski, although only one new faculty position was approved. The SSHE administration encourages larger class sizes and fewer permanent professors. The program, with about 35 students, was moved into the Department of Geosciences in October 2014 and renamed the Institute of Science and the Environment in order to "encourage the research component," says Dr. Jennifer Temchak, department chair. The department has an actve speaker series, Earth Day activities, and "supports both sides of fossil fuel energy—we are neither for nor against it," says Dr. Temchak. The department, which now has an environmental issues course, has not received grants from the industry. However, the reality is that energy companies and their lobbying groups may eventually fill a hole created when Gov. Corbett slashed higher education funding and the system's administration at that time refused to protect academic integrity in the state-owned universities.

With the election of Dr. Tom Wolf as governor, tuition was frozen and the educational budget increased for the 2015–2016 fiscal year.

The Community College of Philadelphia (CCP)—assisted by the state's Department of Labor and Industry and the Marcellus Shale Coalition—created the Energy Training Center in November 2012 to offer certificate and academic programs for workers either already employed in the fossil fuel industry or intending to enter jobs that provide services to Marcellus Shale companies. The Marcellus Shale Coalition gave CCP an initial $15,000 grant for scholarships.

In a news release loaded with pro-Corbett and pro-industry appeal, college president Stephen M. Curtis announced, "The goal is to support the supply chain now serving energy companies and offer specialized career training that connects residents to the high-pay, high-demand career paths."[1769]

Creation of the Center "is short-sighted and foolhardy [since] we now know that shale gas drilling actually accelerates climate change," warned Margaret Stephens, associate professor of environmental conservation and geography.

Several faculty questioned the accelerated pace in which the program was initiated and the failure to inform most staff and faculty more than a day before the public announcement. Dr. Miles Grosbard, head of the Department of Architecture, Design and Construction, charged:

> "Normal college procedures for instituting new academic curricula were completely sidestepped. There is no information available about the proposed unit's mission, student audience, administrative structure, budget, facilities or educational objectives, apparently because none exists. Moreover, $15,000 is an impossibly tiny endowment to even begin a training center."[1770]

John Braxton, an ecologist and assistant professor of biology, said CCP "must not be used as a PR puppet for shale gas fracking companies," accurately noting that the fracking industry "got a free publicity ride" by the administration's hasty decisions.

Within two weeks of CCP's announcement, the faculty union (AFT Local 2026), which represents the college's 1,050 faculty and 200 staff, condemned the decision to establish the Center "without the consideration or approval of the faculty, and with total disregard for established College procedures for instituting new academic curricula." In a unanimous vote by the Representative Council, the faculty declared, "the natural gas drilling . . . industry and peripheral and related industries present unacceptable dangers and risks to public health, worker safety, the natural environment, and quality of life." The faculty called for the college administration to "sever all ties to the Coalition, halt the implementation of any workforce training efforts related to the shale-gas fracking industry, and insist that any energy training programs be disassociated from the fracking industry and decided through the normal College governance process." In contrast to pushing training and education to benefit the fracking industry, the faculty urged the college "to expand its initiatives and offerings in clean, green energy, and environmental career fields."[1771] Curtis left CCP in Summer 2013; the proposed program was never developed.

ENDOWED AND SOLD OUT

Lackawanna College, a private two-year college in Scranton, Pa., became a prostitute in April 2014.

The administration doesn't think of themselves as prostitutes. They believe they are doing a public service. Of course, call-girls also believe they are doing a public service.

Lackawanna College's price is $2.5 million.

That's how much Cabot Oil & Gas paid to the School of Petroleum and Natural Gas, whose nine-building campus is in New Milford in northeastern Pennsylvania. On the School's logo are the words, "Endowed by Cabot Oil & Gas Corporation."

That would be the same Cabot Oil & Gas Corporation that by the time it endowed the program had racked up more than 500 violations since it first used horizontal fracking to extract gas in 2008.

That would be the same company that was found to be responsible for significant environmental and health damages in Dimock, Pa.

It's the same company that managed to keep Vera Scroggins, a peaceful grandmother and anti-fracking activist, not only off its property but away from Susquehanna County's recycling center, a hospital, grocery stores, restaurants, several public roads, and 40 percent of the county where Cabot has mineral-rights leases.

Lackawanna College proudly claims its Cabot-endowed School is "focused on its vision of becoming a nationally-recognized, first in class program in the field of petroleum and natural gas technology." There is no question the School is fulfilling its promise. A $500,000 outdoor field laboratory simulates a working gas field; all students are required to complete internships. The School's mission includes creating "a campus that is focused and dedicated to the oil and gas industry."

Richard Marquardt, the School's executive director, has B.S. degrees in petroleum engineering and business management, as well as a long history of work in the industry. The eight other full-time faculty also have engineering degrees and significant industry experience. Fifteen adjunct faculty also have significant industry experience.

The 150 full-time students major in one of four programs—petroleum and natural-gas technology, natural-gas compression technology, petroleum and natural-gas measurement, and petroleum and natural-gas business administration.

397

Admission to the School's academic programs "is highly competitive," with students needing a strong science and math background prior to acceptance, says Marquardt. The students earn an associate in science degree upon completion of the two-year program. "It is focused on a very specific market," says Marquardt, providing personnel at a level between the vocational training programs and the B.S. in engineering programs. The placement rate is over 90 percent, he says.

In their fourth semester, students take a course in "Leadership, Ethics, & Regulations," which, according to the course description, explores "the holistic environment in which the Petroleum and Natural Gas industry operates, including the effect of corporate leadership on the company's credibility and reputation; real world ethical issues . . . and the relationship of the industry to federal, state, and local governments, including regulatory agencies."[1772]

An associate's degree doesn't mean the students, no matter how prepared they are to work in the shale-gas industry, will be exposed to the issues, reports, and scientific studies that suggest fracking causes significant environmental and health problems, major concerns of those who oppose the process of horizontal fracking. After all, Cabot wasn't going to invest in a college program that presented all sides of the issues. Nor is Cabot likely to invest anything more if the college expands its program to require that students also take classes in renewable energy, and the health and environmental effects of fracking. Other major gas and oil companies and suppliers—including Anadarko, Baker Hughes, Chesapeake Energy, Halliburton, Noble Energy, Southwestern Energy, and Williams Mid-stream—have also contributed scholarships, equipment, and funding to the college to make sure the students receive what may be one of the nation's best possible educations to be prepared to work in the gas fields. They didn't put money and resources into a program that would ask some of the most important questions—"What are the major effects to the health and environment from what we are doing?", "What should we be doing to develop new technology that doesn't threaten the health and safety of the people?", and "Is fossil fuel the best way to assure the production of energy?"

FRACKING PENNSYLVANIA STATE UNIVERSITY

Two of the reasons Pennsylvania has no severance tax and one of the lowest taxes upon shale gas drilling are because of a fawning corporate-friendly legislature and a pair of research reports from Penn State, a private state-related university that receives about $300 million a year in public funds.

Opponents of the tax cited a Penn State study that claimed a 30 percent decline in drilling if the fees were assessed, while also touting the economic benefits of drilling in the Marcellus Shale. What wasn't widely known was that the lead author of the study, Dr. Timothy Considine, "had a history of producing industry-friendly research on economic and energy issues," according to reporting by Jim Efsathioi Jr. of *Bloomberg News*.[1773] The Penn State study was sponsored by a $100,000 grant from the Marcellus Shale Coalition, a coalition of about 300 energy companies which says it provides "in-depth information to policymakers, regulators, media, and other public stakeholders on the positive impacts responsible natural gas production is having on families, businesses, and communities across the region."[1774] Dr. William Easterling, dean of Penn State's College of Earth and Mineral Sciences, said Considine's study may have "crossed the line between policy analysis and policy advocacy."[1775]

The second Penn State research report used by the legislature to justify low taxes and an all-out assault on the environment, with the belief that gas drilling would produce jobs and boost the economy, concluded:

> "Dissolved methane did not increase at fracked sites and was not correlated to the distance to the nearest Marcellus well site. . . .
>
> "Results of the water quality parameters measured in this study do not indicate any obvious influence from fracking in gas wells on nearby private water well quality. Data from a limited number of wells also did not suggest a negative influence of fracking on dissolved methane in water wells."[1776]

However, the study's authors also pointed out:

"[I]it is important to note that this study largely focused on potential changes within a relatively short time period (usually less than six months) after fracking occurred, given the timeline of the project's funding. More detailed, longer-term studies are needed to provide a more thorough examination of potential problems related to fracking, and to investigate changes that might occur over longer time periods."[1777]

The Penn State study was funded by the Center for Rural Pennsylvania, a legislative agency of the Pennsylvania General Assembly, which was pro-fracking.

The Marcellus Center for Outreach and Research (MCOR), a part of Penn State, announced that with funding provided by General Electric and ExxonMobil—which donated a combined $2 million to Penn State, the University of Texas, and the Colorado School of Mines—it would offer a "Shale Gas Regulators Training Program."[1778] The Center had previously said it wasn't taking funding from private industry. However, the Center's objectivity may have already been influenced by two people—Gov. Tom Corbett, who sat on the university's board of trustees, and billionaire Terrence (Terry) Pegula, former CEO of East Resources, which he had sold to Royal Dutch Shell for $4.7 billion in July 2010. On the day Pegula donated $88 million to Penn State to fund a world-class ice hockey arena and support the men's and women's intercollegiate ice hockey team, he said, "[T]his contribution could be just the tip of the iceberg, the first of many such gifts, if the development of the Marcellus Shale is allowed to proceed."[1779] At the groundbreaking in April 2012, Pegula announced he increased the donation to $102 million.[1780] MCOR has a staff of five professionals, and provides numerous short courses and seminars, and conducts research about "energy and energy independence including methods of extraction, alternative fracturing stimulation methods and materials and water treatment and disposal."[1781]

The Shale Technology and Education Center (ShaleTEC) program at the Pennsylvania College of Technology (Williamsport, Pa.), a branch of Penn State, was established in 2008 "to serve as the central resource for workforce development and education needs of the community and the oil and natural gas industry," according to its website.[1782] The program is a collaboration between the college and Penn State Extension. All courses at ShaleTEC, none of which carry academic credit, "may be customized to meet your company's needs."[1783] Many courses are half-day or full-day lectures/seminars, primarily in Natural Gas, Supervisory/Leadership, and Emergency Responder Training programs; there are four full-time and one part-time faculty, all with significant industry experience. In its first four years, the program had more than 10,000 class registrations, according to David C. Pistner, director of energy initiatives. The college also offers several degrees in fields allied to the natural gas industry.

Although ShaleTEC was developed to assist the natural gas industry, and to take advantage of the temporary boom, it has tried to mitigate some of the wildly optimistic claims by the industry and Pennsylvania politicians:

> "No one can accurately estimate how long the drilling phase will last within the Marcellus and Utica plays, but estimates range from 10 to 70 years which in part reflect uncertainty created by future fluctuations in commodity prices, economic conditions, and technological changes among other variables. A number of drilling scenarios are possible for future shale gas development, and they include a relatively quick flurry of activity that subsides when drilling moves to another location, high-intensity drilling that jumps from hotspot to hotspot and moderate and sustained drilling across the Appalachian Basin lasting for decades. Each development scenario changes the direct workforce requirements and opportunities for business development and entrepreneurship."[1784]

The 10 to 70 years of active drilling were hyper-inflated estimates from the industry; the reality is drilling has already begun to decline in the Pennsylvania part of the Marcellus Shale. However, Penn State's own words about the uncertainty of drilling operations didn't preclude it from accepting a $10 million grant from General Electric in September 2014 to create The Center for Collaborative Research on Intelligent Natural Gas Supply Systems, which it declared to be a "new innovation center focused on driving cutting-edge advancements in the natural gas industry." Penn State president Eric

Barron declared the Center "will aim to produce tangible benefits to the natural gas industry as well as the communities impacted by that industry, from the points of extraction right through to the energy that reaches consumers."[1785]

Any question about Penn State's academic integrity on fracking issues was quashed in Fall 2015 when it bought more than 7,000 copies of Russell Gold's pro-fracking book, *The Boom: How Fracking Ignited the American Energy Revolution and Changed the World,* as the sole text for the Penn State Reads program required of all freshmen.

A university press release explained that the program, in only its second year, was established to create "a communal experience that aims to facilitate intellectual engagement inside and outside the classroom, as well as to stimulate students' critical thinking skills."[1786] The university also sponsored an essay contest—each of the two winners would receive a $100 Amazon gift card. There was no mention of who the judges were or if any had affiliations with the oil and gas industry.

Barry Bram, senior associate director of student unions, one of the two Penn State administrators in charge of the program, said the selection of the topic was because, "[F]racking is an issue very relevant to Pennsylvania communities, and we thought it would be good to engage students, faculty and staff about something that's going on right within our home state." Dr. Jacqueline Edmondson, associate vice-president and associate dean of undergraduate education, the other person in charge of the program, explained, "Fracking itself could touch upon almost every discipline at Penn State, so we're eager to engage a large swath of the university."[1787]

Large chunks of university news releases emphasized the book was balanced, and that the selection committee vetted the book and its author through some faculty members in the College of Earth and Mineral Sciences.[1788] There was no mention if any faculty in biology, chemistry, sociology, social work, or the environmental or health sciences read or nominated the book.

Although *The Boom* is well-written and had good reviews from mainstream media, it definitely leans one direction, just as Gold's employer, *The Wall Street Journal*, owned by Rupert Murdoch, definitely leans in one direction. The book itself focuses more on business issues and benefits of fracking, while providing cursory looks at numerous other critical issues, including the politics of fracking, climate change, and the effects of fracking upon the environment and public health.

There are several questions about the selection of the book and the topic. With thousands of fiction and non-fiction books that present critical social issues written by Pennsylvanians or with Pennsylvania themes to choose from, why did Penn State select this book and this issue? The university could have selected books written by Pennsylvanians James Michener, John O'Hara, Gertrude Stein, John Updike, or John Edgar Wideman , all of which would allow students to discuss relevant social issues. Penn State could have selected David DeKok's true-life murder mystery, *Murder in the Stacks*, which has a Penn State focus, and raises numerous social and political issues, many of which predate the recent Jerry Sandusky scandal. Penn State could have selected works by Pennsylvania journalist Ida Tarbell whose in-depth investigation of Standard Oil in the early part of the twentieth century uncovered greed, corruption, and the problems of monopolies. Penn State could also have selected works by Pennsylvanian Rachel Carson, whose *Silent Spring* spurred the environmental movement and the creation of the EPA.

The acceptance of significant outside funds, some sloppy and industry-tainted research, the establishment of industry-funded centers of instruction and research, and the requirement imposed on all freshmen to read and explore fracking, burnished Penn State's focus upon fossil fuel exploration and development, while negating potential research about the advantages of renewable energy.

Integrity Issues at Other Universities

Horizontal fracking to extract shale gas "is a wonderful gift that has arrived just in time," say Dr. Richard Muller, professor of physics at the University of California, and Elizabeth Muller, executive director of Berkeley Earth. The Mullers argue, "Environmentalists should recognize the shale gas revolution as beneficial to society and lend their full support to helping it advance."[1789] The Mullers are principals of the China Shale Fund, which is trying to get China to develop shale gas drilling; they would get financial compensation if China moves from coal to shale gas technology.[1790]

A 205 page July 2015 report from West Virginia University concluded the Utica Shale could contain almost 800 trillion cubic feet of recoverable natural gas,[1791] about 20 times the USGS estimate. The study was funded by 15 energy companies.

Dr. Timothy Considine, whose research at Penn State revealed conflicts-of-interest, continued his optimistic industry-friendly research as director of the Center for Energy Economics and Public Policy at the University of Wyoming. Among his observations for a June 2013 study were that in the Rocky Mountain region, oil and gas development in the decade beginning 2013 "could generate almost $10.6 billion in value added per annum[,] support more than 87,000 job equivalents, and generate more than $3 billion in tax and royalty payments per year." He concluded, "[T]here is considerable upside potential from developing oil and natural gas on federal lands."[1792]

The following year, in two separate studies, one funded entirely by the oil and gas industry,[1793] Dr. Considine determined that allowing drilling off the eastern seaboard would add $11–60 billion in economic values and generate $2–12 billion in tax revenue,[1794] and fracking the agriculture-rich Monterey Shale in central California could add as many as 557,000 jobs a year and increase the domestic product by as much as $63 billion.[1795] Acknowledging that environmental damage could amount to $4.3 to $12.8 billion per year, Dr. Considine still concluded, "[O]nce both the benefits of increased production and the costs of environmental damage are considered, the net economic benefits of developing California's oil and gas resources are between $7 [billion] and $51 billion per year—a significant stimulus for the state's economy."[1796] Less than a week after the research was published, the Energy Information Administration cut its estimate of potential recoverable gas in the Monterey Shale to only 600 million barrels, a 96 percent reduction,[1797] rendering Dr. Considine's best-case scenarios moot.

Shortly after the inaugural of Dr. Tom Wolf as Pennsylvania governor in January 2015, and with a possibility of a severance tax on natural gas production, Dr. Considine, with funding from the American Petroleum Institute, claimed the tax would significantly reduce well production, income, and jobs.[1798]

In 2015, with numerous oil and gas exploration companies defaulting on loans or going bankrupt, three professors from the Harvard Business School determined fracking "is one of the single-largest opportunities to change the trajectory of the U.S. economy and the prospects for the average American in the coming decades."[1799] Two of the three authors work for the Boston Consulting Group, which received $648,875 for "advocacy" consulting from the Western States Petroleum Association, according to an investigation by the Public Accountability Initiative.[1800]

Luring Students to Support Fossil Fuel Energy

Several public school districts have courses in energy, but Houston's Energy Institute High School is the only one to specialize in fossil fuel development. The $37 million school will enroll 1,000 students by 2017.

The American Petroleum Institute (API) and America's Natural Gas Alliance (ANGA), in cooperation with EdVenture Partners, conduct separate contests for college students to showcase fossil fuel energy. Each contest gives cash awards of $5,000, $3,000, and $1,000 to the teams' colleges.

The API contests between 2011 and 2015 drew 101 teams from 61 colleges, according to to EdVenture Partners, which provides guidance for each team. The challenge and winning teams were: 2015—"Conduct research to identify the level of awareness and perceptions of oil and natural gas careers within the target audience. . . . Develop and execute a campaign that educates the audience on the wide variety of jobs available in the oil and natural gas industry and drives them to [a specific website] . . . Submit a report that showcases your campaign's highlights, results and recommendations for future

API engagement."; 2014—"Research the level of energy awareness and issues identified as important. Based on Research, each team identified a target market that offered the greatest willingness to learn about energy issues and the value of oil and natural gas."; 2013—"Outline a complete redesign plan including website discovery and website design of the Adventures in Energy Website."; 2012—"Develop a most compelling and resonant message possible to generate enthusiasm for energy sector career paths as well as deliver this message to the target—college-educated 18–25, majoring in relevant engineering, science, accounting, and communications programs."; 2011—"[Develop] a comprehensive integrated marketing communications plan that will help the target market understand the current situation of energy issues and that will raise awareness of energy issues in the target market, including challenges and opportunities facing the U.S. in meeting future energy demand."

The ANGA/EdVentures competitions required student teams to "design and implement a marketing campaign that raises awareness that natural gas is being produced safely and responsibly and is a safe, reliable, abundant and affordable domestic source of energy that will meet future demand, create jobs, promote energy independence and enhance national security." The ANGA competitions drew 31 teams from 25 different colleges, according to EdVenture Partners.

There is nothing wrong or unethical about a company or organization sponsoring a competition. There is nothing wrong or unethical about students entering that competition. And, there is nothing wrong or unethical about a co-sponsor of that competition assisting students. Where the problem occurs is that the competition, with its tight focus and direction, encourages student teams (and their faculty advisors), in exchange for awards and financial gain, to look at only one side of a controversial issue. It's possible those teams already leaned toward the direction of the challenge parameters. However, it's also possible the requirements helped move those teams to the wishes and philosophies of Big Energy, leading not to objective research and analysis but to the teams becoming a part of the industry's public relations campaigns.

CHAPTER 21
The Revolving Door

The Pennsylvania Experience

Because Pennsylvania is the center of the Marcellus Shale, the most productive of the gas-producing shales,[1801] what happens in that state is indicative of a nationwide trend in possible corporate-political-regulatory collusion.

After slightly more than two years as head of the Pennsylvania DEP—praised by the industry, condemned by environmentalists—Michael Krancer left to again become a partner in the Blank Rome law firm; his assignment was to chair the firm's Energy, Petrochemical and Natural Resources Practice.

According to a Blank Rome news release, he would enhance "the firm's existing energy and public policy talent and advising US and global energy clients in the full range of legal, public policy, government relations, state and federal regulatory, financial, corporate, and labor matters."[1802]

Krancer's comments after leaving DEP and returning to Blank Rome, possibly written by a PR person but attributed to him, suggest a conflict:

> "The rapid move towards American energy self-sufficiency has created new opportunities for our nation. . . . The eyes of the world are on Pennsylvania and Appalachia in particular as a focal point of this paradigm shift in the energy landscape. Blank Rome understands the new energy and natural resources reality in Pennsylvania and throughout the world, and it has tremendous intellectual assets with which to help clients navigate this complicated and evolving industry landscape.
>
> "A key asset of Blank Rome's energy industry practice is our ability to leverage the expertise of practitioners in all of our offices to form cross-border, cross-practice teams. . . . Our lawyers and policy specialists in Houston and Philadelphia,

for example, have enabled us to advise a range of Texas-based oil and gas firms making substantial investments in Pennsylvania. In New York, we advise energy businesses on the structuring of financial derivatives to manage their risk, and in Washington, DC, the firm advises clients on the gamut of federal industry regulation and public policy issues. From every point of our international footprint, from Blank Rome's offices in China and Los Angeles to our presence in Texas, in New York, and in Pennsylvania, where so much of the energy industry is focused on developing new resources, we are well-situated to serve energy clients."[1803]

If Krancer came from the industry, became a regulator and then left after slightly more than two years, was his tenure at the DEP primarily to represent and regulate the environmental interests or to continue to develop a resumé that would impress those of the private sector to which he must have known he would return?

Unlike Krancer who had a solid background in environmental policy and law, E. Christopher Abruzzo, a lawyer who succeeded him in April 2013, had no background in environmental science, but did have close political connections to Gov. Tom Corbett,[1804] and was chief deputy attorney general when Corbett was attorney general. Before the Pennsylvania Senate Environmental Resources and Energy Committee, Abruzzo said that until he became acting secretary, he didn't realize "how challenging it is to lead an agency that, in many respects, is driven every single day by science and facts."[1805] However, it was his opinion of the science of climate change that surprised those who believed someone with an environmental science background should lead the DEP. Abruzzo said that he hadn't "read any scientific studies that would lead me to conclude there are adverse impacts to human beings, animals, or plant life" from climate change.[1806]

Abruzzo's nomination "is a reflection of what I view to be a lack of seriousness with which the governor treats environmental issues," said State Sen. Daylin Leach.[1807]

Abruzzo's tenure as DEP secretary didn't last long. He resigned in October 2014 following disclosures by the attorney general that he was one of about three dozen employees in the attorney general's office, while Corbett was attorney general, to send and receive pornography using their government e-mail accounts on public computers.[1808] Also resigning was Glenn Parno, DEP general counsel, who had also been Gov. Corbett's deputy attorney general. Corbett himself was not implicated.[1809]

Michael Krancer wasn't the only one to jump from being a regulator to representing those firms he once regulated.

Within four months of Krancer leaving DEP, Kevin Sunday, Tom Corbett's deputy press secretary, also left. Sunday, whose primary responsibilities included acting as the principal DEP spokesperson, joined Quantum Communications. A company press release included Sunday's effusive quote: "Pennsylvania is strategically positioned to be the center of the energy world, and I am looking forward to helping Quantum's clients successfully take advantage of that."[1810] Shortly afterward, he went to the Pennsylvania Chamber of Business and Industry.

With the inauguration of Dr. Tom Wolf as governor, Patrick Henderson, Gov. Corbett's energy executive, resigned his $145,000 a year job in January 2015 to become director of regulatory affairs for the industry-sponsored Marcellus Shale Coalition.[1811] That move "elevates cynicism about how government operates," said Barry Kauffman, executive director of Common Cause PA, who asked, perhaps rhetorically, "What was [Henderson] doing in his position of high power to encourage people to hire him after he [left office]?"[1812]

Days before he left office in January 2015, Corbett appointed John T. Hines to the DEP Citizens Advisory Council.[1813] Hines had been a DEP executive deputy secretary before jumping to Shell Oil as a government relations advisor.

The actions by some Pennsylvania regulators and advisors leads to questions about whether they were giving full and complete information to the people or whether, bound by political realities, they were saving their jobs and pensions or planning to enter the more lucrative private sector.

During the tenure of Tom Corbett, transparency in the DEP had become clouded, with some PR staff deflecting questions or creating situations that made it difficult to get accurate and adequate information about fracking. The problem apparently extended over other state agencies. An AP investigation of citizen and media requests for information reveals "previous

409

responses from the Department of Health about the numbers of complaints it has received about drilling and health have been at best confusing and at worst misleading."[1814]

In June 2013, Corbett fired Richard J. Allan, head of the Department of Conservation and Natural Resources, who had sent sexist emails that became public. Corbett replaced Allan with Ellen Feretti, who holds a B.S. in environmental science/ biology, and has a history of environmental work in both the public and private sectors.[1815] Four months later, he appointed Brian Grove to be DCNR deputy secretary for administration. Grove had no experience in environmental science or conservation. However, he did have significant experience as a senior staff member for a Republican state senator and two Republican governors, and came to the DCNR after four years as senior director for corporate development at Chesapeake Energy.[1816] StateImpact reported that at Chesapeake, "Grove repeatedly defended the company against charges of environmental harm."[1817]

An investigation by the non-profit Public Accountability Initiative (PAI) revealed that 45 current and former state officials and senior staff at the DEP as of the end of 2014 had been employed by the energy industry either before or after they left the DEP, that three former Pennsylvania governors (Tom Ridge, Mark Schweiker, and Ed Rendell) "have strong ties to the natural gas industry" and that "Every Secretary of Environmental Protection since the DEP was created has had ties to the natural gas industry."[1818] Some, like Kathleen McGinty and John Hanger, both of whom served in the Ed Rendell administration, have more ties to environmental protection than they do to the natural gas industry—but both believe horizontal fracking is a means to extract natural gas.

"The revolving door trend in Pennsylvania raises questions about whether regulators are serving the public interest or private industry interests in their oversight of fracking [and] raises troubling questions about the incentives that may be guiding public officials' oversight of fracking in Pennsylvania, from governors to DEP secretaries to well inspectors," the 30-page fully-documented PAI report argues.[1819]

The Revolving Federal Door

The problem of the "revolving door" isn't confined just to Pennsylvania.

Mary Landrieu (D-La.), in 18 years as senator, was one of the most effective cheerleaders for the energy industry. She was aggressive in pushing the senate to vote for approval of the Keystone XL pipeline, and was instrumental in pushing for expedited action on behalf of proposed LNG facilities. Within months of losing a bid for her fourth term, she became an executive with the lobbying/law firm Van Ness Feldman, which represents TransCanada, builder of the Keystone XL pipeline. Forbidden by law from active lobbying for two years, Landrieu became a "senior policy advisor," dealing with matters of energy. In a corporate news release, Landrieu stated she "always respected the firm and worked closely with them" while a senator, and acknowledged that as a senator, the company's "substantive and sophisticated approach to important public policy issues in the areas of energy, the environment and natural resources was a major factor in my decision-making process."[1820] As senator, Landrieu earned $174,000 a year plus benefits; at the age of 62 (2018), based solely upon her senate service, she would be entitled to about $78,000 a year in pension and about $24,000 a year in social security benefits. Van Ness Feldman didn't reveal how much more she would earn or what additional benefits she would receive as a member of top management.

John Krohn, who had been the communications director for Energy in Depth (EID),[1821] the industry's information and disinformation PR front, became a communications manager for the federal Energy Information Administration (EIA).[1822] While at EID, Krohn claimed fracking never contaminated water or the air, and that that global warming/climate change, if there is such a thing, is not caused by humans. At EIA, he said not only is there is an "increasing precision and efficiency of horizontal drilling"[1823] but there is a continuing increase of production,[1824] although scientific evidence shows that many of the shale plays are in decline.[1825] Krohn's background was revealed not by establishment media but by Steve DeHorn at DeSmogBlog.[1826]

An investigation by DeHorn and Lee Fang for the *Republic Report* revealed even more revolving door politics at the federal level.

Bill Cooper had been counsel for the House Energy and Commerce Committee, leaving to become a lobbyist for the American Petroleum Institute and then the Center for Liquefied Natural Gas.

David L. Goldwyn held several senior level appointments in the Departments of Energy and State, including developing the Global Shale Gas Initiative for Secretary of State Hillary Clinton. After leaving governmental service, he founded Goldwyn Global Strategies, which identifies itself as "a leading provider of energy sector intelligence, analysis and strategy to Fortune 100 companies."[1827]

David Leiter—who had been chief of staff for Sen. John Kerry and deputy assistant secretary of energy for President Clinton—is president of ML Strategies, a government relations lobbying firm. Among the clients is Sempra Energy.[1828]

Christopher A. Smith became President Obama's Principal Deputy Assistant Secretary for Fossil Energy after several years as a Chevron executive.[1829]

Prior to being appointed by President Obama to be Deputy Assistant Secretary for Oil and Natural Gas, Paula Gant was vice-president of the American Natural Gas Association.[1830]

Many of the former government officials, according to Fang and DeHorn's investigation, are either employed by or have close ties to Cheniere Energy, the first corporation to receive a permit from the Obama Administration to construct an LNG facility.[1831]

Spencer Abraham, a former senator and George W. Bush's Secretary of Energy (2001–2005), developed a corporate energy lobbying firm, taking several Department of Energy officials with him. Several top Abraham aides, report Fang and DeHorn, "now work directly in-house for Cheniere" although Abraham "does not directly lobby for the company."[1832]

Amkit Desai, a senior staffer for both Sens. John Kerry and

Joe Biden, became vice-president for government relations for Cheniere.[1833]

John Deutsch, former director of Central Intelligence, is not only a member of the board of Cheniere, but also a member of the advisory commission for the U.S. Secretary of Energy. To serve on the Commission, Deutsch needed to get a waiver from established federal ethics laws.[1834]

Heather Zichal, Deputy Assistant to President Obama for Energy and Climate Change, resigned in October 2013 after serving almost five years;[1835] within months, she become a member of the board of Cheniere Energy.[1836] DeHorn had noted that Zichal, while a White House staffer, had met twice with several top level Cheniere officials months before leaving federal employment.[1837]

There is nothing wrong with someone jumping between private and public sectors—as long as there are no actual or perceived conflicts of interest. Going from private sector employment into the public sector could mean adding valuable knowledge and experience; it's possible some who moved from the private sector into the public sector wanted to do public service. However, it's also possible some moved into government jobs to influence public policy and use their temporary jobs as stepping stones into even more lucrative private industry jobs.

Going from the public sector to the private sector could raise questions about the loyalty of the individual to the public concerns and needs.

It's possible that staff with industry ties could separate their public jobs from the lure of the more financially lucrative private sector jobs. It's possible that some left the public regulatory agencies, especially the Pennsylvania DEP, because they were frustrated with the declining morale and increased political influence. But it's also possible that some staff in all states realized there were better jobs with better benefits, and that looking to a future in the private sector may have been a good career move.

As long as there is unencumbered movement between private and public sectors, there will always be suspicion and mistrust of a person's motives.

PHOTO: Renard Cohen

Dr. Sandra Steingraber (lower center) at a rally in the rotunda of the state capitol in Albany, N.Y., discusses critical issues about fracking.

PHOTO: *Groundswell Rising*

More than 5,000 persons gathered on the West Lawn of the U.S. Capitol, July 28, 2012, in the first "Stop the Frack Attack," and then marched to the offices of the American Petroleum Institute and the American Natural Gas Association. Media coverage was minimal.

414

CHAPTER 22
The People Push Back

In September 1964, with the rise of the civil rights and anti-war movements, the administration of the University of California at Berkeley, in violation of the First Amendment, declared it would actively enforce regulations against on-campus political speech and fundraising. Three months later, with student protests increasing, and an intractable stand by the Board of Regents, Mario Savio, leader of the newly-formed Free Speech Movement, stood on the steps of Sproul Hall and told more than 4,000 people that the people demanded academic freedom and their constitutional rights of assembly and speech:

> "There's a time when the operation of the machine becomes so odious, makes you so sick at heart, that you can't take part! You can't even passively take part! And you've got to put your bodies upon the gears and upon the wheels, upon the levers, upon all the apparatus, and you've got to make it stop!"[1838]

Mass arrests and student strikes shut down the university, but led to a restoration of constitutional rights and the beginning of similar movements throughout the country.

Defending the Environment and Public Health

The anti-war and civil rights movements of the 1960s had been the base for the environmental movement of the 1970s. In a different era, with a different social cause, the anti-fracking groups help those affected by fracking. From small village meetings, with only one or two members willing to speak against the incursion of the wells, to mass rallies of thousands, protestors are organizing to promote the message that fracking for natural gas can harm public health, animal life, and the environment. Most have websites with extensive information about fracking and an interactive Facebook page; many of the larger groups have blogs and periodical electronic newsletters. The volunteer leaders of the major anti-fracking organizations, many with 501(C)(3) non-profit status, often work the equivalent of a full-time job. All of them follow Susan B. Anthony's rallying cry of more than a century earlier—"Organize. Agitate. Educate."

They organize petition campaigns and lobby for moratoriums; they organize, attend, and speak at conferences, forums, seminars, churches, community events, music festivals, and at governmental hearings; they organize special events, marches, and rallies; they are a vital part of Earth Day every April, and the backbone for Global Frackdown Day every October.

Unable to match the money spent by by the natural gas industry and its front groups, the anti-fracking movement uses social media and mass demonstrations to get its message out. Financial donations, not corporate sponsorships, help create the signs and communications. Unlike the gas companies and trade association front groups, the anti-fracking volunteers do a lot of legwork because, as Jan Milburn of the Westmoreland Marcellus Citizen's Group points out, "We don't have money to take out newspaper and TV ads."

Among dozens of national organizations that oppose fracking are Americans Against Fracking, the Clean Air Task Force, Clean Water Action, EcoWatch, Environmental Defense Fund, Food and Water Watch, Friends of the Earth, Green Party, Natural Resources Defense Council, Nature Conservancy, Oil and Gas Accountability Project, the Sierra Club, and 350.org.

In Iowa, the Allamakee County Protectors are fighting fracking and the destruction of their state by corporations that are mining silica sand. Colorado has more than a dozen active groups. A dozen New Mexico anti-fracking groups are fighting not only fracking but the drought that has furthered problems for the agriculture industry and residents while Big Energy outbids farmers for what water there is.

In August 2012, more than 1,000 individuals marched on

Albany, N.Y., to target Gov. Andrew Cuomo who had given suggestions he might approve fracking for five southern tier counties. The newly-formed coalition, New Yorkers Against Fracking, had helped organize the protest. Founders were long-time activist David Braun of United for Action, Wes Gillingham of Catskill Mountainkeeper, Julia Walsh of Frack Action, and Eric Weltman of Food and Water Watch. The advisory committee members were environmental expert Dr. Sandra Steingraber; Lois Gibbs, who had organized the Love Canal Homeowners Association in 1978 and then created the Center for Health, Environment and Justice, which provided assistance for more than 11,000 grassroots groups; musician/singer Natalie Merchant, who has a long history of environmental and human rights activism; and actor Mark Ruffalo, who has spoken out on numerous radio and TV shows about the destruction of the environment; Ruffalo has also written numerous columns and OpEds, and is a frequent speaker at rallies against fracking and for the environment.

California, which has large oil reserves but no full-scale fracking, has two dozen working groups. New Jersey, affected by what's happening in Pennsylvania and New York, but which doesn't have drilling, has two dozen organizations that are working to inform the state about the effects if corporations develop economical ways to mine gas from the deeper Utica Shale in the north and the smaller Marcellus/Newark Basin in the south. Ohio, on Pennsylvania's western border, has two dozen active groups that oppose the injection wells and the probability that the state's Utica Shale will become profitable for the companies because of the wet gas it produces. Other groups have organized and are fighting fracking in almost every state that is being pockmarked by wells, including Alabama, Arkansas, Idaho, Illinois, Indiana, Michigan, Tennessee, Texas, and West Virginia. Surprisingly, there are almost no groups in North Dakota, which is finding out that wheat and fossil fuel don't mix, but is still enamored of the temporary economic boom.

More than 5,000 marched from the U.S. Capitol to the headquarters of the American Petroleum Institute and the American Natural Gas Association, July 28, 2012.

One hundred activists blocked trucks carrying fracking wastewaster from entering the Greenhunter Water facility in Marietta, Ohio, in February 2013. The activists protested Greenhunter's request to the Coast Guard for permission to ship frackwaste on barges, each carrying 500,000 gallons, on the Ohio River.[1839] Because the Ohio falls into federal jurisdiction, the oversight and regulation of a primary source of drinking water will be significantly higher than if Ohio, West Virginia, or any state along the river monitored the barge traffic.

On March 6, 2013, several hundred Ohioans used Legislator Accountability Day to rally at the statehouse to demand full disclosure of all chemicals and elements in wastewater, and bans on wastewater spread as deicer in roads and on injection wells. Madeline ffitch from Millford, Ohio, outlined the purpose of the rally:

> "Appalachian Ohio is on the front lines of the oil and gas Industry's assault on rural communities. State regulators ignore us, out-of-state industry seeks to prey on us. But we who live in the beautiful hills and valleys of Southeast Ohio are standing up and saying 'no.' We are building a strong coalition of farmers, landowners, nurses, business owners, teachers, families and other rural community members. Appalachian Ohio is proud to send a strong message to industry: We will use every available strategy, including direct action, to stop our communities from being turned into environmental sacrifice zones."[1840]

ffitch, a carpenter and writer, had been charged the previous year with "inducing panic" for having chained herself to two concrete barrels at the entrance to a wastewater injection well. Several police agencies converged at the site, ordered the dozen protestors, spectators, and the media away, and then blocked their view of the protest. ffitch later pleaded guilty to a reduced charge of trespassing. She told Kelly Doran, editor-in-chief of Ohio University's *College Green Magazine*:

> "One of the things about direct action is that it's usually part of a larger campaign with a lot of different tactics going on. The media sort of made it look like it's just me; one lone woman taking a stand . . . [T]here's a lot of different people working on the issue at a lot of different levels. . . . The well

418

that I blocked is still operating right now, it's still running everyday so I was not successful in shutting down that well but I would say that what that action did is bring a lot of attention to the issue and I think it's really important to mention that we've been calling for testing on that frack waste for months. The week after my action they conducted a test from that well site."[1841]

On April 22, 2013, the 43rd annual Earth Day, protestors throughout the country demanded the end of fracking. In Pennsylvania, the Green Party organized rallies in front of all DEP offices. At the Meadville regional office, Karen Bagdes-Canning told the crowd, "We are engaged in a battle for our communities." At the Norristown regional office, Dr. Walter Tsou, one of the nation's leading public health physicians, asked the DEP to conduct "serious scientific studies" on the health impact of fracking. The Green Party demands, all of which were ignored by Gov. Tom Corbett, called for a DEP head who was an environmental expert, a moratorium on permits, full disclosure of water studies by the DEP, and the re-establishment of the DEP Office of Energy and Technology to pursue studies and oversight of renewable energy.

Seven days after the Earth Day rallies, the last day of the Midwest Catholic Worker Faith and Resistance retreat, police arrested 35 protestors in Winona, Minn., and charged them with trespassing after they blocked traffic at two frack sand processing plants and refused to disperse after receiving several warnings. During the previous year, activists held several rallies to protest the destruction of the environment by sand mining, and the shipment of silica sand to sites that use it as proppants for fracking. Matthew Francis Byrnes, one of those arrested, in a letter to the editor of the *Winona Daily News*, wrote:

"According to the Gospel, Jesus kicked the money changers out of the Temple. If there was a frac sand and fracking boom in Palestine in the first century, it's likely that he would have stood in front of trucks, too."[1842]

The protests may have had an effect upon Winona's city council. In November 2013, Winona became the first American city to monitor air pollution from frack sand mining, following a recommendation by Dr. Holly Lenz, an assistant professor of public health at Winona State University.[1843]

Near Morgantown, W.Va., about 100 protestors in July 2013 blocked two entrances of Momentive Specialty Chemicals to protest the manufacture of resin-coated proppants, and the state's plan to allow fracking as early as 2015.[1844] After several hours of non-violent protest, police gave the protestors from Croatan Earth First a half-hour to disperse, and then arrested 12 for trespassing, resisting arrest, and disorderly conduct.[1845]

Global Frackdown Day II, Oct. 19, 2013, brought anti-fracking activists together in 250 separate actions in 28 states and several foreign countries. In New York City, more than 1,000 people rallied to urge Gov. Andrew Cuomo to continue the state's moratorium; in Oakland, Calif., about 200 persons urged Gov. Jerry Brown to issue a moratorium.

In Pittsburgh, Pa., Global Frackdown Day was the second day of a four-day Power Shift conference that drew 7,500 youth from around the country to learn more about environmental issues, including renewable energy sources, and to protest fossil fuel energy.

In France, which bans fracking, thousands gathered in several cities to reinforce their demands that they also wanted assurances there would be no experimental drilling.

In Romania, the Social-Democrats came to power in 2012 on a promise to ban shale gas drilling after the Democratic-Liberal government supported the policy.[1846] However, following extensive lobbying by U.S. Secretary of State Hillary Clinton and U.S. ambassador Mark Gitenstein, according to Mariah Blake in *Mother Jones Magazine,*[1847] the Romanian parliament voted against a proposed fracking moratorium; the government then signed a 30-year lease agreement with Chevron. Thousands of Romanians, many of them farmers, supported by the Romanian Orthodox church, protested Chevron Romania's planned drilling. The protests in Summer 2013 led Chevron to temporarily suspend drilling. A Chevron executive told the newspaper *Adevarul:*

"We have a motto we work by—'Do it safely or don't do it!' During the [protest] in Pungesti we concluded that these operations cannot be done safely. Thus, we announced a suspension of the works. This because the area is not safe for our workers, nor for the villagers. This situation will last until we and the local authorities will reach the conclusion that the area is safe. The operations were not shut down, only delayed."[1848]

Chevron resumed the exploration for shale gas after the new government sent in the national police to suppress the citizens' rights of assembly and freedom of expression. However, the extensive fighting in the Romanian parliament and the massive citizen protests may not have been necessary. In November 2014, Prime Minister Victor Ponta said:

"It looks like we don't have shale gas. We fought very hard for something that we do not have. I cannot tell you more than this but I don't think we fought for something that existed."[1849]

Frackdown was organized by Food and Water Watch in 2012, with 180 organizations participating. In 2013, there were more than 350 organizations involved in the largely peaceful protests.[1850]

Two days after Global Frackdown II, an environmental protest closed downtown streets and sidewalks in Pittsburgh, Pa. The 1,000 protestors, many of whom were in Pittsburgh for the last day of the industry's Power Shift conference, wanted the county government to stop plans to allow fracking in Deer Lakes Park. Police arrested seven for blocking the entrance of a bank that provided loans for mountaintop mining operations.

THE PEOPLE'S CLIMATE MARCH

The Sunday morning political talk shows didn't cover it. The evening news on the three major broadcast networks and three news cable networks barely mentioned it. Newspapers, although they had copy from the wire services, generally ignored it or downplayed it.

"It" was the "People's Climate March." More than 310,000 people[1851] spread across about 25 blocks in New York City, Sept. 21, 2014, the last day of a four-day "fractivista convergence," organized by Bill McKibben and 350.org, and supported by more than 1,500 organizations.[1852] Among those in the march were Al Gore, Mayor Bill de Blasio, U.N. Secretary-General Ban Ki-Moon, and hundreds of celebrities.

The People's Climate March was the largest gathering of environmental activists in history, with more than 2,600 events in 162 countries,[1853] but the establishment mass media didn't have the time, resources, or desire to give it adequate coverage. Had they chosen to devote their resources to informing the people about the march and global warming, the media would have begun their coverage on Sept. 18 when several people from the U.S. and South America spoke about the effects from fracking. The next day, they could have attended a six-hour series of workshops, sponsored by the Rainforest Action Network. On Saturday, they could have attended several more information and workshop sessions at several venues in Manhattan, where they would have learned more about fracking, climate change, renewable energy, and ways to protect the environment.[1854] And yet, as they did on numerous anti-war marches during the Vietnam War and the two Gulf wars, the establishment media—the media that Tea Party right-wingers call the "liberal" media, even the "socialist" media—generally gave the four-day awareness conference little attention.

However, there were exceptions. *TIME, Newsweek*, the *New York Times*, the *Washington Post*, the London *Guardian*, BBC, NPR, Al Jazeera America, and hundreds of alternative online and print publications gave significant coverage of the workshops and the march. Equally important in the emerging digital age, there were more than 630,000 social media posts, according to data compiled by 350.org.[1855]

For their part, the right-wing media—some newspapers and numerous radio/TV talk shows—did did give the march and events some coverage. The *Washington Examiner* reflected the conservative views:

> "Climate protestors descended upon Manhattan on Sunday to vent their frustrations about a world they don't understand. On display at the People's Climate March were ideology and idealism conspiring at the expense of common sense." [Sept. 24, 2014][1856]

Pennsylvania State Sen. Tim Solobay, defending Big Energy, declared he was frustrated "when people spin and challenge every bit of information and action out there with the sky-is-falling mentality." He said protestors "enjoy spreading fear and uneducated comments," and erroneously stated, "A majority of the negative voices out there are paid activists [who] do nothing but spread false rumors and scare people."[1857]

Stephen Moore, chief economist for the conservative Heritage Foundation, told a conference of the Marcellus Shale Coalition in September 2014, "The disinformation and propaganda machine against what you do is frightening." He added the campaign against fracking "may have been instigated by out-side agitators." It was a claim echoed by thousands in the industry.

With absolutely no proof, Moore was referring to the possibility that Russia and the oil-rich oil countries, and not millions of Americans, were behind the anti-fracking campaign. Russia's Gazprom is the world's largest natural gas distribution company, and many in the U.S. oil/gas industry believe Gazprom or Russian president Vladimir Putin wanted to increase Russia's share and domination of the natural gas industry by closing down American natural gas production.

Six months after the MSC conference, Harold Hamm, billionaire CEO of Continential Resources, and with U.S. Oil profits in decline, continued the drum beat that the Russians and American fracktivists were somehow in collusion. "Russia's spent a great deal of money over here to cause a panic in the United States over fracking to stop it, because suddenly their market share is going away," Hamm told Forbes Media.[1858]

The same gaseous windbags who tried to discredit American environmentalists by linking them to Russia also blamed the Arab countries for opposing horizontal fracking because they were making money off oil and didn't want competition. None acknowledged that the Arab countries have been far ahead of the United States in the development of renewable energy, knowing that fossil fuel contributes to global warming and is not infinite.

PROTESTING THE KEYSTONE XL PIPELINE

During the decades when vertical fracking was the primary method to extract gas and oil, there was little public opposition. Almost negligible was any opposition to the problems of waste management, compressor stations, and aging pipelines. But in the decade after horizontal fracking became the dominant method to extract oil and gas, opposition increased not only to focus upon the environmental and health effects of horizontal fracking, but also against all operations connected with the extraction, storage, and transportation of fossil fuel. The uncompleted Keystone XL pipeline became a focus for pieline opposition.

On Feb. 17, 2013, 35,000–40,000 people gathered at the National Mall in Washington, D.C., to demand President Obama block the completion of the Keystone XL pipeline. The Forward on Climate rally was organized by the Sierra Club and 350.org. The mostly peaceful demonstration blocked a street in front of the White House. Among the 48 arrested were former NAACP president Julian Bond, Robert F. Kennedy Jr., Daryl Hannah, author/activist Bill McKibben and Sierra Club executive director Michael Brune and president Allison Chin. Brune's and Chin's arrests underscored a philosophical change in the Sierra Club's long-standing opposition to civil disobedience.

The day after the Forward on Climate rally, activists in Pennsylvania locked themselves to the entrance gate of the Delaware State Forest near Milford to keep pipeline workers from cutting down trees and clear paths for the construction of a segment of the Tennessee Pipeline that would carry Marcellus Shale gas through several states. Affected was a 130 acre swatch. For two days, 7 a.m. to 4:30 p.m. each day, the protestors maintained their vigil, successfully turning away crews hired to clear the forest. "Our goal was to stop the cutting until March 31," says Alex Lotorto, one of the protest organizers. March 31 was a date that had long ago been set by the federal government, under authority of the Migratory Birds Act, to stop all construction and cutting; if work had continued beyond that date, the return migration of brown bats would have been affected. But there would be tree cutting before then, the result of an injunction requested by Kinder Morgan, operator of the pipeline.

In April, two men were arrested near Holdenville, Okla., after chaining themselves to equipment at a construction site for the Keystone XL pipeline.[1859]

Washington, D.C., metro police arrested 398 students who were at the White House, March 2, 2014, to protest the proposed completion of the pipeline. More than 1,000 students, many wearing mock HazMat suits and carrying signs protesting fracking and the completion of the pipeline, were from 80 colleges.[1860] They began their march at Georgetown University and walked a mile to Lafayette Park near the White House. Hundreds then used plastic handcuffs to attach themselves to the fence surrounding the White House[1861] or lay down on the street. The students were booked, fined, and released.[1862]

On a 160-acre farm near Neligh, Neb., more than 8,000 persons paid $50 each in September 2014 to hear the music of Willie Nelson and Neil Young, to hear speakers opposed to fracking, and to rally against the completion of the Keystone XL pipeline. Joe Duggan of the *Omaha World-Telegram* portrayed Art and Helen Tanderup, on whose farm the Harvest of Hope concert was held, as having "never imagined that they would play a key role in helping pipeline opposition . . . Perhaps that's because they come off less as activists than typical retirement-age farmers."[1863] The Tanderups went from being neutral about the pipeline when approached by a land agent wanting an easement on their property to run the pipeline, to being active opponents after doing research about the tar sands oil being shipped in the 36-inch diameter pipeline. Even after TransCanada agents dropped six-figure rights agreements upon the residents to convert them from remaining neutral or opposed to the pipeline to having them sold out solely for the money, the Tanderups refused to accept payment. "They could offer me $10 million, and we're not going to take it," Art Tanderup told Duggan.[1864] About 400 of the 500 land owners in the Nebraska section of the proposed pipeline had already accepted payments.

Among those at the concert were Sioux from South Dakota and Ponca from Oklahoma; the pipeline crosses tribal lands. The income from the concert, about $300,000 after expenses, was distributed to Bold Nebraska, the Indigenous Environmental Network, and the Cowboy and Indian Alliance, three groups that oppose the pipeline's completion.[1865] "These boots and moccasins are going to stop this pipeline,"[1866] said Jane Kleeb, the concert organizer and executive director of the environmental action group, Bold Nebraska.

"Were not just standing here complaining about problems, but finding solutions." Neil Young told reporters before the concert.[1867]

THE MILLIONAIRES PROTEST

Rex W. Tillerson, like many of his neighbors of Bartonville, an affluent rural community in northeastern Texas, was upset with his city council. That's not unusual. Many residents get upset at their local governing boards. And so they went to a city council meeting in November 2013 to express their concerns that the council was about to award a construction permit.

The residents were upset that the Cross Timbers Water Supply Corp. planned to build a 160-foot tall water tower. That tower would be adjacent to an 83-acre horse farm Tillerson and his wife owned, and not far from their residence. The residents protested, and then filed a suit to stop construction. The lead plaintiffs were Dick and Susan Armey; Dick Armey, conservative and opposed to much of the environmental regulation, was House Majority Leader in the 1990s. The tower the millionaires opposed would store water to be sold to companies that needed it for high-volume horizontal fracturing.

The residents weren't concerned about the health or environmental impact of fracking. They were upset that construction of the water tower would impact their views. They argued there would be excessive traffic and noise during construction and after the tower was built. They complained that the water tower would lower property values.

Tillerson isn't your typical resident. He's the CEO and the chairman of the board of ExxonMobil, the third largest corporation in the world, and the company that leads all others in exploring, drilling, extracting, and selling oil and gas. It's also a company that has had more than its share of political, social, and environmental problems. Tillerson was an engineer when the Exxon Valdez fouled the western shore of the United States in 1989. By 2004, he was the company's president.

In 2013, Tillerson earned $40.3 million in compensation, including salary, bonus, and stock options. His company that year had $438.26 billion in revenue; profit was $32.58 billion, down 27.4 percent from the year earlier.[1868]

When you have that much money, every million or so dollars matters, especially if a large ugly tower impacts not just your view but your quality of life and the value of your property. It's a problem that affects millionaire owners of oil conglomerates but, apparently, not the corporations they run, which don't seem to care about the view, the noise, the property values, and the health and environmental effects on hundreds of thousands of residents in their path to excessive profits.

There is something more about Rex W. Tillerson. He's proud of his association with the Boy Scouts. He's a former Eagle Scout and was president of the national Boy Scouts of America. (Both the Boy Scouts and ExxonMobil have their headquarters in Irving, Texas.)

The 12th part of the Scout Law is to be reverent. A widely-accepted interpretation of that law, according to Scouting Trail, is: "As a Scout experiences the wonders of the outdoors, stormy weather and calm blue skies, pounding surf and trickling streams, bitter cold and stifling heat, towering trees and barren desert, he experiences the work of God. . . . We need to play the role of steward rather than king—tending and caring for our world instead of taking all we can for our own comfort."[1869]

Protesting the construction of a water tower because it might lower property values, even for selfish purposes, is Tillerson's right as a citizen. But, destroying God's world to maximize profits is not his right.

PROTECTING SMOKEY—AND FRACKING

Lori LaRoe, an environmentalist and artist from Brooklyn,

N.Y., had an idea. Energy companies wanted to frack public land. The Obama administration, although more environmental friendly than the Bush–Cheney administration, was still opening federal lands to oil and gas companies. And so LaRoe looked at one of the most famous Americans to get out the message that fracking should not occur on public lands. She drew a bear's head, put a Smokey Bear hat on it with the words "No Fracking" and a message, "Only you can prevent faucet fires." Because of the reach of social media, LaRoe was soon putting the modified Smokey image onto posters, T-shirts, and tote bags, and sending them throughout the world.

"This is the radicalization of Smokey the Bear," LaRoe told reporter Peter Rugh. "Smokey wants to fight the corporations and protect the air and the water and the plants and the animals and the people," said LaRoe.[1870]

The U.S. Forest Service also wanted to protect something, specifically Smokey's likeness. LaRoe received a cease-and-desist order. Failure to agree could cost her up to six months in prison and a $150,000 fine.

"Any time anybody uses Smokey's image for anything other than wildfire prevention," a Forest Service official told Rugh, "it confuses the public. What we're trying to do is keep Smokey on message."

LaRoe's message is just as important—stop fracking public lands. Against possible legal action, she continues to produce her message, hoping that if it comes to court, the law will recognize that message as protected free speech and acceptable parody.

A Few Case Studies of Anti-Fracking Groups

DAMASCUS CITIZENS FOR SUSTAINABILITY

Sometime in 2007, gas company landmen came into rural Pennsylvania, promising riches in exchange for mineral rights. For most of the landowners, struggling with an economy that was at the edge of sinking deeper than the Marcellus Shale, they accepted the promises that would help them pay their debts, improve their farms, and give them financial security.

Few had heard about the Marcellus Shale or high-volume hydraulic horizontal fracturing. Even fewer knew anything about the effects of fracking upon the environment and health.

"We were alerted to what was going on by some friends in other areas," says Barbara Arrindell, "and felt we had to learn what's happening in other places and act on it before we were also affected."

The "we" included Arrindell, an engineer and ceramic glass artist; Joe Levine, an architect; Jane Ciphers, a teacher; and several other friends in Wayne County's Damascus Twp., on the western edge of the Delaware River in the northeastern part of the state. Pat Carullo, an activist and graphics designer, co-ordinated media and became the new organization's webmaster. Carullo had moved from Staten Island to Wayne County after 9/11.

By 2008, Damascus Citizens for Sustainability (DCS) became the first anti-fracking organization in the eastern part of the United States.

"We poured ourselves into this issue," recalls Arrindell, "because we realized if we didn't simultaneously scream about what was going to happen and also educate the public, the media, policy makers, and regulatory agencies that we would not be able to do anything about it, that we would not be able to live in our homes."

In response to the oil and gas industry claiming that fracking wasn't responsible for increased methane levels in home water wells, DCS volunteers developed a procedure to measure methane levels prior to drilling.[1871]

"Just as we were informed by people living in the Barnett Shale as to what was happening, we hope to provide information to others in areas where drilling is just starting to let them know what they're looking at." There are now more than 5,000 persons on the DCS mailing list, and a website[1872] that has extensive information about fracking and its effects.

One of the people DCS influenced was Josh Fox, whose family lived in Milanville in Damascus Twp. "We woke him up," says Arrindell. That wake-up call led to *Gasland*, and Fox dedicating the ground-breaking film to the Damascus Citizens.

Although education has been the primary focus of DCS, the organization also tracks methane emissions. "We're looking at baseline data before drilling, wellhead emissions, and fugitive leaks," says Arrindell.

GAS FREE SENECA and WE ARE SENECA LAKE

The Finger Lakes is a series of 11 long and narrow lakes between Ithaca and Rochester, N.Y. The region is one of the nation's leading tourism areas, with extensive water recreation, waterfalls, hiking trails, museums, art and photo galleries, colleges, farmer's markets, boutiques, and outdoor malls. Almost 100 wineries and vineyards, all taking advantage of the "lake effect" and good soil, make the area the largest wine growing region in New York.

It's also where Inergy Midstream Partners, one of the nation's largest independent gas storage and transportation companies, paid U.S. Salt about $65 million in August 2008 to acquire 576 acres on the southwestern side of Seneca Lake, the largest of the Finger Lakes, and plans to spend an additional $40–50 million to convert abandoned salt caverns into what may become the largest gas storage facility in north-eastern United States.[1873]

The Finger Lakes region is also where Dr. Joseph Campbell and Yvonne Taylor in 2011 organized Gas Free Seneca, a grassroots organization of several hundred residents, more than 180 businesses, and more than 5,000 persons who signed a petition to state regulatory agencies asking them to block Inergy's plans, citing geological data that the caverns are structurally deficient and that the corporations' plans would disrupt the health, environment, and economy of the region. Within a three week period in July 2012, says Taylor, "Inergy had two equipment failures resulting in brine spills which killed trees and vegetation along the hillside." The company also received permission from the state to discharge 44,000 pounds of chloride into Seneca Lake per day. That lake provides drinking water to more than 100,000 individuals.

Gas Free Seneca has raised objections that there are two major problems that need to be addressed: potential cavern collapse and potential cavern leakage through faults in walls, roofs, and the floor. It is a concern that the state's Department of Environmental Conservation is also looking into, especially

430

since the previous owner abandoned the salt cavern after the roof collapsed in an earthquake. But, the facts of that roof collapse and other issues of the integrity of the caverns are protected as a "trade secret," a ruling issued by the DEC in November 2010. "There's a lot of information about these caves that we can't get our hands on," says Dr. Campbell.

If the entire project is approved, trains carrying liquid propane and liquid butane would cross a 77-year-old train trestle that spans the Watkins Glen Gorge. Because propane and butane are heavier than air, "if there were an accident and one of the cars breached, the gas would form a dense, explosive vapor cloud and funnel right down to the gorge and into the heart of Watkins Glen" where it could ignite merely by someone lighting a cigarette, says Dr. Campbell.

"This type of industrialization will destroy the local, sustainable economy that brings several hundred million dollars into New York on an annual basis," says Taylor, who points out that the storage of natural gas liquids (NGLs) will bring active burning flare stacks, a six-track train depot that would have the capacity to load and unload 24 rail cars every 12 hours, a truck depot (which would add an additional eight trucks per hour to narrow roads), open brine pits, four compressor stations, and pipelines.

Dr. Campbell says he has "never seen this kind of activity; drilling went on non-stop for about 15 months." That drilling was solely to test the caverns and upgrade them to accept NGLs when the state issues its permits.

In October 2013, Inergy was merged into Crestwood Midstream Partners, a corporation based in Houston, Texas. If past practices are any indication, Crestwood will receive property tax benefits. In other places where Inergy had built, it often violated state and local regulations and freely used the power of state-approved eminent domain rules to seize private property.[1874]

To raise funds for expert witnesses and miscellaneous expenses, Gas Free Seneca held what Taylor describes as a "very lucrative" fundraiser in August 2013; several dozen individuals and businesses contributed art, crafts, and gift certificates for a silent auction. Fundraisers are necessary for citizen groups opposed to fracking and its numerous auxiliary enterprises—they don't have the financial support of billion dollar corporations.

The Federal Energy Regulatory Commission (FERC) received more than 31,000 comments, most of them opposed to Inergy/Crestwood's plans. However, in May 2014, FERC approved the application of the Arlington Gas Co., a Crestwood subsidiary, to store more than two billion cubic feet of gas.[1875] The company plans to expand its storage to more than 10 billion cubic feet of gas, according to filings with the Securities and Exchange Commission. The caverns will have more product in the warm months that coincide with the tourist season.

In a statement to DC Bureau, an independent news service, Taylor said she believes, "The DEC and the EPA are not in place to protect the environment or care for individuals and small businesses so much as they are there to promote heavy industry and protect large corporations."[1876]

Neither Dr. Joseph Campbell nor Yvonne Taylor is a long-time activist. Dr. Campbell, a chiropractor, says he had a "limited role" in successfully opposing Chesapeake Energy from creating a wastewater disposal injection well near Keuka Lake, but the probable destruction of a scenic area for corporate profit caused him to become more active. Taylor, a speech and language teacher, was previously active in stopping gas drilling in the Finger Lakes National Forest on the east shore of Seneca Lake. However, they and thousands of residents and businesses, in the few years since Gas Free Seneca was formed, have become adept at not only maneuvering through governmental regulations and procedures but also campaign organization and media relations. They recognize they are a David to Crestwood's Goliath; they also recognize that, sometimes, David wins. In this case, David did not win. In September 2014, FERC, which seldom denies applications presented to it, ruled that Crestwood Midstream could proceed with construction of facilities for storage of 2.0 billion cubic feet of methane gas.[1877]

Two dozen persons in March 2013 linked arms and blocked the driveway of a gas compressor station owned by Inergy near Seneca Lake, N.Y. Their home-made banner—"OUR FUTURE IS UNFRACTURED. WE ARE > [GREATER THAN] DIRTY INERGY"—explained the reason for the protest. A rally at the park near

Watkins Glen Village Marina, organized by Gas Free Seneca, brought almost 300 persons—including numerous owners of small businesses in the tourist/recreational area—who would exercise their free speech rights to explain why they opposed the development of the salt caverns.

Dr. Sandra Steingraber told *EcoWatch* why the protest was necessary:

> "It is wrong to bury explosive, toxic petroleum gases in underground chambers next to a source of drinking water for 100,000 people. It is wrong to build out the infrastructure for fracking at a time of climate emergency. It is right for me [to] come to the shores of Seneca Lake, where my 11-year-old son was born, and say, with my voice and with my body, as a mother and biologist, that this facility is a threat to life and health."[1878]

She said it was her first act of civil disobedience. It was also the first time she was arrested.

Police arrested 12 who were part of the driveway blockade. Most paid fines. Three—Dr. Steingraber; Michael Dineen, a farmer; and Melissa Chipman, a business owner—did not, and were sentenced to spend 15 days in jail, the maximum time allowed by law.

At a press conference before the sentence was imposed a month after the blockade, Dr. Steingraber told a rally of about 150 persons:

> "[T]his act of civil disobedience is a last resort for me. Prior to this, I and other community members have taken every legal avenue to raise the serious health, economic, and environmental concerns associated with the Inergy plant. However, time and again, we've been deterred from participating in the decision-making process . . .
>
> "It is my belief that paying trivial fines does not excuse the crime of salting the lake. And it's because I have such a high respect for the rule of law that I will be choosing not to pay a fine for my act of trespassing and will instead show responsibility by accepting a jail sentence."[1879]

In courtroom statements,[1880] Chipman and Dineen spoke about protecting the environment and public health:

433

"I believe it is wrong for corporations and people with lots of money/power to harm the earth and the beings living on it! The government has done nothing to protect the earth and the people they are supposed to represent, even though we have written letters, given them scientific evidence and held peaceful protests. They are not hearing us; therefore, I committed civil disobedience as a last resort to stop them from destroying my environment. And furthermore, I would rather eat bread and water now than have no bread and toxic water later!" –Melissa Chipman

"I just want you to know that I do not take this step lightly. I'm 64 years old, my wife and I have a small farm in Seneca County. We grow organic grains and maintain a large garden we use to feed our daughters families and ours. Our garden is irrigated with lake water. I believe the Inergy gas storage complex will, at best, damage the community, and has the potential to do catastrophic damage. Important information has been kept from the public with the DEC's cooperation. I did this to attempt to protect the community when all other means have failed." —Michael Dineen

The three served eight days of the 15 day sentence and, like hundreds who are members of Gas Free Seneca, would continue to protest. But two politicians added an outrageous demand. In an OpEd column for the Corning (N.Y.) *Leader*, journalist Peter Mantius revealed the politicians wanted reimbursement for the $1,600 that Schuyler County paid for the incarceration of the three citizens who exercised their rights of civil disobedience, pleaded guilty, and accepted their sentence. "Your group may want to do the responsible thing and take up a collection to reimburse county taxpayers," Schuyler County legislator Barbara Halpin wrote to Yvonne Taylor. The chairman of the county legislature, Dennis Fagan, called the $1,600 "a relatively minor amount," but argued, "it is galling that we have to pay for their political statement." Mantius observed:

"Since 2009, when . . . Inergy applied for a state permit to store liquid petroleum gas, or LPG, in salt caverns next to Seneca Lake, [Halpin and Fagan] have blocked their ears to noisy dissent and averted their eyes from unpleasant facts.

Meanwhile, they've held fast to their faith in the company and its state regulators."[1881]

"If Schuyler officials continue to shut their ears and eyes," Mantius concluded, "they better start budgeting for more orange jumpsuits."

In the week beginning Oct. 29, 2014, Schuyler County (N.Y.) sheriff's deputies arrested 25 persons, and charged them with trespassing for blocking the gates at the Finger Lakes salt caverns. Several were also charged with disorderly conduct.[1882] By the end of 2014, there were more than 160 arrests.

COVE POINT PROTESTS

About 1,000 protestors marched in withering heat and humidity from the National Mall to the offices of the Federal Energy Regulatory Committee (FERC) on a Sunday in July 2014 to protest what appeared to be a rubber-stamp approval for the Cove Point (Md.) LNG plant.[1883] The next day, 24 were cited and released with a $50 fine for peacefully blocking the entrances to FERC for about 90 minutes.[1884] Almost no media covered the protest; a few more covered the arrests. The protestors objected not only to the future awarding of a permit for Cove Point, but the problems with LNG. More than 150,000 signed petitions to stop the construction, citing several environmental and health concerns.[1885] If the Cove Point facility received all approvals and began shipping LNG, among the myriad problems, as pointed out by the citizens group We Are Cove Point, would be that the plant would emit about 20.4 tons of pollutants and 3.3 tons of greenhouse gases every year, that it would destroy forests and waterways, disturb the balance of nature, increase truck traffic in a residential area while leading to more accidents and slower emergency response times, and create noise pollution that could be diminished only by building a 60-foot high barrier.[1886] The protestors also objected to the Calvert County commissioners waving $506 million in property taxes over a 14 year period,[1887] while requiring that Dominion create only 25 permanent jobs.[1888]

Although the protests had brought the issue into the news, it was a legal decision that temporarily stopped progress of the plant. The Calvert County commissioners, who had willingly signed a non-disclosure agreement with Dominion Resources in August 2012,[1889] undoubtedly in violation of state freedom of information rules, had fast-tracked Dominion's requests for zoning variances. However, in August 2014, Circuit Judge James P. Salmon ruled the zoning exceptions to benefit one industry violated state law.[1890] The following month, FERC approved the plant without requiring an environmental impact statement. That decision was "the result of a biased review process rigged in favor of approving gas industry projects no matter how great the environmental and safety concerns," said Mike Tidwell, director of the Chesapeake Climate Action Network.[1891]

Protests continued through much of 2015. In May, Beyond Extreme Energy, an anti-fracking group, organized a media pseudo-event when almost 100 protestors, carrying signs and banners, blocked FERC entrances; the protest drew media attention when a woman, suspended by a harness, swung from an 18-foot high metal tripod to call out reasons why FERC should stop permits for further pipeline and LNG construction.[1892] The protestors left after several warnings, but no arrests by Department of Homeland Security officers.

RESPONSIBLE DRILLING ALLIANCE

No one expected 500 people to pack a meeting of the Department of Conservation and Natural Resources (DCNR), June 3, 2013, in Williamsport, Pa. But there they were. Young adults and retirees. Former gas company workers and environmentalists. Professionals and tradespeople. Liberals, Conservatives, and people affiliated with no political party. Almost all of them were at a public hearing the DCNR didn't even want to have. All of them were there to protest what would be the destruction of the Loyalsock State Forest.

Had not the Responsible Drilling Alliance (RDA) of Williamsport pushed for the hearing and had it not pushed for overflow protest at that hearing, the DCNR, as politicized as the state's Department of Environmental Resources, would probably have reviewed Anadarko's application, made a few changes and suggestions, and allowed drilling. But that was not going to happen.

"We took out ads, networked all over the state with other groups, and used social media" to alert the people to what was happening in the forest, says Ralph Kisberg, one of RDA's founders. The RDA, Marcellus Earth First!, and Shale Justice organized additional rallies to try to protect the forest.

The RDA began in 2009 as drillers began expanding their operations into Lycoming County. The four RDA founders were Kisberg; Robbie Cross, owner of a company that manufactures outdoor recreation gear; Jon Vogle, professor of art at Lycoming College; and Janie Richardson, a housewife. Kisberg, who managed public lands for native Americans, had lived much of his life in Colorado, "where we saw a lot of the problems from oil and gas extraction." Now in Williamsport and the owner of a used book store, Kisberg says the four "met informally for awhile before we realized we needed to form a non-profit organization" to help explain fracking to the people, and to demand "responsible drilling" from the industry.

Informing the public should have not been as hard as the founders experienced. "The local media seemed so biased to promoting fracking; they and the Chamber of Commerce and local politicians seemed to be cheerleaders," says Kisberg, frustrated that the people who should have been looking out for the public welfare "were looking only at the economic benefit." He says, "It was hard to find anything other than a positive spin" to fracking.

In September 2009, the RDA took out a full page ad in the Williamsport *Sun–Gazette*. That ad, which explained fracking's effects upon water and air, helped counter that positive spin by business, media, and politicians.

With numerous public events and a weekly newsletter with a circulation of about 1,200, the RDA slowly cut into that cheer-leading for more fracking, slowly informing the people of Lycoming County, who by 2013 had significant ties to the gas industry, that despite the economic benefits of a "boom," the Industry had responsibilities to protect the people and their environment. It was this focus that brought a standing-room-only crowd of people to the DCNR hearing.

"Keep It Wild!" the crowd chanted. "Keep It Wild!" was the message on protest signs painted environmental green.

The protestors knew there would be drilling; they didn't

know how bad it would be until one of the DCNR panel members revealed that Anadarko Petroleum planned to place 26 well pads and four compressor stations into the forest, after fragmenting it with almost 16 miles of new roads, 34 miles of pipelines, and several towers as high as 200 feet.

DCNR secretary Richard Allan, clearly upset by the turn-out, explained that Anadarko planned to fragment only two percent of the 18,870 acres it and the International Development Corp. "owned," only to hear outrage from the crowd that knew the proposed drilling would impact wildlife, destroy vegetation, significantly reduce recreational opportunities, and undoubtedly create environmental and health problems. To jeers, hoots, and outrage, Allan claimed his department didn't even have full control over the surface rights of the 25,000 acre forest.

But how much contempt the gas drillers and the Tom Corbett administration had for the public's rights became apparent in a request filed in January 2013 by Citizens for Pennsylvania's Future. PennFuture filed a Right-to-Know request for information about Anadarko's plans to deforest Loyalsock. Nine months later, the Governor's Office of General Counsel finally provided the requested documents, much of it redacted, including information about the seismology of the forest, and proposed roads, well pads, and pipelines.[1893] The redacted information was whatever Anadarko claimed to be proprietary information and protected under Act 13. "Why black out the entire table of anticipated disturbances . . . when this information was given by the DCNR at the June meeting; does it imply that the numbers are different?" asked RDA newsletter editor Ann Pinca.[1894]

By the end of Summer, Anadarko had staked out well pads and roads, but still hadn't begun drilling. When that happens, RDA and hundreds of people and dozens of organizations also plan to be in the forest, respectful of the wildlife, streams, and vegetation, furious at the destruction of the land.

MARCELLUS PROTEST

June 2010. Like most Pennsylvanians, Gloria Forouzan had never heard of fracking. But she knew there were people called

438

"landmen" who were working Pittsburgh and Allegheny County. "As we found out more, we became more concerned," she says. At first, a few friends and neighbors met in kitchens and dining rooms; soon, they were meeting in church basements. "It all happened so quickly," she remembers.

"Quickly" happened because Facebook and Twitter, each about four years old, were becoming communication vehicles that would unite sub-cultures, groups, and movements.

"I knew nothing about social media when we started," says Forouzan, "but I had to learn it to get message out."

Within months, the message was being heard throughout the state, and Marcellus Protest would be ready for its first public rally, Nov. 3, a counter-protest to the second annual Developing Unconventional Gas (DUG) East conference that would draw 2,000 industry professionals and politicians, and more than 200 exhibitors, to the city's Convention Center. DUG used a wide variety of media to promote its conference; Marcellus Protest used social media and volunteers to distribute home-made flyers to libraries and other places that had public bulletin boards. Keynoting the DUG convention was Karl Rove, who had been George W. Bush's senior advisor and deputy chief of staff. Speakers for the counter-rally were scientists, environmentalists, those affected by fracking, and people from the creative arts.

Like many in Marcellus Protest, Forouzan had been active in numerous social issues, "but none of us had done anything like this before." She worried about everything that could go wrong, about everything that was not done that should have been done. "If we get 50 people here, I'd be happy with that," she recalls. She didn't have to worry about being just happy. Not long before the rally was to begin, Forouzan heard drums and whistles from a makeshift rag-tag marching band, saw giant puppets on wooden stakes, and what would become more than 500 people behind banners and signs that read, "You Can't Drink Money," ready to march across the Rachel Carson Bridge into downtown Pittsburgh and circle the Convention Center. "I started crying. I was so incredibly moved," she remembers. The rally would be the largest public anti-fracking protest in the state, covered by several newspapers and television stations.

The following week, Pittsburgh City Council member Doug

Shields, who had been active in the march, introduced the state's first resolution to place a moratorium on fracking. The council unanimously passed that resolution.

Marcellus Protest has about 4,700 Facebook "likes," about 2,000 Twitter followers and, like many anti-fracking groups throughout the country, has a newsletter and an interactive calendar that lists anti-fracking activities. It sponsors health symposia; its volunteers speak to anyone who wants information or who needs help dealing with the DEP and corporate drillers. It has also helped start other anti-fracking groups. "We help connect people," says Forouzan, "our networks have expanded, and we can connect with activists throughout the world."

Like the first day it was founded, Marcellus Protest doesn't have officers, but it has a lot of volunteers. "If you want to do something, you do it and get help," Forouzan says. For many of the volunteers, "the work is all-consuming, but while it's easy to despair, we don't ignore the small victories; they are what sustain us."

WESTMORELAND MARCELLUS CITIZEN'S GROUP

With assistance from Marcellus Protest and the Mountain Watershed Association, 15 people met in Delmont, Pa., in September 2010 to form the Westmoreland Marcellus Citizen's Group. Four months later, newly-inaugurated Tom Corbett officially began his push to open up Pennsylvania to fracking.

"That became our first priority, to tell people about the effects of fracking," says Jan Milburn, one of the founders and president of the Citizen's Group. "We wanted people to know what the proposed legislation would do to our state and to the people," she says.

Getting the people to come to informational meetings was relatively easy; getting elected public officials to attend the meetings was impossible. It is a problem faced by anti-fracking groups throughout the country. "We called them, we sent them personal letters, we did everything we could," says Milburn, "but they didn't want to hear what we were saying." Most of the county's politicians apparently had already decided that the possible economic benefits from fracking, which they heard

440

from other politicians and the gas industry, were far more important than anything about health, safety, and the environment.

And so Milburn and members of the Citizen's Group went to meetings of 15 of the 21 townships where leasing was an issue to try to explain the issues of the forthcoming Act 13 and of local, health, and environmental concerns, and why elected officials shouldn't blindly accept fracking as an economic savior. And that's when they learned another reality. "Some of the council members and supervisors welcomed us and allowed us to speak, but many weren't receptive and refused to allow us to talk about the issues," says Milburn. The state's Sunshine Law requires public bodies "to provide a reasonable opportunity [for citizens] to comment on matters of concern, official action or deliberation which are or may be before the board or council prior to taking official action."[1895] The Citizen's Group persisted, continuing to attend meetings of public officials, continuing to speak at any club, church, or organization that wanted to hear something other than the PR claims of the Industry and the politicians who had fallen into the web of political donations and economics.

Jan and Jack Milburn had moved their family from North Huntingdon, a suburban city about 20 miles southeast of Pittsburgh, to Ligonier in Westmoreland County because their two children had asthma and "because the air was cleaner." But clean air may no longer be why people move into Westmoreland County. "We're now facing compressor stations, flaring, and fugitive emissions from gas pipes and wells," says Milburn. Until the people see what it can do," says Milburn, "there won't be a dramatic change of attitude."

'WE'RE ALL IN THE SACRIFICE ZONE'

The Marcellus Shale doesn't extend into south-central and southeastern Pennsylvania, but that doesn't mean the eight million residents of the largely urban and suburban region aren't affected by fracking. The region, downstream of the most active fracking sites, sits on top of a complex series of pipelines, and co-exists with compressor stations and a deep-water port in Philadelphia that will allow multi-national corporations to

441

ship liquefied natural gas. It is in this region that Lehigh Valley Gas Truth, Berks Gas Truth, and Protecting Our Waters are active in the fight against fracking.

"We wanted to have the backs of those in the drilling region at the time, but realized we might be in industry's path ourselves one day," says Karen Feridun, a former corporate librarian who organized Berks Gas Truth in November 2010. The path came faster than any of the BGT leaders thought. "Our first taste of the industry was an incident at an existing compressor station during Hurricane Sandy," says Feridun. Within six months, she says BGT was "still fighting to hold accountable the company that owns the compressor, and we found ourselves in fights against two pipelines and a natural gas power plant."

Feridun and BGT have provided clean water to residents impacted by fracking, and organized numerous panels, special events, and concerts, while also writing dozens of articles and letters-to-the-editor. She and BGT, by 2013 one of the nation's most forceful and effective anti-fracking volunteer groups, became the force that rallied other organizations and innumerable activists to successfully convince the Democratic State Committee to disregard the party's political leaders and recommend a moratorium of shale gas drilling until such a time came when it could be shown to be safe.

Like Karen Feridun, Julie Ann Edgar spends much of her time working to eliminate fracking. She never expected to be an activist; it just slowly happened. She had been environmentally concerned, but says she was "way too absorbed to care much about world affairs, corporatism, and extreme fossil fuel extraction." And then *Gasland* changed her life. "By the end of that movie, I knew from what I had seen of corporate greed and hubris that those who were awake and able needed to take action," she says, "and answer the call to create a cleaner world, and to expose this behemoth destroyer of communities and ecosystems before Pennsylvania became unbearable." She joined the Lehigh Valley Gas Truth, formed in October 2010 by Cathy Frankenberg, program organizer of Clean Water Action (CWA). Two years later, after Frankenberg left CWA, Edgar became director of LVGT. Edgar estimates she has worked more than 5,000 volunteer hours to learn about fracking and to help inform the people. "It has been a very long journey," she says, "but I am more in touch with my heart now than I have ever been before."

Conferences

If nothing else shows the differences between the corporate world of fracking and the grassroots world of those who protest the destruction of the environment and the health problems fracking causes, it could easily be seen in how each group handles its conferences.

At the first Shale Gas Insight conference, September 2011 in Philadelphia, former Pennsylvania Gov. Tom Ridge called for natural gas to be "at the forefront of the energy revolution." Natural gas extraction through fracking, said Ridge, "should lead the energy revolution, and be the foundation of the 21st century."[1896] As expected, Ridge, who was also the country's first secretary of homeland security, argued it was a matter of "national security" to reduce importing oil and to increase domestic gas production. Ridge, a lobbyist for the Marcellus Shale Coalition, which sponsored the conference, also praised gas field job creation and condemned government subsidies for renewable energy. Gov. Corbett, less than a year in office, and former Gov. Ed Rendell, who worked with and for the natural gas industry, also spoke at the conference to praise gas extraction. That conference drew more than 1,500 politicians and oil/gas executives and professionals, as well as hundreds of protestors.

For the second annual conference, Corbett, now the keynote speaker, brought cheers from the audience when he declared, "We are advancing, even in the face of unreasoning opposition. Our opponents agree that we can land a rover on Mars, but they can't bring themselves to think that we can safely drill a mile into our own soil."[1897] If Corbett had said nothing else about fracking and natural gas extraction in two years, those words to the industry should have been a warning that as the state's chief executive his loyalties were with the industry and not with the people. Those words explained all anyone needed to know about why the Department of Environmental Protection, now a politicized regulatory agency, was being attacked for failing to adequately regulate the natural gas industry.

By the third conference in 2013, keynoter Newt Gingrich, former speaker of the House of Representatives, who less than two years earlier had called for the elimination of the Environmental Protection Agency,[1898] now told the cheering industry-and-politician crowd, "[W]e'll see almost inevitably the weight of technological and scientific progress grind down the regulators."[1899] Much of the conference focused upon what the Industry thought was over-regulation. Although there were significant technological advances to create hydraulic fracturing, journal articles by independent scientists revealed the myriad health and environmental problems associated with fracking. Further, grinding down regulators should not be hailed as an achievement—the purpose of Pennsylvania's Department of Environmental Protection and the federal Environmental Protection Agency, as their names suggest, is to protect the environment, and not to accede to industry demands and overlook violations of law. That third conference was also a subtle recognition that natural gas fracking in Pennsylvania's Marcellus Shale may not be as lasting or as economically beneficial as all the speakers once believed. The new name was the "Conference," dropping the word "gas" from its title.

Gov. Corbett and former Gov. Ridge were keynote speakers for the the 2014 conference; other keynoters included the CEOs of XTO Energy, MarkWest Energy, and EQT Corp., plus Fox News pundits Sean Hannity and Dana Perino, who had been George W. Bush's press secretary. Hannity's speech was sponsored by Consol Energy; Perino's speech was sponsored by Range Resources.[1900] At the conference, as on their TV shows, both praised the oil and gas industry, while attacking government and anti-fracking activists.

A major theme of the September 2015 conference was equating energy independence with national security. Former New York City mayor Rudy Giuliani was the keynoter. Among other speakers were Michael Krancer and Mike Turzai, speaker of the Pennsylvania House of Representatives.

Unlike industry conferences that have corporate CEOs and well-known politicians and FOX commentators as speakers, anti-fracking conferences bring together a mixture of scientists, health care professionals, environmentalists, activists, and those who have been harmed by the effects of fracking. There are often folk musicians and others from the creative and performing arts.

Unlike industry conferences, where registrants pay as much as $300 a night for a hotel room, and deduct thousands of dollars of travel, meals, and other expenses, registrants at anti-fracking conferences often stay in homes of friends, fractivists, or sponsoring churches and synagogues.

Unlike Industry conferences, which distribute full-color flyers and pamphlets, anti-fracking conferences rely upon social media, email lists, and newsletters.

Freedom from Fracking—its motto is to "educate, organize, and agitate!"—is the counter-conference to the Shale Insight Conference.

Unlike the Shale Insight Conference, which charges up to $1,075 to attend,[1901] registration for Freedom from Fracking was only $20 in 2011 and 2012; in 2013, there was no charge. At the Marcellus Shale Coalition conferences, each of the 150 exhibitors paid $3,500–$4,500 for a 10-foot by 10-foot booth.[1902] There is no cost to display literature or recruit members at Freedom from Fracking. At the Marcellus Shale Coalition conferences, sponsors paid $5,000–$50,000;[1903] there were 45 sponsors at the 2013 conference, and 36 the following year.[1904] Major sponsors for Freedom from Fracking were Berks Gas Truth, Clean Air Council, the Delaware Riverkeeper Network, PennEnvironment, and Protecting Our Waters. Together they paid about $2,000. More than 60 "endorsers" are primarily environmental organizations, including the Sierra Club and local and regional anti-fracking groups. Another difference is media coverage. Newspapers and television carried daily coverage of the Shale Insight Conference. "We contacted the media but got little response," says Karen Feridun, "and we had people who were living with the effects of fracking willing to talk, and the press didn't bother to come." The *Philadelphia Inquirer* did send a photographer, and the Pennsylvania Cable Network (PCN-TV) sent camera crews to broadcast all events.

Freedom From Fracking's first counter-convention in 2011

was held at the Congregation Rodeph Shalom, with about 200 participants.

The 2012 conference began with two sessions on peace-keeping and the Constitutional rights of protestors. On the third day, 1,500 protestors attended a rally and march, stopping at the campaign headquarters of President Obama, PNC Bank (which heavily invested in the shale gas industry), the Philadelphia office of Gov. Corbett, and the Pennsylvania Chamber of Commerce, both in same office. Among the speakers at the rally were Dr. Sandra Steingraber, Maya van Rossum, Deborah Rogers, Dr. Stephen Cleghorn, Josh Fox, Iris Marie Bloom, Karen Feridun, and other leaders of the anti-fracking movement. The protest in Philadelphia wasn't the only one. Two weeks later was Global Frackdown Day where the voices of the fractivists were heard in small towns and urban areas in the United States and overseas in more than 100 events organized by Food and Water Watch.

For the third conference, Freedom From Fracking held a two-day water drive to collect jugs of clean water to distribute to those whose water had been polluted by fracking. "People would come by, hear what we had to say, and then go to a store and buy a gallon of water," says Karen Feridun. The Shale Insight Conference had no charity events.

An all-morning session at the College of Physicians and Surgeons on the fourth day of the conference focused upon health effects from fracking. John Fenton of Pavillion, Wyo., a farmer whose water was polluted by fracking, was the keynote speaker of the afternoon session. Iris Marie Bloom, director of Protecting Our Waters, told about trains from the Bakken Shale in North Dakota coming into the Philadelphia Energy Solutions refinery "on an elevated, old, shaky-looking track and whistles all night long, so this is a really new development."[1905] There was no protest or march the third year, prompting some media and the shale gas industry to declare the anti-fracking movement was dying. However, the anti-fracking movement is getting stronger, with almost half of all Americans now calling for a halt to fracking, and support for fracking declining, according to a poll published by the Pew Research Center in September 2013.[1906] The reason why there wasn't a rally was "because it costs a lot of money and energy to put together a rally and march, and we wanted to put our resources on other areas that are critical," says Feridun.

There was no 2014 Freedom from Fracking Conference in the Philadelphia metropolitan area because the industry moved its Shale Insight Conference to Pittsburgh, about 270 miles away in the southwestern part of the state.

Dozens of other conferences—all of them low-budget and focused on grassroots education and activism—bring unity and purpose to the anti-fracking movement.

Divesting Fossil Fuel Stocks

Socially conscious churches, universities, non-profit organizations, and local governments have begun to divest themselves of fossil fuel stock.

DIVESTMENT BY RELIGIOUS INSTITUTIONS

The World Council of Churches, which represents about 590 million Christians in 520,000 congregations,[1907] decided in July 2014 that to continue to hold fossil fuel stock would compromise its ethics, and recommended the 349 member denominations consider divesting oil and gas stock.

The Church of England, a member of the World Council of Churches, represents about 25 million members. The church, which has an endowment of about £5.2 billion (about US $8.5 billion), divested all tar sands oil and thermal coal from its portfolio in April 2015. The Church of England still has about £100 million invested in Shell and about £91 million in BP. "Companies like the oil and gas majors are significant players in the business world and political world, and we'd like them to be part of a constructive call for the transition to the low-carbon economy," said Edward Mason, head of the investment portfolio. However, he cautioned, "If engagement with companies isn't productive, the policy does make clear divestment is there as a last resort."[1908]

Six of the eight Anglican dioceses of New Zealand and Polynesia, and four dioceses in Australia divested their portfolios of fossil fuel stock.[1909]

The Religious Society of Friends (Quakers) in England began

to divest in October 2013. The church says that investment in fossil fuel extraction "is incompatible with [the] commitment made in 2011 to become a sustainable low-carbon community" and to encourage the United Kingdom to develop "an energy system and economy that does not rely on fossil fuels."[1910]

In the United States, the United Church of Christ and the Unitarian Universalist churches became the first denominations to begin to divest themselves of fossil fuel stock.[1911] Both denominations have a long history of fighting for social justice. However, the UCC pension board is not divesting fossil fuel stocks. That Board "sees itself as ensuring a mandate to provide adequate funds for retirement, as opposed to the church philosophy of social justice," says the Rev. William Thwing, an active environmentalist and opponent to fracking.

Several synods of the Evangelical Lutheran Church of America have passed resolutions asking the national board to divest themselves of fossil fuel stocks.

Two weeks after Pope Francis issued the Catholic church's first encyclical on the protection of the environment, the Episcopal church leadership approved divestment of the $380 million in church holdings. The divestment doesn't affect the church's $9 billion pension fund or the $4 billion that individual parishes and dioceses control;[1912] however, the leadership did ask congregations to consider eliminating fossil fuel stock and replacing it with renewable energy stock.[1913]

Also divesting are several other denominations, including the Society of Friends.[1914] Students at the Quaker-run George School, a high school in Newton, Pa., convinced the administration in May 2015 to divest its $150 million endowment of all stock owned by coal companies. The George School became the first secondary school to eliminate fossil fuel stock.[1915]

Several religious groups that have shares in Chevron asked the corporation in 2011 to go "above and beyond regulatory requirements" to protect the environment and public health. Chevron flippantly dismissed the request. Chevron claimed, it "is already committed to meeting or exceeding all applicable laws and regulations [and the suggestions] would merely duplicate Chevron's current efforts and thus would be a waste of stockholder money."[1916]

An organization of Catholic priests asked the ExxonMobil shareholders at the annual meeting in May 2015 to put a climate-change expert on its board. The Board opposed that resolution, which got only 21 percent approval by others in attendance or who had submitted their proxies.[1917] All environmental resolutions, including those to reduce greenhouse gases and to look into renewable energy, were also defeated. In clinging to the corporation's determination to remain in fossil fuel extraction and distribution, CEO Rex Tillerson told a cheering audience that renewable energy wasn't profitable, and because of that, "We choose not to lose money on purpose."[1918]

The Union Theological seminary, with a $108.4 million endowment, became the nation's first seminary to divest itself of fossil fuel stock. The Rev. Dr. Serene Jones, the seminary's president, explained the decision:

> "As vulnerable communities have been swallowed by rising shorelines, as potable water has become a commodity of increasing rarity, as hundreds of thousands of people have been killed by violent weather, it is ever clear that humanity's addiction to fossil fuels is death-dealing—or as Christians would say, profoundly sinful."[1919]

DIVESTMENT BY COLLEGES AND UNIVERSITIES

College students have been active in pushing their institutions to eliminate fossil fuel stocks from their portfolios. The result has been an awareness of a social issue that was not seen since students pressured their colleges to divest funds in tobacco companies and in corporations that dealt with the apartheid government of South Africa.

"If we don't deal with climate change now, we consign our grandkids to an unlivable planet," said Unity College president Dr. Stephen Mulkey.[1920] Unity, which became the nation's first college to divest its endowment portfolio of fossil fuel stock,[1921] specializes in environmental and natural sciences.

Among three dozen colleges and universities that are committed to eliminating fossil fuel stock from their investment portfolios are Foothill–DeAnza Community College, Green Mountain College, Humboldt State University, San

Francisco State University, the University of Dayton, and Hampshire College, which in 1977 in protest of apartheid policies, was the first U.S. college to eliminate all stocks related to South Africa.[1922]

Pitzer College, a liberal arts college in Southern California, divested about $4.4 million of its $5.5 million in fossil fuel stock by the end of 2014, and pledged to work to eliminate most of the rest of the stock in fossil fuel industries. The college has about a $124 million endowment.[1923] "I don't think that anyone who favors divestment is arguing that the institutions' sale of the fossil fuel company stock is going to have much impact, if any, on either the stocks or the companies themselves," Donald P. Gould, chair of the Pitzer College portfolio, told the *New York Times*. However, he said, "Divestment seeks to work indirectly on these companies by changing the conversation about the climate."[1924]

Stanford University announced in May 2014 it would no longer hold stock in the coal industry, but did not include divesting oil and gas stock in its $18.7 billion endowment fund.[1925] About 300 Stanford University professors published an open letter to the administration in January 2015 to request the university divest itself of all fossil fuel stock. Among the signers were Drs. Elizabeth A. Hadley, senior associate vice-president; Donald Kennedy, former Stanford president; Roger Komberg, Nobel Prize winner in chemistry; and Douglas Osheroff, Nobel Prize winner in physics.[1926]

Syracuse University, in March 2015, announced it would no longer include fossil fuel stocks in its investment portfolio.[1927] The decision by the Board of Trustees followed an 18-day sit-in four months earlier and public information campaign by Divest SU, a student organization founded in 2012.

Harvard University, whose $32.7 billion endowment is the largest among educational institutions, doesn't plan to eliminate fossil fuel stocks from its portfolio. Dr. Drew Gilpin Faust, Harvard president, told the *New York Times* that in spite of wide-scale student protests, eliminating fossil fuel stocks is not "warranted or wise," and that the university's portfolio is "a resource, not an instrument to impel social or political change."[1928] However, the response by 235 faculty, in a letter of support to Divest Harvard, pointed out the fallacy of the administration—"If the Corporation regards divestment as 'political,' then its continued investment is [itself] a similarly political act, one that finances present corporate activities and calculates profit from them."[1929]

David Crane, CEO of NRG, one of the nation's leading energy providers, said he didn't "relish the idea that year after year we're going to be educating a couple million kids from college, who are going to be American consumers for the next 60 or 70 years, that come out of college with a distaste or disdain for companies like mine."[1930]

DIVESTMENT BY FOUNDATIONS AND GOVERNMENTS

The Norwegian parliament, in June 2015, ordered Norges Bank Investment Management, which manages the country's $890 billion pension fund, to divest all stocks from companies that have at least 30 percent of their revenue from coal or coal-related energy. The effect will be the elimination of stock worth about $8.7 billion from 122 companies, including 49 based in the United States, according to Greenpeace.[1931] However, not affected are oil and gas stocks.

The Rockefeller Brothers Fund, founded on income from oil exploration and development, declared in October 2014 it would divest fossil fuel stock. Stephen Heintz, president of the $860 million philanthropic foundation, said his executives had already divested its portfolio of coal and tar sands stocks. The Foundation will invest in more renewable energy companies.

Several dozen U.S. cities and counties, as early as 2012, have begun the process to divest themselves of fossil fuel stocks and to urge their independent pension boards to divest.

San Francisco in April 2013 began divesting about $580 million of fossil fuel stock, and by 2019 will stop purchasing stock of any company associated with fossil fuel exploration and development.[1932]

Seattle, Wash., Mayor Mike McGinn in December 2013, a month before leaving office, asked the city's pension board, which oversees a portfolio of about $1.9 billion, to "begin exploring options for moving existing investments from fossil fuel companies."[1933] The city had $17.6 million invested with ExxonMobil and Chevron.

The other eight cities of the first ten to divest fossil fuel stocks are Richmond and Berkeley, Calif.; Boulder, Colo.; Ithaca, N.Y.; Rochester, Minn.; Eugene, Ore.; and Bayfield and Madison, Wisc.[1934]

Even if all major religious institutions and university foundations removed fossil fuel stock from their portfolios, the effect would be more symbolic than destructive to the continuation of fossil fuel energy. Profits might be a bit lower, but there will still be investors, mutual funds, corporations, and financial institutions that would invest.

Collateral Damage in the Marcellus Shale

On Feb. 18, 2012, the residents of the Riverdale Mobile Home Village in Jersey Shore, in north-central Pennsylvania, found out after reading a story in the Williamsport *Sun–Gazette* their village had been sold and would be demolished. The 32 families of the village, most with limited incomes, paid their rents and fees on time; they kept up the appearances of their trailers and the land around it. They worked their jobs; they survived. But now they were told they would have two months to leave. It was abrupt. Business-like.

The new owner was Aqua–PVR, a partnership of Aqua America and the smaller Penn Virginia Resource Partners. Aqua America provides water to about 2.8 million residents and corporations in nine states, including about 1.4 million in Pennsylvania.[1935] It had received permission from the Susquehanna River Basin Commission (SRBC) to withdraw up to three million gallons of water a day for 15 years from the West Branch of the Susquehanna; the families of the mobile home village would just be in the way. The company planned to initially spend about $20 million to build a pump station and create an 18-mile pipe system to provide fresh water to natural gas companies that use hydraulic fracturing.[1936] The pumping station would be between two others, both owned by Anadarko Petroleum, both taking water from the Susquehanna.

Most residents had only a vague knowledge of fracking. Deb Eck says she "learned real fast" what it was, and what it was doing to the people and the land. Kevin June says the residents of Riverdale "have a lot more knowledge now." Eck, manager of a retail store in nearby Williamsport, and June, a disabled auto body repairman, would become leaders of a protest movement.

Aqua dangled a $2,500 relocation allowance to those it was evicting. However, the cost to move a trailer to another park was $6,000–$11,000, plus extra for skirting, sheds, and any handicap-accessible external ramps. About a third of the trailers couldn't be moved. Even if all could be moved, there were few places that would take the other families. The parks want the newer trailers, but most parks were full.

Kevin June was constantly on the move, going from trailer to trailer to try to keep the residents informed, to help families who were abruptly evicted, because the lives of his neighbors mattered.

There's Betty and William Whyne. Betty, 82, began working as a waitress at the age of 13 and now, in retirement, makes artificial Christmas trees. At the time of their eviction, she had a cancerous tumor in the same place where a breast was removed in 1991. William, 72, who was an electrician, carpenter, and plumber before he retired after a heart attack, goes to a dialysis center three times a week, four hours each time. They brought their 12-wide 1965 Fleetwood trailer to the village shortly after the 1972 Agnes flood. Like the other residents, they couldn't afford to move, and they couldn't find adequate housing. They earned $1,478 a month from retirement, only $252.17 above the federal poverty line.[1937] One son is in New Jersey, one is in Texas; the Whynes don't want to leave the area; they shouldn't have to.

There's April and Eric Daniels. She's a stay-at-home mom for their two children; he's a truck driver for Stallion Oilfield Services, delivering water to natural gas companies in Pennsylvania and wastewater into Ohio. April Daniels had grown up living in a series of foster houses, "so I know what it's like to move around, but this was my first home, and it's harder for me to leave."

Doris Fravel, 82, a widow on a fixed income of $1,326 a month, lived in the village 38 years. She was proud of her 1974 12-wide trailer with the tin roof. "I painted it every year," she says. Eight months earlier, she paid $3,580 for a new air conditioner; she recently paid $3,000 for new insulated skirting. The trailer has new carpeting. Unlike most of the residents, she found housing—a $450 a month efficiency, far smaller than her mobile home. She sold or gave away most of what she owns. "I never knew I would ever have to leave," she says, but she still wants to "see one of those gas men come to my door—and I'd like to punch him in the shoulder."

Few lots are available, apartments are too expensive because landlords inflated rents to take advantage of the gas company workers, most residents don't qualify for a mortgage, and there are waiting lists for senior citizen and low-income housing.

Kevin June went to State Rep. Garth Everett "to ask what he could do to help, but his secretary just coldly told me there was nothing that could be done because whoever owns a property can do with it what he wants to do." He never saw the state representative.

During the week Aqua–PVR issued eviction notices, its parent company issued a news release, boasting that its net income in 2011 was $143.1 million, up 15.4 percent from the previous year.[1938] Two months later, Penn Virginia Resource Partners bought Chief Gathering, an energy pipeline company, for $1.06 billion.[1939] But, for some reason, Aqua–PVR just couldn't find enough money to give the residents a fair moving settlement. "They just expected us to throw our homes into the street and live in tents," says June.

The eviction united the residents, but by the middle of May, most reluctantly took the $2,500 relocation allowance and left.

"I was scared of what they would do, because they had the lawyers and the money," says Fred Kinley, a 79-year old Air Force veteran and retired sheet metal worker.

Other residents stripped their trailers of pipes, fixtures, porches, and anything that could be taken and used in a new home, anything that could be sold, anything other residents might be able to use.

For years, village residents had formed a loose neighborhood; they were friends and acquaintances. They had picnics and, like residents of all neighborhoods, they sometimes had arguments or just didn't associate with other residents. But, the stress of dealing with Aqua and the forced evacuation led to an increase of rumor and innuendo, with some of the residents verbally challenging others. Everyone was frustrated and tired, their emotions bled raw by dealing with a corporate entity and issues they didn't fully understand.

Kevin June had become adept at how to work with social media. Alex Lotorto, a grounds technician and volunteer, set up a website, SaveRiverdale.com, with information and photos. The pleas and stories drew assistance from individuals and several anti-fracking groups. Soon a nation was able to see micro-documentaries on YouTube, posted by several amateur videojournalists.

A few readers of the Williamsport *Sun–Gazette* wrote letters to the editor or posted their comments to the online site of the newspaper to show their support for the residents of Riverdale.

However, in the center of the Marcellus Shale boom, most of the neighbors and readers of the *Sun–Gazette* condemned the residents who protested their sudden eviction.

Hiding behind anonymous screen names, the writers could have been among thousands of employees and managers who had temporarily moved into the area. They could have been the landlords who raised their rents three and four times to take advantage of the incoming gas field workers. They could have been those who leased part of their land to the oil companies. They could also have been the business owners who profited because of selling products to the workers. But most of them condemned the residents.

Linhk48, who posted several dozen times, believed "the new owner's only obligation is to give you notice to vacate. He is under absolutely no obligation to subsidize your move, allow you to live rent free until you move, or hire professionals to help you with relocation. Anything he does is a generosity and SHOULD be appreciated!"

No2spanish believed, "[T]his is why people in mobile homes should save their money—instead of spending it on booze or drugs."

Czkb217 thought the police or National Guard could move in, and advised the residents, "SO just pack your stuff and MOVE, you are now breaking the law." It's doubtful any of the commentators knew Pennsylvania state law that establishes legal processes that must be met to evict persons from their homes. Czkb217 also believed the residents should have gotten together and bought the park. Since most of the families lived slightly above the poverty line, they probably didn't have an extra $550,000 plus lawyer fees and closing costs laying around. Nevertheless, Czkb217 believed the residents should "Just man up and put your big boy panties on and MOVE." He objected that his taxes supported some of the residents who used Legal Aid, which receives state and federal funds to assist the impoverished. John Person with the Williamsport office of North Penn Legal Services and Kevin Quisenberry of the Community Justice Project in Pittsburgh assisted the low-income residents of the village; Jonathan Butterfield of the Williamsport law firm of Murphy, Butterfield and Holland assisted *pro bono* for those residents who didn't qualify for legal services.

Many of those who attacked the residents and defended corporations probably believed they were good Christians; they attended church regularly and, in one of the more conservative and highly Christian parts of the state, praised God publically.

However, the Rev. Dr. Leah Schade, who held an interfaith service at the village, doesn't see them as good Christians. "Those criticizing the residents profitted from the way things are or they were so insulated from the pain and suffering the people are undergoing that they are unable to respond with compassion," says Rev. Schade, pastor of the United in Christ Lutheran Church in nearby Lewisburg. "As a Christian," she says, "I make a decision to do what Jesus calls us to do—to minister to those most vulnerable and resist the powers and the principalities that seek their own self perpetuation and their own profit." The teachings of Jesus, she says, "would tell us that what is happening to these families isn't right. He would ask, 'Who controls the resources; who does not?' The residents and the surrounding ecosystem are the disempowered ones."

The situation in Riverdale was becoming a public relations and operations disaster for Aqua–PVR. It was David vs. Goliath, and no corporation wants to be seen as Goliath. Some Aqua executives had to have felt conflicted, maybe even

456

confused—they had bought the land in good faith, and couldn't understand why 32 families living in trailers would cause such a problem to a huge corporation that just wanted to build a pump station. Aqua America may have reasoned it wasn't even the one that was fracking the earth; it was merely pulling out water to send to those companies that did the fracking. What Aqua–PVR was doing was relatively clean. Even if it had consequences for the marine life and vegetation, a governmental body had approved its application to withdraw up to three million gallons a day.

THE OCCUPATION OF RIVERDALE

On May 31, 2012, the final day residents were legally allowed to remain in Riverdale, seven families remained in what was left of their micro-village. They were joined by 50 anti-fracking activists who showed up to begin a vigil. "We asked the residents about their concerns, explained what we wanted to do, and made sure what we wanted to do would help the residents and not cause them further harm," says Alex Lotorto.

The next day, the day Aqua said it would begin construction, the protestors blockaded the two entrances to the park. The barriers included old washing machines, tires, cinder blocks, and fiber board, anything the protestors could take out of the trailers that were abandoned. What informally became known as the Occupation of Riverdale began Friday, June 1, 2012.

"When we first heard about the protest," says Eric Daniels, "we [he and his wife, April] weren't very happy because anti-fracking protestors would throw rocks at my truck." However, Daniels soon realized these protestors weren't involved in destruction of property. "After sitting and talking with them· the first day, I realized they were working class people, just like we are. They were here to help us, brought food, and shared their lives with us." During what would be a two-week protest, many of the protestors maxed out their credit cards to buy food and supplies for themselves, the residents who remained in the village, and former residents.

Online comments by readers of the *Sun-Gazette* identified the protestors as "environmental activists" and "out-of-town activists," never acknowledging that the landmen and most gas company workers were from out of state.

Bobbie2 called the scene a "liberal zoo . . . a veritable microcosm of the liberal social system." Joe123 called the protestors "unorganized morons," and decided the residents "are on display by 'Fame Seekers', like trick-monkeys in a circus." Many, who had never been to the village, called the protestors unwashed hippies who were living off welfare and the government. However, most protestors had jobs, and came to the village on their days off and in the evenings. Some were students; some were retired.

Tom Shepstone of Energy in Depth (EID) engaged in a continuing campaign to try to discredit the protestors. On the EID website, Shepstone claimed, "While the residents of the Riverdale park may think they have found supportive allies, in actuality they found a convenient partner that will be on to their next headline grabbing effort in quick fashion likely leaving Riverdale behind when this occurs."[1940]

Eric Daniels, who was still hauling wastewater to Ohio, began questioning the impact of fracking, and by now knew the protestors may have been right in how they were drawing attention to the problems. He ripped the roof off his trailer to make signs. Matthew West, a sculptor and digital fabricator, helped the residents create many of the signs, banners, and mobile billboards that would go into the barrier.

The signs were placed adjacent to, but not blocking, the main entrance into the village. The residents and protestors had become closer during the week, but "on that day is the day we [the residents and protestors] bonded, and solidified the sense of community we had all worked to grow," recalls Dr. Wendy Lee, a protestor and university professor of philosophy.

The Riverdale Mobile Home Village was becoming a battle zone. In addition to the barriers, there were now a security tent, a command/media trailer, a trailer with a large red cross painted onto it that served as an infirmary, and a trailer that served as a community kitchen and a place that protesters could get naps and hot showers. The command trailer and one other trailer also served as a place protestors could sleep.

Tuesday, June 12, 2012, was overcast with intermittent drizzles. Many of the protestors wore light rain ponchos. Only a cold heavy rain could have made this summer day any uglier than it was. Most had known they would eventually be forced to end the protest; most didn't expect it to be this day.

On site to "neutralize" the protestors were six guards from Huffmaster Security, identified on its web site as a company that is the "leading provider of strike management solutions."[1941]

Shortly after noon, about two dozen State troopers, local police, and representatives of various companies working in the Marcellus Shale showed up at Riverdale, determined to end the protest and secure the property. The protestors had no plans to become violent; the police had no reason to know that.

About 1:30 p.m., a State Police trooper asked Deb Eck's help to end the protest. "We want to resolve this peacefully without arrests," she remembers the trooper telling her. But, she also remembers being told that the police had every intention to arrest and jail every one of the protestors.

With the police determined to handcuff and remove every protestor, Eck says she saw no way the arrests would continue to help their cause. "The money everyone, including the protestors, raised for us should now be used to help the residents, those who had already left Riverdale and the ones who remained, and not for bail money," she says.

Throughout the protest, the State Police didn't confront the protestors, but showed up now and then to assure there would be no violence or destruction of property. This time, the police told the protestors they could continue their protest, but not the occupation. They had to take down the barriers and if they wished to exercise their First Amendment rights, it would have to be on a berm outside the village.

About 2 p.m., Deb Eck asked the protestors to end their resistance to avoid being arrested. Dr. Lee says the protestors "were prepared to sit down, arm-in-arm in defiance of the police" and continue the protest. But the moment Eck, representing the remaining residents, asked the protestors to "stand down" they were no longer invited guests and now would be arrested as trespassers. The residents and protestors hugged each other; the 12-day occupation was over.

"We had no leverage," says Alex Lotorto. Goliath was finally victorious. The dividing line between residents and a corporation had been breeched. A few, protestors and residents, openly wept. The longest around-the-clock continuous protest against fracking and for the residents was now over.

PHYSICAL DESTRUCTION OF A COMMUNITY

At 7 a.m. of what would have been the fourteenth day, demolition crews came into the village to take down the barricades and some of the trailers.

Aqua America got what it wanted, but it was still giving residents problems. It demanded everyone who left their trailers in the village to give Aqua America clear title. It forbid any of the former protestors from coming into the village to help the residents move. Aqua America also demanded that in addition to excluding the former protestors, only family members could help, and directed Huffmaster Security to record the license plates and car descriptions of anyone who tried to help.

The residents who had stayed and negotiated with Aqua America received additional financial compensation above the $2,500 that Aqua America had originally promised. However, Aqua America forced the residents to sign a non-disclosure agreement that forbid them from talking about that settlement. The settlement required the residents not to sue Aqua America, and never speak against Aqua America for any reason at any time or else face legal action. The secret settlement is also believed to have included individual payments of about $12,000, but only for the seven families who had remained in the village after June 1. Those who left before June 1 received only $2,500. The lawyers and the seven families who stayed at Riverdale after June 1 had tried to get Aqua America to give full compensation to all families who left before June 1, but the billion dollar corporation wouldn't yield.

"I didn't want to sign the agreement," says Deb Eck who was upset that the non-disclosure agreement removed her rights of free speech. "That's when I was told," she says, "that if any one of the seven families didn't sign, then no one would get anything. It was all or none." Aqua representatives, says Eck, told us if we didn't sign, they would also go after everyone, including those who took the $2,500 payment and left." Eck

460

believed it was an idle threat, just "a lot of gesturing," but she signed the agreement. "I didn't want the other six families to lose what we have worked for," she says. My pockets were heavier, but I wanted to puke."

Where Riverdale once stood, construction lights turned the night into daylight, natural vegetation was cut down, and wildlife was displaced.

A one-story 5,468 square foot green brick and concrete building that houses the pump station now sits adjacent to the West Branch of the Susquehanna River. Beginning in February 2013, the pump station began taking up to its allotted three million gallons of raw water from the river every day and sending it through 54 miles[1942] of 12-inch diameter pipes to Range Resources and other companies that need water. The pipeline, Aqua America boasts, replaces about 6,000 water truck trips per day.

Revenue for Aqua America in 2012 was $757.8 million, a 10.3 percent increase from 2011. Net income was $196.6 million, an increase of 37.4 percent.

A STORY OF PEOPLE

The story of Riverdale began with the people; it ends with the people.

Kevin June, who first united the residents, was hospitalized several times for bleeding ulcers. He still collects disability. "I can't see myself ever recovering," he says. With the settlement from Aqua America, he bought a 1988 14-by-70 mobile home with a shingle roof "in real good condition" at a nearby trailer park. Around his trailer he planted and tends to several flower gardens, and beams with pride when he hears others tell him it's the nicest yard in the park. The new trailer park, says June, "is nice, the neighbors are friendly, but it isn't like Riverdale."

Nine other families moved into that same park. They dipped into savings, retirement accounts, or borrowed from their families to be able to afford the move and the $100 a month higher rent.

Other residents eventually found spaces at other trailer

parks, often renting smaller trailers; others are living in efficiency apartments and in the homes of what are now extended families; a couple of families, after years of owning their own homes, are living in senior citizen apartment buildings.

Fred Kinley, 15 years after his retirement, wants to move to senior citizen housing, but his $1,500 a month pension is too high to give him a priority on the governmental lists. And so he waits, living in a rented trailer in a non-descript trailer court, where there is no playground, no picnic area, no access to the river, and no well-kept paved roads. Traffic and wind from the dirt roads throw dust into his trailer; it's just a fact of his existence.

In February 2013, Stallion Oilfield Services, the Houston-based corporation that promotes itself as "the largest provider of auxiliary rentals and services for oil and gas operations in the domestic United States,"[1943] laid off 32 drivers, including Eric Daniels. The layoffs were attributed to reduction in the need for trucks because of the Aqua America pumping station. Even if offered work in the industry, Eric Daniels says he won't take it.

Deb Eck and her two 11-year-old daughters, who moved their trailer, live in a small no-name trailer cluster about five miles north of Riverdale. She pays $165 more per month in lot rental.

Drifters and managers from the gas drilling industry and their vendors come into Eck's store. "I try to be polite," she says, "but it's hard." The workers buy windshield washer anti-freeze by the case, even in warm weather. "I asked one why so much antifreeze," says Eck, "and he said they add it to the lines to keep the drills from freezing. They also buy a lot of liquid soap for the same reason." All of that stuff, she says, somewhat sadly, "is poison. It goes into our ground." But it's their purchase of bottled water, often clearing the shelves of all bottles and cases, that brings out her strongest anger. This isn't the water they use to frack the wells, this is drinking water. She says she wants to yell at everyone who buys water by the case, "You come here and poison our water, destroy our land, and you take our clean water." She doesn't say it, but the rage builds every time she politely sells them cases of water.

For five months, the struggle against eviction had brought the residents closer than they had ever been; those five months also tore them apart. There are still bad feelings between those who took the $2,500 and left by June 1, and those who stayed, became part of a larger protest, and eventually got larger settlements. But for most, the residents understand why some took the money and left, and why other stayed. The 32 families that were a small community have moved on, their community now destroyed.

"Once in awhile, in town I'll see one of the residents; we'll chat," says Eck. "Sometimes we'll send messages on the computer," she says.

But it's not like it ever was.

CODA

In May 2012, the Pennsylvania House of Representatives passed an amendment to the Manufactured Home Community Rights Act to benefit residents of mobile home parks who are forced to move.

None of the families from Riverdale Mobile Home Village can take advantage of the legislation.

History will record they were just collateral damage.

The 'Call' and the 'Effect'

Louis W. Allstadt, former executive vice-president of Mobil Oil, retired in 2000, before hydraulic horizontal fracturing became the common process to extract gas. He had promoted fossil fuel development; he also opposes fracking. "Allstadt told Ellen Cantarow of *TruthOut:*

> "It will take masses of people demanding action from politicians to offset the huge amount of money that the industry is using to influence lawmakers, a world-scale version of those standing-room-only town meetings. Something has to wake up the general public. It will either be education from the environmental movements or some kind of climate disaster that no one can ignore."[1944]

Jonathan Wood, who authored a major report on the anti-fracking movement for Control Risks, a consulting firm that leans heavy to the use of fracking to extract oil and gas, readily acknowledged the effectiveness of grassroots social activism. "The oil and gas industry has often failed to appreciate social and political risks, and has repeatedly been caught off guard by the sophistication, speed and influence of anti-fracking activists," wrote Wood.[1945]

The effect of anti-fracking protests is reflected in a series of independent polls. In March 2013, about 48 percent of Americans favored fracking, while 38 percent opposed it, according to the Pew Research Center. By the end of 2014, only 41 percent favored fracking, while 47 percent opposed it, the first time more people opposed fracking than supported it. The Pew poll also revealed that fracking tended to be supported by men (52 percent support; 40 percent oppose), by those over 50, those in the South (45–42 percent), and by Republicans (62–25 percent). Those opposing fracking tended to be women (54 percent opposed, 31 percent support), by those under 50, by college graduates (49–38 percent), by those in the Northeast (48–37 percent), Midwest (47–39 percent), and West (54–38 percent), and Democrats (59–29 percent). In all cases, support for fracking declined from March 2013 to November 2014.[1946]

The enthusiasm and determination of activists doing what they can to protect the environment and public health—while demanding that politicians, governments, and corporations think about the future and not act for immediate gratification—has only grown stronger as more people learn the facts. The movement is all that Louis W. Allstadt had hoped for, and Jonathan Wood recognized. It is one of the major social movements of the past few decades.

PHOTO: Wendy Lee

Protestors at the Riverdale Mobile Home Village

The peaceful occupation of the Riverdale Mobile Home Village brought attention to the rights of mobile home tenants.

PHOTOS: Wendy Lynne Lee

The destruction of the village was because companies drilling in the Marcellus Shale needed millions of gallons of water per well.

PHOTO: Doug Crawford/Red Rock Pictures

Debra Anderson interviews occupational toxicologist
Daniel Thau Teitelbaum for her film, *Split Estate.*

PHOTO: Lorenzo Ciniglio

The Strike Anywhere Performance Ensemble spoofed the
Talisman Terry frackasaurus in the "Same River" multi-media tour.

CHAPTER 23

Framing the Message in 30 Frames per Second— and Other Media

Films and Video

GASLAND and *GASLAND II*

The presence of methane in drinking water in Dimock, Pa., was the basis of Josh Fox's advocacy documentary, *Gasland*, which received the Sundance Special Jury Prize and then began airing on HBO in June 2010.

Robert Koehler, writing in *Variety*, called it "one of the most effective and expressive environmental films of recent years. [It] may become to the dangers of natural gas drilling what *Silent Spring* was to DDT."[1947] The 107-minute film received an Academy Award nomination in 2011 for Outstanding Documentary; Fox was nominated for best documentary screenplay by the Writers Guild of America. He also received an Emmy for non-fiction directing, and was nominated for outstanding writing for non-fiction programming and outstanding cinematography. The film was nominated for exceptional merit in nonfiction filmmaking. *Gasland* was also honored by the Big Sky Documentary Film Festival and the Environmental Media Awards. Hundreds of activist organizations have shown the film, many with Fox as guest speaker.

Fox's interest in fracking had intensified when a natural gas company offered $100,000 for mineral rights on 21 acres of property his family owned in Milanville, in the extreme northeast part of Pennsylvania, about 60 miles east of Dimock. Angela Monti Fox, Josh Fox's mother and a social justice activist dating to the 1960s, told the Moms Clean Air Force, she would defend her house against the shale gas industry:

> "If we lose in the Upper Delaware River, I will be the first one to be arrested. I know my kids would be very upset, but I would do it, there's no question. I would rather sit in jail than watch them come down my road and destroy that house. That's my home. We built that house with our own two hands."[1948]

The natural gas industry launched a major public relations campaign opposing *Gasland*. Energy in Depth sent a letter (Feb. 1, 2011) to the Academy of Motion Pictures Arts and Sciences, asking the Academy to deny the film honors because, "Although we believe the film has value as an expression of stylized fiction, the many errors, inconsistencies and outright falsehoods catalogued in the appendix attached to this letter . . . cast serious doubt on GasLand's [*sic*] worthiness for this most honored award, and directly violate both the letter and spirit of the published criteria that presumably must be met by GasLand's competitors in this Category."[1949] Attached to the letter was a seven page list of what EID believed were inaccuracies and distortions.[1950] Fox responded to EID's attacks in a 39-page rebuttal.[1951]

Rex W. Tillerson, CEO of ExxonMobil, declared, "The 'Gasland' movie did more to set us back in this endeavor [to discount fear about fracking] than anything else out there, and yet every aspect of that movie has been completely, scientifically debunked."[1952]

The *Pittsburgh Tribune–Review*, which had called Al Gore's *An Inconvenient Truth* "claptrap," now called Gasland "tommyrot."[1953]

Gasland is a film "Joseph Goebbels would have been proud," said Teddy Borawski, chief oil/gas geologist for the Pennsylvania Department of Conservation and Natural Resources, who called *Gasland* "propaganda" and compared Fox, whose father and paternal grandparents survived the Holocaust, to the Nazi propaganda minister.[1954] Within a week, a contrite Borawski issued a public apology, acknowledging he "used very poor judgment" and that his remarks "do not truly reflect the person that I am or my understanding of the atrocities committed by the Nazis."[1955]

John Hanger, Gov. Ed Rendell's DEP secretary who was interviewed by Fox, opposed *Gasland*. In his blog for Feb. 23,

2011, Hanger claimed the film "presents a selective, distorted view of gas drilling and the energy choices America faces today [and] seeks to inflame public opinion to shut down the natural gas industry."[1956]

At an industry conference to develop public relations strategies "For Engaging The Public On A Positive Image For The Industry," Jeff Eshelman, EID executive vice-president, acknowledged EID follows "Josh Fox, wherever he goes."[1957]

On his Facebook Page, July 17, 2012, State Rep. Jesse White wrote that following a screening, "Industry representatives told the 211 people in attendance the gas industry is going to start getting more aggressive in attacking people who don't agree with them, and that because their facts are totally right, there can be no middle ground in the debate."[1958]

Josh Fox followed up *Gasland* with an 18-minute video, *The Sky is Pink*, released in June 2012. The video, a condensed version of *Gasland* with new footage, was a plea to Gov. Andrew Cuomo to continue the moratorium on fracking in New York. But, it would be Fox's next major documentary that would again ignite environmentalists, while receiving vicious condemnation from conservative politicians and front groups for Big Energy.

The forthcoming release of *Gasland Part II* was announced by a one-minute movie trailer,[1959] posted on YouTube in November 2012. The video, which matched Hollywood production standards, ominously cast fear as the driving theme of what could be a horror film. But this wasn't a trailer produced by Fox or his production company, International WOW Co., or HBO, which planned to release the two-hour documentary in Summer 2013.

The trailer was a well-written biting satire produced by Energy for America and funded by the American Energy Alliance, Americans for Prosperity, and the Institute for Energy Research, front groups for the oil and gas industry, which have ties to the ultra-conservative billionaire Koch Brothers, according to *Sourcewatch*.[1960]

The two-hour sequel premiered in April 2013 at the Tribeca Film Festival in New York City's Greenwich Village. The film continued the story of politicians who willingly praised fracking for what was perceived as its economic benefits, but countered by families who were damaged by actions of the drillers and by the health and environmental consequences of fracking. For the sequel, Fox checked out fracking across the country and in Queensland, Australia.

Gasland Part II, said Fox:

> "is about who gets to tell the story. Do the oil and gas companies get to tell it? Do reporters get to tell it? Do the people who are experiencing it get to tell it? In the film, we tried to take a look at whether or not we have true democratic procedure when it comes to oil and gas."[1961]

Fox told *Rolling Stone* he needed to make the sequel because, "The story wasn't over. We wanted to track whether or not there would be change, and what, if anything, was in the way of that. We also wanted to examine how this crisis is being handled by the government."[1962]

Unlike *Gasland*, which had a one week theater run, *Gasland Part II* did not go into theaters, nor did Fox tour the film as extensively as he did *Gasland;* he relied upon local organizations and colleges paying the screening fee to show the film locally.

By the time *Gasland II* premiered, Fox had already established a strong relationship with a national public relations firm that specializes in working with producers to get publicity for their films and with celebrities to get them awards. However, the agency claimed it was "unable" to answer most questions about the film, including budget and tour dates.

As expected, Energy in Depth attacked *Gasland II*. A year before the film was released, Lee Fuller, EID executive director, had sent a letter to Sheila Nevins, president of documentary and family programming for HBO.[1963] In that letter, Fuller reiterated EID's concern that *Gasland* contained significant "questionable assertions and claims," and that "the content of *Gasland 2* will suffer from similar flaws." Fuller then issued a veiled threat:

"Although we are hopeful that Mr. Fox will include our recommendations, his active participation in rallies and other events urging bans on hydraulic fracturing—all since the release of *Gasland*—leads us to believe that *Gasland 2* will not be an objective presentation of the facts, but rather a narrative that works backwards from a preordained conclusion—the complete opposite of what your subscribers would expect when tuning in to watch an HBO documentary."[1964]

Upon the film's release, EID expanded the attack:

"[T]he main challenge was manifest: Regain the public's trust by discarding hyperbole and laying out the challenges and opportunities of shale development as they actually exist in the actual world. In short, do everything a documentary filmmaker *should* do, but which he chose not to do in *Gasland*.
"Unfortunately . . . Josh eschewed that path entirely, doubling down on the same old, tired talking points, and playing to his narrow base at the exclusion of all others."[1965]

In an 8,000 word rebuttal to *Gasland II*, EID's Steve Everley pulled apart the film; sometimes, he correctly found errors, both of omission and commission; sometimes he cherry-picked information to enhance the energy industry's arguments. For anyone who doesn't know much about the Marcellus Shale or fracking, EID's counter-arguments make sense, and appear to strike a knockout blow to Fox's credibility; for those who are biological, chemical, and environmental experts, there is more to the story.

The reality is that both Fox and EID could be fairly accused of picking data to meet pre-determined conclusions.

Perhaps the most honest review of *Gasland II* was published by *Indiewire*, a respected on-line newspaper that looks at the independent film industry. Eric Kohn, senior editor and the newspaper's chief film critic, noted:

"Fox's exposition is a cluttered, scattershot affair that shifts from one location and case study to another with little narrative fluidity, but the collage holds together mainly due to his dark wit, snappy editing [by Matthew Sanchez] and musical cues that give the message an added kick."[1966]

Gasland, a documentary, grossed \$30,846[1967] in its first (and only) month of theatrical release; the rest of the income is primarily from television and video sales and rentals. The first major motion picture to focus upon the natural gas industry brought in about \$173,915 in box office receipts the first week of a limited release beginning Dec. 28, 2012.[1968] By the end of its first week of general release (Jan. 6, 2013), *Promised Land* had grossed \$4.4 million from 1,676 theaters, before it quickly faded from general release, grossing about \$7.6 million in four weeks.[1969] Although the stars had made several TV talk show appearances before the film's release, the failure of both the production company and studio to do a full promotion campaign, especially during the post-release period, contributed to the low box office.

With a \$15 million budget,[1970] *Promised Land* was written by Matt Damon and John Krasinski, based upon a story by Dave Eggers, and directed by Gus Van Sant. The movie stars Damon as a sweet-talking salesman from a natural gas driller who is trying to lease drilling rights from families in a town hit hard by the recession; Krasinski is an environmentalist, and Hal Holbrook is a science teacher who is knowledgeable by a lifetime of experience. The movie was partly filmed in western Pennsylvania in Spring 2012.

"We wanted to show how it [fracking] tears apart local communities and subverts democracies and corrupts political leaders and eviscerates all the things that Americans value," Damon told Robert F. Kennedy Jr.[1971]

Even before its release, *Promised Land* was attacked by a "rapid response team" from Energy in Depth, and by a horde of conservatives, including Fox News,[1972] the *National Review,*[1973] and the website, Breitbart.com.[1974] Most of the objections focused upon claims that the movie, a drama not a documentary, is an amalgamation of factually-inaccurate liberal bias that claims fracking is unsafe. Cherry-picking wisps of information from a few research studies and an orchard of half-truths from the natural gas industry, the right-wing press claimed not only is there no danger in the water, but that any claims about environmental damage and health problems are exaggerations meant solely to fuel a liberal bias.

The movie "lacks any substantive scientific evidence illustrating the alleged dangers of fracking [and] falls back on many conventional anti-capitalist themes," wrote Dr. Bill Bennett,[1975] conservative political commentator and former secretary of education in the Reagan administration. Bennett's solution to the energy problem is "freeing up more state and federal lands for drilling [and] cutting unnecessary, burdensome restrictions."

Energy in Depth created a Facebook website, "The Real Promised Land," to present what it called:

> "[R]eal life narratives of how farmers, homeowners and business owners, living on shale gas deposits have been able to realize the benefits of natural gas development in their towns and communities. Their genuine experiences with hydraulic fracturing reveal that the process isn't unregulated or dangerous, but rather a means for economic growth and job creation, in areas of the country that would have otherwise been economically barren."[1976]

Filmmakers Phelim McAleer and Ann McElhinney also became part of the attack upon both *Promised Land* and *Gasland*, buying space on a billboard near Rock Hill, N.Y., to promote their own film while also declaring: "Matt Damon: The water has been on fire since 1669." The reference was to methane-based fires dating into Colonial America. Calling Damon a liar, McAleer told *Politico:*

> "This stuff can get into the water supply naturally. . . . [H]undreds of years ago, native Americans named three different locations 'Burning Springs.' Quite simply, this is because of the naturally occurring gases, like methane, that are found beneath the earth's surface and that can rise to the surface with water. This pre-dates fracking and, in fact, there is no evidence that suggests that fracking has led to an increase in this phenomenon."[1977]

The McAleer/McElhinney attack, of course, was a half-truth, echoing Industry propaganda. It is true there was methane that burned centuries before fracking. But it is also true that methane did not migrate into numerous wells and kitchen faucets until after horizontal fracking began.

Nevertheless, *Promised Land* justifiably received mixed reviews from media critics, primarily because of its descent into clichés in characterization and a few weakness in the script. But, there was another problem that hovered over the film.

Three months before the film's release, the conservative Heritage Foundation, which had said the creators "have gone to absurd lengths to vilify oil and gas companies," now revealed that financing came from a wholly-owned subsidiary of the United Arab Emirates, "a member of the Organization of Petroleum Exporting Countries (OPEC), [which] has a stake in the future of the American fossil fuel industry." Thus, the Heritage Foundation concluded, the reason to fund *Promised Land* was "slowing the development of America's natural gas industry" to protect the petroleum industry.[1978] The story generated hundreds of negative stories by conservative commentators and others who favored fracking. One of those was John Hanger.

Hanger's pro-Industry review said the movie is "grating, arrogant, elitist [and] sends the horribly unfair, false message that those who sign drilling leases typically are greedy, stupid, and waste their gains on conspicuous consumption like sports cars." Several times in his review, Hanger referred to the film's "Persian Gulf investors." The development of gas production and distribution in the United States, said Hanger, "threatens the power of oil dictators in the Middle East and Putin's Russia that has a near monopoly on supplying gas to Europe. These oil and gas oligarchs, therefore, use their bulging purses to assault shale gas production that could mean new gas production in many countries and a lessening of their geopolitical power."[1979]

What the Heritage Foundation, Hanger, and others didn't reveal was that neither the UAE nor its media subsidiary funded the film. In 2004, billionaire Jeff Skoll, the first president of eBay, formed Participant Media to produce socially relevant films. Among them were *An Inconvenient Truth, American Gun, Fast Food Nation, The World According to Sesame Street, Syriana,* and *Good Night and Good Luck.* In 2008, Image Nation of Abu Dhabi invested $250 million into the company to assist in production of several films of social significance.[1980] Image Nation Abu Dhabi did not have editorial control and, according to Participant Media, the investment "covers all qualifying Participant narrative films regardless of genre or subject matter."[1981]

474

Damon said the first time he and Krasinski "were aware that Image Nation was involved with our movie was when we saw the rough cut and saw their logo."[1982]

However, there was another "bulging purse." The Marcellus Shale Coalition produced and placed a 15-second ad before every showing of *Promised Land* in about three-fourths of all Pennsylvania theaters.[1983] What the Commission may not have factored into what would normally be a brilliant marketing strategy was that most of the audience were probably already opposed to fracking, weren't susceptible to the politically biased negative reviews, and saw the movie to get reinforcement for their own beliefs. The fact the movie had artistic flaws did not negate the reality that anti-fracking arguments were receiving national attention and discussion.

COUNTERING *GASLAND* AND *PROMISED LAND*

The environmentalists and fracktivists had the documentaries *Gasland* and *Gasland II,* and the feature film *Promised Land* as their beacons to enlighten a nation about fracking. Big Energy countered with two films—*Truthland* and *FrackNation.*

Truthland,[1984] a 35-minute film, claimed Fox was "a spoiled avant-garde showman from New York City."[1985] That pretend-documentary focused upon Shelly Depue, a farmer from Susquehanna County, Pa., who went on what was billed as an independent national fact-finding tour; she concluded that fracking is safe. Although the film emphasized Depue was not paid, it skirted the issue that wells were put onto her farm, for which she and her family would receive several forms of income. As *Littlesis.org,* the website for the Public Accountability Initiative, noted in its own investigation, "[T]he film and its 'full-scale website and social media campaign' [pushed by EID] was planned from start to finish by the natural gas industry."[1986] The film was partially funded by a $1 million grant from the American Natural Gas Alliance to Chesapeake Energy.[1987] The original website[1988] was registered in the name of Chesapeake Energy on Feb. 1, 2012, but later re-assigned to the Independent Petroleum Association of America. *Littlesis* pointed out that the film, with all the characteristics of an infomercial, "was made by a high-profile ad firm based in

475

Hollywood, CA [Strategic Perceptions, Inc.] and promoted by a Washington DC media relations expert."[1989] John Krohn, communications manager for Energy in Depth, readily acknowledged he "Created and oversaw completion of multiple special projects including the film 'Truthland.'"[1990] Several EID senior staff had previously been employed by Republican members of Congress, according to Dory Hippauf,[1991] whose "Connect the Dots" series, published in *Common Sense 2*,[1992] shows relationships between the production/PR company, EID, and other natural gas industry front groups, industry funding, and politics.

FrackNation, a 77-minute film directed and produced by conservatives Phelim McAleer and Ann McElhinney, reported what they believed was the truth about the benefits of fracking, and to debunk claims made in *Gasland*. McElhinney said her team "talked to people across the country in areas where fracking is happening in their backyard. They all told us the same thing: *Fracking is safe and it's saving our community, but the media is ignoring us.* We were willing to listen."[1993] Funded by $212,265 in Kickstarter donations,[1994] with much of the donations from individuals associated with energy companies,[1995] *FrackNation* premiered on the AXS cable network in January 2013.

Among their previous films was *Not Evil, Just Wrong* (2009), which tried to debunk the Oscar-winning *An Inconvenient Truth* (2006), David Guggenheim's 100-minute documentary of Al Gore's quest to alert the world to climate change. Guggenheim won the Academy Award for best feature documentary; Gore and the Intergovernmental Panel on Climate Change shared the 2007 Nobel Peace Prize.

CRITICALLY-ACCLAIMED LESSER-KNOWN FILMS

Although *Gasland* and *Promised Land* received the most media hype, other films are at least as strong in presenting the issues about fracking.

Fracking Hell: The Untold Story (2010),[1996] an 18-minute video short, is an investigation into how fracking affects the people and environment in the Marcellus Shale. Released in

2010, not long after fracking began to expand in Pennsylvania, the film focuses upon Bradford County, with interviews of local residents affected by the process.

There's Truman Barnett, who says that before the rig and wells came, "The only thing you heard at night-time was your heartbeat. Now it's just totally devastated here. Inside my home you can hear and see the pictures vibrate on the walls." He shows the journalists a pond that once had plant life and turtles, now all killed by whatever was in the drilling fluids and waste water. "Our drinking water and our house has high concentrations of lead, they've told us not to drink it and don't bathe in it," Barnett says.

There's Carolyn Knapp who says, "As an organic farmer I don't feel that they should be allowed to put chemicals into my ground. Chemicals that I feel can do harm to my family, to the people around me, and I feel violated."

Among others interviewed were Lou Allstadt, former senior vice-president at ExxonMobil, who called proposed regulations that allow wells to be drilled 150 feet from fresh water "just insane"; and Dr. Tony Ingraffea, who says Americans have "purposely polluted large quantities of fresh water with chemicals that do not belong in the human environment."

Fracking Hell first aired in 2010 on Link TV, which produces global news and documentaries. The video was a collaboration of Link's Earth Focus team and the United Kingdom's Ecologist Film Unit. The network has also produced and aired specials about fracking's effects in England, Poland, and South Africa.

In Potter County, in the north-central part of Pennsylvania, is a triple divide. From this point, about 2,530 feet above sea level, are the headwaters of three small creeks that lead to the Genesee River, which flows north to Lake Ontario and the St. Lawrence River; the Allegheny River, which flows into the Ohio and then the Mississippi rivers before emptying into the Gulf of Mexico; and the West Branch of the Susquehanna, which flows into the Susquehanna and then into the Chesapeake Bay and then into the Atlantic Ocean.

Why this triple divide is important is because there are 71 wells, which had 189 environmental violations between Jan. 1, 2009, and Dec. 31, 2013,[1997] in this 17,000-resident rural county.

Pollution from fracking operations in Potter County could easily affect millions of Americans, as well as the ecosystem of three major watersheds in the eastern United States.

To tell the story of people impacted by fracking in Potter County, and how the DEP and drillers handled violations, journalists Joshua B. Pribanic and Melissa Troutman created a 90-minute video documentary, *Triple Divide*. The film was first shown at the Coudersport Theater in March 2013 and then became available on DVDs for public discussion groups.

Groundswell Rising: Protecting Our Children's Air and Water (2013) combines history with social activism. The 70-minute documentary was produced by Matt Cohen and Renard Cohen, whose award-winning documentaries often appeared on PBS. Mark Lichty, who has a long history of social justice, served as executive producer. Associate producers were Dave Walczak and Anne Marie Lauharn. The Cohens focus upon the effects of fracking upon the people living in the Marcellus Shale:

> "We meet parents, scientists, artists, teachers, clergy, community organizers and business leaders who are convinced that this unproven form of gas extraction is a serious health and environmental risk. Driven by a deep moral conviction, we see how they are standing up to one of the world's most powerful industries.
>
> "The film takes us back to the groundswells of the past such as the civil rights movement, women's rights, the denormalization of cigarette smoking, and the environmental movement, for connection and inspiration. And looking forward we see how the resistance to fracking is part of a growing movement towards dealing with climate change and making the transition to sustainable energy. The notion that burning more fossil fuels as a bridge to a cleaner planet is explored.
>
> "We will see how an industry rich with political connections managed to slip into a position of almost untouchable power and how at-risk communities have come together to fight back. And we will see how not protecting the cyclical flow of air and water puts us all at risk, no matter where we live."[1998]

The L.A. Weekly noted, the film "is ultimately a rousing, convincing rallying cry that the little guys, working in numbers, can triumph."[1999] *The Hollywood Reporter* said, "You're likely to be moved by the film's portraits of grassroots activists managing to make their voices heard despite the opposition of major corporations and the big money at their disposal."[2000]

The producers of *Dear Governor Cuomo* promote the 75-minute film as "a cross between 'The Last Waltz' and 'An Inconvenient Truth,' equal parts message and music."[2001] *The Last Waltz* is Martin Scorsese's documentary of The Band's last concert in 1976. Although *Dear Governor Cuomo* doesn't match either film, it is a powerful advocacy documentary that is screened at innumerable conferences and rallies. The base of the film is a star-studded concert in Albany, N.Y., that rallied anti-fracking activists to try to convince Gov. Cuomo there were significant reasons to continue the state's fracking moratorium. Protestors, scientists, environmentalists, and persons affected by fracking are interspersed with the concert. The film was written and directed by award-winning journalist and nature filmmaker Jon Bowermaster, with concert footage by Academy Award-winning documentary director Alex Gibney, and music direction by Natalie Merchant.

The producers of *Switch—to a Smarter Energy Future* (2012) promote their 98-minute film as "the acclaimed agenda-free energy documentary, embraced across political lines."[2002]
Mostly, they're right. A talented production crew under the direction of veteran filmmaker Harry Lynch went to 11 countries and interviewed more than 50 experts over a period of two years to get the story of energy production in all its forms. All forms—fossil fuel and renewable energy—are considered, with the benefits and problems of each carefully explained.
John Anderson of *Variety* sums up the *Switch* structure as:

> "Explore one energy source at a time, examine its pros and cons, forecast its future effectiveness, and give a rational evaluation of where the world is heading. It isn't sensational, but it is intelligent, and should be a draw to [audiences] with green thoughts in scattered theatrical play."[2003]

Michael O'Sullivan of the *Washington Post* says the film is "a thorough and sober-minded documentary about the past, present, and future of energy, [that] is refreshingly free of hot air."[2004]

Switch is narrated by geologist/professor Dr. Scott W. Tinker who went to the sites and co-wrote the film with Lynch. "Tinker comes across as affable, reasonable, and unfailingly curious," wrote Mark Feeney of the *Boston Globe*.[2005]

Aware of the controversy in energy dependence and energy independence, the producers of *Switch*—which has been a featured selection in several film festivals and has been screened at hundreds of colleges and community organizations—assured the media and their audiences:

> "We pushed ourselves to look beyond assumptions and remain as objective as possible. Interviewees were selected for their deep knowledge and respected positions, not for their support of any issue or agenda, and are a broad mix of industry, government, academic and NGO [non-governmental organization] leaders, from differing political backgrounds. The information in this project includes the opinions of these diverse experts along with years of research and learning from the field.
> "The end product has been embraced by energy companies and environmental groups, government agencies and leading universities, general audiences and most reviewers. However, because it does not overtly advocate for or condemn any resource, it has also angered some who do, including anti-nuclear protesters, renewable-only promoters, fossil fuel lobbyists, and some movie critics."[2006]

As much as Lynch and Tinker wanted to make the film as objective as possible, there are several holes that skew the film.

Although energy from coal, with both its advantages and disadvantages, is discussed, not included is the current process of strip mining and its destruction of the environment. There is also no rebuttal to Chesapeake CEO Aubrey McClendon's explanation of how safe fracking is.

Although there may not be a direct correlation between the praise of fracking and the absence of the rebuttal, both the director and narrator had to be aware that most of the trustees of the American Geoscience Institute, a major co-sponsor of the project, are either former or current executives of energy cor-

480

porations. Perhaps, Lynch and Dr. Tinker too readily accepted McClendon as an expert.

The absence of alternative and anti-establishment views, says the *Globe*'s Mark Feeney "does seem a bit odd." Because of that omission:

> "The film's generally upbeat attitude begins to feel a bit hollow. Nuclear-power opponents will likely leave the theater shaking their heads. And opponents of hydraulic fracking . . . will likely leave shaking their fists. The more important energy becomes in our lives . . . the more controversies it inspires. The way 'Switch' manages to avoid controversy is both an impressive achievement and perplexing."[2007]

The Ethics of Fracking (2014) is a 37-minute documentary, directed and produced by Scott Cannon for the Gas Drilling Awareness Coalition. The film is an overview of the process, combined with video clips from gas fields, animation, charts, and interviews with scientists, religious leaders, and those directly affected by fracking. Cannon, a former TV videographer who owns his own production company, made the film available at no cost to non-profit organizations and colleges because, "It's important that we get the message out about what fracking is, and what it can do."

From 1947 through 1995, Arkansas wanted to be known as "Land of Opportunity." Part of that campaign was to encourage business to come to the state to take advantage of the natural resources. One of those resources was natural gas in the Fayetteville Shale, which couldn't be economically extracted until horizontal fracking was developed. In October 2004, Southwestern Energy announced it acquired 575,000 acres and had begun drilling test wells.[2008]

In 2011, after a series of fracking-related earthquakes hit the Fayetteville Shale, Emily Lane and her brother, Sam, and his wife, April, began to investigate the effects of fracking upon the people of Arkansas. The result is a comprehensive website (www.arkansasfracking.org) and a 96-minute film, *Land of Opportunity* (2014) that explores fracking's effects on the people in the Fayetteville shale.

"We ran through our savings," says Lane, who has an MFA

in film making, "but we knew this was important." It was also a problem because the gas industry was seen as an economic boost to the state. "A lot of friends and family were not supportive of what we were doing," says Lane, whose family has lived in Arkansas for six generations. Also not supportive were the media—"We got almost no coverage," she says—and the Hot Springs Film Festival, which declined to accept the film because, as the Lanes were told, it was an election year and the film was "too controversial."

Even with minimal support, the film is being seen in college classes as far away as Oklahoma because, says Emily Lane, "they are also being affected by earthquakes caused by fracking."

Disruption (2014), a 52-minute film, directed and produced by Kelly Nyks and Jared P. Scott, was released two weeks before the Sept. 21, 2014, People's Climate March. The film premiered simultaneously in at-home and public screenings, mostly to promote the March. *Disruption*, according to 350.org, the non-profit environmental organization that created the march in New York City, is the story of "the dangerous environmental tipping points after which the entire climate system could spiral out of control, as well as the need for a mass social movement to disrupt the status quo and business-as-usual approach which is inhibiting the bold actions necessary to protect the planet's future."[2009] The film features commentary by scientists, environmentalists, and social activists, as well as behind-the-scenes footage of preparations for the march.

Faith and Fracking (2015) looks at the responsibility of mankind to be stewards of the earth. The 15-minute documentary was directed by David Braun, and produced by Braun, Jessica Wohlander, and Diana Mayoral.

With housing scarce and rents for substandard apartments extraordinarily high in the boom-town of Williston, N.D., the Rev. Jay Reinke, pastor of the Concordia Lutheran Church, allowed nomadic oil company workers to sleep in the church. *The Overnighters* (2014), a 102-minute documentary by Jesse

Moss, focuses upon Rev. Rinke as he extends what he believes is Christian charity and compassion, while his congregation, neighbors, and even local media question his motives and the character of those he's trying to serve, many of whom have criminal backgrounds and are suspected of being part of the rising crime rate. Avoiding judgment about the effects of fracking, Moss weaves a tale that parallels the California Gold Rush of 1848 and almost every other energy boom. Several media critics, including those from the *San Francisco Chronicle*, *Boston Globe*, and the *Chicago Tribune* called the film one of the best documentaries of the year. The *New York Times* called the film a "mediation on the meaning of community and the imperative of compassion."[2010] The *Los Angeles Times* called it "exceptional. . . . A film of disquieting moral complexity."[2011] The *Washington Post* noted it was a revelation "that masterfully explores so many modern-day conundrums."[2012] Moss was nominated for best director by the Directors Guild of America; the film was placed on the "short list" by the Academy of Motion Picture Arts and Sciences for an Academy Award in documentary film production. Like most documentaries, *The Overnighters* had negligible theater books and revenue, and was seen primarily at film festivals and by viewers at home and club showings. In June 2015, PBS aired the documentary in its POV series.

Un-earthed (2014) is a 90-minute documentary from South Africa. Jolynn Minaar and her team interviewed more than 400 persons in South Africa, the United States, and Canada to present an overview of fracking—and to conclude, based upon almost two years of research—that fracking harms the environment and poses health risks.

An Autumn Diary, produced by Phillip Davison, is a 30-minute story of fracking issues in England, as seen through the transformation of an elderly resident of Sussex. The film, which focuses upon the effects of fracking in a countryside, is available at no cost on DVD. It's often screened for community audiences; the program usually includes audience discussion and speakers who discuss the issues of fracking.

Three major Canadian films—*White Water, Black Gold; The Tipping Point: The Age of the Oil Sands*; and *Shattered Ground*—expose the truth about oil and gas exploration.

For three years, David Lavelle, a former teacher/counselor, mountaineer, and hiking guide, researched the tar sands of eastern Alberta, Canada, asking questions and learning about the effects that tar sand mining has upon what is identified as the third largest watershed in the world.[2013] His journey had begun with a hike to Mt. Snow Dome (elevation: 11,339 feet), in the Columbia Icefield on the Alberta and British Columbia border, and continued along the Athabasca River to Fort Chipewyan in the oil fields. Lavelle chose this journey because water—about three million barrels a day—is used to mine the tar sands oil.

The result of his questions is *White Water, Black Gold*, an 83-minute point-of-view documentary, narrated by actor Peter Coyote. Lavelle said he made the film "to tell the story of water and how the tar sands are impacting an element essential to all life on this planet."[2014]

The film, completed in March 2010, became a hit in film festivals and at dozens of public screenings by environmental groups. It first aired on the cable network Free Speech TV in May 2013. *White Water, Black Gold* was named the best Canadian film at the Vancouver International Mountain Film festival, winner of the John Muir Award of the Mountain Film Festival, and received the platinum award of the Oregon Film Festivals.

Lavelle concludes that mining and transporting tar sands oil should be stopped because of its health and environmental effects.

The Tipping Point and *Shattered Ground* first aired on the Canadian Broadcasting Corp. series, "The Nature of Things." Both films received critical acclaim; both were attacked by the oil and gas industry. The Canadian Association of Petroleum Producers, in a 1,100 word selective-truth rebuttal, called both films "a slanted approach and one-sided narrative to suit a pre-conceived story line."[2015]

The Tipping Point, a two-hour documentary produced by Tom Radford and Niobe Thompson, first aired in January 2011.

A CBC news release explained the reason and conclusions of the research that led to *Tipping Point*:

"For years, residents of the northern Alberta community of Fort Chipewyan, down the Athabasca River from the oil sands, have been plagued by rare forms of cancer. They were concerned that toxins from oil sands production might be to blame. Industry and government, meanwhile, claimed production in the oil sands contributed zero pollution to the Athabasca River.

"But in 2010, new and independent research measured pollution in waters flowing through the oil sands and discovered higher-than-expected levels of toxins, including arsenic, lead and mercury, coming from industrial plants. Leading the research was renowned freshwater scientist Dr. David Schindler. At the same time, the leaders of tiny Fort Chipewyan took their battle to the boardrooms of global oil companies, demanding change.

"Leading the campaign was Dene Elder Francois Paulette, whose battles with Ottawa a generation ago launched the era of modern land claims. From New York, to Copenhagen, to Oslo, to the oil sands themselves, our camera followed Paulette on his relentless search for allies. When he finally enlisted the support of Avatar director James Cameron, Paulette created a storm of controversy for the Alberta's oil sands industry.

"By the end of 2010, Schindler's alarming discovery of toxic pollution and the media attention Cameron's visit had raised was putting federal and provincial environmental policy under serious pressure. Separate reports by Canada's Auditor General, the Royal Society of Canada, and a panel of experts appointed by . . . Environment Minister Jim Prentice revealed a decade of incompetent pollution monitoring, paid for by industry, in Alberta's oil sands."[2016]

Leif Kaldor and Leslea Mair spent 15 months touring the gas and oil fields of Texas, Colorado, Alberta, and British Columbia to learn and record the story of fracking. Their one hour film, *Shattered Ground*, first aired on CBC in February 2013.

"There are very strong opinions on both sides of the issue, and very little middle ground," said Kaldor, so "we felt there was a need for a film that would give people a clear understanding of the process, the latest science, and how fracking is playing out on the ground, so they'd be better able to make up their own minds."[2017]

In Texas, Kaldor interviewed Calvin Tillman, the former mayor of DISH, which began experiencing fracking and health problems as early as 2005; in Erie, Colo., he interviewed those who suspected—and whose views were later confirmed by the Colorado School of Public Health—that intestinal problems among children may be traced to the presence of gas wells.

Kaldor suggests that, based upon his interviews, "If we are making decisions that will affect our future, then we should look for answers first."

LOOKING AT OTHER PARTS
OF THE FRACKING NIGHTMARE

Split Estate, a 76-minute documentary, focuses upon families of Garfield County, Colo., who, says director/producer Debra Anderson, "struggle against the erosion of their civil liberties, their communities and their health" after the natural gas industry moved into the San Juan Basin. The film, which premiered October 2009 on the cable network Planet Green, received an Emmy for Outstanding Individual Achievement in Craft for Research; researchers were Mitchell Marti and Matt Vest; writers were Joe Day, Avery Garnett, and Jean Wendt; Ali MacGraw was the narrator.

"Even reasonably knowledgeable viewers are likely to come away with a heightened understanding of both the politically privileged position of our nation's extraction industries and the role that concerned citizens can play in holding those Industries accountable," wrote Dr. Christopher H. Foreman, professor of public policy at the University of Maryland.[2018]

Bill Richardson, governor of New Mexico at the time *Split Estate* was produced, called the film, "an eye-opening examination of the consequences and conflicts that can arise between surface land owners . . . and those who own and extract the energy and mineral rights below. This film is of value to anyone wrestling with rational, sustainable energy policy while preserving the priceless elements of cultural heritage, private enterprise above-ground, and the precious health not only of people but the land itself."[2019]

The importance of silica sand in the fracking process, as well as its health and environmental dangers, is told in *Frac Sand Land*, a 45-minute documentary by Robert Nehman; and *The Price of Sand*, a 57-minute documentary by Jim Tittle.

In October 2012, Nehman learned that Minnesota Sands was planning to build a sand mine in Allamakee County, in the extreme northeast part of Iowa. The people of the upper Midwest accepted the presence of sand mines; they had been part of the economic landscape for decades. When the residents questioned Rick Frick, owner of Minnesota Sands, he brushed off their concerns, telling them, "It's just sand."[2020] But it wasn't "just sand," but a sand mine that would destroy much of the scenic river valley to provide silica sand to companies that were fracking the earth to get natural gas.

Within hours of learning of Minnesota Sands' proposal, Nehman became a co-founder of Allamakee County Protectors. He says he "never wanted to be a protestor, I just wanted to protect the environment." He also never planned to do a video. With only a digital video camera, a home computer, and the desire to show what the proposed fracking sand mine would do to destroy the beauty and historic features of the county, Nehman went into central Wisconsin to talk with residents who were affected by silica mining and to show the devastation to the environment. "I wanted to bring back their stories," he says, "and understood the power that a video could have."

With raw footage of the working mines and of numerous Wisconsin residents affected by the around-the-clock mining operations, including hundreds of trucks going into and out of the mine each day, Nehman became proficient in digital videography and editing, and added locally-produced music to the mix.

A decorative house painter, Nehman says he devoted 70 to 90 hours a week to the project because in winter "there isn't that much work, and my customers who did have work for me also knew how important this project was to all of us." For two months, Nehman spent most of his time filming, editing, and producing *Frac Sand Land*. "It took over my life," he says.

Beginning in January 2013, Nehman has shown *Frac Sand Land* once or twice a week to county boards, clubs, and organizations. "When they could see what large sand pits could do to the environment and the people's lifestyle, even though the company and the industry kept pushing they were adding jobs to the economy, the people understood the reality," he says.

Jim Tittle also used the power of video to bring reality to the people of Minnesota.

In 2011, Windsor Permian bought about 150 acres of land adjacent to the home of Tittle's mother, Barbara, who lived near Red Wing, Minn. The purchase was kept secret for months. When the plans became public, the residents formed an opposition group, held a series of public meetings, and secured a temporary county moratorium on frack sand mining.[2021]

It also led Tittle to spend 18 months researching silica sand and its impact upon the people and environment—"They told me stories—intense truck traffic, plummeting property values, toxic silica dust—a catalog of complaints that surprised me with its variety and intensity." He concluded his story wasn't of environmentalists vs. Big Business, nor conservatives vs. liberals, but of "very large corporations coming into a rural area and extracting something valuable in a reckless way."

But the presence of the sand pit near economically-depressed Red Wing also led to a split within the people of the county, one based upon economics vs. health and environment. Those who saw economic opportunity wore neon green T-shirts with the message "SAND=JOBS" silkscreened on the back. Those opposed countered with their own buttons and posters. Tittle interviewed those on both sides, including government officials, business owners, and those who worked in the industry. A professional videographer, whose work appears on several major cable networks, Tittle had been hired by the Al Franken senatorial campaign in 2008 to produce short videos. "I knew about YouTube," he says, "but I didn't know until then how much power it has."

During the Summer of 2012, Tittle posted 15 sand mining videos, each 30 seconds to nine minutes long. The YouTube videos gave credibility to his larger project and helped stimulate donations for the film's production; a CrowdSource campaign brought in about $5,000 in donations; he added several thousand dollars of his own funds.

For video of the sand pits, he resorted to aerial photography, creating a mount to keep the camera from being jarred or covered by splashes of airplane oil. Photo stills from the fly-overs were used on billboards and posters; most residents had never seen the enormity of an open pit sand mine. For micro-scopic pictures of silica sand, he brought samples to a professor who helped him use a microscope with an attached camera. For music, he sponsored a song writing contest that drew 10 entries; the music of many of the finalists is heard in the film. For color correction—balancing the color of the different shots from different cameras—he would have had to pay a production house about $900 an hour; Ditch Edit of Minneapolis waived its fees. For the final editing, he hired an editorial consultant, but did most of the editing himself.

The Price of Sand, Tittle's first full-length documentary, premiered March 22, 2013, at the Sheldon Theatre in Red Wing. On May 1, 2013, it opened at the Minneapolis–St. Paul International Film Festival.

Books

Among major books to explore fracking are those from Michelle Bamberger and Robert E. Oswald (*The Real Cost of Fracking*, 2014), Walter M. Brasch (*Fracking Pennsylvania: Flirting With Disaster*, 2011; and *Collateral Damage in the Marcellus Shale*, 2013), Chris Faulkner (*The Fracking Truth*, 2014), Russell Gold (*The Boom: How Fracking Ignited the American Energy Revolution and Changed the World*, 2014); John Graves (*Fracking: America's Alternative Energy Revolution*, 2013), Richard Heinberg (*Snake Oil: How Fracking's False Promise of Plenty Imperils Our Future, 2013*), Michael D. Holloway and Oliver Rudd (*Fracking: The Operations and Environmental Consequences of Hydraulic Fracturing*, 2013); Margaret Jackson (*Unconventional Oil and Gas Development: Environmental and Public Health Requirements, Risks and Size of Shale Resources*, 2013), Greg Kozera (*Just the Fracks, Ma'am: The TRUTH about Hydro-fracking and the Next Great American Boom*, 2012), Seamus McGraw (*The End of Country: Dispatches from the Frack Zone, 2012*), Bill Powers and Art Berman (*Cold, Hungry and in the Dark: Exploring the Natural*

Gas Supply Myth, 2013), Alex Prud'homme (*Hydrofracking: What Everyone Needs to Know*, 2013), Dr. Vikram Rao (*Shale Gas: The Promise and the Peril*, 2012), Frank R. Spellman (*Environmental Impacts of Hydraulic Fracturing*, 2012), Nadia Steinzor (*Blackout in the Gas* Patch, 2014), Alan Tootill (*Fracking the UK*, 2013), Dylan Weiss (*Sebastian's Tale*, 2015), Tom Wilber (*Under the Surface: Fortunes, and the Fate of the Marcellus Shale*, 2012), Nadia Steinzor and Bruce Baizel, *Wasting Away* (2015), and Gregory Zimmerman (*The Frackers: The Outrageous Inside Story of the New Billionaire Wildcatters*, 2013).

Faulkner is CEO of Breitling Energy, and the recipient of the honor of Oil Executive of the Year from the American Energy Research Group. In speeches and news releases, Faulkner proudly refers to himself as the "Frack Master," and his book reflects an overwhelming industry bias. *The Fracking Truth* was packaged and published by Winans Kuenstler Publishing, which charges a minimum of $65,000 for services that include ghostwriting, book production, and promotion, according to its website.[2022]

Dr. Stephen Chu, former U.S. secretary of energy, calls Russell Gold's book, "a compelling account of the last half century of natural gas technology development." Gold is a *Wall Street Journal* reporter, and focuses upon fracking from the business viewpoint.

John Graves, a financial analyst, believes fracking is beneficial to America's energy independence. On his book website, Graves, who uses the industry-preferred term, "fracing," rather than "fracking," writes:

> "Local communities reap significant rewards when the E&P [energy and production] guys come to town. This may sound odd: movies like 'Gasland' and the new 'Promised Land' make it appear as if the poor people in rural America are being scammed once again by rich corporations. Nothing could be further from the truth. Locals are no fools when it comes to negotiating. The lease income they receive creates wealth where none existed. It allows them to support their families, to offer them more than they had. Tax revenue enhances the quality of life for the community. These big rigs tear up roads, destroy commons and create havoc—for a few weeks. Then

they leave, many having repaired the damage; all having paid taxes to do so.

"Consumers also win. Gas prices stabilize as local crude replaces imported. Electricity prices may decline, as may heating oil, propane and butane. Natural gas powers much public transportation today and is increasing its presence in the trucking industry. Beyond fleet application, cars can run on natural gas, as they do in many other nations. In a decade, the gasoline car may be out, as the CNG—compressed natural gas—car becomes popular. The fuel is far less expensive and the engines run better and longer. You would fuel up each evening in your garage.

"Fracing offers both amazing opportunities and serious challenges."[2023]

Greg Kozera, an engineer who spent more than 35 years in the oil and natural gas industry, generally discusses fracking from an anecdotal point of view. While trying to present "just the facts," *Just the Fracks, Ma'am* favors fracking. Like many in the Industry, he praises fracking as being safe since "it has been used successfully for more than 60 years," but doesn't reveal that horizontal fracturing presents far greater problems than vertical fracturing. The book was praised by Barry Russell, president of the Independent Petroleum Association of America, who called it, "a great first read for those wanting to understand hydraulic fracturing."[2024]

Barbara McGraw, a widowed resident of Dimock, Pa., was one of the first to sign a lease to allow Cabot Oil to drill on her property. *The End of Country* is her son's story about the effects of fracking upon the people of the extreme northeast part of Pennsylvania. *Booklist* said that Seamus McGraw's research and personal experiences "led to this impressively detailed, highly engaging look at issues of energy policy, economics, and sociology that arose when a bucolic town was suddenly faced with the 'traveling circus' of energy exploration. . . . [This is] a completely engaging look at how energy policy affected a quiet, rural town." *Kirkus Reviews* called the book, "An unusual—and successful—marriage of memoir and investigative journalism." Tom Brokaw suggested, "This cautionary tale should be required reading for all those tempted by the calling cards of easy money and precarious peace of mind."[2025]

Bill Powers and Art Berman's *Cold, Hungry and in the Dark* focuses upon the gas industry's supply-and-demand trends. They argue that shale promoters, "relying on faulty science, bought-and-paid-for-white papers masquerading as independent research and 'industry consultants' . . . have vastly overstated the viable supply of shale gas resources for their own financial gain."

Dr. Vikram Rao, whose Ph.D. is in engineering, is executive director of the Research Triangle Energy Consortium, and former senior vice-president and chief technology officer of Halliburton. He is also a commissioner on North Carolina's Mining and Energy Commission. *Shale Gas* attempts to present all sides of the issue, and has drawn praise from environmentalists. "Rao makes clear that he is an advocate of fossil fuel production—but wants us to do it intelligently, and doesn't think that current market and regulatory structures will get us there. This is a healthy antidote to the frequently sloppy coverage in the media," observed Carl Pope, former executive director of the Sierra Club.[2026] Dr. Joseph Strakey, former chief technology officer of the National Energy Technology Laboratory, called the book "an outstanding contribution to the literature on this extremely important fuel and its implications for the energy future of the United States."[2027] However, Dr. Rao's focus is obvious—as a former Halliburton executive and as a commissioner on a state regulatory agency, he is a strong advocate for fossil fuel development.

One of the strengths of Frank R. Spellman's *Environmental Impacts of Hydraulic Fracturing* is his extensive discussion of the technology, including the tools and chemicals used in the process, with an emphasis upon the environmental effects. A retired Naval officer, university professor of environ-mental health, and the author of several books about the environ-ment, Spellman recommends numerous ways to mitigate the damage from fracking, while also accepting that natural gas development could make the United States energy inde-pendent. "As a practicing environmental professional," writes Spellman, "I recognize that we must achieve a balance between protecting the air, water, and soil, and ecosystems that life on earth depends on and utilizing the natural resources that earth possesses."[2028]

Steinzor, an environmental scientist with Earthworks, examines, "The consequences of prioritizing industry expansion without an equal commitment to protecting the public."[2029] Her conclusions in *Blackout in the Gas Patch* are based upon what she says is an "operational record of 135 wells and facilities" in Pennsylvania, as well as in-depth interviews with persons affected by nearby drilling.

Alan Tootill is a technical writer and novelist whose first non-fiction book, *Fracking the UK*, looks at the issues and effects of fracking, with an emphasis upon the United Kingdom. Tootill concludes, based upon his studies of fracking in the United States and the geology and environment of the United Kingdom, that "shale gas fracking, Coal Bed Methane (CBM) and Underground Coal Gasification (UCG) are entirely inappropriate for the UK, that they are inherently unsafe, that they would hasten our decline into climate change chaos and should be unconditionally opposed."

Gregory Zimmerman, a business reporter for the *Wall Street Journal*, explores how several energy company executives overcame industry doubts about the economic value of horizontal fracturing to bring additional billions to their companies. *Kirkus Reviews* praised *The Frackers* as, "A fascinating study of American entrepreneurial culture and the modern robber barons who succeeded in creating an energy revolution."[2030]

Mainstream Newspapers

The Williamsport (Pa.) *Sun–Gazette* is the only daily newspaper in Lycoming County, which has the third highest number of permitted wells of any county in the Marcellus Shale.[2031] Like most newspapers in areas where gas drilling occurs, it tries to cover a technology and social issue that few Americans had heard about before 2008. Like most newspapers, it also believes fracking contributes to the economic benefit.

Sun–Gazette editorials claim it's possible to balance economic benefits with environmental and public health concerns:

> "A variety of studies indicate there is little or no danger of groundwater being contaminated by chemicals used in fracking, as the industry points out."[2032] [Nov. 19, 2012]

"We believe most drilling companies are inherently respectful of the environment to begin with. We will say what we've been saying for years: It is possible for the natural gas boom and a healthy environment to coexist."[2033] [March 22, 2013]

"As for the environment, the vigilance must never end but to date there have been few missteps of significance. That part of the equation is within the control of the industry and governmental agencies and if there is a sincere partnership between the two, energy production and the environment can coexist.

"The industry has its critics who will never be satisfied. So be it. But any objective analysis would grade the presence of the natural gas drilling boom as a plus for our region."[2034] [Oct. 30, 2013]

The *Pittsburgh Tribune–Review*, about 170 miles to the southwest of Williamsport, has not been silent about its support for gas drilling:

"Smart energy diplomacy, for government and industry alike, is the safe, responsible extraction of Marcellus shale natural gas—not flaming claims that defy scientific substantiation." [March 19, 2012][2035]

"The Obama Administration must resist environmentalist supporters' calls for carbon taxes, cap-and-trade legislation, rules against drilling on public lands, anti-fracking laws and opposition to the Keystone XL pipeline. The Obama White House's love for winner-picking, taxpayer-funded "green" energy boondoggles doesn't help, either.

"Proven, safe and economically sensible technologies for shale oil and gas extraction present an opportunity to strengthen America in a way unthinkable not long ago. If the Obama administration prevents U.S. energy independence, its policies will cost this nation dearly by weakening it immeasurably." [November 16, 2012][2036]

"Marcellus shale drilling in Pennsylvania is exceeding advocates' expectations for economic and energy benefits— and doing so safely, refuting environmental extremists' alarmism.

"Safe, responsible shale gas extraction is more than living up to its promise of a brighter future for Pennsylvania and for America." [August 21, 2013][2037]

However, the Wilkes–Barre (Pa.) *Times–Leader*, six years after fracking came to the Marcellus region, argued: "Safeguarding the state's air and water—especially as it might pertain to residents' physical well-being—should be atop every lawmaker's priority list. And yours, too." [Sept. 30, 2014][2038]

The *Salt Lake City Tribune*, in one of the most conservative states in the country, editorialized against a coalition of counties to spend millions of dollars on infrastructure development that would directly benefit the oil and gas industry. It charged that the coalition was "making dangerous moves toward betting the future of their jurisdictions, their constituents and the land they hold in trust for future generations on the dirty boom-and-bust of the 20th Century fossil fuel economy." [Oct. 27, 2014].[2039] One month later, the *Tribune* editorialized: "The eagerness with which Utah accepts its role as the favorite lab rat of the fossil fuel industry should be troubling to everyone who lives here." [Nov. 24, 2014][2040]

The Fayetteville (N.C.) *Observer*, three months before fracking began in North Carolina, did not oppose fracking, but warned:

> "With so many environmental regulatory issues, government leaders see the primary customer as business and industry, whose interests appear to rank ahead of the health and safety of the people of North Carolina.
> "If North Carolina has an abundant gas resource, companies will want to mine it and will work with our regulations—which really should be strong. We don't need to put out an "I'm easy" sign." [March 16, 2015][2041]

The Wilmington (N.C.) *Star–News* urged stronger regulation:

> "[S]tate lawmakers have an obligation to protect public health and the environment and to hold energy companies responsible for any damage they do. Anything less is a breach of their duties to the people." [Jan. 29, 2015][2042]

The Frederick (Md.) *News–Post* argued for an outright ban on fracking:

> "We believe the issue of fracking has been studied enough, and enough evidence exists of the illnesses, air pollution and

water contamination in other states that Maryland should go ahead and ban fracking altogether, rather than just stall it." [Feb. 12, 2015][2043]

The *Boulder* (Colo.) *Daily Camera*, while accepting the industry's claim of economic benefits from fracking, still looked at the critical issues, including the recent legislation in Texas to remove local control over fracking:

"Fracking has been a boon to the Colorado economy and a major contributor toward meeting America's goal of energy self-sufficiency. While we join many critics in looking forward to the day when clean energy takes over from fossil fuels, we recognize we are not there yet. But that does not mean that the state needs to be in bed with the industry it is supposed to be regulating. . . .

"Colorado has a history of greater concern for the environment and for local control than the governor [John Hickenlooper, a former oil company geologist] seems to appreciate. If the industry wins from the courts the right of implicit state preemption in all matters relating to oil and gas development, we hope the governor's declaration of surrender on behalf of those who would regulate fracking more stringently proves premature." [June 9, 2015][2044]

The *San Francisco Chronicle*, recognizing there is still a lot about the fracking chemical soup that isn't known, urged the Legislature and state agencies to demand "more data from oil companies about their activities." [July 12, 2015][2045]

However, throughout the country, more newspapers seemed to favor fracking, citing economic benefits. Shortly before Illinois finalized regulations for oil and gas fracking, the (Champaign, Ill.) *News–Gazette* claimed:

"North Dakota is perhaps the best example of what fracking can do for a state's economy. Thanks to fracking, its biggest problem is a shortage of workers to fill available jobs, a supply/demand issue that has driven wages sky-high. At the same time, a gusher of new revenue is flowing into the state's treasury.

"Does Illinois wish to take similar advantage of this opportunity? Or will it continue to slow-walk the permitting process to the point that developers redirect their efforts to states where their capital and their jobs will be welcome?" [Aug. 31, 2014][2046]

The *Houston Chronicle* reflected the position of the oil and gas industry, as well as most Republican legislatures:

> "For decades, traditional liberal groups had little respect for our Texan attitude of praise and reverence for the oil and gas industry. Where they saw pollution, we saw profit.
> "Fracking has provided a means to mighty ends. Ambitious politicians should treat it with the respect it deserves." [Jan. 6, 2015][2047]

The Las Vegas *Review–Journal* editorialized, "Nevada needs more good-paying jobs and more industry to diversify and grow its economy. Fracking fits the bill."[March 30, 2014][2048]

Editors say they impose a rigid wall between the newspaper's editorial opinions, often formulated by publishers, and its coverage of news. It's an ideal, but often doesn't reflect reality. Reporters, like workers in every industry, know what their boss believes and, whether consciously or sub-consciously, even if no verbal direction has been given, may follow that direction. Publishers and newspaper executives are often members of the local Chambers of Commerce, most of which praise fracking for its economic benefits. Sub-editors may assign, or not assign, stories based upon perceptions of the newspaper's editorial philosophy; they will decide where to place stories and how large a headline to give it, based upon editorial philosophy. Reporters who see that editors give certain stories "bigger play" may increase their reporting in that area; reporters who see some of their stories "spiked," may look elsewhere for stories. It's known as "the socialization of the newsroom."

Another factor that influences news content is the severe downsizing of news staffs that has been occurring for more than a decade as publishers fight to hold onto what is left of once-hefty 20 and 30 percent profits. Among newsroom beats that have been diminished or eliminated are environmental and science reporting. When those stories do appear, they're usually from wire services or syndicates.

497

Membership in the Society of Environmental Journalists is one way to track the effects of downsizing. Between 1995 and 2013, membership in SEJ increased from 1,080 to 1,232; however, those who indicated they were primarily employed on newspapers declined from 339 to 187, according to Beth Parke, SEJ executive director. Not all environmental writers are SEJ members, and some SEJ members include environmental writing as only a part of their overall responsibilities.

Many newspaper editors say they don't cover the natural gas industry and its controversies because their circulation areas don't have active wells, but fail to recognize not only do pipelines cross their areas and trains carrying volatile gases and crude oil roll past farmlands and into the cities but that air and water pollution and health problems do not stop at county borders. When stories are assigned, more often than not to general assignment reporters or reporters whose specialty is covering the business community, the result can be an aggregation of facts that follow a "he said/she said" formula—record the stories and quotes issued by the gas industry or their front organizations and, if there's time before the deadline, also find someone who says something different. Often, leaders of the anti-fracking movement aren't quoted for rebuttal statements; anti-fracking rallies and conferences become invisible to the mass media.

The keynote speaker for the opening night of the annual convention of the Society of Environmental Journalists (SEJ) in September 2014 was a BP executive who attacked "opportunistic" environmentalists and reporters who sensationalize the news. Geoff Morrell, senior vice-president of communications and external affairs for BP in the United States, discussed the corporation's response to problems at the Deep Water Horizon rig in 2010 that led to the deaths of 11 employees and spilled almost 180 million gallons of oil into the Gulf of Mexico. He acknowledged BP's role in the spill, cited numerous examples of how the Gulf was returning to normal, and argued, "There needs to be less sensationalism and more balance and context to tell the whole story of the health of the Gulf." He said "advocacy groups . . . are pushing a narrow one-sided perspective [that paints] an incomplete and inaccurate picture," which places the blame on BP for "all environmental problems afflicting the Gulf. Activists, he said, "ignored key facts."[2049] (The day after Morrell's speech, U.S. District Judge Carl Barbier declared BP had sacrificed safety for "profit-driven decisions," its actions in the Gulf oil spill were "reckless [and] the result of gross negligence or willful misconduct."[2050])

Lt. Gen. Russel Honore (USA-ret.), who had directed military relief efforts following Hurricane Katrina in 2005, didn't buy what Morrell was selling. Between the time Morrell addressed the SEJ membership and the decision by Judge Barbier, Gen. Honore told the journalists the oil and gas industry "hijacked our damn democracy. They lobby. They write the laws." He told the journalists that although fossil fuels are needed for energy, "It doesn't give the industry the right to destroy where we live." Turning to journalistic coverage, Gen. Honore said, "There's never a time the world needs you more to shed light on environmental problems. Do your damn job!"[2051]

Several reporters and editors have sliced through the spin and tried to get the truth behind the impact of fracking.

Don Hopey, president of the Society of Environmental Journalists, 2013–2014, has been reporting about environmental issues for the *Pittsburgh Post–Gazette* since 1993. The *Post–Gazette* has been a leader in reporting news and features from the Marcellus Shale. The Pipeline project was begun in February 2011 "as a specialty news website that employs multimedia, social media and interactive maps and that curates daily coverage from the PG and other news organizations to provide an authoritative resource for Marcellus Shale news and information."[2052] The year after the project was created, the Pipeline team won the Scripps Howard Award for Environmental Reporting and the Best Specialty Site award from the Online News Association. Erich Schwartzel, who headed the team in its first two years, received the G. Richard Dew Award, which the Pennsylvania Newspaper Association calls its "most prestigious honor for outstanding journalism." Schwartzel was honored for an investigation of Chesapeake Energy and its founder, Aubrey McClendon. Schwartzel, who had worked at the *Post-Gazette* as a business reporter for four years, went to the *Wall Street Journal* in 2013 to become a West Coast entertainment writer. He was replaced by Anya Litvak; like Schwartzel, she covers the business side of the industry. Michael Sanserino, primarily a sports reporter, also covers shale gas issues.

At the Scranton (Pa.) *Times–Tribune*, Laura Legere, a general assignment reporter, became that newspaper's primary reporter on issues related to the shale gas industry. Legere left the newspaper in 2013 to become a freelancer and work with State-Impact NPR. She was replaced by Brendan Gibbons, a recent college graduate with a B.S. in science journalism and a concentration on conservation who left two years later for a job with the San Antonio (Texas) *Express–News*.

The Wilkes–Barre (Pa.) *Times–Leader* also supports fracking and natural gas "as part of the energy mix," but raised questions about the health and environmental effects. In a February 2015 editorial, the newspaper charged:

> "In their zeal to take advantage of a new, abundant energy source, all but a few of our leaders largely turned their backs on our area and let it be mined—literally—for its resources. Health and other consequences be damned.
>
> "We wonder if future residents of Northeastern Pennsylvania will look back on how our leaders have handled natural-gas drilling at the dawn of the 21st century and ask themselves if history simply repeated itself. . . .
>
> "In our state's zeal to go after The Next Big Thing, it rushed into horizontal fracking and is only now beginning to learn about the consequences (good and bad). . . . Aside from ongoing discussions about how much to tax the industry, there hasn't been a sufficient examination of what this still-emerging energy boom will do to Pennsylvania (to help and to hurt it) and how this revolution should be conducted.
>
> "Why aren't lawmakers, for instance, raising red flags as companies proceed with plans in this region to lay down a sprawling network of natural-gas pipelines, including some lines blueprinted to go through former mining areas that have a history of subsiding?
>
> "We're pleased that natural-gas exploration has provided a much-needed boost to the region's economy, but we're disap-

pointed that has happened because the majority of lawmakers eager for a win have rubber-stamped projects.

"Today, American natural gas—as part of a larger portfolio of energy sources—can help to end our dependence on foreign oil. However, the fossil fuel found below Pennsylvania should be extracted only with proper oversight. And safety guidelines. There simply are not enough regulators tasked with safeguarding our soil, water and air to guarantee that the current way of doing things is the safe way of doing things."[2053]

That editorial would have been more effective, and more powerful, had it been published three or four years earlier, but it does reflect that the establishment news media, not just in the Marcellus Shale but throughout the country, have slowly begun to understand that horizontal fracking carries significant environmental and health effects.

Alternative Print and Online Media

MINING FACTS TO GET TO THE TRUTH

Lisa Song, a reporter for *InsideClimate News (ICN)*, was sitting in the reading room of the Texas Commission on Environmental Quality (TCEQ). She was there to look at hundreds of documents related to fracking in the Eagle Ford and Barnett shales. Near her was a TCEQ paralegal whose job was to make sure Song didn't take too many notes, and to keep her from photographing or scanning any documents.

ICN had previously asked for copies of the documents; TCEQ told the online newspaper the cost would be about $3,400. However, the TCEQ said if someone from *ICN* went to Austin and looked at the documents, there wouldn't be any charge. When Song got to the room, she was told that taking excessive notes, typing information into her laptop computer, photographing the documents from her cell phone or scanning them onto a thumb drive was the same as copying, and the TCEQ would impose the fee.

Most state and local governments charge 15–50 cents a page, and add a reasonable labor charge to copy public records.

Often, the fees are less or even waived for news organizations. Anyone—private citizen, corporate employee, lobbyist, or reporter—can make notes and even photograph or scan the documents without costs. But, this is Texas, and this is the TCEQ. David Hasemyer, and Elizabeth McGowan had previously won the 2013 Pulitzer Prize for national reporting for a series of *ICN* articles that began with an investigation of a million gallon tar sands oil spill into the Kalamazoo River and then expanded to look at the problems in the nation's oil and gas pipelines and governmental regulation. And now Song, Hasemyer, and Jim Morris of the Center for Public Integrity were the primary reporters of what would be a 42-part series that looked at fracking, air quality, health issues, and politics in Texas.

The 20-month investigation, which involved more than two dozen reporters and editors, had begun in Spring 2013. "Very little reporting was done about air pollution in the shale," says Hasemyer, "but the emissions from air poses a greater threat to humans than water pollution," which, at the time, was the subject of several research studies.

"When we started to look into the issue," says Song, "we did cursory interviews, and we didn't see any newspapers doing anything." David Sassoon, *ICN* founder and publisher, believes Texas media did not investigate the problems of air quality caused by fracking because not only do the current media not cover stories that are "long detailed investigations that take up a lot of space," but that the subject may have been "too hot to handle."

What the reporters experienced in Texas trying to get information was similar to what reporters in Pennsylvania and other states experienced. Rather than making public records and information available to the public, the TCEQ and Railroad Commission executives may have believed that preventing information from getting to the people was its priority. Often, say the three primary reporters, agency officials instead of directly answering questions sent them to obscure websites to make them find the answers that were buried deep in the data.

After the TCEQ and the Texas Railroad Commission routinely denied requests for public records, the reporters filed more than 50 requests with the state attorney general. "Within days," says Hasemyer, the attorney general would order the agencies to give us the records.

State regulators also refused to allow the reporters to go to drilling sites, and refused numerous requests for interviews. "We wanted to have phone interviews," says Song, "but they wanted written questions and we got written answers." She says the reporting team "had no idea if the response was from the scientists or from the PR people," and had difficulty verifying the accuracy of the information they received. Song says the team made repeated requests to meet with TCEQ personnel, and were finally allowed a one-hour meeting with the chief toxicologist, two of his colleagues, and a TCEQ PR person. When the reporters later wanted to ask follow-up questions and to verify earlier information, standard journalistic practice, they were told that the reporters already had the interview and would not be allowed to talk with the state employees again.

When Hasemyer called inspectors at home, after being refused permission to talk with them at work, he received a strong rebuke from a PR person who ordered him never to call staff at home again. Hasemyer says he was trying to get information about "overworked staff who were trying to inspect and regulate more than 7,000 wells." That same PR person, says Hasemyer, stepped between him and chairman Bryan Shaw, who was in a hallway before a public hearing, and blocked him from an interview. Public Relations also blocked him from talking with a mid-level TCEQ executive. Hasemyer had found an email from that official who had suggested there could be significant air emissions problems in the Eagle Ford shale. Hasemyer says he called the agency, talked to a receptionist, asked to talk with the official—"and after about a three-minute delay, while I thought she was getting the official, the PR person answered my call" to say the official wasn't available.

It was even more difficult for The Weather Channel (TWC) to get interviews. The six-person TWC team found that no regulators wanted to talk with them. The Texas Railroad Commission did make commissioner David Porter available for an interview in San Antonio, but for only 10 minutes. When the team arrived for the interview, they were told Porter was ill and wouldn't be available.

The investigation revealed there was significant air pollution, some of it far in excess of the baseline for lethal doses. One story reported that three facilities in the Eagle Ford shale generated more air pollution than a Houston oil refinery. The team interviewed numerous residents and analyzed data that could have suggested a high correlation between fracking-caused air pollution and significant health problems that included leukemia, rashes, headaches, and liver, kidney, skin and respiratory problems. The TCEQ emphatically denied "chemicals are being emitted at levels high enough to cause adverse health effects."[2054] However, Lisa Song says, based upon the team's investigation, the TCEQ "knows almost nothing about air quality in the drilling areas."

The reporting also revealed why the TCEQ "knows almost nothing" about air quality and refused to acknowledge a fracking/health correlation. About one-fourth of the legislators or their spouses had financial interests in the Eagle Ford shale drilling. In a six year period, beginning in 2008, the Legislature cut the TCEQ budget by one-third and funding for air monitoring equipment by half.

Energy in Depth (EID) attacked the reporters for journalistic impropriety and factual inaccuracy; it attacked their employers for accepting funding from anti-fracking activists. As part of a vigorous 4,200-word attack, EID's Steve Everley claimed *InsideClimate News* and the Center for Public Integrity showed "a willingness [to] present false information as part of its interviews, or to deliberately withhold information about their affiliations from interviewees." Everley also claimed, "The tone of coverage is also tilted toward the sensational," and concluded, "[I]t's difficult to see how ICN and CPI are not themselves 'intertwined' with a different type of industry: the anti-fracking movement."[2055]

In response, David Sassoon pointed out:

> "We believe we've aroused the group's displeasure because our work shines an unwelcome spotlight on these toxic air emissions and the manner in which they are released, with

little regulation or regard for neighboring homes and communities. . . .

"Energy in Depth did not dispute the evidence we presented. Instead, it published a litany of allegations charging journalistic malfeasance. Not one of the allegations touched on the substance of our reporting . . .

"It is not unusual for powerful industries to go on the attack to undermine unfavorable news reports. Even though Mr. Everley speaks as if he is protecting the public from misrepresentations by media with an agenda, in fact he is protecting the bottom line of the oil and gas industry, which pays his salary. . . .

"No such thing as an 'anti-fracking industry' exists, and Everley provides no evidence that would pass muster in an honest newsroom that it does. Instead, he manufactures an imaginary public enemy, a rival 'industry,' no less, that is out to do harm."[2056]

Specifically addressing EID's claims of the reporters' lack of journalistic ethics, Sassoon fired back: "Our reporters do not conceal their identities, nor do they employ other questionable reporting techniques, and no credible evidence exists to the contrary."[2057]

Unlike the general news media that can boast of circulations everywhere from a few thousand to millions, specialized online sites have smaller audiences. Sassoon, who has spent almost three decades in journalism, says there wasn't much viewership on the site, "but we're not chasing traffic but trying for impact."

That impact was felt by other media capsulizing the information documented by the series and by numerous radio and TV hosts interviewing the principal reporters.

The series influenced other states to look into issues of air quality; it also invigorated the anti-fracking movement. "Impacts to water had dominated the discussion in the Marcellus for years, but the Eagle Ford series proved that air impacts were significant and just as much cause for concern as water impacts are," says Karen Feridun, founder of Berks Gas Truth. Pennsylvania, says Dory Hippauf of the Gas Drilling Awareness Coalition, is still being touted as the New Texas by the industry, their front groups, mainstream media and politicians, but 'Big Oil, Bad Air' revealed what the industry doesn't want people to know—drilling and fracking is anything but clean or good for the environment and it's killing people—and forced Pennsylvanians to ask if they want to become a 'New Texas'"

"Big Oil, Bad Air" won first-place awards in journalistic excellence competition from Editor and Publisher, the Association of Healthcare Journalists, the Society of Professional Journalists, and the National Press Foundation. It also won the prestigious Kevin Carmody Award for Outstanding In-depth Reporting from the Society of Environmental Journalists (SEJ). In announcing the award, SEJ stated:

> "The authors created this compelling package with dogged data work, comprehensive interviews, and an obvious drive to shed light on the problems that residents had described, and that politicians, regulators and companies had largely ignored. . . .
> "The reporting here is solid, and the multimedia work is both captivating and thorough. The organizations proved their commitment to the inquiry by keeping at it, story after story. This commendable work upholds the best traditions of investigative reporting."[2058]

More important than awards, the series may have led Texas to install a $122,000 air pollution monitor in Karnes County, one of the most heavily-drilled parts of the Eagle Ford Shale. It was only the sixth air pollution monitor in the 20,000 square mile shale; the other five were installed at the fringe areas of the shale, far from the heaviest drilling activity.

Unfortunately, Hasemyer doesn't believe the series had much impact upon TCEQ, the Railroad Commission, the state legislature, or the oil and gas industry. Their reaction, he says, "was as if our stories never happened; people suffering didn't faze them." The regulatory agencies, says Hasemyer, "are still not responsive to the people."

It is no different in Texas than in other parts of the country where politicians and the business community flaunt economic benefits while ignoring environmental and health impact, and where the major mass media don't have the time, resources, or desire to investigate the critical social issues and, often, tend to accept what they are told by the industry.

PROPUBLICA AND SOCIAL MEDIA

Long before the establishment media figured out what fracking is and what its impact could be, Abrahm Lustgarten, a *ProPublica* journalist, was researching and writing about the natural gas industry. His articles and analyses provide a substantial base for understanding the effects of fracking. In 2009, he was honored with the George Polk Award for Environmental reporting for a series of articles about the dangers of natural gas drilling.

Dozens of online-only newsletters and mini-newspapers, representing a variety of editorial philosophies, focus upon the shale-gas industry. Many are "aggregator sites"—a collection of truncated stories with links to their original sites, including print and web. Although many newsletters are either issue-neutral or produced by oil and gas companies and their associations, most are produced by the anti-fracking movement, which has become adept at the use of digital technology and social media to meet the three basic requirements of the mass media—to inform, entertain, and persuade.

More than 400 websites, blogs, and digital newsletters, written and edited by volunteers, fill the gap from the mainstream media. Some have readership in the double digits; some in the thousands.

Tracking, sorting, and condensing information from the establishment media, the shale gas industry, and the anti-fracking community is Judy Morrash Muskauski, a retired teacher and school librarian from Baltimore, Md. For persons in the anti-fracking movement, "Fracking in Northeast Pennsylvania," a Facebook page that can be accessed with Muskaski's permission, is one of the most important resources. "I log on about noon and just keep going on and off all day and night," says Muskauski, who often puts in 10–12-hour days to deliver an average of 75 messages—news, features, events, opinion articles, and action items, which originate throughout the United States and in several foreign counties. Muskauski and assistants Dory Hippauf and Debbie Ziegler Lambert not only receive and verify story accuracy, they also serve as a resource for the movement. "If I don't check my email regularly during the day," says Muskauski, "I will usually have at least 300 emails waiting for me."

Radio and Television

ADVERTISING

Between 7 and 8 p.m., every Tuesday, KDKA-AM, Pittsburgh, Pa., aired "The Marcellus Shale Hour." KDKA, one of the most powerful stations in the United States, is owned and operated by the CBS Radio Network. The show, hosted by KDKA staff member Robert Mangino, was professionally produced and presented significant information about natural gas and fracking—with one major problem: the co-host was Mark Pitzarella, official spokesperson for Range Resources. Mangino and Pitzarella seldom disagreed on the importance of natural gas and methods used to extract it. KDKA, which has a rate of $3,000 per week for that time slot, called the show "education-based."

"It's hard for the truth to be voiced when industry has the money to buy time to advertise their toxic operations to try and justify what they are destroying—our environment and our democracy," says Briget Shields, one of the leaders of the anti-fracking movement in the Pittsburgh area.

KDKA, and the three other CBS radio stations in Pittsburgh, hosted a one-day "Marcellus Shale Festival" in August 2013. The festival, which had an attendance of about 2,500,[2059] included music, speeches by politicians, and the Pittsburgh premiere of the pro-fracking film *FrackNation*, with discussion led by producers Phelim McAleer and Ann McElhinney. Co-sponsors were several trade associations and energy companies.[2060] "We promoted it as a Marcellus Shale Festival, and the town-hall meeting was about how private companies and the public sector were working together," CBS Radio senior vice president Michael Young told the *Pittsburgh City Paper*, but claimed the stations' reporters could cover fracking issues objectively.[2061] Nevertheless, environmentalists and others opposed to fracking were denied the opportunity to distribute literature or make their voices heard at the festival.

"KDKA programming does not present a balanced viewpoint and is considered by many listeners to be nothing more than an advertisement for the shale gas industry," says Jan Milburn, president of Westmoreland Marcellus Citizen's Group.

News operations in all media try to maintain a separation from advertising, but when a major advertiser is buying time immediately following a newscast, as was the case with "The Marcellus Shale Hour," it's sometimes difficult not to be influenced, even if there are no memos from management.

Every month since May 2011, WNEP-TV, the ABC affiliate in northeastern and central Pennsylvania, airs a half-hour "Power to Save" program, sponsored by Cabot Oil and Gas and PPL, one the nation's largest energy distribution companies. Among almost 200 segments is a wide range of local news, including recycling, preventing creek erosion, lowering thermostats in winter, using rain barrels to save water, and a university that saved money by switching from coal to natural gas. Depending upon the story, the benefits of using natural gas can be non-existent, subtle, or overt. One story even focused upon a major florist wholesaler that switched from gas and oil to wood chips for energy, saving about $150,000 a year.

Energy From Shale began running 30-second TV ads in 2015; its focus, as presented by the owner of a coffeehouse and bakery in Washington, Pa., was that fracking has been done safely for 60 years, and she had no problems with the quality of food she served.

NEWS AND FEATURES

The Allegheny Front, which produced its first radio show in 1991, is one of the nation's primary media operations to focus upon environmental issues. The five-person staff distribute 3–7 minutes stories, news updates, and 30- and 60-minute specials to public radio stations in Ohio, Pennsylvania, and West Virginia.

The first radio segments had focused upon river pollution and sewer overflow in the greater Pittsburgh, Pa., area. With better awareness and remedial measures, combined with the loss of the steel mill and mining industries, the rivers became less polluted. But, then, the fossil fuel industry moved into the Marcellus Shale, and water and air pollution increased.

Reid Frazier began working as a freelance reporter in 2006, and became full-time in 2011, primarily to cover the expanding fracking operations and their effects.

"When I first started covering fracking, I was very frustrated," says Frazier, "because the information I wanted from the DEP was only available from a file review." The files, says Frazier, "were confusing, the information was buried, and it was difficult to get clear answers." Filing right-to-know requests for information often denied to him usually took at least six weeks, he says. About 2013, it became a little easier to get information, "but it was still difficult," he says. However, he says, it was easier to get information from the state's Department of Conservation and Natural Resources.

Even with the difficulties getting information from official sources, Frazier and the other staff have been able to get technical and scientific data, interview numerous sources, produce hundreds of stories and, says Frazier, explain the impact "in a way that the people could understand what the issues are." Because of that, Allegheny Front reporters have won numerous awards from the Pennsylvania Associated Press Broadcasters Association (PAPBA), the Pennsylvania Association of Broadcasters (PAB), the Press Club of Pennsylvania, the Radio Television Digital News Association, the Radio and Television News Directors Association, and the Society of Environmental Journalists (SEJ). In 2015, Frazier earned several awards for his coverage of the fracking industry. He and Josh Raulerson of WESA-FM (Pittsburgh), who combined for a five-part series about the influence of gas industry money in state politics, earned first-place for news documentary from the PAPBA; he and Matt Richmond of WSKG-FM (Vestal, N.Y.) took first place honors in public affairs reporting from the PAB for their six-part series documenting wastewater and pollution caused by gas drillers. In awarding Frazier honorable mention for beat reporting in 2015, the SEJ pointed out "Reid's straight forward, multi-faceted radio reports on fracking and the loss of tax credits for wind energy showed an adroit use of public records and the ability to tackle both politics and science with skill."[2062]

One of the reasons why Allegheny Front stories may have had a heavy impact upon the public is because unlike other establishment media, environmental reporting is its focus, and not just an occasional part-time assignment in an industry that has rapidly downsized its news operations and increased entertainment reporting, soft features, and rewriting press releases. Another reason may be because there are no demands upon showing increased profits by cutting back on certain expenses and coverage—the mini-network is funded by both listener donations and several non-profit endowments, including those of the Heinz and Mellon foundations.

StateImpact Pennsylvania is a joint reporting project of WITF-FM (Harrisburg, Pa.), WHYY-FM (Philadelphia, Pa.), and National Public Radio. In 2013, the Alfred I. DuPont–Columbia University Awards Center honored StateImpact with its Silver Baton award for excellence. The reason for the award was because the project:

> "showed the significant impact of natural gas drilling on Pennsylvania residents, and is an important model for reporting on local issues. Reporters Susan Phillips and Scott Detrow covered the public policy, fiscal and environmental impact of the state's booming energy economy, with a focus on Marcellus Shale drilling. Their broadcast reports were heard on public radio stations across Pennsylvania and on a dedicated web site featuring multimedia, data-driven stories."[2063]

A few commercial radio stations (including WKOK-AM, Sunbury Pa.) broadcast occasional segments devoted to fracking issues, including guests who represent a variety of opinions. However, if there is news about the shale gas industry, most radio stations rely upon the AP radio wire.

Most television coverage about fracking is superficial, usually 30–60 second packaged sound-bites with some narrative. Analysis and in-depth reporting is almost non-existent in most markets. However, in September 2013, the Pennsylvania Cable Network (PCN), which had devoted several hours of live coverage of hearings, speeches, and energy conferences, created "Energy

Month: From the Wellhead to the Marketplace." The 37 separate programs included tours of gas fields and manufacturers, panel discussions, coverage of the Freedom From Fracking and Shale Gas Insight conferences, and "On the Issues" half-hour discussions with regulators, industry representatives, and anti-fracking activists.

Although some radio and television stations ran stories about fracking that were not industry-inspired, it was David Letterman on CBS-TV's "Late Night" who helped make the issues known throughout the country. In a two-minute commentary in July 2012, Letterman said:

> "Here's what I know about fracking:
> "The greedy oil and gas companies of this country have decided that they can squeeze every last little ounce of oil and gas out of previously pumped wells by injecting the substrata of our planet with highly toxic, carcinogenic chemicals, which then seep into the aquifer and hence into the water supply of Americans.
> "The Delaware water gap has been ruined. The Hudson Valley has been ruined. Most of Pennsylvania has been ruined. Virginia, West Virginia has been ruined. Colorado has been ruined. New Mexico has been ruined.
> "I'm no expert on fracking history and regulations, but this sounds like a typical 'compromise' working in the extractor's favor.
> "They're poisoning our drinking water and the EPA said, 'You know what? You no longer have to comply with EPA standards for stuff you put into the water.' So the greedy oil and gas companies said, 'Great, let's go crazy,' and then some states are saying, 'No, we have transparency laws,' so the oil and gas companies say, 'Okay, we'll tell you everything but two percent of what we're putting into your tap water.'
> "And that's supposed to make us feel better.
> "Ladies and gentlemen, we're screwed!"[2064]

Letterman was attacked by numerous industry-friendly PR people and commentators, most of whom called him ill-informed, while laying out a series of half-truths of their own; they touted the numerous benefits of shale production in advancing the economy, but mentioned nothing about documented effects upon the environment or health.

Two years after the Letterman rant, and two days before the November 2014 midterm elections, *The Simpsons*, one of the best satires in TV history, attacked fracking. In "Opposites A-Frack," written by Valentina Garza, the evil Montgomery Burns secretly fracks under Springfield. When Marge Simpson finds out, Burns has the solution to stop the city-wide panic—he employs Homer to explain why fracking is good, and hands out checks to residents to quell their fears. By the time the animated episode ends, even Homer is questioning the value of fracking, but concludes it's bad only if it's in his own back yard. The episode, in the series' 26th year, was seen by about 4.2 million viewers.[2065]

Advertising

Long before the people began to understand the negative effects of fracking, the fossil fuel industry realized it had to get its message out early.

In 1991, the coal industry tested advertising in three cities to "reposition global warming as theory," rather than scientific reality.[2066] Seven years later, several individuals representing conservative organizations and the shale gas industry met at the headquarters of the American Petroleum Institute to plan a $5 million advertising campaign that underscored what it believed were benefits of shale gas and oil while debunking global warming as scientific certainty. However, after details within an eight-page memo were leaked to the mass media, the plan was abandoned.[2067]

Near the end of 2009, the American Natural Gas Association (ANGA) bought about $80 million in advertising. "In public opinion research conducted earlier this year," according to *AdWeek*, "ANGA found that natural gas wasn't well understood by Americans and wasn't part of the nation's energy conversation."[2068] The multi-platform campaign used the tagline: "America's new Natural Gas. Cleaner, smarter energy."

Multi-million dollar ad campaigns soon enriched the stagnant or declining net profits of the mass media during the economic recession that had begun in 2008. ExxonMobil, in 2012, secretly delivered $2 million to the Independent Oil & Gas Association of New York,[2069] to be spent for pro-fracking advertising in a state where a moratorium blocked drilling into the gas-rich Southern tier. The campaign focused upon local residents wishing to take advantage of lucrative leasing deals; the message was, "We've Waited Long Enough."[2070]

In the Baltimore and Washington, D.C., area, Food & Water Watch sponsored a 30-second radio ad, narrated by actor Ed Norton, during the Maryland legislature's discussions about a possible statewide moratorium. The ad drew numerous calls to the governor and members of the legislature to place a moratorium on fracking, says Emily Wurth, water and program director for Food & Water Watch.

In North Carolina, which passed pro-industry fracking legislation in 2014, the American Petroleum Institute and the Natural Resources Defense Council launched a series of opposing radio and TV ads during the general election cycle. The NRDC spent more than $500,000 for the ads that singled out a half-dozen candidates as "The Fracking Crew."

The American Petroleum Institute bought air time on local stations in Colorado, Michigan, Ohio, Pennsylvania, and Washington, D.C during SuperBowl XLIX in February 2015. The 30-second ads continued the myth that fracking has been safe for 65 years. The ads for Colorado, Michigan, and Ohio focused on energy independence from foreign oil; the ad that ran in Pennsylvania focused on economic benefits. The ad that ran in the nation's capital concluded that fracking was, "Supporting millions of new jobs, billions in tax revenue, and a new century of American energy security." The cost of the ad on WRC-TV (D,C.), an NBC owned-and-operated station, was $100,000.[2071] Other prices were not disclosed by API or the stations. (A 30-second ad broadcast on all stations during SuperBowl was about $4.5 million.)

Although StateImpact Pennsylvania presents a daily stream of objective news and features, NPR itself has come under attack for airing pro-fracking advertising-like spots distributed by America's Natural Gas Alliance (ANGA). "All you hear on NPR all day, every day is how gas is 'natural' and fracking is 'safe,' and you should just 'think about it' at the ANGA website of the same name," says Drew Hudson of Environmental

514

Action.[2072] David Braun of Americans Against Fracking says he believes NPR "was much more fair before it began taking oil and gas industry money."

Some NPR affiliates agree with Hudson and Braun.

WUNC (North Carolina Public Radio) stated it "complained to NPR about the copy and about how this advocacy group [ANGA] is represented through the announcements."[2073]

Brian Sickora, president of WSKG-FM/WSQX-FM, which broadcasts into New York's southern tier counties on nine separate frequencies, said he was "very disappointed that NPR continues to accept ANGA funding." In a letter to the editor of the Oneonta (N.Y.) *Daily Star*, Sickora stated his station doesn't accept ads or sponsorship from either the gas industry or anti-fracking groups, and asked NPR "to reconsider its decision to accept this funding [because] it puts WSKG in a very difficult position."[2074]

The Federal Trade Commission (FTC) has specific rules governing ad content. Under the Federal Trade Commission Act, ads must be "truthful and non-deceptive," based upon evidence, and cannot be unfair.[2075] Ads from both the industry and anti-fracking movements often push the limits, sometimes crossing over. However, FTC authority is limited, especially in political and social message advertising. Commercial speech, once outside First Amendment protection, is now largely covered by the First Amendment. In *Edenfield v. Fane* (1993), the Supreme Court ruled:

> "The commercial market place, like other spheres of our social and cultural life, provides a forum where ideas and information flourish. Some of the ideas and information are vital, some of slight worth. But the general rule is that the speaker and the audience, not the government, assess the value of the information presented. Thus, even a communication that does no more than propose a commercial transaction is entitled to the coverage of the First Amendment."[2076]

Unlike the United States, where the First Amendment gives the media broad protection against governmental inteference and censorship, other countries have governmental agencies that have the power to interfere, especially if a story or ad is incorrect. In England, the Advertising Standards Authority assures citizens that advertising is truthful. Chris Faulkner, CEO of Breitling Energy, which is headquartered in Dallas, Texas, wrote an ad in the form of an open letter to the people of England; that ad/letter appeared in the *Daily Telegraph* of London in February 2014. Faulkner claimed there were numerous advantages of natural gas, including "decades worth of natural gas," "reducing greenhouse emissions by replacing coal with natural gas for energy," that natural gas extraction would lower energy prices, and that fracked gas in 2013 helped keep England from a catastrophe. The ASA investigated all claims made by Faulkner, determined there were numerous false statements, and ordered his corporation to cease placing the ad; more important, it cautioned him that he needed evidence to justify claims made in other ads. The company's response was that Faulkner was "simply sharing his views and experience to bring some balance to the debate on hydraulic fracking in the UK, a debate which has been dominated for many months by the sometimes outrageous claims made by opponents of the fracking process."[2077] The ASA didn't order anti-fracking activists to pull any ads.

The Power of a Word

The oil and gas industry has retreated from its entrenched position to have the public delete the "k" in "fracking," and write it as "frac'ing" or "fracing." Those who have been the strongest advocates for fracking have scorned and mocked those who place the "k" in the word. The problem is that without the "k," the word sounds like "frasing."

The first use of the word "fracking" can be traced to an oil and gas journal article in 1953;[2078] for most of the next five decades, the industry accepted that term and its spelling.

As hydraulic horizontal fracturing became a standard to extract gas and oil about 2008, anti-fracking activists began using the word in campaigns that slyly bordered on the obscene—"Frack off!" and "No Fracking Way!" Research by Gregory FCA, a pro-fracking PR agency, revealed that by the end of 2010, less than two years after fracking began in the

Marcellus Shale, the public had begun to develop negative attitudes about the process. Gregory Matusky, the firm's founder and president, reported, "Positive public sentiment to Marcellus Shale development in both traditional media and social media is slowly eroding, as some environmentalists make Marcellus Shale their cause célèbre, and as the media take a more negative turn in their reporting." Database analysis revealed that within the year, media perception of the process had fallen from +3.1 to −0.3 on a scale of +5 to −5; social media perception had fallen from a high of 4.0 to 1.1.[2079]

The industry, faced by being the brunt of a series of near-obscene jokes, dug in and demanded that "unconventional drilling" or just "horizontal fracturing" were the acceptable terms. But, if "fracking" had to be used in print, the preference was for "frac'ing" or "fracing." Most dictionaries—including the *Oxford English Dictionary*[2080] and *Merriam–Webster*[2081]—use the word "fracking" as the preferred and acceptable term.

In September 2014, the Marcellus Shale Coalition (MSC) became proactive with a series of newspaper, radio, TV, and YouTube ads. The ads were revealed at the annual Shale Insight conference, sponsored by the MSC. The fractivists "tried to hijack that word and paint it as something negative," David J. Spigelmyer, MSC president, told the *Philadelphia Inquirer*. He said the use of the "k" was "our effort to take that word back."[2082] Randy Cleveland, XTO Energy president, told the conference that where people said "frac'ing," the industry thrived, but where they said "'fracking,' we have difficulty." The PR and advertising campaign, said Cleveland, is "to regain the high ground."[2083]

In the newspaper ad, the word "fracking" is used five times; the largest word in the ad is "JOBS"—the ad emphasized job creation, using the inflated and discredited number of 240,000. Anchoring the ad was a new aphorism: "FRACKING: ROCK SOLID FOR PA."[2084] In radio and TV ads, a girl says, "Fracking rocks! My dad does it." At the conclusion of a three-minute YouTube video, in which a series of rumors was replaced by a series of half-truth "facts," one of the narrators tells the audience, "Fracking, a good word," and concludes with the newly-created motto.[2085]

YouTube

Thousands of videos—shorts and segments to full-length films—have been posted on YouTube by the energy industry, news and entertainment media, scientists, and anti-fracking activists. Most of the videos have been posted by those opposed to fracking. Among those videos are numerous interviews with persons directly affected by fracking and recorded lectures and discussions by scientists and activists. As expected, the production qualities vary from rough "home movie" quality to Hollywood production standards.

The Delaware Riverkeeper Network created weekly short video interviews with experts who present the truth about fracking issues. The Shale Truth interviews, says Maya van Rossum, the Delaware Riverkeeper, are:

> "an effective way to share new insights and technical information with those who are just beginning to learn about shale gas development as well as those who have been immersed in the subject for years. . . . We thought that short, focused pieces that pull out key issues which may have slipped under the radar for folks, including discussions about how to most effectively talk about shale gas development, would be a meaningful contribution to the learning and dialogue. We are careful to hit new and meaningful areas of focus from the ecological impacts, to the economic ramifications to the better energy alternatives that could and should make shale gas development a bottom rung choice for energy."

Among anti-fracking videos with the most views are the *Gasland* trailer; *Light Your Water on Fire*, a 1:45 minute slice that shows a woman lighting her kitchen faucet; *Fracking Hell*, which first appeared on cable TV; and *Fracking: Things Find a Way*, a two-minute animated video from *EarthJustice*. *Animation of Hydraulic Fracturing*, a six minute-30 second documentary about fracking, with both high-quality production standards and an industry slant, was produced by Marathon Oil Corp; by June 2015, it had about 880,000 views.[2086] A2L Consulting produced an easy-to-understand eight minute animated video, *Fracking Explained with Animation* but with a pro-industry bias. Nevertheless, both the Marathon and A2L videos give good overviews of the process, even if they soft-sell the effects.

Music

When corporate America wishes to promote or protest something, it gathers its money and sprinkles it upon politicians. To influence the public and what the people think are opinion leaders, it will turn to the mass media, sometimes to place ads, often to spray PR releases and "information bulletins," which much of the media, downsized and desperate for content to fill diminished news holes, will often run unquestioned.

When social movements want to get their message to the people, they turn to the social media, hold rallies and concerts. And so it was that on a rainy May 15, 2012, New Yorkers Against Fracking staged a 90-minute rally on the capitol lawn in Albany, and then a concert at The Egg at the Empire State Plaza. Hosts were Oscar®-nominated actor Mark Ruffalo and Oscar®-winner Melissa Leo, both of whom live in New York. Among the dozen musicians were Tracy Bonham, Joan Osborne, and Natalie Merchant, as well as rallying speeches by Dr. Sandra Steingraber and the stories of several persons who were directly affected by the health and environmental effects of fracking.

"We are not the sum of dollars and cents, but people of flesh and blood who have entrusted the folks here in Albany to safeguard our health and our common good," said Ruffalo. "We are here to celebrate our strengths and bear witness to the devastating public health and safety issues that surround hydrofracking," said Ruffalo to sustained cheers—"We are here to lift up the scientists and their better judgments against the paid-for political science of the oil and gas industry."[2087]

But it isn't the star-studded concerts that are the basis of the music of the social movement, but the songs written and recorded by activists, most of whom donate their time and creativity at everything from small coffeehouse appearances to mass rallies for social justice.

Dozens of songs give power to the movement, to unite the people and tell a story in about three minutes. Many of the songs have been recorded and placed as mini-documentary music video productions, often with state-of-the-art computer animation, onto YouTube, where thousands not dozens of people can hear and appreciate the power of protest.

Robin and Linda Williams created an upbeat foot-tapping four-minute bluegrass song, "We Don't Want Your Pipeline," to protest the proposed 550-mile 42-inch diameter pipeline planned from West Virginia through North Carolina. Dominion's Atlantic Coast Pipeline, when approved and built by 2019, will carry about 1.5 billion cubic feet of methane every day. The Williamses, professional singer/musicians for more than four decades, are active in the Augusta County (Va.) Alliance; a few residents had suggested the Williamses write a protest song. Another member of the Alliance videotaped the song and put it onto YouTube. In July 2015, about a year after the song's premiere, Robin and Linda Williams headlined the Music for the Mountains festival, which brought in about $20,000 for the anti-pipeline campaign. "We all love their song," says Alliance co-chair Nancy Sorrells, "because it hits all the different points that we are trying to make about why this pipeline is so wrong."

Neal Young's, "Who's Gonna Stand Up?" is a four-minute rock-and-folk song that was recorded in 2014 with 60 musicians and a 30-member choir.

Kris Kitko's upbeat "Frack Pit Love Song" is a sweetly sarcastic look at the lies the industry tells about the safety of wastewater. Kitko, from North Dakota, donates proceeds of her songs to Bakken Watch, an anti-fracking organization.

Anne Hill of Bethlehem, Pa., is a singer/actress/writer/musician who has a master's in social work and a long history of folk music and activism. She wrote "The Trade" at the request of the Lehigh Valley, Pa., Quakers. She asks several rhetorical questions that reveal the lies of the industry—"What good is a job with no safe place to settle? What good is a big paycheck without clean water to drink?"

There's also the country beat of Corey Koehler's "Frac Sand Blues"; Judith Van Allen's up-tempo "No Frackin'," set to the music of the Negro spiritual, "Oh, Freedom"; and David Rovics, whose "No Frackin' Way" tells about a landmen who promises riches, but whom he calls "a corporate crook."

In New York's Otsego County, the anti-fracking group Friends of Butternuts recorded Kathryn Gibson's lyrics to

Stephen C. Foster's "Oh, Susanna," to beg "Governor Cuomo, don't bring no fracking here, don't pollute the air and water, for the future we do fear."

Excellent computer-generated animation enhances an easy-rap beat on "My Water's on Fire Tonight" to explain what fracking is, and what its effects are. The song is a product of Studio 20 NYU, in collaboration with ProPublica. Vocals and lyrics are by David Holmes and Niel Bekker; music is by Holmes and Andrew Bean; animation was created by Adam Sakellarides and Lisa Rucker.

Sean Lennon, Yoko Ono, and Artists Against Fracking created "Don't Frack My Mother," a three-minute music video with a Bob Dylan spirit that explains what fracking is, and why we need to protect Mother Earth. Because of the creators' celebrity, the video has had more than 200,000 thousand YouTube views,[2088] and Lennon and Ono have performed it on several late night television shows.

In the fall of 2014, before Denton, Texas, voters were about to ban fracking, three young women (Niki Chochrek, Angie Holliday, and Tara Linn Hunter), part of Frack Free Denton, produced "Fracking is Your Town's Best Friend" to the tune of "Diamonds are a Girl's Best Friend." Later, in response to the Texas legislature, which was about to overturn the will of the voters, the Frackettes created a five-minute parody of "Come to the Cabaret." In this song-and-dance version, "The Death of Democracy," the Frackettes sang about how Texas was "bought and paid for" by the oil and gas industry.

The United States isn't the only country that has anti-fracking music. From Australia, Joel Kalma raps "Fracking Fluid," backed by a combo and his father, Ariel, playing the didgeridoo. From England comes the Grass's powerful protest, "We Don't Want Your Fracking Well." From Ireland comes "Frackin' Devil," by Colin Beggan and Frank Malloy. From South Africa, Ben Ulric created "Frackin' Blues," recorded in Afrikaans, with English subtitles over photos of the soon-to-be-fracked Karoo Desert.

Theatre and Multimedia

"Same River," produced by New York City's Strike Anywhere

Performance Ensemble, toured the Marcellus region between summer 2010 and summer 2013 to explain the problems of fracking. The multimedia production combined a structured script with improvisational acting, dance, music, and stage lighting and video which are displayed on several surfaces. The production "was a reflection of the places we perform," says Leese Walker, artistic director and Ensemble producer. Each performance was an extended three-act experience. The first act was a local work of art that hung in the lobby; Act 2 was a 75-minute structured improv; Act 3 was a town hall discussion. "Our intent," says Walker, "was not only to address a sociopolitical issue but also to initiate dialogue, to spur the people to think, talk, and act."

Strike Anywhere followed up "Same River" with "Farce of Nature," which premiered at Jacqueline Kennedy Onassis High School (Manhattan, N.Y.) in May 2015. With the same structure as "Same River," "Farce of Nature" looked at climate change. In additional to the professional actors, the cast included students and the school's teachers.

Miscellaneous Media

APPEALING TO CHILDREN

With a donation of $1.2 from energy giants EQT Corp. and Energy Corporation of America, the Clay Center for the Arts and Sciences of West Virginia created a mobile exhibit to showcase what it believes are the benefits of natural gas and horizontal fracking.

"Power Your Future," packed into a truck, is targeted to elementary and middle school students in West Virginia, Ohio, and Pennsylvania. The mini-museum on wheels, says the Clay Center, "uses interactive games and activities to take visitors on a journey through the exploration and extraction of natural gas, the engineering and technology of processing it, and the many uses of this important natural resource." The pre-teen visitors can not only "Put [their] skills to the test in an interactive game show, create a musical composition of seismic sounds, [but also] reclaim the land and return it to its natural beauty after drilling is completed."[2089]

522

There is nothing in the exhibit about alternative renewable forms of energy or the health and environmental consequences of natural gas extraction. However, Lloyd G. Jackson II, chairman of the museum's board, says, [T]he Center will continue to bring balanced and fact-based educational programming to inform our citizens and to educate our children."[2090] Jackson, in an OpEd in the West Virginia *Gazette–Mail*, cited part of the Governor's STEM Council Report, claiming because "Marcellus activity in West Virginia is anticipated to last for decades," it's important that that "this increase in natural gas production requires a steady flow of workers who are skilled in Science, Technology, Engineering and Mathematics (STEM) in the Mountain State. The West Virginia education system has the challenge of preparing the needed STEM-skilled workers to fill these positions."[2091]

Jackson, who served 12 years as a state senator, is president of the Jackson Gas Co.

Chesapeake Charlie, an orange-tinged cartoon beagle who wore an American flag on his blue gas field jumpsuit, was Chesapeake Energy's tool to lure children into believing that "because natural gas is clean, affordable, abundant and American, it's good for us and our country." The 14-page coloring book, designed by freelancer Kev Brockschmidt, also told children that natural gas "doesn't pollute the air and helps protect our environment." Chesapeake Energy also transformed the all-American patriotic Charlie into a costumed mascot for fairs and picnics.

Talisman Terry the Fracasaurus was Talisman Energy's hero in a 24-page coloring book to help indoctrinate children into why gas drilling and fracking is beneficial. Smiling animals, beautiful forests, and rainbows, all of which children were expected to color, provide the background for the to-be-colored drilling sites. The booklet was distributed throughout northeastern Pennsylvania in 2009 and much of 2010.

Terry became extinct when a newspaper, a politician, and a satirist protested the depiction of fracking that showed a better world after fracking than before. The newspaper was the *Pittsburgh Post–Gazette*, which summed up the essence of the coloring book as "drilling is smart, safe and American [while] glossing over the environmental and economic controversies that have surrounded drillers tapping the Marcellus Shale rock formation for lucrative pockets of gas."[2092] The politician was Rep. Ed Markey (D-Mass.), who said the problem with the book "is that unless you are a 'FRACK-A-SAURUS' named 'Talisman Terry,' this [idyllic] world doesn't exist."[2093] The satirist was Stephen Colbert, who skewered Talisman and Terry in a five-and-half minute segment on "The Colbert Report" in July 2011. Colbert, a liberal who portrayed a bumbling conservative talk show host on the cable network Comedy Central, said he supported hydrofracking, which he said was "like giving the Earth an Alka-Seltzer if the Alka-Seltzer shattered your internal organs so oil companies could harvest your juices." Terry, said Colbert, "is a dinosaur and he's encouraging us to use the remains of his own dead relatives to heat our homes." Colbert added a few pages that were not in the company's booklet, and concluded his own additional pages by showing Terry, now consumed by grief at exploiting his ancestors, who steps into a shower, lights a cigarette, and commits suicide from the flames that came from methane in the water.[2094] Shortly after the segment aired, Talisman discontinued distributing the coloring book.

Anadarko Energy apparently didn't learn the lessons from the Fracasaurus meltdown. For the 2013 Little League World Series in South Williamsport, Pa., it created a metal pin that 11- and 12-year-olds could buy and trade. The Anadarko Energenie was a round-headed comic blue figure who wore a hard hat, had a baseball in his right hand, a mitt covering his left hand, and was depicted on a baseball diamond that was framed by two bats. Covering the bottom one-third of the pin was "Anadarko" in bold letters, a rough depiction of a rig, and "Williamsport, Pa."

The marketing and public relations campaign of the natural gas industry and its many front groups is far more sophisticated than pins and coloring books. America's Natural Gas Alliance spent about $80 million in 2009 to implement its programs.[2095] Part of that $80 million went to Hill + Knowlton, the largest PR agency in the world. For more than a decade, the PR agency's primary mission was to assure Americans there were no links between cigarettes and cancer. Its mission for the gas industry included assuring Americans that fracking, like smoking unfiltered cigarettes, was safe and beneficial.

ExxonMobil, faced by a continuing moratorium on fracking in New York state, spent $2.1 million in 2012 to develop a media campaign to try to convince the people of the state's southern tier, which borders Pennsylvania, of the benefits of fracking. The Elmira *Star-Gazette* explains why ExxonMobil spent so much: "In 2008, XTO Energy—now a subsidiary of Exxon—struck a $110 million deal to lease the oil-and-gas rights to about 46,000 acres of land in eastern Broome and Delaware counties in the Marcellus Shale region."[2096]

PSEUDO-EVENTS

Pseudo-events are often necessary to attract media attention. Members of Rising Tide Chicago dressed as elves, tied a red bow to an eight-foot mock rig and placed it on the front lawn of the home of Illinois Gov. Pat Quinn, Dec. 23, 2013. "If Gov. Quinn and the other people that have opened up our state to fracking had to live next to fracking and had to obtain their water from a well I think they would not bring fracking to our state," said Mike Durshmid, one of the elves.[2097] State police dismantled the fake rig; they made no arrests.

In South Burlington, Vt., five women staged a 90 minute "knit-in" in the lobby of Vermont Gas Systems in July 2014. The company had planned a gasline extension that would have impacted their property. Vermont Gas's response was to try to talk with the women in private; the women wanted the discussion to be public. Police arrested them for trespassing.[2098]

BILLBOARDS

On highways throughout the gas shale regions are billboards that preach the advantages of natural gas. The message from gas and oil companies, their trade associations and front groups is that natural gas drilling provides jobs, and is cheaper and cleaner than other forms of energy.

A pro-fracking front, which calls itself Big Green Radicals, developed an unusual tactic to scare the public into believing that environmentalists were the problem, not the solution. On billboards in Colorado and Pennsylvania, it placed an image of Lady Gaga with a piece of meat on her head. The message was, "Would you take energy advice from a woman wearing a meat dress?" On another billboard, it placed a picture of Robert Redford, known for his environmental activism, and the message, "Demands green living. Flies on private jets." The website identifies the motives of Big Green Radicals:

> "A web of wealthy foundations and individuals bankroll these radical activists, making the Sierra Club, the Natural Resources Defense Council, and Food & Water Watch among the most powerful (and radical) voices pushing the green agenda.
> "These activists strive to end the use of both traditional fossil fuel energy sources—coal, natural gas, and petroleum—and several carbon emission-free energy sources not deemed 'green' enough—nuclear energy and hydropower. By focusing solely on wind and solar, Big Green Radicals will eliminate the energy sources that provide 95 percent of our current electricity needs, raising prices dramatically and reducing reliability."[2099]

Big Green Radicals is an Astroturf Group, an organization created by the oil/gas corporations to appear to be grassroots activism. Behind Big Green Radicals is Richard Berman, CEO of Berman and Co., which identifies itself as a "dynamic research, communications, advertising, and government affairs firm [that blends] aggressive, creative thinking with functional expertise to achieve extraordinary results for our clients."[2100] The company often represents clients that are opposed to unions, animal rights, and environmental issues. At the June 2014 annual meeting of the Western Energy Alliance, Berman told fracking industry executives that they had to exploit fear and greed, and not be afraid to dig up embarrassing background information about environmentalists to promote their message, that they could "either win ugly or lose pretty." In his one-hour speech, which was secretly taped and provided to the *New York Times*, Berman disclosed one of his company's strategies:

"People always ask me one question all the time: 'How do I know that I won't be found out as a supporter of what you're doing?' We run all of this stuff through nonprofit organizations that are insulated from having to disclose donors. There is total anonymity. People don't know who supports us."[2101]

Occasionally, Berman's tactics, including the use of the "false prophet," have been exposed, but usually long after they have succeeded in delivering the message. Several newspapers published pro-fracking OpEds by Anastasia Swearingen, identified as a senior research analyst for the Environmental Policy Alliance. Using conclusions provided by Dr. Tim Considine, known for optimistic projections about the benefits of fracking, Swearingen attacked the procedures of the Bureau of Land Management in awarding permits to drilling companies:

"Even as the BLM slow-walks oil and gas permits, it fast-tracks green energy projects to appease the Obama administration's environmental allies. . . .

"Thanks to BLM and radical green groups, oil and gas production is declining on federal lands amid the largest domestic energy boom in U.S. history. Technological advancements in hydraulic fracturing and horizontal drilling have made America the world's largest combined oil and gas producer, surpassing Russia and Saudi Arabia. This boom, however, is occurring only on lands not controlled by federal bureaucrats.

"The truth is that BLM and radical environmentalists have prevented Western states from reaping enormous economic benefits."[2102]

Swearingen was employed by Berman; the Environmental Policy Alliance, with the convenient acronym EPA, was nothing more than a front group created by Berman.

The greater problem is not that the article was nothing more than a well-written PR release, but that newspaper editors did not verify the information or the source prior to publishing it.

As those in the anti-fracking movement developed better revenue sources and became more media-sophisticated, they also developed messages to place along highways. However, the gas industry has used intimidation and fear to eliminate messages they disagree with.

Craig and Julie Sautner moved to Dimock, Pa., in March 2008. By September, a month after drilling began near their property, the water turned yellow and cloudy.[2103] Cabot Oil and Gas denied responsibility, but provided a 550 gallon water buffalo. Three years and innumerable problems later, Catskill Citizens for Safe Energy bought space on an outdoor billboard on Route 29 near Dimock. The billboard showed a pitcher of yellow-brown water, a list of chemicals in wastewater, and the message, "Fix It!" Cabot, whose name was not mentioned in the billboard ad, claimed the water was safe, met all federal guidelines, and objected. That billboard message lasted two days. John Krohn of Energy in Depth told Staci Wilson of the *Susquehanna Independent*, that the owner of the property who leased land to Park Outdoor Advertising "did not appreciate or welcome the message at all." Wilson reported that Bill Kelley, owner of two rental companies, provides equipment to the natural gas industry.[2104]

Rebecca Roter, with financial assistance from friends and other anti-fracking activists, kept a billboard message for two years in northeast Pennsylvania before Park Outdoor Advertising cancelled the message. That message showed three glasses of discolored water from Susquehanna, Bradford, and Washington counties. The message was, "Would you drink this gasfield tap water?" Prominent on the billboard was the slogan, "Water is Life."

Roter tried again in 2012. The new billboard would have shown a picture of a gasfield flare and the message, "Keep Our Families Safe." Included was a phone number for an EPA tip line. Roter told the Philadelphia *City Paper* that Park's president had e-mailed her, "We cannot accept controversial advertisement requests," and later told her that the flare and the word "safe" were controversial.[2105]

In Watkins Glen area of New York, Gas Free Seneca wanted to put up two billboards that showed Seneca Lake and a message, "Save Seneca Lake." It was a plea to continue to keep fracking out of the state. Park Outdoor Advertising also rejected that ad. In a letter to Kerry Leipold, general manager of the Finger Lakes Division, Yvonne Taylor stated:

"Everything seemed to be going along fine in acquiring ad space, until it was learned that we are a non-profit organization with a mission to preserve and protect the region. Then we were told that we were not permitted to do business with you.

"We represent 155 area businesses, and are incredulous that we would not be permitted to place the image attached in a few locations this summer."[2106]

Park Outdoor Advertising never responded, says Taylor.

On U.S. Route 36, near Coshocton, Ohio, Mike Boals paid about $1,000 to put messages on two billboards in July 2014. One sign told motorists that injection wells "pump POISONED WATER under the feet of America's citizens." The second sign quoted Biblical scripture of people dying from bitter waters. The messages remained on the boards two months until Boals was forced to remove them. Buckeye Brine, a Texas-based corporation that operates a nearby injection well, sent Boals a letter demanding he remove the signs, claiming they defamed the company and an individual who managed the well and owned the land where the injection well was drilled. "So I sent them a notice that I wouldn't remove the sign," says Boals. Buckeye then contacted the owner of the signs, threatening court action. The owners, a husband and wife who farm and own a miniature golf course, supported Boals, but when threatened with court action, and extensive legal fees to defend themselves, reluctantly asked Boals to remove the signs a month before his three-month campaign would have ended. Although the billboard messages came down, Boals says in other ways he is "pursuing the right to tell the public the truth."

Several individuals and groups have been successful in placing billboard messages along highways.

Along the Major Deegan Expressway in the Bronx, N.Y., Yoko Ono and Sean Lennon of Artists Against Fracking placed a billboard in November 2012 with the words, "IMAGINE THERE'S NO FRACKING" in black letters on a blue background.

In East Columbus, Ohio, Radioactive Waste Alert posted a picture of a girl guzzling from a bottle that has a radioactive waste label. The message is "Don't frack my water, Protect Columbus."[2107] Jen French of WSYX-TV interviewed several persons who were concerned about easy permitting for drilling waste to be brought and disposed in Ohio. However, the Ohio Department of Natural Resources told WSYX there was no connection between fracking and drinking water problems.

Faced by opposition, the Industry became bolder in its attacks. The Pennsylvania Independent Oil and Gas Association (PIOGA) placed a series of billboards on the Pennsylvania Turnpike south of Pittsburgh. PIOGA at first tried to hide behind a veil of anonymity, but finally revealed itself after an Associated Press investigation.[2108]

One billboard message, oozing green letters, told motorists, "YOU'VE BEEN SLIMED!" It throws out the words "Distortions," "Half-Truths" and "Lies." A nearby billboard, with the same design, suggests motorists get the truth from the website nogreenslime.com. That website reveals the intent:

"Green Slime is the propaganda being thrown at you and your neighbors every day by 'environmental activists' to condemn and vilify the revolution in energy development that is taking place in the United States. . . .

"We're pointing a light on the deliberate misinformation, distortion and junk science from groups that are trying to stop the growth of an industry that is doing a number of important things."[2109]

Ironically, the same arguments have been said for years by those who have challenged the statements made by front groups and trade associations of the oil and gas industry.

CHAPTER 24
From 'Boom' to 'Bubble' to 'Bust'

The energy exploration companies received billions of dollars in loans from Wall Street financial institutions to lease land, drill exploratory wells, buy equipment, and drill and process the gas. The *New York Times* in October 2012 reported:

> "After the [mortgage] financial crisis, the natural gas rush was one of the few major profit centers for Wall Street deal makers, who found willing takers among energy companies and foreign financial investors.
>
> "Big companies like Chesapeake and lesser-known outfits like Quicksilver Resources and Exco Resources were able to supercharge their growth with the global financing, transforming the face of energy in this country. In all, the top 50 oil and gas companies raised and spent an annual average of $126 billion over the last six years [2006–2012] on drilling, land acquisition and other capital costs within the United States, double their capital spending as of 2005, according to an analysis by Ernst & Young."[2110]

In November 2014, Saudi Arabia decided not to lower production of crude oil, which would have kept prices stable. It was a deliberate move to lessen competition. The result of the Saudi decision was the cost per barrel dropped from above $100 to about $50 by the end of 2014. By the end of Summer 2015, prices dropped to below $40 a barrel, a 60 percent slide that led to significantly lower prices at the gas pumps, and to a panic among investors, financial institutions, and corporations drilling and planning to export natural gas, and their myriad suppliers. The sudden and unexpected drop in oil prices could cause a "shock large enough to trigger the next wave of defaults," according to an analysis from Deutsche Bank.[2111] The Saudis "have no incentive to lower supply to defend the price of crude oil

531

. . . so [they] are not going to rescue the market," predicted Bob Tippee, editor of the *Oil & Gas Journal*.[2112]

Illinois, which had completed regulations for companies that wished to conduct fracking, and which hired 31 persons to regulate the practice, found the plunge of oil prices and the glut of natural gas was too much for any company to apply for a permit to drill into the New Albany basin.

Gazprom, the world's largest supplier, was forced to drop its own prices to be competitive, and has been developing plans to provide gas to Europe and Asia, especially China where American gas is headed, at a price that makes it uneconomical to do long-term contracts.

Royal Dutch Shell, by the end of 2014, had invested about $24 billion into unconventional shale development in North America. It "did not exactly play out as planned," said Peter Voser, the company's CEO.[2113] Shell took a $2.1 billion tax loss for 2013, primarily because it had come into the Marcellus Shale development late, and was consumed by the glut that forced the price of natural gas to all-time lows.

Although the value assigned to the 200 largest fossil fuel companies is about $4 trillion, there is almost $1.3 trillion in outstanding debt, according to the Carbon Tracker Initiative, an investment analyst.[2114] Between 2010 and 2014, debt doubled, while revenue increased only about 5.6 percent, according to an analysis of 61 drillers by *Bloomberg News*.[2115] Between June 2014 and the beginning of 2015, the ten largest oil and gas corporations lost about $200 billion, according to a CNNMoney analysis.[2116]

WBH Energy, which drilled for oil and gas primarily in Texas, declared debt of between $10 million and $50 million, and filed for bankruptcy protection in January 2015.[2117] Two months later, Quicksilver Resources, with $2.35 billion in debt and only $1.21 billion in assets, filed for Chapter 11 bankruptcy. Sabine Oil & Gas, with assets of about $2.5 billion and debt of about $2.9 billion, filed for bankruptcy in July.[2118] Samson Resources defaulted on a $110 million interest payment in August, and filed for Chapter 11 bankruptcy protection. Samson, according to analyst Wolf Richter, used about $4.1 billion, accumulated significant debt when it "went on to drill this cash into the ground to produce lots of natural gas and sell it below cost, losing money all along." Richter says the company's cash "is running out, and new cash to drill into the ground isn't readily forthcoming."[2119] Other major gas and oil companies that declared bankruptcy were American Eagle Energy, BPZ Resources, Dune Energy, and Saratoga Resources.

Range Resources had a $301 million loss in the third quarter of 2015; one year earlier, it had a third quarter net income of $146 million. It also cut the number of rigs from 15 to six in the Marcellus Shale.[2120]

Chevron reported a net profit of $571 million for the second quarter of 2015, down 90 percent from its second quarter profit of $5.67 billion a year earlier. The company cut $5 billion in capital expenditures in 2015, and expects to make significant spending reductions in 2016 and 2017, according to Pat Yarington, Chevron chief financial officer.[2121]

Among oil and gas companies that defaulted on bank loans in 2015 were Halcon Resources, Midstates Petroleum, RAAM Global Energy, American Energy Partners' Woodford unit in northern Oklahoma, Venoco, and Warren Resources.[2122]

Nineteen oil and gas stocks "are toxic to your portfolio" and in danger of declaring bankruptcy, according to energy analyst David Fessler. At the top of his list were Antero Resources ($4.14 billion debt), Energy XXI ($3.84 billion debt), EV Energy Partners ($1.15 billion debt), EXCO Resources ($1.55 billion debt), and Exterran Partners ($1.23 billion debt).[2123]

Two of the world's largest banks, both with London headquarters, have warned investors against keeping or purchasing fossil fuel stock.

Mark Carney, governor of the Bank of England, the United Kingdom's official bank to help set monetary policy, warned that if the world is determined that it not reach the critical point of being unable to reverse global warming, then the "vast majority of reserves are unburnable."[2124] HSBC, a British bank with total equity of more than $190 billion,[2125] warned customers that fossil fuel companies may become "economically non-viable," according to reporting in *Newsweek*. HSBC proposed its investors in fossil fuels do one of three things— eliminate all fossil fuel stock, remove coal and oil from their portfolios, or wait and see what develops. However, according to the HSBC report published in May 2015, those who keep their fossil fuel stocks "may one day be seen to be late movers, on 'the wrong side of history.'" [2126]

"We are all losing our shirts today. We're making no money," said Rex Tillerson, CEO of ExxonMobil.[2127]

"Because of the intricate financial deals and leasing arrangements that many of them struck during the boom, they were unable to pull their foot off the accelerator fast enough to avoid a crash in the price of natural gas," wrote *New York Times* reporters Clifford Krauss and Eric Lipton in October 2012.[2128] The energy exploration and distribution companies, they noted, "are committed to spending far more to produce gas than they can earn selling it. Their stock prices and debt ratings have been hammered."[2129]

The *Guardian* reports, "Ratings agencies have expressed concerns, with Standard and Poor's concluding that the risk could lead to the downgrading of the credit ratings of oil companies within a few years."[2130]

Deborah Rogers further explains how financial institutions contributed to, and may have caused, the problem, even without the Saudi strategy:

> "By ensuring that production continued at a frenzied pace, in spite of poor well performance (in dollar terms), a glut in the market for natural gas resulted and prices were driven to new lows. . . .
>
> "It is highly unlikely that market-savvy bankers did not recognize that by overproducing natural gas a glut would occur with a concomitant severe price decline. This price decline, however, opened the door for significant transactional deals worth billions of dollars and thereby secured further large fees for the investment banks involved. In fact, shales became one of the largest profit centers within these banks in their energy . . . portfolios since 2010. The recent natural gas market glut was largely effected through overproduction of natural gas in order to meet financial analyst's production targets and to provide cash flow to support operators' imprudent leverage positions.
>
> "As prices plunged, Wall Street began executing deals to spin assets of troubled shale companies off to larger players in the industry. Such deals deteriorated only months later, resulting in massive write-downs in shale assets. In addition, the banks were instrumental in crafting convoluted financial

products such as VPP's (volumetric production payments); and despite . . . the obvious lack of sophisticated knowledge by many of these investors about the intricacies and risks of shale production, these products were subsequently sold to investors such as pension funds. Further, leases were bundled and flipped on unproved shale fields in much the same way as mortgage-backed securities had been bundled and sold on questionable underlying mortgage assets prior to the economic downturn of 2007."[2131]

Another problem is that Big Energy and the world's major financial institutions may have inflated the value of recoverable gas. "The financial crisis has shown what happens when risks accumulate unnoticed," said Dr. Nicholas Stern, one of the world's leading economists and professor at the London School of Economics.[2132]

The low prices gave both residential and business consumers a false sense of financial dependence upon natural gas as a primary heating fuel. It has, ironically, also benefitted the natural gas industry. Instead of pushing for more renewable energy, consumers attacked both coal and oil heating, and bought into natural gas arguments about natural gas being a clean, safe way to reduce American dependence upon foreign oil.

The economic outlook for tight oil production is only slightly better. Tight oil is mined from shale or tight sandstone, and often contains gas. The largest fields—Bakken and Eagle Ford—are expected to be economically productive until about 2019 and then, according to the Post Carbon Institute, "will collapse back to 2012 levels." Tight oil production, according to the Institute, "will be a bubble of about ten years' duration."[2133]

Matt Kelso of fracktracker.org said the data confirms, "The industry is ramping down permitting, drilling and production efforts."[2134] Many companies are "holding gas back in anticipation of a better price, and if that happens, they can turn the wells back on very quickly," said Dr. Terry Engelder, professor of geosciences at Penn State, and a booster of fossil fuel extraction.[2135] Dr. Sergei Komlev, head of export contracts and pricing for Gazprom, told the AP, "We do not expect the currently abnormally low prices in the USA to last for long."[2136]

A side effect of the falling oil prices is that America's conspiracy theorists and conservative media pundits have blamed environmentalists for that decline, falsely claiming that by massive protests to horizontal fracking, they were nothing more than dupes for foreign energy companies that saw American natural gas and oil production as competition.

Searching to Keep Their Market Share

To compensate for overproduction, mild winters, competition from the coal industry, and the forced reduction in oil prices by the Saudis, yet trying to repay their heavy debts, natural gas energy companies developed a four-part strategy: downsize operations, reduce production, look for better revenue-producing areas, and export gas.

DOWNSIZING

When the industry first went into the shale fields, it came with promises of riches to leaseholders, jobs to the unemployed, and increased income to suppliers. However, the promises by the industry, and disseminated by politicians and chambers of commerce, never met the expectations.

EXCO Resources in February 2013 cut 70 of its 120 person staff in its Warrendale, Pa., office.[2137] Antero Resources laid off 250 landmen in West Virginia;[2138] Chesapeake Energy, once the leading player in the Marcellus Shale, in 2014–2015 terminated almost 2,000 employees nationwide,[2139] about 36 percent of its peak workforce in 2010, and eliminated most of its community relations personnel in Pennsylvania.[2140] Consol Energy laid off about 600 employees and cut health benefits for retirees.[2141] Chevron laid off 2,100 contractors and employees in 2015.[2142]

Schlumberger laid off about 20,000 workers (15 percent of its workforce),[2143] and cut spending to $3 billion in 2015 from $4 billion the previous year;[2144] Baker Hughes laid off 13,000;[2145] Halliburton laid off 18,000 workers, about one-fifth of its work force.[2146] The cutbacks affected not just the three largest providers of oil and gas services, but all of their suppliers. Layoffs exceeded 200,000 by the end of October 2015, according to *Forbes*.[2147]

Not only are jobs leaving the shale fields, but small business owners—from those who own restaurants, motels, and apartment complexes, and who had raised their rents three or four times what they were in 2010 to take advantage of the higher-paid workers, to those who provide supplies in the Marcellus Shale—have seeing sudden profits turn into crushing losses.

In addition to the immediate effects, the layoffs will likely affect the programs at colleges and universities, which created certificate and degree programs to meet industry claims and expectations.

REDUCING PRODUCTION

To correct for overproduction, fossil fuel companies reduced production to try to drive prices back up. Many corporations by the end of 2014, less than five years after many had moved into the shales, began cutting their drilling operations; others abandoned drilling in the gas shales, burdened by debt to the lending institutions. Range Resources, which spends about 95 percent of its drilling budget in Pennsylvania, cut capital spending to $870 million in 2015, down from $1.3 billion the year before.[2148] Chesapeake Energy, the nation's second largest producer of natural gas, sold 1,500 wells and physical assets, and leases to about 413,000 acres in the Marcellus and Utica shales to Southwestern Energy for $4.975 billion at the end of 2014.[2149] Continental Resources, which uses horizontal fracking primarily to extract oil, lowered its drilling budget by 40 percent in 2015,[2150] and reduced its rigs in the Bakken and other northern shales from 50 to 31.[2151] EQT, which has significant drilling ooperations in the Marcellus Shale and Permian Basin, cut its 2015 budget 18 percent, to $2.05 billion.[2152] Consol Energy began selling off $1.55–$2.3 billion in assets at the end of 2015 and stopped all drilling until at least 2017.[2153]

Hess Corp. and Newfield Exploration, two of the majpor drilling companies, terminated about 1,500 leases covering about 100,000 acres in northeast Pennsylvania during the first quarter of 2015.[2154] Hess Corp. cut its 2015 budget in the Shale to $1.8 billion, 18 percent less than its $2.2 billion budget the year before.[2155]

Continental Resources cut its budget to $2.4 billion for 2015, down from $4.6 billion the year before.[2156] Cabot Oil and Gas cut its 2015 budget to $900 million, down from the $1.6 billion it reported in October 2014 that it planned to spend in 2015. Cabot recorded net income loss of about $2.8 million for the first three quarters of 2015.[2157] The cutback by the major drillers forced thousands of suppliers to cut back their workforce and their budgets.

In January 2012, there were 114 gas rigs in Pennsylvania;[2158] by January 2015, there were only 62,[2159] seven of which were vertical rigs.[2160] Nine months later, the count dropped to 33 rigs. The number of gas rigs throughout the nation by December 2015 was 192,[2161] down from 439 in January 2013.[2162] In the Bakken Shale, which produces mostly crude oil, the rig count dropped from 202 in January 2012 to 61 in December 2015.[2163] Throughout the country, the number of oil and gas rigs dropped from 1,920 in December 2014 to 744 by December 2015.[2164]

The Industry "over-fracked and over-drilled," Matthias Bichsel, projects and technology director for Royal Dutch Shell, said in an interview with the *Financial Post*. In the United States, he said, "The reservoirs don't need that many wells. The reservoirs don't need that many stages of fracks, because not all the pieces of the rocks are as good." He then gave a dose of reality to those who were pushing horizontal fracking throughout the country:

> "We only talk about the Bakken, Eagle Ford, and the Permian in West Texas, and the Marcellus—we never talk about the basins that have not worked. We have some areas that are simply not as good as others."[2165]

CHANGING STRATEGIES

To accomplish the third part of their strategy, energy companies began to move back into oil production. Talisman moved much of its operation to the Eagle Ford Shale in south Texas at the beginning of 2012;[2166] Royal Dutch Shell, which had a $26.6 billion profit in 2012,[2167] took a $2 billion write-off on its shale assets in the United States, and announced it was moving into more productive oil-rich areas.[2168] WPX Energy announced in November 2014 it would not drill new wells in Pennsylvania, and was concentrating upon oil and wet gas fields in Colorado, North Dakota, and New Mexico;[2169] it sold 46,700 acres and 63 wells, its assets in northeastern Pennsylvania, to Southwestern

Energy for $300 million,[2170] and most of its assets in southwestern Pennsylvania for about $200 million.[2171]

Some corporations began to move into the deeper Utica Shale in Ohio, mining the more lucrative wet gas than the dry gas in the Marcellus Shale. Mike Knapp, president of Knapp Acquisitions & Production, explains the differences between wet and dry gas:

> "Natural gas is a gas comprised of multiple hydrocarbons, the most prevalent being methane. The higher the methane concentration, the 'drier' or 'colder' the gas is. Other constituents of natural gas are evaporated liquids like ethane and butane, pentane, etc. We refer to these collectively as natural gas liquids (NGLs), or 'condensates'. The higher the percentage of NGL's, the 'hotter' or 'wetter' the gas is. NGL's must be stripped out of the gas before it can be put in a pipeline and used. Ethane, which is prevalent in Western PA wet gas, is the feedstock for Ethylene, which is what we use to make plastics."[2172]

As wet gas begins to dominate production, economic benefits of dry gas will diminish, says Knapp:

> "Whether you are in a wet gas or dry gas area is going to have a huge impact on the value of your lease. Right now, the NGL's are worth considerably more than dry gas. In some areas, the value of the gas is more than doubled because of the NGL's. . . . Dry gas areas are dead as a doornail right now for leasing, but wet gas areas are seeing nice offers. With the impressive . . . results companies have been having in the Utica in Ohio with oil production (which is far more profitable than wet or dry gas) dry gas areas have been reduced to a distant third tier. Dry gas areas will not be in high demand for a long time, possibly decades. That is not to say that they will not be drilled . . . but companies will not be competing and landowners shouldn't expect to see the huge up front bonuses (that they did a few years ago) again any time soon. With the low price of gas, it's simply not economical to pay out thousands of dollars per acre just to be able to pull a rig on the property to spend millions to drill a well that will barely make a profit at these prices. Wet gas area landowners have . . . more leverage."[2173]

A partnership of three companies—Fossil Creek Ohio, Gastar, and Stone Energy—began drilling into the Utica Shale in Ohio and West Virginia in 2014, drilling down about 11,000 feet before turning pipes horizontally more than 4,000 feet.[2174] In October 2013, Reuters reported that a firm run by Aubrey McClendon, disgraced former CEO of Chesapeake Energy, had raised $1.7 billion, and planned to acquire about 110,000 acres in the Utica Shale. The firm, according to Reuters, planned to have 12 rigs by 2017.[2175]

However, successful drilling in the Utica Shale may not be an option. Numerous petroengineering studies and test drillings show that with natural gas prices hovering between $3 and $4 MCF the cost to recover gas in the Utica Shale may be impractical. The rock is much denser and recoverable gas is much deeper than in the Marcellus Shale. Devon Energy sold 157,000 acres in the Utica Shale to concentrate on oil recovery in other shales. Philip Weiss, an oil and gas analyst, told *Bloomberg News*, "The results [in the Utica Shale] were somewhat disappointing. Early data show it's not as good as we thought it was going to be."[2176]

INCREASING PROFITS BY EXPORTING GAS

The fourth strategy is to export gas, hoping to reap higher profits. Liquefied Natural Gas (LNG) is natural gas that has been super-cooled to −259 degrees Fahrenheit, which reduces its volume to 1/600. This allows it to be transported by ships and then returned to gas by the host countries.

The Fukushima Daiichi meltdown in Japan in March 2011 made countries re-evaluate their dependence upon nuclear energy. "Given the uncertain prospects of nuclear power in Japan, the political drive to clean up the Chinese energy system and the acute energy shortages in India, Asia is intensely looking for energy supplies," Maria van der Hoeven, executive director of the International Energy Agency, told a business audience in Calgary in August 2012.[2177]

The principal advantage to U.S. companies is that they can sell gas overseas at three to five times the price in the U.S. This would lead to an increase in domestic prices as the glut from overdrilling is reduced and the amount of gas that is economically feasible to mine is diminished. It would also significantly cut hope that natural gas consumption in the U.S. would reduce the nation's energy dependence on foreign oil.

Production of natural gas is expected to exceed consumption, rising from about 5.0 trillion cubic feet in 2010 to about 8 trillion cubic feet in 2020, according to estimates by the U.S. Energy Information Administration.[2178] This should lead to the U.S. becoming "a net exporter of [LNG] in 2016" to take advantage of lower gas prices in the U.S. while being able to command higher prices overseas, according to the EIA.

Projections for 2035, according to the EIA, are for 13.6 trillion cubic feet of production a year. A $5.6 billion expansion of the Panama Canal, scheduled for completion by the end of 2015, will cut transportation costs; super-tankers loading LNG in the Gulf or East Coast ports will reach Asian markets in about 25 days, down from a current average of 41 days.[2179]

In November 2011, Aubrey McClendon, at that time CEO of Chesapeake Energy, said he wanted "the right to export natural gas, but I am hopeful we never do." He said he hoped "in the next four years that we embrace natural gas for transportation so we don't need to export it outside the country."[2180] Two months later, Mike Stice, Chesapeake senior vice-president, declared, "Chesapeake Energy wants to export LNG."[2181]

American corporations are spending several billion dollars to move natural gas from domestic use to foreign consumption.

Cheniere Energy of Houston spent more than $12 billion[2182] to convert its Sabine Pass terminal in Louisiana from import to export. Cheniere and the BG Group signed a 20 year $8 billion agreement to export 3–5 million tons of LNG a year.[2183]

In Pascagoula, Miss., Kinder Morgan plans to convert its $1 billion import facility into an export facility, hoping to get federal permission to transport LNG to countries that don't have free trade agreements with the United States. The cost of conversion could be about $8 billion.[2184]

Sempra Energy's $6 billion export terminal in Louisiana will be completed in 2017. Sempra expects to ship about 1.7 billion cubic feet of LNG per day on 20-year contracts, with one-third of production going to Mitsubishi and one-third going to Mitsui, both Japanese-owned companies. Sempra believes it will earn

$300 million a year in after-tax profits.[2185]

Dominion Resources is budgeting about $3.8 billion to convert its 1,100 acre LNG terminal at Cove Point, Md., in a residential area, from import to export.[2186] The terminal, built in the 1970s to import gas from Algeria,[2187] is almost idle, processing only enough imported LNG to keep it from deteriorating; the conversion is expected to be completed in 2017. Dominion plans to deliver up to one billion cubic feet of LNG, primarily from the Marcellus Shale, to 20 nations, including India and Japan. Among the companies that plan to ship their shale gas output to Cove Point is Cabot Oil and Gas, which has a 20 year contract with Pacific Summit Energy of Japan to provide 350,000 million BTUs daily.[2188] The Cove Point terminal, if operating at maximum production, will emit about two million tons of greenhouse gases per year.[2189]

Freeport LNG of Houston has contracts with several overseas companies. From its Quintana Island, Texas, facility, Freeport will export up to 2.2 million tons of LNG per year to Toshiba of Japan and SK E&E of South Korea, 2.2 million tons per year to Osaka Gas and Chubu Electric Power, and 4.4 million tons per year for BP Energy. All exports are long-term contracts up to 20 years.[2190]

Sunoco Logistics Partners is spending about $2.5 billion for its Mariner East 2 pipeline to transport gas from the Marcellus fields to Philadelphia for processing, and then onto ships to Asia and Europe. The 350-mile pipeline will carry about 275,000 barrels (about 11.5 million gallons) of natural gas liquids per day.[2191]

Under the Natural Gas Act mandate, American companies may not export gas to countries that do not have free trade agreements with the U.S. unless the Department of Energy determines the exports are in the public interest. Most applications have been deemed to be in the public interest, including those from Cheniere, Freeport, and Lake Charles Exports, which is building a $2 billion facility in Louisiana to export up to two billion cubic feet of LNG a day to non-trade partners.[2192]

The oil and gas industry, shut out from fracking in New York, has been lobbying to build LNG facilities in that state. However, New York banned LNG facilities in 1973. Even proposed regulations that apply to truck fueling stations drew several hundred protestors to a public hearing in October 2013.[2193]

However, long-term profits from shipping LNG may not be as high as some experts and investors may believe. Gazprom, the Russian-owned world's largest gas supplier, has every reason to keep its own prices below what the American companies can charge. Russia can provide gas by pipeline to much of Europe and Asia. American companies have to charge higher rates because of the cost of overseas shipping. With prices of domestic-produced gas low because of overdrilling, the American-based companies can't show a profit domestically; the increased costs to build terminals, and to refine and send gas by ship could reduce long-term advantages of shipping LNG.

PROBLEMS WITH LNG

Peer-reviewed research studies reveal that spills and flaring make LNG about 30 percent worse than conventional natural gas in its effects upon climate warming.[2194] LNG is also both flammable and highly toxic. Transported by 1,000-foot long ships, LNG explodes if it spills into water.[2195] Terrorist or military attacks against LNG ships will immediately kill the crew and much aquatic life for at least three miles, and leave lasting pollution.[2196] The volatility on land is just as great. If spilled on the ground, it will form a cloud that can result in asphyxiation not only of nearby residents but also those below where the cloud may drift.

"Local officials and community groups have challenged numerous LNG infrastructure proposals on the grounds that they may represent an unacceptable risk to the public," according to an analysis of LNG problems by the non-partisan Congressional Research Service.[2197]

These problems, combined with the reality that the presence of large underground storage and LNG terminals, will extend the use of fossil fuel energy at the expense of renewable energy, have led to numerous public protests against awarding federal permits for LNG facilities.

More than 100 of the nation's leading physicians, scientists, and environmental engineers petitioned President Obama in

December 2012 to reconsider increased exporting LNG. The petition pointed out the "opening of LNG export facilities would serve to accelerate fracking in the United States in absence of sound scientific assessment, placing policy before health."[2198]

Dr. Seth B. Shonkoff, an environmental scientist and executive director of the group that had petitioned the President, asked:

> "Why would the United States dramatically increase the use of an energy extraction method without first ensuring that the trade-off is not the health of Americans in exchange for the energy demands of foreign nations? The only prudent thing to do here is to conduct the needed research first."[2199]

Even if the U.S. rethinks approvals for overseas distribution, natural gas producers may find that estimates of overseas purchases may not match optimistic predictions. China, India, and most European countries have been developing non-fossil green energy sources faster than the United States.

Once looking as a savior for the industry faced by numerous problems, including its own greed that led to overproduction, exporting LNG has become a less attractive option. Excelerate Energy stopped all progress on its application to establish and export LNG from a terminal in Lavaca Bay, Texas. The company had planned to ship as much as 5.5 million tons of LNG, beginning in 2017.[2200] In a notice to the Federal Energy Regulatory Commission, Excelerate declared the fall in crude oil process led it to a "strategic reconsideration of the economic value of the project."[2201] A Reuters News analysis of LNG export plans, partially based upon the Excelerate decision, "bodes ill for thirteen other U.S. LNG projects, which have also not signed up enough international buyers, to reach a final investment decision." Acording to Reuters:

> "Prices that LNG projects can charge for long-term supply are falling from historic highs as new producers crowd the market, which is already oversupplied due to slowing demand and rising output that has seen spot Asian LNG prices halve this year.
> "At the same time, major consumers from Japan to South Korea and China are seeking to offload some of their long-term LNG supply commitments, contributing to the glut."[2202]

Conclusion

The fracking boom is like the housing bubble. At first, the availability of mortgages looked like a boom. However, a combination of greedy investors and lending institutions with almost no governmental oversight, combined by a client base of ordinary people who were lured into buying houses with inflated prices they couldn't afford, led to the Great Recession. Those who didn't learn from the housing bubble guaranteed the fracking boom would become a fracking bubble.

Oil and gas prices are cyclical; but if the investors and industry plan to allow the market to "correct itself," the result will be an earlier bust than what should be expected in energy development and production.

The boom led to a bubble, which is leading to a bust.

In January 1901, Beaumont, Texas, had a population of about 9,000. Two months later, with oil discovered in what became known as the Spindletop Oil Boom, the population was 30,000. The discovery of oil led to a higher crime rate and extensive fraud by hucksters selling leases.

Pilots André Borschberg and Dr. Bertrand Piccard in Abu Dhabi, January 2015, two months before the first part of the around-the-world flight began.

PHOTOS: © 2015 Solar Impulse 2 project

Solar Impulse 2, with André Borschberg at the controls, about to touch down in Kalaeloa, Hawaii, on July 3, 2015.

PHOTO: Itamar Grinberg

The Arava Power Co. installed a 4.95 MW solar field at the Kibbutz
Ketura in southern Israel. The kibbutz is part of the Green Kibbutz
Movement, and houses the Arava Institute for Environmental Studies.

PHOTO: David DeKok

Centralia, Pa., shows the contrast between fossil fuel and renewable energy.
Beneath the village, an anthracite coal mine fire has been burning since 1962,
forcing more than 1,000 people to leave; the population is now 10. On hills above
the village is the Locus Ridge Wind Farm, built by Iberdrola Renewables.

CHAPTER 25
A Brief Look into the Future

The continued push for fossil fuel development, and more than $4 billion in governmental subsidies, slows the development of renewable energy, while escalating the problems associated with climate change and brings the world closer to a time when global warming is irreversible.

Scientific projections show that under the strictest regulation for air emission, surface temperature is likely to rise at least 0.5 degrees Fahrenheit in the 21st century; with minimal restrictions, the rise is expected to be between 4.7 and 8.5 degrees Fahrenheit.[2203] An increase in carbon dioxide emissions from 400 parts per million to 450–600 parts per million, which is possible under current projections, will make global warming irreversible, according to research published in the *Proceedings of the National Academy of Sciences*.[2204]

About 12.5 trillion watts (TW) of power is the maximum energy consumption at any moment, according to the U.S. Energy Information Administration. By 2030, the maximum power necessary will be about 16.9 TW.[2205] Fossil fuels will not be able to provide that coverage.

Most gas shales are now in decline. The Haynesville shale in Louisiana and Texas, which began producing fracked gas in 2008, peaked in 2012 as the nation's biggest shale gas play; by the end of 2014 it was producing less than half its production from two years earlier.[2206] To maintain the same production of natural gas that existed at the end of 2013, geologist David Hughes predicts that 7,000 new wells a year, at a cost of about $42 billion, is necessary; shale oil production will require 6,000 new wells a year, at a cost of about $35 billion.[2207] His data analysis of 65,000 wells concludes:

"[P]rojections by pundits and some government agencies that [horizontal fracking] can provide endless growth heralding a new era of 'energy independence,' in which the U.S. will become a substantial net exporter of energy, are entirely unwarranted based on the fundamentals. At the end of the day fossil fuels are finite and these exuberant forecasts will prove to be extremely difficult or impossible to achieve."[2208]

The Marcellus Shale boom may last only until 2016–2017 before declining and Big Energy eventually abandons it for the more lucrative oil-producing shales or, if it can develop better technology, the Utica shale and the "wet gas" it can produce. Only five shales produce 80 percent of all natural gas.[2209]

The oil and gas industry is facing an even greater burden from a coal industry that has no plans to fade away. Coal generates about half of all electricity in the U.S.[2210] Between 2000 and 2010, coal as a source of electricity rose more than 56 percent throughout the world, with a prediction of an increase of 40 percent between 2010 and 2020.[2211]

Helping to keep coal as a major energy source is the development of clean coal technology, which reduces but doesn't eliminate the emission of nitrogen oxide, carbon, sulfur dioxides, and other particulates and gases that had once made coal the dirtiest of all energy sources and a major contributor to ozone depletion and the destruction of the environment. Even the use of CCS (carbon capture and sequestration), which isolates carbon emissions and buries them, is still in research, rather than full production.[2212] The use of coal as a source of energy is almost as great as the use of gas, nuclear, water, oil, solar, and wind power combined. However, oil, coal, and gas resources are finite—no dinosaurs are left to willingly sacrifice their lives to produce fossil fuels—and alternative non-fossil fuel sources will increase during the next two decades. No matter how clean coal, oil, or gas energy become, the processes to mine and refine them are dirty, and workers are exploited.

The Sierra Club argues that the country needs "to leapfrog over gas whenever possible in favor of truly clean energy [wind, solar, water]. Instead of rushing to see how quickly we can extract natural gas, we should be focusing on how to be sure we are using less—and safeguarding our health and environment in the meantime."[2213] Millions throughout the world agree.

Renewable Energy

The Eco-Justice Programs of National Council of Churches, in March 2012, recognized that the earth can no longer sustain itself on fossil fuel development:

> "We have the technology to produce energy in renewable or sustainable ways, using God's gifts of sun, wind, and water. Shifting away from an energy economy based on scarce fossil fuels to more renewable and site-specific energy technologies allows us to embrace an economy of abundance. Renewable energy sources such as solar, wind, geothermal, and hydro-electric may also have negative impacts on God's Creation; however, renewables generally produce less pollution than tradition carbon-based energy sources and, unlike more traditional sources of energy, they will be available as long as Earth endures. . . .
>
> "Renewable energy offers the potential for local energy development. In the U.S. this has meant new economic opportunities for struggling rural communities, which are often the best locations for solar and wind farms. In less developed countries renewable energy has the potential to allow off-the-grid and mini-grid systems in rural communities to support necessities such as irrigation and food preservation, as well as aiding in the development of a robust economy."[2214]

Cyril L. Scott, the Rosebud Sioux tribal president, affirming the beliefs of the Great Sioux Nation and most Native American cultures, declared:

> "The Lakota people have always been stewards of this land. We feel it is imperative that we provide safe and responsible alternative energy resources not only to Tribal members but to non-Tribal members as well. . . . We need to start remembering that the earth is our mother and stop polluting her and start taking steps to preserve the land, water, and our grandchildren's future."[2215]

The mainstream media, after first ignoring alternative energy, and then casting doubts upon the use of renewable energy as prime energy sources, have slowly accepted the reality that renewable energy may be part of the solution to reducing global warming, while weaning the United States off fossil fuel. The *Detroit Free Press* pointed out:

> "Wind energy and recycling hold as much promise, if not more, to create new investment and jobs over several decades as does drilling for fossil fuels. Also, an unprecedented construction boom to upgrade or replace coal-fired power plants means billions of new investment and thousands of construction jobs." [Oct. 12, 2014][2216]

In more than 30 nations, renewable energy accounts for at least 20 percent of energy production, with the European Union countries planning for a 20 percent use by 2020.[2217] Among the countries in which renewable energy accounts for at least half of energy production, according to data compiled by the Renewable Energy Policy Network, are Iceland (100 percent), the Congo (100 percent), Mozambique (100 percent), Zambia (100 Percent), Malawi (98 percent), Norway (97 percent), Costa Rico (91 percent), Brazil (89 percent), Cameroon (88 percent), Ethiopia (84 percent), Colombia (80 percent), Austria (75 percent), Venezuela (73 percent), Somalia (69 percent), New Zealand (65 percent), Hondourus (65 percent), Canada (63 percent), El Salvador (63 percent), Sudan (63 percent), Peru (57 percent), Switzerland (57 percent), and Sweden (54 percent).[2218]

There are several kinds of renewable energy:

• Bioenergy is energy produced from plants and recently deceased plants and animals. Examples include timber, charcoal, and wood chips, which produce heat. Biodiesel fuels to power motors and engines can be produced from vegetable oils and recycled grease. Corn and sugar cane can also be used to create fuels for transportation.

• Hydropower energy is generated by dams and from the oceans. Three Gorges Dam on the Yangtze River in Hubei Province, China, is the world's largest dam, generating almost 100 terrawatt hours (TWH) in 2014.[2219]

• Geothermal energy is the conversion of both the radioactive decay of minerals and the energy from the earth's core. Examples include natural hot springs.

The development of solar and wind energy is proving to be the most popular methods to wean the planet from fossil fuels.

SOLAR ENERGY

Tokelau, an independent territory of New Zealand, is a small three island archipelago of about 1,400 residents about 300 miles north of American Samoa in the South Pacific. In October 2012, it turned off the last of its diesel generators and became the first country to use solar power as its only energy source.

New Zealand loaned Tokelau $7 million (about $5.8 million U.S. dollar equivalent) to purchase and install the 4,032 solar panels, 392 inverters, and 1,344 batteries.[2220] The territory had been spending about $1 million a year for diesel fuel.

"We know that the longer the fossil fuel industry gets its way, the worse climate change will be, and the more sea-level rise will threaten our islands," wrote Mikalele Maiava, one of its residents.[2221]

On March 2, 2013, residents of 15 Pacific island nations called for an end of fossil fuel energy. Maiava explained why:

> "[We are] laying down a challenge to the fossil fuel industry. It is their coal and oil and gas vs. our future. They cannot both coexist. And it is our future that has to win.
>
> "In this moment, and in the years to come, we need you to walk beside us. Because we live far away from the mines and power plants that threaten our future, we need the world's solidarity. . . . [S]tand with us during this weekend of Pacific Warrior climate action!
>
> "We want to show the world that people from countries and cultures everywhere are standing with us—the Pacific Warriors—in the fight against climate change."[2222]

About 9,300 miles west of the Polynesian nation of Tokelau is Saudi Arabia. Beneath its sands is an estimated 262.6 billion barrels of oil, the largest reserve in the world.[2223] But, Saudi Arabia isn't counting on oil as its sole form of energy. By 2032, Saudi Arabia plans to have 41,000 megawatts of solar energy, enough to generate about one-third of the nation's electricity. Nuclear, wind, and geothermal energy would provide an additional 21,000 megawatts.[2224] The project is expected to cost about $109 billion,[2225] and save about 523,000 barrels of oil a day.[2226]

On Saudi Arabia's northern border is the United Arab Emirates (UAE). Beneath its sands is the world's seventh largest oil reserve.[2227] About 75 miles south of Abu Dhabi, the capital of the seven principalities that make up the UAE, are 285,000 parabolic mirrors that generate 100 megawatts of energy, enough to meet the energy needs of 20,000 homes.[2228] The solar power plant is expected to reduce carbon dioxide emissions by about 175,000 tons a year.[2229]

By 2020, almost half of Morocco's electricity will come from wind, hydroelectric, and solar energy sources.[2230] The 500,000 mirror Noor Solar Energy field, which will be the world's largest concentrated solar panel field,[2231] is expected to produce 500 MW of energy by 2018, enough to provide electricity to about one million homes.

China and Pakistan are teaming up to build a $46 billion solar project in the Cholistan Desert, to be completed by the end of 2017, that will produce about 1,000 MW of electricity, enough to power about 320,000 homes.[2232]

Arava Power built Israel's first solar energy plant in 2011, providing energy for the Kibbutz Ketura; the profits are donated to charity.[2233]

BrightSource Energy is building the Ashalim plant, also in the Negev Desert, that will generate solar energy for about 40,000 homes by the end of 2016.[2234] By 2020, Israel plans to generate about 10 percent of all its power from solar energy.

India, a world leader in wind-generated power,[2235] is likely to develop what will become the world's largest solar energy system. The first phase, being developed on a 23,000 acre site,[2236] is expected to be completed by 2016, and will provide about one gigawatt of power.[2237] The 30-year plan will allow India to provide four gigabytes of electricity, about one-eighth of its energy needs, through solar power.[2238]

The Jasper Solar Plant in South Africa, generating 180,000 megawatts of power from its 10,000 heliostat mirrors, is Africa's largest solar plant. Completed in 2014, Jasper can provide energy for about 80,000 homes.[2239]

The 3,500 pound Solar Impulse 2, powered entirely by 17,000 solar cells on its 208-foot wing, with Swiss pilots Andre

Borschberg and Dr. Bertrand Piccard alternating in the cockpit, completed about half of a 13-leg 22,000 mile around-the-world trip in August 2015. During that time, it stayed aloft 117 hours, 52 minutes while crossing the Pacific from Japan to Hawaii.[2240] The rest of the flight resumes in late Spring 2016. By May 2012, Germany's solar energy plants produced as much power as 20 nuclear power plants;[2241] by 2050, Germany plans to have about 60 percent of its energy provided by renewables, and reduce greenhouse emissions by as much as 95 percent.

In the United States, solar energy projects are expanding; both employment and taxes paid by solar corporations to local and state governments are increasing. The presumed advantages of fracking, including job growth, are daily being muted by the realities of renewable energy beginning to overtake fossil fuel energy.

About 600,000 homes and businesses were powered by solar energy by the end of 2014, according to the Solar Energy Industries Association.[2242] By the end of 2016, solar energy in the United States is expected to be the primary energy source for about one million homes, and is expected to replace about 50 million tons of harmful carbon emissions.[2243] Solar energy at that time is expected to be at least as cheap as fossil fuel and nuclear energy in 47 states, assuming the United States maintains a 30 percent tax credit, according to an analysis by the Deutsche Bank.[2244] The average low cost of solar energy was 5.6 cents a kilowatt hour by the end of 2014, compared to natural gas and coal at 6.6 cents, according to a study produced by Lazard, one of the world's leading financial asset management firms.[2245]

For the consumer, solar panel costs plunged more than 50 percent since 2008, and almost 99 percent less since they were introduced in the 1970s.[2246] With a 30 percent federal tax credit per home system, and local and state subsidies, the cost to purchase panels and have professionals install them can be about $10,000 for the average house that uses about 3 kilowatts per year;[2247] savings exceed costs within about five years on average, according to the U.S. Department of Energy.[2248]

Between 2009 and the end of 2015, employment in the solar industry increased by 90 percent, to about 210,000 jobs,[2249] 25,000 more than construction jobs in the fossil fuel industry.[2250]

In late 2014, the U.S. began a six year program to train 50,000 American veterans for jobs in the solar industry.[2251] President Obama had taken the initiative to establish the training at military bases after Republican members of Congress blocked efforts to create realistic job programs, and clung to fears that renewable energy would crowd out fossil fuel development. By executive order, the President directed all federal agencies to reduce greenhouse gas emissions by 40 percent during the decade beginning in 2015, and to work together to increase renewable energy. In May 2015, the Federal Aggregated Solar Procurement Project established a common purchasing program for nine federal sites in northern California. When fully implemented, the program will affect about a half-million buildings that annually use about 57 billion kilowatt hours of electricity; the use of renewable energy and a common purchasing agreement, will reduce costs by about $1 billion from the current $5 billion a year, according to the EPA.[2252] In addition to linking federal agencies, the President also announced the federal government was providing $68 million for 540 renewable energy projects, 240 of them solar projects, in rural areas.[2253]

Between 1984 and 1991, Luz International, an Israeli company, built nine solar plants in California's Mojave Desert before reorganizing as BrightSource. Its more recent projects are at Hidden Hills in southeastern California, which provides 500 megawatts of energy to about 178,000 homes;[2254] and the Ivanpah Dry Lake project, which provides about 377 megawatts of energy to about 140,000 homes.[2255]

In Gila Bend, Ariz., about 70 miles southwest of Phoenix, on a three square mile farm is the world's largest parabolic trough plant. Each of the 3,200 mirrors is 500 feet long, 25 feet wide, and 10 feet high, generating 280 megawatts of power to steam turbines, enough to provide electricity to 70,000 houses.[2256] The plant was built by Abengoa, a Spanish corporation, with a $1.45 billion loan from the Department of Energy,[2257] and went online in Summer 2013. Abengoa reports it employed more than 2,000 persons at the peak of construction, and purchased more than $966 million in supplies and components from 165 companies, most of them from the United States, over a three year period. The company guarantees loan repayment and $420 million in taxes over a three decade period.[2258]

In the Carrizzo Plains in central California, the $2.5 billion Topaz Solar Farm went online at the end of 2014. The 9.5 square mile project, with nine million solar panels, can provide as much as 550 megawatts of energy, enough to power about 160,000 homes. Just as important is that by not producing greenhouse gases, it keeps 377,000 tons of carbon dioxide, which would be produced by fossil fuel energy, from being released into the atmosphere. The project provided $192 million in wages for about 400 construction jobs over a three-year period, $52 million income for local suppliers, and about $400,000 a year in new tax income, according to project contractor First Solar.[2259]

In February 2015, the Desert Sunlight solar project in southern California's Riverside County, east of Los Angeles, became the world's largest solar farm, producing enough solar energy to power about 160,000 homes. That project was partially funded by a $1.5 billion loan guarantee from the Department of Energy; the federal government expects a profit of $5–6 billion.[2260] Also in Riverside County, Bright-Source and Abengoa teamed up to develop the Palen solar thermal power system; when completed in 2016, it will provide 500 megawatts of power for about 200,000 homes.[2261] Nearby is the $1.1 billion Blythe Solar Power Project, which will provide about 1,000 MW of energy from a 6,000 acre site when the four adjacent plants are completed in 2018.[2262]

By 2016, MidAmerican Solar's Solar Star plant in central California is expected to eclipse both the 550 megawatt Desert Solar project and 500 megawatt BrightSource projects to become the world's largest solar operation, producing 579 megawatts of energy. The three-year project, according to MidAmerican, will require about 650 construction jobs and will reduce carbon dioxide emissions by about 570,000 tons a year.[2263]

In Nevada, Crescent Dunes is a 110 megawatt 360,000 mirror solar energy plant that can power 75,000 houses.[2264] The plant holds about 70 million pounds of molten salt at 550 degrees Fahrenheit that will store solar energy to be used any time of the day or night, says Kevin Smith, a mechanical engineer and CEO of SolarReserve, the company that built and is maintaining the operation. SolarReserve has built plants capable of producing more than 5,000 megawatts of renewable energy using solar thermal and photovoltaic technology.

Near South Buffalo, N.Y., SolarCity is building the largest solar panel manufacturing plant in the Western Hemisphere. SolarCity plans to hire about 1,000 employees[2265] to construct the 2.5 million square foot plant, scheduled for full production in 2017. The company predicts it will employ about 5,000 persons in manufacturing, sales, and support services.[2266]

Lancaster, Calif., a city of about 160,000 in Los Angeles County, was the first American city to require all new houses to produce a minimum of one kilowatt of solar energy, beginning Jan. 1, 2014. With the development of solar energy, other cities may follow Lancaster's example. Even if no other city requires new houses to have a minimal amount of renewable energy, homeowners themselves are beginning to realize the benefits of solar energy, which can be considerably less expensive than fossil fuel energy and significantly less of a problem to the environment and health than fossil fuel energy.

Hundreds of schools, government agencies, hospitals, and companies—from "mom-and-pop" operations to major corporations—are increasing their use of solar energy. The three main campuses of the Community College of Baltimore County, Md., installed 16,500 solar panels on the roofs of carports that protect about 1,400 parking spaces. The panels are generating about 6.5 million KW hours of energy every year.[2267]

By the end of 2014, the United States was generating almost 22.4 million gigawatts of solar energy,[2268] enough to power about 4.5 million homes. That amount is expected to increase significantly as more solar farms are built, as more Americans take advantage of a 30 percent tax credit before it expires at the end of 2016—and as science develops more efficient ways to capture solar energy. One of those ways is the use of a bionic artificial leaf. Drs. Joseph Torella and Christopher Gagliardi developed a procedure to split hydrogen and oxygen in artificial leaves; they then added an organic product that changes carbon dioxide and the resulting hydrogen into isopropanol, essentially converting sunlight into a liquid form.[2269]

Frank Shuman, an inventor from Philadelphia, produced a

solar motor in 1907. Four years later, in the Sept. 30, 1911, issue of *Scientific American*, Shuman predicted:

> "Where great natural water powers exist, sun power cannot compete; but sun-power generators will, in the near future, displace all other forms of mechanical power over at least 10 per cent of the earth's land surface; and in the far distant future, natural fuels having been exhausted it will remain as the only means of existence of the human race."[2270]

Within two years, Shuman and a team of construction engineers built the world's first solar generating plant at the edge of the Nile, using solar energy to pump 6,000 gallons of water a minute into adjacent cotton fields.[2271]

The development of solar energy, blocked by the fossil fuel industry for more than a century, could eventually become the primary energy force Frank Shuman had believed. By 2050, solar energy will become the world's leading source of electricity, according to projections by the International Energy Agency.[2272]

WIND ENERGY

For more than seven millennia, wind energy was the primary source of energy for ships and for producing grains. However, in the past century, the fossil fuel industry, aided by numerous politicians who have taken campaign funds, has been vigorous in casting doubt upon the emergence of wind energy as a viable method of generating energy. The attacks have been that wind energy causes health problems, damages the environment, and lowers property values. These pseudo-science claims are little more than a smokescreen to obscure the attacks made upon the fossil fuel industry; they are countered by numerous research studies that show wind energy is an effective method to produce clean, renewable, energy, with no significant impact upon health or the environment.[2273] Even the scare-mongering claims that wind energy kills birds and bats is muted by the reality that high-tension wires, cats, pesticides, vehicle impacts, hunters, impacts with windows, and oil and waste-water spills kill about 1.5–2.0 billion birds and bats a year, about 10,000 times more than wind turbines.[2274] Equally important, fossil fuel extraction and global warming have a greater effect upon bird/bat mortality than all turbines.[2275]

China has a maximum capacity of 115,000 megawatts of wind energy, more than all nuclear energy in the United States. It plans to increase maximum wind power under ideal conditions to 200,000 megawatts by 2020.[2276]

Denmark, with the addition of 3.6 megawatts of offshore wind turbines in early 2013, has more than a gigawatt of wind energy, providing about one-third of all power to its citizens.[2277] Denmark plans to increase wind power capacity to provide about one-half of all domestic energy needs by 2021.[2278]

Wind power in Spain barely edged out nuclear power as the main source of energy in 2013. Wind energy accounted for about 20.9 percent of all power, according to Red Eléctrica de España (REE).[2279] Nuclear energy accounted for 20.8 percent, followed by coal energy (14.6 percent), hydroelectric power (14.4 percent), and solar energy (3.1 percent). Wind power, according to REE, produced about 54.5 GWh of energy.

By the end of 2014, renewable energy—primarily wind energy, but also including solar and hydroenergy—produced about one-third more electricity in Scotland than any other source of energy, according to the United Kingdom Department of Energy and Climate Change.[2280]

By the end of 2012, there were 1,662 off-shore wind turbines, generating enough electricity to power 10 million homes in 10 European countries, according to the European Wind Energy Association (EWEA). The EWEA believes that by 2020, off-shore wind turbines could provide electricity to about 39 million homes if countries continued to push for wind energy and rely less upon fossil fuel energy.[2281] With political and economic support, according to an EWEA analysis, "The energy produced from turbines in deep waters in the North Sea alone could meet the EU's electricity consumption four times over."[2282]

Among hundreds of major corporations that have begun using renewable energy is Ikea, which invested about $1.7 billion to provide non-fossil fuel energy for its buildings. That investment includes 314 offsite wind turbines and about 700,000 solar panels.[2283] Both Ikea and its non-profit foundation have also donated about $1.13 billion to assist Third

World countries to transition from fossil fuels to renewable energy. About half of the funding is for wind energy.[2284]

Wind power in 36 states provided about 4.5 percent of electricity generation in the United States in 2014.[2285] The largest wind farm in the United States is the 300 turbine Alta Wind Energy Center in central California.[2286] The plant, owned by Terra-Gen Power, can produce as much as 3,000 MW of power, while reducing carbon emissions by more than 5.7 tons;[2287] Terra-Gen says the construction and operation of the farm will provide more than 3,000 jobs.[2288]

Wind turbines provide about one-fifth of all power in Iowa and South Dakota,[2289] and almost one-fourth of all power in Kansas.[2290]

Off the coast of Castline Harbor, Maine, the nation's first off-shore wind turbine began producing power in June 2013. The 65-foot tall three-blade floating turbine was conceived and built by a partnership of University of Maine scientists, staff at the Maine Maritime Academy, and the Cianbro Corp.[2291]

Burlington, Vt., in September 2014 became the largest U.S. city to receive all of its power from renewable energy. The 42,000-population city receives its energy from wind, water, and biomass; the biomass energy comes from a generating station that uses residue wood chips.[2292]

Kodiak Island, in southern Alaska, went almost 100 percent wind turbine energy in 2015. The 15,000-population island had been burning almost three million gallons of diesel a year. "Stable electricity rates have . . . brought in more construction, expanded the fishing industry, and brought in more jobs and tax revenue," said Darron Scott, CEO of the Kodiak Electric Association.[2293]

Construction of a 102 turbine wind farm on 22,000 acres near Elizabeth City, N.C., will provide about 204 MW of energy, enough to power about 60,000 homes, according to developer Iberdrola Renewables of Spain.[2294] The project, which began in August 2015, will use towers as high as 460 feet, about 200 feet taller than most turbines. Iberdrola Resources is also building a 38-turbine wind farm, capable of producing 76 MW of energy on 11,000 acres near Lamar, Colo. When completed in 2017, the farm will will provide enough wind energy

to power about 30,000 homes.[2295] The company is also producing wind energy in several other states.[2296]

An extensive analysis by Andrew Menaquale for Oceana, the
world's largest organization devoted to studying and preservation of ocean life, suggests a "modest and gradual development of offshore wind on the East Coast could generate up to
143 gigawatts of power [by 2035], which is enough to power
over 115 million households." Menaquale's analysis also
suggests that wind power development could create about
91,000 more jobs than offshore oil drilling, and provide more
energy than all possible recoverable oil and gas along the
Eastern shore.[2297]

A Need for a More Aggressive Energy Policy

Renewable energy sources accounted for 49.81 percent of
new domestic energy capacity in 2014, slightly higher than the
48.65 percent for natural gas, according to the Federal Energy
Regulatory Commission. Overall, renewable energy was 16.63
percent of all energy sources.[2298]

About three-fourths of Americans want to see more development of solar energy, according to a Gallup poll conducted in
March 2013. That same poll also revealed that Americans want
to see more emphasis on wind energy (71 percent), natural gas
energy (65 percent), oil (46 percent), nuclear power (37 percent)
and coal (31 percent). However, that same poll revealed that
persons who consider themselves Democrats or Independents
support renewable energy in greater numbers than do Republicans who as a group tend to support fossil fuel energy. About
78 percent of Republicans support placing more emphasis on
natural gas exploration, extraction, and development, significantly higher than do the Democrats (59 percent) and Independents (62 percent).[2299]

"[W]e have wasted a lot of time that should have gone into
seriously looking into and developing alternative energies,"
Louis W. Allstadt told *TruthOut*'s Ellen Cantarow in July 2013.
Allstadt, a former executive vice-president of Mobil Oil who
had direct supervision over exploration and production in the
western hemisphere before retiring in 2000, argues:

"[W]e need to stop wasting that time and get going on it. But the difficult part is that the industry talks about, well, this is a bridge fuel [that] will carry us until alternatives [are developed] but nobody is building them. It's not a bridge unless you build the foundations for a bridge on the other side, and nobody's building it."[2300]

George P. Mitchell, whose company had developed horizontal fracking, believed that extraction of natural gas had a place in the world's energy consumption. But, in an exclusive interview with the *Economist* slightly more than a year before his death, he also believed:

"We need to ensure that the vast renewable resources in the United States are also part of the clean energy future, especially since natural gas and renewables are such great partners to jointly fuel our power production. Energy efficiency is also a critical part of the overall energy strategy that our nation needs to adopt."[2301]

Two American scientists believe renewable energy could provide all of the planet's needs by 2030 if world leaders and corporations were willing to make the investment. The plan would eventually cost less per year than fossil fuel energy, take up less land than currently used for oil and gas drilling, and would release no carbon emissions into the atmosphere, say Drs. Mark Z. Jacobson, professor of civil and environmental engineering at Stanford University, and Mark A. Delucchi, research scientist at the University of California at Davis.

"A combination of wind, concentrated solar, geothermal, photovoltaics, tidal, wave and hydroelectric energy could more than meet all the planet's energy needs, particularly if all the world's vehicles could be run on electric batteries and hydrogen fuel cells," says Dr. Jacobson.[2302] The world consumption would be lowered to about 11.5 TW, they say, with a reduction to 1.8 TW from 2.8 trillion watts in the United States. The reduction in required usage is possible, they say, because eliminating fossil fuel dependence increases energy efficiency.

They rejected use of nuclear energy because:

"[I]t's not carbon-free, no matter what the advocates tell you. Vast amounts of fossil fuels must be burned to mine, transport and enrich uranium and to build the nuclear plant. And all that dirty power will be released during the 10 to 19 years that it takes to plan and build a nuclear plant."[2303]

They call for 3.8 million wind turbines and 720,000 wave converters, which would account for about half of the earth's energy needs; the turbines would take up about one percent of the earth's land surface, with the space between turbines "used for agriculture or ranching or as open land or ocean." The rest of the energy requirements would be provided by 90,000 solar plants and rooftop photovoltaic systems; 490,000 tidal turbines, 5,350 geothermal plants, 900 hydro-electric plants, and 720,000 wave converters. The cost, say Drs. Jacobson and Delucchi, would be about $100 trillion over a 20 year period.[2304] Within that cost would be increased employment. The obstacles, they say, "are primarily political, not technical."[2305]

The National Renewable Energy Laboratory, a part of the Department of Energy, is less optimistic. But it does suggest, "Renewable electricity generation from technologies that are commercially available today, in combination with a more flexible electric system, is more than adequate to supply 80% of total U.S. electricity generation in 2050 while meeting electricity demand on an hourly basis in every region of the country."[2306]

A significant increase of non-fossil fuel energy would produce a cleaner fuel source, reduce workplace accidents and substantially negate the health and environmental effects of fossil fuel energy; it would also increase employment, one of the basic reasons why politicians say they like natural gas drilling.

Governments, which have helped subsidize fossil fuel energy development for more than two centuries, will have to change their focus. Worldwide, governments contributed about $523 billion to the fossil fuel industry in 2011, but only $88 billion to renewable energy development, according to data from the International Energy Agency.[2307] A decrease in fossil fuel subsidies with a corresponding increase in renewable fuels subsidies will show that governments are willing to put aside the politics of energy and focus upon the health, environment, and safety of the people.

Conclusion

On Nov. 6, 2015, with about 14 months left in his presidency, Barack Obama rejected the construction of the northern segment of the Keystone XL pipeline. In a seven minute speech, the President stated the pipeline "would not serve the national interest of the United States [because it] would not make a meaningful long-term contribution to our economy . . . would not lower gas prices for American consumers [and that] shipping dirtier crude oil into our country would not increase America's energy security." Permitting the pipeline, said the President, "would have undercut" the country's leadership in reducing climate change.[2308]

The President's decision, after a seven year review by the Department of State, infuriated conservatives, Congressional Democrats from the gas-producing shale states, the oil/gas industry, Chambers of Commerce, and corporations. House Speaker Paul Ryan (R-Wisc.), said the President's decision "isn't surprising, but it is sickening."[2309] About 56–65 percent of Americans supported construction of the pipeline, according to several independent national polls.[2310]

Liberals and environmentalists praised the decision. Rejecting the pipeline gave the President **"new stature as an environmental leader," said Bill McKibben, one of the leading opponents of fracking and the pipeline.**

Financial analyst Deborah Rogers told the *Village Voice:*

> "After a decade of fracking, we're beginning to be able to show that, without a doubt, this was simply a very well-orchestrated public relations campaign. . . . There is gas there, but is there as much as they said? No. Are we gonna see the economic stability they promised? The answer is no." . . .
>
> "Shale gas was supposed to be this economic powerhouse for the next 40 years, they said. It didn't even work out in the past seven. And it's the same story in every other state. Unfortunately, that's just how the game is played."[2311]

By September 2014, about 51 percent of Americans opposed increased use of horizontal fracking, according to a poll conducted by the Pew Research Center. More important, about two-thirds of scientists who were members of the American Association for the Advancement of Science opposed continued use of horizontal fracking.[2312]

When the history of natural gas exploration is finally written, the story will be that for a few years it was a cheaper, cleaner-burning energy source than coal or oil, that it temporarily helped some people in rural areas, and brought with it some well-paying jobs.

But history will also record that natural gas and oil drilling using hydraulic horizontal fracturing contributed to global warming. It will also record that, as with every other energy industry, it will have taken everything it could from the earth, left rusting and deteriorating pipelines and cement casings, thousands of abandoned wells dotting the landscape, with many of the wells possibly leaking methane and wastewater. History will also record the lure of immediate gratification led politicians to heed the call of the sirens of oil and gas industry lobbyists and willingly accept political donations that led them to sacrifice their citizens' health and the environment.

PHOTO: Scott Cannon/Gas Drilling Awareness Coalition

Members of the Gas Drilling Awareness Coalition and the Damascus Citizens for Sustainability are part of the People's Climate Change Rally in New York City; Sept 21, 2014.

ENDNOTES

[1] http://www.scientificamerican.com/article.cfm?id=safety-first-fracking-second

[2] http://www.eia.gov/forecasts/aeo/pdf/0383er(2011).pdf

[3] http://www.eia.gov/analysis/studies/worldshalegas/

[4] http://www.nofrackingway.us/2014/09/04/popping-the-shale-gas-bubble/

[5] http://truth-out.org/news/item/28406-russia-blamed-us-taxpayers-on-the-hook-as-fracking-boom-collapses

[6] http://www.netl.doe.gov/File%20Library/Research/Oil-Gas/shale-gas-primer-update-2013.pdf

[7] http://www.energyfromshale.org/hydraulic-fracturing/marcellus-shale-gas

[8] http://www.postcarbon.org/reports/DBD-report-FINAL.pdf

[9] http://www.netl.doe.gov/File%20Library/Research/Oil-Gas/shale-gas-primer-update-2013.pdf

[10] http://nyshalegasnow.blogspot.com/2010/11/contentious-bedrock-photos-of-marcellus.html

[11] http://www.portal.gov.on.ca/drinkingwater/stel01_049392.pdf

[12] http://www.njgeology.org/enviroed/newsletter/v2n1.pdf

[13] http://www.nps.gov/frhi/parkmgmt/upload/GRD-M-Shale_12-11-2008_high_res.pdf

[14] http://www.netl.doe.gov/File%20Library/Research/Oil-Gas/shale-gas-primer-update-2013.pdf

[15] http://www.eia.gov/forecasts/aeo/er/

[16] http://www.growthstockwire.com/3148/Shocking-Gov-t-Statistics-Every-Investor-Should-See

[17] https://www.eia.gov/tools/faqs/faq.cfm?id=427&t=3

[18] http://www.cantonrep.com/article/20140602/News/140609939

[19] http://www.netl.doe.gov/File%20Library/Research/Oil-Gas/shale-gas-primer-update-2013.pdf

[20] http://prezi.com/k6khssankkmr/copy-of-untitled-prezi/

[21] http://www.ewg.org/analysis/usgs-recent-earthquakes-almost-certainly-manmade

[22] http://yosemite.epa.gov/sab/sabproduct.nsf/c91996cd39a82f648525742400690127/d3483ab445ae61418525775900603e79/$file/draft+plan+to+study+the+potential+impacts+of+hydraulic+fracturing+on+drinking+water+resources-february+2011.pdf

[23] http://www.netl.doe.gov/File%20Library/Research/Oil-Gas/shale-gas-primer-update-2013.pdf

[24] http://www.chk.com/investors/documents/latest_ir_presentation.pdf

[25] http://www.forbes.com/sites/christopherhelman/2013/07/27/father-of-the-fracking-boom-dies-george-mitchell-urged-greater-regulation-of-drilling/

[26] http://www.usnews.com/science/news/articles/2012/09/23/decades-of-federal-dollars-helped-fuel-gas-boom

[27] http://thebreakthrough.org/archive/interview_with_dan_steward_for

[28] http://bigstory.ap.org/article/decades-federal-dollars-helped-fuel-gas-boom

[29] *Ibid.*

[30] http://www.economist.com/blogs/schumpeter/2013/08/interview-george-mitchell

[31] http://extremeenergy.org/about/what-is-extreme-energy-2/

[32] http://thinkprogress.org/climate/2013/10/26/2841841/15-million-americans-live-near-fracking/

[33] http://www.netl.doe.gov/File%20Library/Research/Oil-Gas/shale-gas-primer-update-2013.pdf

[34] http://www.aga.org/Kc/analyses-and-statistics/statistics/annualstats/appliance/Documents/Table10-4.pdf

[35] http://en.wikipedia.org/wiki/Natural_gas#Domestic_use

[36] http://www.businesswire.com/news/home/20120718005840/en/GE-Researchers-Developing-At-Home-Refueling-Station-NG

[37]http://www.popularmechanics.com/cars/how-to/maintenance/should-you-convert-your-car-to-natural-gas

[38]http://www.sciencemag.org/content/343/6172/733.summary?sid=715a8fd8-f722-471b-b729-3e2c823086a9

[39]http://www.nytimes.com/natural-gas.html?_r=0

[40]http://energy.nationaljournal.com/2010/09/natural-gas-a-fracking-mess.php

[41]http://bigstory.ap.org/article/israel-faces-geopolitical-tangle-natural-gas

[42]http://bigstory.ap.org/article/calif-coastal-panel-takes-offshore-fracking

[43]http://blogs.seattletimes.com/politicsnorthwest/2014/07/31/inslee-signs-letter-opposing-coastal-oil-exploration/

[44]http://www.cnn.com/2015/05/22/us/california-oil-spill/

[45]http://www.cnn.com/2015/05/20/us/california-oil-spill/

[46]http://www.bbc.com/news/world-us-canada-32821384

[47]http://money.cnn.com/2015/05/21/news/california-pipeline-spill-plains/

[48]http://www.nytimes.com/2011/02/27/us/27gas.html?_r=3&pagewanted=all&

[49]http://www.netl.doe.gov/File%20Library/Research/Oil-Gas/shale-gas-primer-update-2013.pdf

[50]http://nyagainstfracking.org/no-compromise-on-an-independent-comprehensive-health-impact-assessment/

[51]http://energy.nationaljournal.com/2010/09/natural-gas-a-fracking-mess.php

[52]http://www.dnr.state.mn.us/lands_minerals/silicasand.html

[53]http://www.acs.org/content/acs/en/pressroom/newsreleases/2014/august/a-new-look-at-whats-in-fracking-fluids-raises-red-flags.html

[54]http://democrats.energycommerce.house.gov/sites/default/files/documents/Hydraulic Fracturing Report 4.18.11.pdf

[55]http://www.dec.ny.gov/data/dmn/rdsgeisfull0911.pdf

[56]http://www.truth-out.org/news/item/20039-fracking-unfocus-how-the-epas-long-awaited-hydraulic-fracturing-study-could-miss-the-mark

http://www2.epa.gov/sites/production/files/2015-03/documents/fracfocus_analysis_report_and_appendices_final_032015_508_0.pdf

[58]http://www.halliburton.com/en-US/ps/stimulation/fracturing/cleanstim-hydraulic-fracturing-fluid-system.page

[59]http://www.damascuscitizensforsustainability.org/wp-content/uploads/2012/11/PSECementFailureCausesRateAnalysisIngraffea.pdf

[60]http://www.texastribune.org/2013/05/24/texas-railroad-commission-adopts-well-construction/

[61]https://www.onepetro.org/journal-paper/SPE-106817-PA

[62]http://www.watertowndailytimes.com/article/20111213/OPINION02/712139975

[63]http://www.halliburton.com/ps/default.aspx?pageid=4272

[64]http://catskillcitizens.org/learnmore/PSECementFailureCausesRateAnalysisIngraffea.pdf

[65]http://www.cnn.com/2012/08/29/us/new-york-fracking-artists-protest/index.html

[66]http://www.pnas.org/content/111/30/10955

[67]http://www.magazine.columbia.edu/print/1091

[68]http://www.magazine.columbia.edu/print/1091

[69]http://fractoids.blogspot.com/

[70]http://www.desmogblog.com/%E2%80%98energy-depth%E2%80%99-was-created-major-oil-and-gas-companies-according-industry-memo

[71]*Ibid.*

[72]http://www.environmentamerica.org/sites/environment/files/reports/EA_FrackingNumbers_scrn.pdf

[73]http://www.marcellusgas.org/well_details.php?well_id=2500&wellsitename=DCNR95%2050027#2500

[74]http://banmichiganfracking.org/?p=1483

[75]https://today.duke.edu/2015/09/frackfoot

[76]http://dailyitem.com/0100_news/x1284938395/Susquehanna-River-Basin-Commission-approves-water-use-for-drilling
[77]http://www.environmentamerica.org/sites/environment/files/reports/EA_FrackingNumbers_scrn.pdf
[78]http://www.dec.ny.gov/energy/75370.html.
[79]http://grist.org/news/marcellus-shale-fracking-wells-use-5-million-gallons-of-water-apiece/#comment-1104605063
[80]http://uk.reuters.com/article/2013/05/20/us-usa-water-idUKBRE94J0Y920130520
[81]http://www.ceres.org/resources/reports/hydraulic-fracturing-water-stress-water-demand-by-the-numbers/view
[82]http://www.jpl.nasa.gov/news/news.php?release=2014-242;http://onlinelibrary.wiley.com/doi/10.1002/2014GL061055/pdf
[83]http://www.slate.com/articles/technology/future_tense/2014/07/lake_mead_before_and_after_colorado_river_basin_losing_water_at_shocking.html
[84]http://online.wsj.com/article/SB10001424052748703739204576228823641659148.html#articleTabs%3Dcomments
[85]http://gov.ca.gov/docs/4.1.15_Executive_Order.pdf
[86]http://www.californiansagainstfracking.org/content/californians-against-fracking-releases-new-data-analysis-oil-industry-california-wastes-2
[87]http://www.latimes.com/local/california/la-me-drought-oil-water-20150503-story.html
[88]http://money.cnn.com/2012/08/10/news/economy/kansas-oil-boom-drought/index.html
[89]http://www.nytimes.com/2012/09/06/us/struggle-for-water-in-colorado-with-rise-in-fracking.html?pagewanted=all
[90]http://www.nytimes.com/2012/09/06/us/struggle-for-water-in-colorado-with-rise-in-fracking.html?pagewanted=all
[91]http://www.environmentamerica.org/sites/environment/files/reports/EA_FrackingNumbers_scrn.pdf
[92]http://www.ceres.org/resources/reports/hydraulic-fracturing-water-stress-growing-competitive-pressures-for-water/view
[93]http://www.forbes.com/fdc/welcome_mjx.shtml
[94]http://www.theguardian.com/environment/2013/aug/11/texas-tragedy-ample-oil-no-water
[95]http://truth-out.org/news/item/16338-the-mines-that-fracking-built-part-two
[96]http://www.xcelenergy.com/Safety_&_Education/Nuclear_Safety/About_Nuclear_Energy/Prairie_Island_Nuclear_Generating_Plant
[97]http://www.r-cause.net/uploads/8/0/2/5/8025484/frackingwithpropane1.pdf
[98]*Ibid.*
[99]http://www.reuters.com/article/2013/05/28/us-california-oil-insight-idUSBRE94R0CO20130528
[100]http://www.gao.gov/assets/590/588205.txt
[101]http://www.earthworksaction.org/files/publications/WastingAway-FINAL-lowres.pdf
[102]http://smartenergyuniverse.com/utility-news/923-epa-issues-final-air-rules-for-the-oil-and-natural-gas-industry
[103]http://www.huffingtonpost.com/robert-f-kennedy-jr/fracking-natural-gas-new-york-times-_b_1022337.html
[104]http://www.fairwarning.org/2011/04/list-reveals-3200-uncapped-abandoned-wells-in-gulf-of-mexico/
[105]http://www.fairwarning.org/2011/04/list-reveals-3200-uncapped-abandoned-wells-in-gulf-of-mexico/
[106]http://www.texasvox.org/abandoned-oil-wells-dangers-pose/
[107]http://www.huffingtonpost.com/2012/09/27/new-york-oil-and-gas-drilling_n_1917659.html
[108]http://www.theguardian.com/environment/2014/sep/18/pennsylvania-abandoned-fracking-wells-methane-leaks-hidden

[109]http://ir.eia.gov/ngs/ngs.html
[110]http://www.eia.gov/pub/oil_gas/natural_gas/analysis_publications/ngpipeline/undrgrn
d_storage.html
[111]http://theadvocate.com/home/4227993-125/scientists-give-sinkhole-insight
[112]http://www.nola.com/environment/index.ssf/2012/08/sinkhole_neighbors_dont_know_
w.html
[113]http://theadvocate.com/news/4282059-123/more-land-and-trees-fall
[114]http://wgno.com/2015/08/03/the-bayou-corne-sinkhole-formed-3-years-ago/
[115]http://www.sunherald.com/2012/11/17/4310905/vent-wells-burning-gas-from-
louisiana.html
[116]http://www.examiner.com/article/sinkhole-strong-chemical-stench-on-bubbling-bayou
[117]http://na.unep.net/geas/archive/pdfs/GEAS_Nov2012_Fracking.pdf
[118]http://na.unep.net/geas/archive/pdfs/GEAS_Nov2012_Fracking.pdf
[119]http://www.jlcny.org/site/index.php/news-articles/32-frontpage/1704-hydrofracking-
brings-prosperity-landowners-told
[120]http://www.cbsnews.com/8301-505123_162-43041239/schlumberger-smith-deal-a-
11b-bet-on-hard-to-reach-oil-and-gas/
[121]http://stateimpact.npr.org/pennsylvania/2011/09/12/can-pennsylvanias-state-forests-
survive-additional-marcellus-shale-drilling/
[122]http://uncoveringpa.com/visiting-penn-brad-oil-museum
[123]http://www.enopetroleum.com/earlyoilpennsylvania.html
[124]http://explorepahistory.com/story.php?storyId=1-9-20&chapter=1
[125]http://www.eia.gov/tools/faqs/faq.cfm?id=69&t=2
[126]http://www.nytimes.com/2011/03/09/opinion/lweb09gas.html
[127]http://www.pacourts.us/assets/opinions/Supreme/out/J-127A-D-2012oajc.pdf?cb=1
[128]http://www.eia.gov/electricity/monthly/epm_table_grapher.cfm?t=epmt_1_12_a
[129]http://heartland.org/issues/environment
[130]http://www.gpo.gov/fdsys/browse/collectiontab.action
[131]http://www.epa.gov/air/caa/
[132]http://cfpub.epa.gov/npdes/cwa.cfm?program_id=45
[133]http://water.epa.gov/lawsregs/rulesregs/sdwa/index.cfm
[134]http://www.epa.gov/lawsregs/laws/rcra.html
[135]http://www.epa.gov/superfund/policy/cercla.htm
[136]http://www.nytimes.com/2011/03/04/us/04gas.html?pagewanted=all
[137]http://www.rff.org/rff/documents/rff-dp-01-38.pdf
[138]*Ibid.*
[139]http://www.gao.gov/new.items/d09872.pdf
[140]http://permanent.access.gpo.gov/lps21800/www.epa.gov/safewater/uic/cbmstudy.html
[141]http://www.earthworksaction.org/files/publications/Weston.pdf?pubs/Weston.pdf
[142]http://www.gpo.gov/fdsys/pkg/PLAW-109publ58/pdf/PLAW-109publ58.pdf
[143]http://water.epa.gov/grants_funding/dwsrf/index.cfm
[144]http://www.google.com/url?sa=t&rct=j&q=&esrc=s&frm=1&source=web&cd=1&ved=
0CCUQFjAA&url=http%3A%2F%2Fwww.wildwatch.org%2FBinocular%2Fbino25%2F
Hydro-fracturingImpactonWildlif.doc&ei=neRlT4T-DYmJgwfws7XKAg&usg=
AFQjCNHhsrEhZunrz78hXtCTrLMJ0PFXog&sig2=0imb2JYsl
[145]http://grist.org/?p=143090&preview=true
[146]http://www.huffingtonpost.com/robert-f-kennedy-jr/fracking-natural-gas-new-york-
times-_b_1022337.html
[147]http://www.forbes.com/sites/christopherhelman/2012/07/19/billionaire-father-of-
fracking-says-government-must-step-up-regulation/
[148]http://www.economist.com/blogs/schumpeter/2013/08/interview-george-mitchell
[149]http://www.scientificamerican.com/article.cfm?id=safety-first-fracking-second
[150]*Ibid.*
[151]http://thomas.loc.gov/cgi-bin/bdquery/z?d111:H.R.2766:

[152]http://thomas.loc.gov/cgi-bin/bdquery/z?d111:S1215:
[153]http://www.govtrack.us/congress/bills/112/hr1084
[154]http://www.govtrack.us/congress/bills/112/s587
[155]http://thomas.loc.gov/cgi-bin/bdquery/D?d113:4:./temp/~bdKuDz:@@@L&summ2=m& | /bss/ |
[156]https://beta.congress.gov/bill/113th-congress/house-bill/1921?q=%7B%22search%22%3A%5B%22hr+1921%22%5D%7D
[157]http://www.cbo.gov/sites/default/files/cbofiles/attachments/02-22-ARRA.pdf
[158]http://www.state.gov/secretary/20092013clinton/rm/2010/04/140286.htm
[159]http://belfercenter.ksg.harvard.edu/files/The Geopolitics of Natural Gas.pdf
[160]http://www.bakerinstitute.org/publications/EF-pub-HKSGeopoliticsOfNaturalGas-073012.pdf
[161]http://www.bloomberg.com/news/2013-04-30/gazprom-2012-profit-drops-9-5-on-decline-in-natural-gas-demand.html
[162]http://www.cbc.ca/m/touch/news/story/1.2649241
[163]http://stateimpact.npr.org/pennsylvania/2012/01/24/energy-in-tonights-state-of-the-union/
[164]http://www.whitehouse.gov/the-press-office/2012/04/13/statements-president-s-executive-order-supporting-safe-and-responsible-d
[165]http://www.whitehouse.gov/the-press-office/2012/04/13/statements-president-s-executive-order-supporting-safe-and-responsible-d
[166]http://thehill.com/blogs/e2-wire/e2-wire/221429-industry-groups-applaud-obamas-natural-gas-executive-order
[167]http://www.whitehouse.gov/state-of-the-union-2013
[168]http://www.whitehouse.gov/the-press-office/2013/06/25/remarks-president-climate-change
[169]http://www.ewg.org/about-us
[170]http://www.ewg.org/news/news-releases/2012/01/25/presidents-speech-misses-mark-addressing-concerns-over-fracking
[171]http://oilpro.com/post/11535/read-oil-industry-response-to-obama-administration-new-hydraulic
[172]http://www.greenpartywatch.org/2013/06/26/green-shadow-cabinet-obamas-climate-proposals-fall-dangerously-short-ignore-time-critical-opportunity-to-revive-economy/
[173]*Ibid.*
[174]http://www.washingtonpost.com/blogs/the-fix/wp/2013/06/20/boehner-thinks-new-white-house-climate-rules-would-be-crazy-but-obama-may-not-have-a-choice/
[175]http://www.infrastructurereportcard.org/
[176]http://www.infrastructurereportcard.org/a/#p/overview/executive-summary
[177]http://www.sanders.senate.gov/newsroom/recent-business/war-on-jobs
[178]http://ww2.epa.gov/carbon-pollution-standards/clean-power-plan-proposed-rule
[179]http://www.pennlive.com/midstate/index.ssf/2014/06/five_things_to_know_about_the.html
[180]https://www.youtube.com/watch?v=UIUgYtuPaLM
[181]http://www.usnews.com/news/blogs/Ken-Walshs-Washington/2013/06/26/gop-brands-obama-climate-plan-a-war-on-coal
[182]http://thehill.com/policy/energy-environment/223298-mcconnell-priority-is-to-get-the-epa-reined-in
[183]http://www.scribd.com/doc/229509188/Time-16-June-2014
[184]http://thedailyreview.com/news/yaw-to-co-chair-state-senate-coal-caucus-1.1557988
[185]http://www.senatorsolobay.com/solobay-named-co-chair-of-bi-partisan-coal-caucus
[186]http://www.greenwisebusiness.co.uk/news/climate-change-campaign-to-block-barack-obama-plan-to-cut-emissions-from-power-plants-4318.aspx#.VB7zaJRdW1o
[187]http://www.huffingtonpost.com/elliott-negin/google-quits-alec-but-che_b_6030760.html

[188]http://www.theguardian.com/environment/2014/may/02/barack-obamas-emissions-plan-comes-under-new-line-of-attack

[189]http://www.mcall.com/news/breaking/mc-gas-drilling-value-20120505,0,5829301.story

[190]http://www.villagevoice.com/2012-09-19/news/boom-or-doom-fracking-environment/4/

[191]http://youngphillypolitics.com/topics/natural_gas_drilling

[192]http://www.magazine.columbia.edu/print/1091

[193]http://www.bloomberg.com/news/2014-02-24/wells-that-fizzle-are-a-potential-show-stopper-for-the-shale-boom.html

[194]http://www.marcellusgas.org/graphs/PA#aver_pro

[195]http://www.examiner.com/article/drilling-permits-decline-sharply-for-the-pennsylvania-marcellus-formation

[196]http://www.ihs.com/images/Americas-New-Energy-Future-Mfg-Renaissance-Main-Report-Sept13.pdf

[197]http://www.post-gazette.com/stories/business/news/shale-study-promises-huge-gains-33-million-jobs-by-20-701889/#ixzz2e1fv8APu

[198]*Ibid.*

[199]http://www.businessweek.com/ap/2012-07-19/us-chamber-touts-benefits-of-pa-dot-gas-drilling

[200]http://pipeline.post-gazette.com/news/archives/24704-chamber-officials-lobby-to-keep-shale-costs-low

[201]http://www.opensecrets.org/news/2013/07/billion-dollar-baby-us-chamber-is-first-to-hit-lobbying-milestone/

[202]https://www.pachamber.org/newsroom/articles/2012/PA Chamber president helps launch national Marcellus Shale campaign.php

[203]http://www.energyindepth.org/tag/marcellus-shale-coalition/

[204]http://marcelluscoalition.org/2012/09/msc-president-in-philadelphia-inquirer-marcellus-shale-transforming-pa/

[205]http://philadelphia.cbslocal.com/2013/06/14/critics-challenge-gov-corbett-at-fracking-industry-meeting-in-philadelphia/

[206]http://thirdandstate.org/2013/november/morning-must-read-setting-record-straight-employment-fracking

[207]http://keystoneresearch.org/sites/default/files/KRC_JobRanking_Sept.pdf

[208]http://www.bls.gov/eag/eag.pa.htm#eag_pa.f.P

[209]http://www.bls.gov/eag/eag.tx.htm

[210]http://www.bls.gov/eag/eag.nd.htm

[211]http://business.financialpost.com/2013/10/18/u-s-has-overfracked-and-overdrilled-shell-director-says/?__lsa=d589-16d8

[212]http://www.post-gazette.com/stories/local/marcellusshale/chamber-officials-lobby-to-keep-shale-costs-low-645818/

[213]http://www.earthworksaction.org/2010summit/Panel7_DeborahRogers.pdf

[214]*Ibid.*

[215]http://notohydrofracking.blogspot.com/2011/10/economic-assessment-of-hydrofracking-dr.html

[216]http://www.catskillmountainkeeper.org/wp-content/uploads/2012/01/JMB+Critique+of+PPI+Jan+2+2012.pdf

[217]http://notohydrofracking.blogspot.com/2011/10/economic-assessment-of-hydrofracking-dr.html

[218]http://www.cbsnews.com/videos/extra-meet-the-shaleionaires/

[219]http://www.wsj.com/articles/SB10001424052702304071004579407572017346590

[220]http://thetimes-tribune.com/news/business/chesapeake-to-pay-7-5-million-to-settle-post-production-cost-suit-1.1544612

[221]http://www.star-telegram.com/2013/06/04/4910170/range-resources-settling-lawsuit.html

[222]http://aese.psu.edu/research/centers/cecd/publications/marcellus/marcellus-shale-land-ownership-local-voice-and-the-distribution-of-lease-and-royalty-dollars/view
[223]http://public.econ.duke.edu/~timmins/w18390.pdf
[224]http://blog.360mtg.com/?p=2407
[225]http://www.magazine.columbia.edu/features/summer-2012/gas-menagerie
[226]http://www.catskillmountainkeeper.org/our-programs/fracking/whats-wrong-with-fracking-2/mortgage-problems/
[227]http://www.dutchnews.nl/news/archives/2013/07/rabobank_will_not_finance_shal.php
[228]http://www.desmogblog.com/2012/09/21/oil-and-gas-leases-create-conflicts-fema
[229]http://www.americanbanker.com/issues/178_218/fracking-boom-may-cause-mortgage-headaches-for-banks-1063561-1.html
[230]http://www.nationwide.com/newsroom/071312-FrackingStatement.jsp
[231]http://www.fema.gov/pdf/nfip/manual201205/splashscreen.pdf
[232]http://www.desmogblog.com/2012/09/21/oil-and-gas-leases-create-conflicts-fema
[233]http://www.desmogblog.com/2012/09/21/oil-and-gas-leases-create-conflicts-fema
[234]http://www.desmogblog.com/2012/09/21/oil-and-gas-leases-create-conflicts-fema
[235]http://transportation.house.gov/news/documentsingle.aspx?DocumentID=391079
[236]http://bigstory.ap.org/article/some-anti-drilling-activists-change-tactics-tone-0
[237]http://www.blm.gov/wo/st/en/prog/energy/oil_and_gas/best_management_practices/split_estate.html
[238]http://www.environmentamerica.org/news/ame/senate-votes-favor-dirty-keystone-xl-tar-sands-pipeline
[239]http://www.whitehouse.gov/the-press-office/2012/03/22/remarks-president-american-made-energy
[240]http://www.law.cornell.edu/supct/html/04-108.ZO.html
[241]http://www.kslaw.com/library/newsletters/energynewsletter/2011/november/article1.html
[242]http://primis.phmsa.dot.gov/comm/reports/safety/Allpsi.html?nocache=1783
[243]http://www.ilr.cornell.edu/globallaborinstitute/research/upload/GLI_Impact-of-Tar-Sands-Pipeline-Spills.pdf
[244]http://www.propublica.org/article/pipelines-explained-how-safe-are-americas-2.5-million-miles-of-pipelines
[245]http://www.sqwalk.com/q/pipeline-companies-seize-land-texas-will
[246]http://www.chron.com/opinion/outlook/article/Railroad-Commission-favoring-pipelines-over-Texas-3706181.php?cmpid=twitter
[247]http://www.youtube.com/watch?v=WKqs7Fc45XM
[248]http://www.cbsnews.com/8301-207_162-57526540/daryl-hannah-arrested-in-texas-protesting-pipeline/
[249]http://www.fwweekly.com/2009/10/14/sacrificed-to-shale
[250]http://www.desmogblog.com/2015/02/08/voices-arlington-texas-unify-protect-environment-and-community-fracking
[251]http://www.reuters.com/article/2012/10/02/us-chesapeake-landgrab-substory-idUSBRE8910E920121002
[252]https://www.texasobserver.org/the-holdouts-three-texas-families-refused-sell-mineral-rights-fracking/
[253]http://www.supreme.courts.state.tx.us/historical/2012/mar/090901rh.pdf
[254]http://uk.reuters.com/article/2009/05/03/btscenes-us-energy-gas-drilling-idUKTRE5422TG20090503
[255]http://www.tctimes.com/news/homeowner-s-property-destroyed-by-enbridge/article_6ca1168e-e89f-11e1-97a0-001a4bcf887a.html
[256]http://www.kwwl.com/story/18864857/montague-couple-fighting-gas-pipeline-extension-route?clienttype=printable
[257]http://www.nj.com/gloucester-news/index.ssf/2012/09/national_park_at_the_center_of.html

[258]http://www.kwwl.com/story/18864857/montague-couple-fighting-gas-pipeline-extension-route?clienttype=printable

[259]http://www.morningstar.com/earnings/PrintTranscript.aspx?id=13725415

[260]http://www.walb.com/story/29098768/palmetto-pipeline-kinder-morgan-cant-take-land

[261]*Ibid.*

[262]https://www.gov.uk/government/consultations/underground-drilling-access

[263]http://www.theguardian.com/environment/2014/oct/07/naomi-klein-uk-fracking-trespass-law-flouts-democratic-rights

[264]http://www.theguardian.com/environment/2014/sep/26/fracking-trespass-law-changes-move-forward-despite-huge-public-opposition

[265]http://news.stv.tv/scotland/293529-snp-condemns-uk-plans-to-allow-fracking-drilling-below-peoples-land/

[266]https://www.govtrack.us/congress/bills/113/hr1944/text

[267]http://ketr.org/post/texas-supreme-court-will-not-review-ruling-favoring-transcanada

[268]http://www.coloradoan.com/apps/pbcs.dll/article?AID=2012305180028

[269]http://www.westernresourceadvocates.org/schooldrill/

[270]*Ibid.*

[271]http://www.coloradoan.com/apps/pbcs.dll/article?AID=2012305180028

[272]http://environews.tv/residents-furious-about-being-force-pooled-into-idahos-first-fracking/

[273]http://www.legis.state.pa.us/cfdocs/billinfo/billinfo.cfm?syear=2009&sind=0&body=H&type=B&bn=977

[274]http://www.pressconnects.com/viewart/20100811/NEWS11/8110334/Rendell-willing-negotiate-gas-pooling-law

[275]http://www.bizjournals.com/pittsburgh/stories/2010/02/08/story12.html

[276]http://www.legis.state.pa.us/cfdocs/billinfo/billinfo.cfm?syear=2011&sind=0&body=S&type=B&bn=0447

[277]http://thetimes-tribune.com/news/forced-pooling-legislation-for-gas-industry-planned-in-pennsylvania-1.885341

[278]http://www.propublica.org/article/forced-pooling-when-landowners-cant-say-no-to-drilling

[279]http://www.legis.state.pa.us/CFDOCS/Legis/PN/Public/btCheck.cfm?txtType=PDF&sessYr=2013&sessInd=0&billBody=S&billTyp=B&billNbr=0259&pn=1290

[280]http://stateimpact.npr.org/pennsylvania/2013/07/09/corbett-signs-controversial-bill-giving-drillers-power-to-pool-leases/

[281]http://triblive.com/business/headlines/4828098-74/hilcorp-pooling-gas#axzz2htpkhR7F

[282]http://www.opensecrets.org/industries/indus.php?Ind=E01.

[283]http://www.opensecrets.org/industries/indus.php?Ind=E01

[284]http://www.opensecrets.org/industries/recips.php?ind=E01&cycle=2012&recipdetail=P&mem=N&sortorder=U

[285]http://www.american.com/archive/2012/august/presidential-power-obama-vs-romney-on-energy

[286]*Ibid.*

[287]*Ibid.*

[288]http://www.opensecrets.org/industries/recips.php?ind=E01&cycle=2008&recipdetail=P&mem=N&sortorder=Uhttp://www.opensecrets.org/industries/recips.php?ind=E01&cycle=2008&recipdetail=P&mem=N&sortorder=U

[289]*Ibid.*

[290]https://www.opensecrets.org/industries/indus.php?ind=E01

[291]http://crew.3cdn.net/29e185cb94de445559_rom6b5st3.pdf

[292]https://www.opensecrets.org/industries/recips.php?cycle=2014&ind=E01

[293]https://www.opensecrets.org/industries/indus.php?ind=E01

[294]http://www.whitehouse.gov/the-press-office/2013/08/23/remarks-president-town-hall-binghamton-university

[295]http://www.opensecrets.org/industries/summary.php?ind=E01&recipdetail=M&sortorder=U&cycle=2012

[296]http://www.opensecrets.org/industries/indus.php?Ind=E01

[297]https://www.opensecrets.org/industries/indus.php?ind=E01

[298]https://www.opensecrets.org/industries/indus.php?ind=E01

[299]http://crew.3cdn.net/29e185cb94de445559_rom6b5st3.pdf

[300]http://www.opensecrets.org/indivs/search.php?name=mcclendon&state=&zip=&employ=&cand=Upton%2C+Fred&c2012=Y&c2010=Y&c2008=Y&sort=N&capcode=c5s6j

[301]http://www.nytimes.com/2012/06/27/science/earth/epa-emissions-rules-backed-by-court.html?_r=0

[302]http://www.cnn.com/2015/07/25/politics/ted-cruz-wilks-brothers/index.html

[303]http://www.cnn.com/2015/06/03/politics/ted-cruz-super-pac-identities/

[304]https://www.opensecrets.org/politicians/summary.php?type=C&cid=N00005582&newMem=N&cycle=2014#cont

[305]http://epw.senate.gov/public/index.cfm?FuseAction=PressRoom.PressReleases&ContentRecord_id=7B837DA0-802A-23AD-43CD-4C7F2458F87B

[306]http://epw.senate.gov/public/index.cfm?FuseAction=Minority.PressReleases&ContentRecord_id=7280f114-802a-23ad-4bb2-797081f70515&Region_id=&Issue_id=

[307]http://thehill.com/blogs/e2-wire/e2-wire/205455-white-house-throws-in-the-towel-on-interior-nominee

[308]http://www.canadafreepress.com/index.php/article/44055

[309]http://www.eenews.net/public/Greenwire/2012/01/20/1

[310]http://www.epw.senate.gov/public/index.cfm?FuseAction=Minority.PressReleases&ContentRecord_id=8aae983e-cb8e-03f6-0743-ce4913c8ca7e

[311]*Ibid.*

[312]https://content.sierraclub.org/press-releases/2013/05/sierra-club-statement-failure-any-republicans-attend-gina-mccarthys

[313]http://www.commoncause.org/atf/cf/%7Bfb3c17e2-cdd1-4df6-92be-bd4429893665%7D/DEEP%20DRILLING%20DEEP%20POCKETS%20NOV%202011.PDF

[314]http://www.huffingtonpost.com/robert-f-kennedy-jr/oil-industry-republicans-congress_b_3252170.html

[315]http://www.whitehouse.gov/the-press-office/2013/06/25/remarks-president-climate-change

[316]http://www.pnas.org/content/early/2013/04/26/1218453110

[317]http://supreme.justia.com/cases/federal/us/558/08-205/

[318]http://www.whitehouse.gov/the-press-office/statement-president-todays-supreme-court-decision-0

[319]http://legaltimes.typepad.com/blt/2010/01/obama-supreme-court-opened-the-floodgates-for-special-interests.html

[320]http://query.nictusa.com/cgi-bin/dcdev/forms/C00504530/828309/sa/ALL

[321]http://blog.seattlepi.com/seattlepolitics/2012/10/28/2-5-million-from-chevron-usa-to-republicans/

[322]http://www.reuters.com/article/2013/03/31/us-exxon-pipeline-spill-idUSBRE92U00220130331

[323]http://ecowatch.com/2013/fast-track-keystone-xl-tar-sands-pipeline/

[324]http://politics.nytimes.com/congress/votes/113/house/2/519

[325]http://www.cnn.com/2015/02/11/politics/boehner-obama-keystone-veto/

[326]http://www.opensecrets.org/pfds/assets.php?year=2010&cid=N00003675

[327]http://priceofoil.org/2015/01/09/vote-analysis-house-proves-serve/

[328]http://priceofoil.org/2015/01/09/vote-analysis-house-proves-serve/

[329]http://priceofoil.org/2015/01/22/bribery-bargain-big-oil/

[330]http://www.desmogblog.com/2014/09/01/labor-day-news-dump-ferc-hands-enbridge-permit-tar-sands-by-rail-facility
[331]http://www.jdsupra.com/legalnews/us-chamber-of-commerce-sues-sec-to-ove-75416/
[332]http://energywatchmrb.blogspot.com/
[333]http://thenetwork.berkeleylawblogs.org/2013/07/09/federal-court-vacates-secs-extraction-payment-disclosure-rule/#more-1857
[334]http://www.prwatch.org/news/2012/10/11802/big-oil-and-us-chamber-fight-keep-foreign-bribery-flourishing
[335]http://search.nola.com/southeast+louisiana+flood+protection+authority-east/1/all/?date_range=all
[336]http://www.legis.la.gov/legis/BillInfo.aspx?s=14rs&b=SB469&sbi=y
[337]http://www.nola.com/politics/index.ssf/2014/06/bobby_jindal_signs_bill_to_kil.html
[338]http://www.nola.com/politics/index.ssf/2014/06/bobby_jindal_signs_bill_to_kil.html
[339]http://www.nola.com/politics/index.ssf/2014/06/bobby_jindal_signs_bill_to_kil.html
[340]http://www.nola.com/politics/index.ssf/2014/06/bobby_jindal_signs_bill_to_kil.html
[341]http://www.wwltv.com/story/news/2014/12/10/tainted-legacy-money-from-oil-and-gas-industry-flowing-into-state-capitol/17670409/
[342]http://www.petition2congress.com/15559/traitorous-lousiana-gov-bobby-jindal-signs-bill-preventing-lawsui
[343]http://subsidytracker.goodjobsfirst.org/prog.php?parent=marathon-petroleum
[344]https://en.wikipedia.org/wiki/Marathon_Petroleum
[345]http://www.theguardian.com/environment/2015/may/12/us-taxpayers-subsidising-worlds-biggest-fossil-fuel-companies
[346]http://www.dispatch.com/content/stories/local/2013/08/01/fracking-lobby-helps-fund-gop-campaigns.html
[347]http://www.commoncause.org/states/ohio/reports/deep-drilling-deep-pockets.PDF
[348]http://www.dispatch.com/content/stories/local/2015/02/22/industry-donations-ruling-tied.html
[349]http://www.dispatch.com/content/stories/local/2015/02/22/industry-donations-ruling-tied.html
[350]http://www.dispatch.com/content/stories/local/2013/08/01/fracking-lobby-helps-fund-gop-campaigns.html
[351]http://www.commoncause.org/states/new-york/research-and-reports/NY_011314_Deep_Drilling_Deep_Pockets.pdf
[352]http://www.commoncause.org/states/new-york/research-and-reports/NY_011314_Deep_Drilling_Deep_Pockets.pdf
[353]http://www.commoncause.org/states/new-york/research-and-reports/NY_011314_Deep_Drilling_Deep_Pockets.pdf
[354]http://www.commoncause.org/site/pp.asp?c=dkLNK1MQIwG&b=8482079
[355]http://www.commoncause.org/site/pp.asp?c=dkLNK1MQIwG&b=8482079
[356]http://www.commoncause.org/site/apps/nlnet/content2.aspx?c=dkLNK1MQIwG&b=5287775&ct=13223851&notoc=1
[357]http://www.nytimes.com/2015/07/23/nyregion/thomas-libous-new-york-state-senator-is-convicted-of-lying-to-fbi.html?_r=0
[358]http://polhudson.lohudblogs.com/2013/03/07/libous-wants-no-senate-vote-on-fracking-moratorium/
[359]http://www.tomlibous.com/index.asp?Type=B_BASIC&SEC={8273960E-5FCD-4429-BAE6-98818C88E676}
[360]http://www.followthemoney.org/entity-details?eid=6641132&default=lawmaker
[361]http://www.commoncause.org/site/apps/nlnet/content2.aspx?c=dkLNK1MQIwG&b=5287775&ct=13223851&notoc=1
[362]http://www.nysenate.gov/senator/catharine-young/bio
[363]http://www.commoncause.org/site/apps/nlnet/content2.aspx?c=dkLNK1MQIwG&b=5287775&ct=13223851&notoc=1

[364]http://www.opensecrets.org/politicians/summary.php?cid=N00030949&cycle=2012
[365]http://www.buffalonews.com/apps/pbcs.dll/article?avis=BN&date=20131106&categor
y=CITYANDREGION&lopenr=131109203&Ref=AR&profile=1175&template=printart
[366]http://www.stargazette.com/viewart/20131108/NEWS10/311080048/Dem-senator-
seeks-probe-GOP-chief-over-fracking
[367]http://www.commoncause.org/states/new-york/research-and-reports/NY_011314_
Deep_Drilling_Deep_Pockets.pdf
[368]http://www.commoncause.org/states/pennsylvania/press/marcellus-gas-industry.html
[369]http://marcellusmoney.org/findings-2/
[370]http://catskillcitizens.org/learnmore/Marcellus-Money-2014-Main-Findings.pdf
[371]https://www.opensecrets.org/politicians/summary.php?cycle=Career&type=I&cid=N0
0024992&newMem=N
[372]https://www.opensecrets.org/politicians/industries.php?cycle=2014&type=I&cid=N00
001489&newMem=N&recs=20
[373]http://www.opensecrets.org/industries/summary.php?ind=E01&cycle=2012&recipdet
ail=H&sortorder=N&mem=Y
[374]https://www.opensecrets.org/politicians/industries.php?cycle=2012&cid=N00025495&
type=I&newmem=N
[375]https://www.opensecrets.org/politicians/industries.php?cycle=2014&type=I&cid=N00
025495&newMem=N&recs=20
[376]http://dailyitem.com/0100_news/x62498341/Barletta-on-shale-caucus-owns-gas-stock
[377]*Ibid.*
[378]*Ibid.*
[379]http://stateimpact.npr.org/pennsylvania/2011/11/10/common-cause-report-details-
campaign-contributions-from-drillers/
[380]http://marcellusmoney.org/sites/default/files/images/Marcellus Money press release
071212_0.pdf
[381]http://www.publicintegrity.org/2012/10/18/11498/pennsylvania-governor-benefited-
untraceable-15-million-donation
[382]http://articles.philly.com/2011-06-29/news/29717481_1_corbett-campaign-tom-
corbett-marcellus-shale
[383]http://www.rollingstone.com/politics/news/the-big-fracking-bubble-the-scam-behind-
the-gas-boom-20120301
[384]http://www.marcellusgas.org/graphs/PA#prodollars
[385]http://www.marcellusgas.org/record_book_co.php?report_type=top_producing_co&cou
nty_id=&num_results=10&muni_id=
[386]http://www.triplepundit.com/2011/03/socially-responsible-investment-firm-divests-
chesapeake-energy/
[387]http://www.marcellusgas.org/record_book_co.php?report_type=top_violations_co&cou
nty_id=&num_results=10&muni_id=
[388]http://wmarcellusmoney.org/candidates
[389]http://marcellusmoney.org/sites/default/files/images/Marcellus Money press release
071212_0.pdf
[390]http://www.marcellusmoney.org/candidate/yaw-gene
[391]http://wnep.com/2013/03/26/lawmaker-denies-conflict-of-interest/
[392]http://wnep.com/2013/03/27/96777/
[393]http://wnep.com/2013/04/04/invite-only-gas-drilling-meeting-excludes-public-media/
[394]http://marcellusmoney.org/candidates
[395]http://old.post-gazette.com/pg/12115/1226722-178-0.stm
[396]http://articles.philly.com/2012-02-06/news/31030487_1_guilty-verdicts-corbett-
bonusgate
[397]http://www.heraldstandard.com/gcm/news/top_stories/deweese-others-speak-out-
against-corbett-plan/article_8f577f62-d7ae-5131-b859-92ff99e0e6f5.html

[398]http://www.alleghenyfront.org/story/flooding-zone-gas-industry-pours-millions-lobbying-pa
[399]http://www.alleghenyfront.org/story/dcnr-ex-chiefs-calendar-shows-gaps
[400]http://www.citizen-times.com/story/opinion/editorials/2014/11/12/teeth-added-fracking-rules/18916133/
[401]http://www.newsobserver.com/opinion/editorials/article10199591.html
[402]http://www.fayobserver.com/news/local/fracking-with-drilling-on-horizon-opponents-prepare-for-fight/article_62076ed6-3fb6-57f4-b6b4-b35e74ccb5ae.html
[403]http://www.fayobserver.com/news/local/fracking-with-drilling-on-horizon-opponents-prepare-for-fight/article_62076ed6-3fb6-57f4-b6b4-b35e74ccb5ae.html
[404]http://www.followthemoney.org/entity-details?eid=5866134
[405]http://greenpeaceblogs.org/2014/05/28/north-carolina-fracking-bills-sponsor-close-ties-oil-gas-industry/
[406]http://www.fayobserver.com/opinion/editorials/our-view-fracking-can-begin-but-safeguards-are-lacking/article_942c7a6a-20e7-521e-a0b2-f2c61ea9e63f.html
[407]http://www.businessweek.com/news/2012-03-29/obama-says-oil-company-profits-justify-ending-u-dot-s-dot-tax-br
[408]http://priceofoil.org/content/uploads/2014/07/OCI_US_FF_Subsidies_Final_Screen.pdf
[409]http://www.businessweek.com/news/2012-03-29/obama-says-oil-company-profits-justify-ending-u-dot-s-dot-tax-br
[410]http://thinkprogress.org/green/2012/03/29/454853/senators-who-voted-to-protect-oil-tax-breaks-received-23582500-from-big-oil/
[411]http://www.whitehouse.gov/the-press-office/2013/06/25/remarks-president-climate-change
[412]http://www.mediamatters.org/research/2012/03/06/energy-experts-debunk-right-wing-defense-of-oil/184371
[413]http://www.factcheck.org/2011/05/playing-politics-with-gasoline-prices/
[414]http://www.shell.us/aboutshell/projects-locations/appalachia.html
[415]http://www.opensecrets.org/lobby/indusclient.php?id=e01&year=2012
[416]http://publicsource.org/investigations/potter-township-forgotten-player-bringing-shell-oil-pa
[417]http://triblive.com/news/2008446-74/jobs-state-plant-permanent-shell-industry-corbett-numbers-tax-billion#axzz2DbxQ6ptB
[418]http://www.shell.us/home/content/usa/aboutshell/media_center/news_and_press_releases/2012/03152012_pennsylvania.html
[419]http://commonsense2.com/2011/12/naturalgasdrilling/connecting-the-dots-the-marcellus-natural-gas-play-players-part-1/
[420]http://www.forbes.com/profile/terrence-pegula/
[421]http://www.buffalonews.com/city/article326485.ece
[422]http://marcellusmoney.org/company/shell
[423]http://www.followthemoney.org/database/StateGlance/contributor_details.phtml?&c=123883&i=33&s=PA&y=2010&summary=1&so=a&filter%5b%5d=pegula&filter%5b%5d=&filter%5b%5d=&filter%5b%5d=#sorttable
[424]http://www.followthemoney.org/database/StateGlance/state_candidates.phtml?s=PA&y=2012&f=G&filter[]=&filter[]=&filter[]=GOVERNOR/LIEUTENANT GOVERNOR#candidateselect
[425]http://sierraclub.typepad.com/compass/2012/02/the-sierra-club-and-natural-gas.html
[426]*Ibid.*
[427]http://www.sierraclub.org/policy/energy/fracking
[428]http://www.texassharon.com/2011/11/28/who-put-the-k-in-fracking-the-truth-the-whole-truth-and-nothing-but-the-fracking-truth/
[1]http://climatechangecommunication.org/sites/default/files/reports/Fracking_In_the_American_Mind_2012.pdf

[429]http://stateimpact.npr.org/pennsylvania/2012/06/29/dep-secretary-krancer-stumps-for-natural-gas-processing-facility-in-delaware-county/

[430]https://stateimpact.npr.org/pennsylvania/2012/06/29/dep-secretary-krancer-stumps-for-natural-gas-processing-facility-in-delaware-county/

[431]http://www.weitzlux.com/Marcus-Hook-Pa_1961415.html

[432]http://epa-sites.findthedata.org/l/495723/Sunoco-Inc-r-And-m-Marcus-Hook-Refinery

[433]http://stateimpact.npr.org/pennsylvania/2012/07/11/despite-refinery-closure-petrochemical-plant-remains-in-delaware-county/

[434] https://www.eia.gov/dnav/ng/NG_SUM_LSUM_A_EPG0_XDG_COUNT_A.htm

[435]http://www.dced.state.pa.us/public/oor/constitution.pdf

[436]https://www.paoilandgasreporting.state.pa.us/publicreports/Modules/DataExports/DataExports.aspx

[437] Theuticashale.com/fracking-our-food-supply

[438]http://stateimpact.npr.org/pennsylvania/tag/tom-corbett/

[439]http://www.eia.gov/tools/faqs/faq.cfm?id=46&t=8

[440]http://articles.philly.com/2012-09-20/business/33978477_1_corbett-pennsylvania-s-marcellus-shale-shale-gas-outrage

[441]http://www.marcellusgas.org/graphs/PA#prodollars

[442]http://files.dep.state.pa.us/OilGas/BOGM/BOGMPortalFiles/OilGasReports/2012/WEBSITE Weekly Report for Last Week.pdf

[443]http://www.portal.state.pa.us/portal/server.pt?open=512&objID=708&PageID=224602&mode=2&contentid=http://pubcontent.state.pa.us/publishedcontent/publish/cop_general_government_operations/oa/oa_portal/omd/p_and_p/executive_orders/2010_2019/items/2011_01.html

[444]http://files.dep.state.pa.us/PublicParticipation/MarcellusShaleAdvisoryCommission/MarcellusShaleAdvisoryPortalFiles/MSAC_Members.pdf

[445]http://blog.shaleshockmedia.org/2012/09/25/marcellus-shale-advisory-commissions-swivel-chairs/

[446]http://www.theintelligencer.net/page/content.detail/id/567362/Pa--Still-Seeking--Cracker-.html?nav=515

[447]http://thepennsylvaniaprogressive.com/diary/3232/tom-corbett-same-old-corruption

[448]http://www.propublica.org/article/corbett-pa-energy-exec-authority-environment

[449]http://www.propublica.org/article/corbett-pa-energy-exec-authority-environment

[450]http://www.dcnr.state.pa.us/news/newsreleases/2011/0311-allan.htm

[451]http://www.gastruth.org/?p=163

[452]http://stateimpact.npr.org/pennsylvania/2013/03/13/corbett-and-his-wife-took-over-15000-in-gifts-from-law-firm-representing-oil-and-gas-industry/

[453]http://www.dispatch.com/content/blogs/the-daily-briefing/2012/12/12-8-12-kasich-ong.html

[454]http://www.linkedin.com/pub/john-hines/55/b3/542

[455]http://marcellusprotest.org/dep-inspectors-limited-propublica

[456]*Ibid.*

[457]http://thetimes-tribune.com/opinion/dep-boss-bows-to-gas-drillers-1.1126421

[458]http://old.post-gazette.com/pg/11123/1143606-503-0.stm

[459]http://www.phillyburbs.com/news/local/courier_times_news/dep-secretary-krancer-hit-on-marcellus-shale-smallmouth-bass/article_446821c4-2d21-55d3-a497-d995674c994a.html?mode=jqm

[460]http://www.marcellusgas.org/drillco/drillco_details.php

[461]http://marcelluseffect.blogspot.com/2014/07/pa-regulators-unprepared-for-rapid.html

[462]http://www.auditorgen.state.pa.us/reports/performance/special/spedep072114.pdf

[463]http://eagleford.publicintegrity.org/

[464]http://www.post-gazette.com/stories/local/state/pa-parks-director-says-he-was-forced-out-by-corbett-administration-656785/

[465]http://blog.pennlive.com/midstate_impact/print.html?entry=/2012/10/state_parks_director_john_norb.html
[466]http://bit.ly/1cqIuOm
[467]http://www.marcellus.psu.edu/resources/PDFs/MSACFinalReport.pdf.
[468]http://change.nature.org/2011/02/10/how-pennsylvania%E2%80%99s-energy-infrastructure-will-affect-hunters-fishers-trout-birds/
[469]http://www.alleghenyfront.org/story/moratorium-discussions-dcnr-raise-questions
[470]http://www.portal.state.pa.us/portal/server.pt?open=512&objID=708&PageID=224602&mode=2&contentid=http://pubcontent.state.pa.us/publishedcontent/publish/cop_general_government_operations/oa/oa_portal/omd/p_and_p/executive_orders/2010_2019/items/2010_05.html
[471]https://stateimpact.npr.org/pennsylvania/2012/02/07/governors-budget-a-mixed-bag-for-conservationists/
[472]http://readingeagle.com/news/article/corbett-order-allows-drilling-below-pennsylvanias-public-lands&template=mobileart
[473]http://www.dcnr.state.pa.us/cs/groups/public/documents/document/dcnr_20029147.pdf
[474]http://marcellusmoney.org/sites/default/files/images/Marcellus Money press release 071212_0.pdf
[475]http://stateimpact.npr.org/pennsylvania/2015/01/29/gov-wolf-bans-new-drilling-in-state-parks-and-forests/
[476]http://stateimpact.npr.org/pennsylvania/2012/02/08/corbetts-budget-would-cut-dep-spending/
[477]http://www.pennlive.com/editorials/index.ssf/2012/04/gov_corbetts_budget_hurts_envi.html
[478]*Ibid.*
[479]http://paindependent.com/2012/03/dep-budget-cut-not-affecting-pa-gas-well-checks/
[480]http://www.nytimes.com/2012/05/15/us/for-oil-workers-deadliest-danger-is-driving.html?_r=3
[481]http://www.auditorgen.state.pa.us/reports/performance/special/spedep072114.pdf
[482]http://www.portal.state.pa.us/portal/server.pt/gateway/PTARGS_0_2_785_708_0_43/http%3B/pubcontent.state.pa.us/publishedcontent/publish/global/files/executive_orders/2010___2019/2012_11.pdf
[483]http://www.businessweek.com/ap/2012-08-13/foes-pa-dot-state-permit-order-threatens-environment
[484]http://www.usatoday.com/money/industries/energy/2011-04-13-pa-gas-drilling-permits.htm
[485]http://www.earthworksaction.org/files/publications/FINAL-US-enforcement-sm.pdf
[486]*Ibid.*
[487]http://earthjustice.org/news/press/2012/groups-urge-penn-governor-to-reverse-policy-that-delays-warning-of-fracking-water-pollution
[488]http://money.cnn.com/2012/05/01/news/economy/fracking-violations/index.htm
[489]*Ibid.*
[490]*Ibid.*
[491]http://www.auditorgen.state.pa.us/reports/performance/special/spedep072114.pdf
[492]http://www.legis.state.pa.us/CFDOCS/Legis/PN/Public/btCheck.cfm?txtType=HTM&sessYr=2011&sessInd=0&billBody=H&billTyp=B&billNbr=1950&pn=3048
[493]http://www.legis.state.pa.us/CFDOCS/Legis/RC/Public/rc_view_action2.cfm?sess_yr=2011&sess_ind=0&rc_body=H&rc_nbr=854
[494]http://legiscan.com/gaits/rollcall/103497
[495]http://www.facebook.com/photo.php?v=10200406919639519
[496]http://marcellusmoney.org/sites/default/files/images/Marcellus Money press release 071212_0.pdf
[497]http://www.americanlegislator.org/2012/03/alec-encourages-responsible-resource-production/

[498]http://ia700809.us.archive.org/26/items/MarcellusShaleCoalitionAssociatedPetroleu
mIndustriesOfPennsylvania/MSCAPILetterToSmithTurzaiEllis01-12-12.pdf
[499]http://www.newpa.com/newsroom/governor-corbett-signs-historic-marcellus-shale-law
[500]http://www.gastruth.org/?p=534
[501]http://www.legis.state.pa.us/CFDOCS/Legis/PN/Public/btCheck.cfm?txtType=HTM&
sessYr=2011&sessInd=0&billBody=H&billTyp=B&billNbr=1950&pn=3048
[502]http://www.examiner.com/article/gov-corbett-says-yes-to-shale-gas-incentives-but-no-to-solar
[503]http://bit.ly/1dpARou
[504]http://bit.ly/17sLwjq
[505]http://www.businessweek.com/ap/financialnews/D9MA9IF80.htm
[506]http://standardspeaker.com/news/downtown-rally-blasts-spending-plan-1.1126526
[507]http://pennbpc.org/sites/pennbpc.org/files/2009-natural-gas-production-ranking-and-
2010-11-drilling-tax-rates.pdf
[508]http://thirdandstate.org/2012/february/pa-marcellus-shale-fee-among-lowest-nation
[509]https://pennbpc.org/look-other-states-shows-marcellus-impact-fee-shortchanges-pennsylvanians
[510]http://www.philly.com/philly/news/politics/state/20120910_ap_padrillingimpactfeerai
sesmorethan200m.html
[511]http://www.statejournal.com/story/18154011/pa-gas-drilling-brought-35-billion-in-
2011
[512]http://www.post-gazette.com/stories/local/region/2002-court-case-proved-windfall-for-shale-drillers-266017/
[513]http://www.pachamber.org/advocacy/priorities/energy_environmental/energy/testimo
ny/pdf/Severance_Coalition_Letter.pdf?1431612316
[514]http://www.pachamber.org/advocacy/priorities/energy_environmental/energy/testimo
ny/pdf/Severance_Coalition_Letter.pdf?1431612316
[515]http://powersource.post-gazette.com/powersource/policy-
powersource/2015/05/18/Wolf-fires-back-at-industry-groups-for-opposition-to-Marcellus-
Shale-severance-tax-Pennsylvania/stories/201505180163
[516]http://www.scribd.com/doc/265778928/Letter-to-Business-Groups
[517] See--https://www.youtube.com/watch?v=cv7TOXl1EKs
[518]http://pennbpc.org/top-five-facts-about-drilling-and-taxes-pennsylvania
[519]http://earthjustice.org/sites/default/files/Dryden-Decision.pdf
[520]https://www.nycourts.gov/ctapps/Decisions/2014/Jun14/130-131opn14-Decision.pdf
[521]http://www.scribd.com/doc/139445873/Land-Use-Laws-and-Fracking
[522]http://marcellusdrilling.com/2014/04/ny-anti-driller-gets-150k-prize-for-
environmental-activism/
[523]http://earthjustice.org/news/press/2014/ny-communities-triumph-over-fracking-
industry-in-precedent-setting-case
[524]http://caselaw.findlaw.com/pa-supreme-court/1144542.html
[525]http://www.puc.state.pa.us/naturalgas/naturalgas_marcellus_Shale.aspx
[526]http://www.alternet.org/story/154459/fracking_democracy:_why_pennsylvania%27s_a
ct_13_may_be_the_nation%27s_worst_corporate_giveaway_?akid=8391.298606.GhNY
U8&rd=1&t=1
[527]http://www.marcellusprotest.org/sites/marcellusprotest.org/files/June_2012.pdf
[528]http://www.spilmanlaw.com/Mobile/Home/Resources/Attorney-Authored-
Articles/Marcellus-Fairway/Pa--Commonwealth-Court-Strikes-Down-Act-13-Zoning
[529]http://www.facebook.com/BerksGasTruth/posts/480779425267417
[530]http://canon-mcmillan.patch.com/articles/commonwealth-court-to-puc-cease-and-
desist-release-the-impact-fee-money
[531]http://stateimpact.npr.org/pennsylvania/2012/10/26/court-bars-public-utility-
commission-from-reviewing-drilling-ordinances/

[532]http://www.post-gazette.com/stories/local/marcellusshale/pennsylvania-submits-its-arguments-in-shale-drilling-law-appeal-651861/

[533]http://www.post-gazette.com/stories/local/neighborhoods-south/local-group-challenges-new-rules-for-shale-gas-industry-651150/

[534]http://www.marcellusprotest.org/content/press-release-pa-supreme-court-ruling

[535]http://www.pacourts.us/assets/opinions/Supreme /out/J-127A-D-2012oajc.pdf?cb=1

[536]http://theberkshireedge.com/physician-flees-tracking-fields-pennsylvania/

[537]http://www.delawareriverkeeper.org/resources/PressReleases/Press%20Release%20Supreme%20Court%20Ruling%2012.19.13%20last.pdf

[538]http://www.delawareriverkeeper.org/resources/PressReleases/Press%20Release%20Supreme%20Court%20Ruling%2012.19.13%20last.pdf

[539]http://www.post-gazette.com/local/2013/12/19/Pennsylvania-Supreme-Court-declares-portions-of-shale-drilling-law-unconstitutional.print

[540]http://www.pennlive.com/opinion/index.ssf/2013/12/editorial_act_13_local_zoning_preemption_unconstitutional_gas_drilling.html

[541]http://marcelluscoalition.org/2013/12/msc-statement-on-pa-supreme-court-ruling/

[542]http://www.senatorscarnati.com/statement-on-supreme-court-ruling-on-act-13/

[543]*Ibid.*

[544]http://www.prnewswire.com/news-releases/commonwealth-files-for-reconsideration-of-recent-supreme-court-act-13-decision-238500971.html

[545]http://marcellusmonitor.wordpress.com/2014/01/02/pa-puc-dep-seek-to-overturn-supreme-court-act-13-decision/

[546]http://marcellusmonitor.wordpress.com/2014/01/02/pa-puc-dep-seek-to-overturn-supreme-court-act-13-decision/

[547]http://www.legis.state.pa.us/cfdocs/billinfo/billinfo.cfm?syear=2011&sind=0&body=S&type=B&bn=1263

[548]http://pubs.usgs.gov/fs/2012/3075/fs2012-3075.pdf

[549]http://www.phillyburbs.com/my_town/palisades/pa-lawmakers-approve-gas-drilling-moratorium-for-bucks-and-montco/article_5ece9717-8d8e-5e87-ae87-2737c134a187.html

[550]http://www.phillyburbs.com/my_town/palisades/pa-lawmakers-approve-gas-drilling-moratorium-for-bucks-and-montco/article_5ece9717-8d8e-5e87-ae87-2737c134a187.html

[551]*Ibid.*

[552]http://www.phillyburbs.com/my_town/palisades/pa-lawmakers-approve-gas-drilling-moratorium-for-bucks-and-montco/article_5ece9717-8d8e-5e87-ae87-2737c134a187.html

[553]http://www.phillyburbs.com/my_town/palisades/pa-lawmakers-approve-gas-drilling-moratorium-for-bucks-and-montco/article_5ece9717-8d8e-5e87-ae87-2737c134a187.html

[554]http://www.buckslocalnews.com/articles/2012/07/01/bristol_pilot/news/doc4ff067a5cea73798138351.txt

[555]http://www.nofrackedgasinmass.org/2015/12/22/historic-los-angeles-methane-leak-puts-natural-gas-emissions-under-scrutiny/

[556]http://www.arb.ca.gov/research/reports/aliso_canyon_natural_gas_leak_updates-sa_flights_thru_dec_12_2015.pdf

[557]http://www.latimes.com/local/lanow/la-me-ln-gas-company-taken-to-task-over-gas-leak-near-porter-ranch-20151124-story.html

[558]http://losangeles.cbslocal.com/2015/12/17/lausd-board-to-consider-relocating-students-from-2-porter-ranch-schools-in-wake-of-gas-leak/

[559]http://tfr.faa.gov/save_pages/detail_5_3842.html

[560]http://www.nofrackedgasinmass.org/2015/12/22/historic-los-angeles-methane-leak-puts-natural-gas-emissions-under-scrutiny/

[561]http://www.epa.gov/ghgreporting/documents/pdf/2010/Subpart-W_TSD.pdf

[562]http://www.netl.doe.gov/File%20Library/Research/Oil-Gas/shale-gas-primer-update-2013.pdf

[563]http://mashable.com/2014/04/08/carbon-dioxide-highest-levels-global-warming/#:eyJzIjoidCIsImkiOiJfbHoxd3RuN3VlYzcxcWWNlZiiJ9

[564]http://thinkprogress.org/climate/2014/04/09/3424704/carbon-dioxide-highest-level/

[565]http://www.economist.com/blogs/schumpeter/2013/08/interview-george-mitchell

[566]http://pubs.acs.org/doi/full/10.1021/acs.est.5b02275

[567]http://www.endocrinedisruption.com/files/Oct2011HERA10-48forweb3-3-11.pdf

[568]http://www.envirobank.org/index.php?sid=5&rec=138

[569]http://www.nytimes.com/2013/07/29/opinion/gangplank-to-a-warm-future.html?_r=0

[570]*Ibid.*

[571]http://www.springer.com/earth+sciences+and+geography/meteorology+%26+climatology?SGWID=0-10009-12-565099-0

[572]http://www.pnas.org/content/early/2014/06/25/1323422111

[573]http://www.nature.com/news/air-sampling-reveals-high-emissions-from-gas-field-1.9982

[574]http://researchmatters.noaa.gov/news/Pages/COoilgas.aspx

[575]http://www.wtop.com/?nid=41&sid=2651787

[576]*Ibid.*

[577]http://co2scorecard.org/home/researchitem/27

[578]http://dataspace.princeton.edu/jspui/bitstream/88435/dsp019s1616326/1/Kang_princeton_0181D_10969.pdf

[579]http://dataspace.princeton.edu/jspui/handle/88435/dsp019s1616326

[580]http://www.desmogblog.com/2014/10/28/when-shale-runs-dry-look-future-drilling-fracking

[581]http://www.investmentu.com/2011/September/natural-gas-flaring.html

[582]http://www.taxpayer.net/images/uploads/downloads/BurningMoney.pdf

[583]http://breakingenergy.com/2015/01/27/capturing-methane-to-fuel-drilling-operations/

[584]http://breakingenergy.com/2015/01/27/capturing-methane-to-fuel-drilling-operations/

[585]http://www.eia.gov/dnav/ng/ng_prod_sum_a_epg0_vgv_mmcf_a.htm

[586]http://www.environmentamerica.org/sites/environment/files/reports/EA_FrackingNumbers_scrn.pdf

[587]http://www.earthworksaction.org/files/publications/Up-In-Flames_FINAL.pdf

[588] See-http://www.expressnews.com/business/eagleford/item/Up-in-Flames-Day-1-Flares-in-Eagle-Ford-Shale-32626.php

[589]http://www.jacksonkelly.com/jk/pdf/C2110376.PDF

[590]http://www.epa.gov/airquality/oilandgas/pdfs/20120417fs.pdf

[591]http://www.epa.gov/airquality/oiland gas

[592]*Ibid.*

[593]http://clerk.house.gov/evs/2014/roll231.xml

[594]http://www.wsj.com/articles/obama-to-cast-climate-change-as-a-national-security-threat-1432126767

[595]http://www.dailymail.co.uk/news/article-1328366/John-Shimkus-Global-warming-wont-destroy-planet-God-promised-Noah.html

[596]http://www.theguardian.com/commentisfree/2012/nov/04/america-theologians-climate-science-denial

[597]http://www.huffingtonpost.com/victor-stenger/global-warming-and-religi_b_864014.html

[598]http://thehill.com/policy/energy-environment/130405-gop-fight-for-energy-gavel-mars-otherwise-seamless-transition

[599]http://crew.3cdn.net/29e185cb94de445559_rom6b5st3.pdf

[600]http://thehill.com/policy/energy-environment/231540-inhofes-gambit-inside-the-senates-big-climate-vote

[601]http://www.mcclatchydc.com/2015/01/21/253927/senate-not-ready-to-tie-climate.html

[602]http://www.nationaljournal.com/twenty-sixteen/for-gop-presidential-candidates-a-slightly-changing-climate-20150208

[603]http://www.mcclatchydc.com/2015/01/21/253927/senate-not-ready-to-tie-climate.html

[604]http://closup.umich.edu/files/ieep-nsee-2015-fall-climate-belief.pdf

[605]http://www.utenergypoll.com/wp-content/uploads/2014/04/October-2015-UT-Energy-Poll-Final2.pdf

[606]http://www.realclimate.org/index.php/archives/2005/01/senator-inhofe/

[607]http://www.nytimes.com/2005/06/08/politics/08climate.html?pagewanted=print&_r=0

[608]http://www.huffingtonpost.com/al-gore/obama-climate-change-speech_b_3498596.html

[609]http://www.whitehouse.gov/the-press-office/2013/06/25/remarks-president-climate-change

[610]http://www.whitehouse.gov/the-press-office/2014/06/25/remarks-president-league-conservation-voters-capital-dinner

[611]http://www.dailykos.com/story/2014/06/14/1307018/-President-Obama-s-remarks-at-UC-Irvine-commencement-ceremony

[612]http://www.theguardian.com/environment/2015/aug/21/obama-green-light-on-shell-arctic-drilling-tensions-high

[613]http://www.washingtonpost.com/news/energy-environment/wp/2015/07/22/obama-administration-greenlights-shell-drilling-off-alaskas-arctic-coast/

[614]https://royalsociety.org/~/media/Royal_Society_Content/policy/projects/climate-evidence-causes/climate-change-evidence-causes.pdf

[615]http://thinkprogress.org/climate/2014/03/20/3416741/climate-scientists-alarmed/

[616]http://skepticalscience.com/97-percent-consensus-cook-et-al-2013.html

[617]http://stateimpact.npr.org/pennsylvania/2014/06/19/former-state-health-employees-say-they-were-silenced-on-drilling/

[618]http://stateimpact.npr.org/pennsylvania/2014/06/19/former-state-health-employees-say-they-were-silenced-on-drilling/

[619]http://stateimpact.npr.org/pennsylvania/2014/06/19/former-state-health-employees-say-they-were-silenced-on-drilling/

[620]http://ecowatch.com/2015/06/30/fracking-health-complaints/

[621]http://ecowatch.com/2015/06/30/fracking-health-complaints/

[622]http://www.legis.state.pa.us/cfdocs/billinfo/billinfo.cfm?syear=2011&sind=0&body=S&type=B&BN=1100

[623]http://www.pennlive.com/midstate/index.ssf/2014/07/pa_didnt_address_fracking_heal.html

[624]http://www.healthyamericans.org/assets/files/TFAH2013InvstgAmrcsHlth05%20FINAL.pdf

[625]http://powersource.post-gazette.com/powersource/policy-powersource/2014/04/04/Pennsylvania-expects-higher-Act-13-impact-fees-in-2014/stories/201404040167

[626]http://thetimes-tribune.com/opinion/make-health-top-priority-1.1726421

[627]http://www.huffingtonpost.com/2014/07/12/pennsylvania-fracking-former-health-secretary_n_5580980.html?utm_hp_ref=green

[628]http://www.ombwatch.org/files/info/naturalgasfrackingdisclosure_highres.pdf

[629]http://www.aspendailynews.com/section/home/158039

[630]http://www.garfield-county.com/public-health/documents/1   Complete HIA without Appendix D.pdf

[631]http://journals.plos.org/plosone/article?id=10.1371/journal.pone.0131093

[632]http://press.endocrine.org/doi/abs/10.1210/en.2013-1697

[633]http://www.endocrinedisruption.com/chemicals.introduction.php

[634]http://cce.cornell.edu/EnergyClimateChange/NaturalGasDev/Documents/PDFs/fracking chemicals from a public health perspective.pdf

[635]http://www.psehealthyenergy.org/data/LNG_SignOnLetterPDF.pdf

[636]http://articles.baltimoresun.com/2012-01-04/features/bal-cdc-scientist-urges-more-gas-drilling-study-20120104_1_shale-gas-drilling-fracking-impacts

[637]http://www.colorado.edu/news/releases/2012/10/02/nsf-awards-cu-boulder-led-team-12-million-study-effects-natural-gas

[638]http://psna.org/2012/06/nurses-promote-healthier-energy-choices/

[639]http://concernedhealthny.org/wp-content/uploads/2014/05/Medical-Experts-to-Governor-Cuomo-May-29FINAL.pdf

[640]http://www.psehealthyenergy.org/data/lettertoGovCuomofinal.pdf

[641]http://www.scribd.com/doc/67626513/Letter-to-Cuomo-on-Fracking

[642]http://www.nysbcsen.org/the-link-between-hydrofracking-and-cancer/

[643]http://www.timesunion.com/default/article/Health-groups-urge-3-5-gas-frack-ban-by-Cuomo-5515089.php

[644]http://www.shalegas.energy.gov/resources/111811_final_report.pdf

[645]http://www.ehjournal.net/content/13/1/82

[646]http://www.epa.gov/airquality/oilandgas/basic.html

[647]http://www.environmentamerica.org/sites/environment/files/reports/EA_FrackingNumbers_scrn.pdf

[648]*Ibid.*

[649]http://www.stateoftheair.org/2012/health-risks/health-risks-ozone.html

[650]http://www.eurekalert.org/pub_releases/2012-03/uocd-ssa031612.php

[651]http://www.sciencedirect.com/science/article/pii/S0048969712001933

[652]http://pubs.acs.org/doi/pdf/10.1021/es405046r

[653]http://www.sustainableotsego.org/Risk Assessment Natural Gas Extraction-1.htm

[654]http://www.timesunion.com/business/article/Air-near-fracking-sites-carries-cancer-risk-5858256.php

[655]http://www.publicintegrity.org/2014/10/07/15890/new-five-state-study-finds-high-levels-toxic-chemicals-air-near-oil-and-gas-sites

[656]http://truth-out.org/news/item/18636-rural-new-jersey-township-fights-ferc-approved-gas-compressor

[657]http://www.epa.gov/iaq/schools/pdfs/publications/iaqtfs_update44.pdf

[658]http://www.ceh.org/making-news/press-releases/29-eliminating-toxics/658-new-report-finds-fracking-poses-health-risks-to-pregnant-women-and-children-

[659]http://dyson.cornell.edu/research/researchpdf/wp/2012/Cornell-Dyson-wp1212.pdf

[660]http://www.aeaweb.org/aea/2014conference/program/preliminary.php?search_string=currie&search_type=last_name&association=&jel_class=&search=Search#search_box

[661]http://www.upmc.com/media/NewsReleases/2015/Pages/plos-one.aspx

[662]http://gdacc.org/tag/drinking-water/

[663]*Ibid.*

[664]http://ehp.niehs.nih.gov/1307732/

[665]http://triblive.com/business/headlines/6772348-74/health-state-studies#axzz3SfdEtQsk

[666]http://www.propublica.org/article/science-lags-as-health-problems-emerge-near-gas-fields

[667]http://www.marcellushealth.org/final-report.html

[668]http://www.ncbi.nlm.nih.gov/pubmed/23552648 *and* http://leanweb.org/our-work/water/fracking/investigating-links-between-shale-gas-development-and-health-impacts-through-a-community-survey-project-in-pennsylvania

[669]http://www.prendergastlibrary.org/wp-content/uploads/2013/03/New-Solutions-23-1-Binder.pdf

[670]http://www.marcellus-shale.us/pam-judy.htm

[671]http://www.marcellus-shale.us/pam-judy.htm

[672]http://protectingourwaters.wordpress.com/2012/08/16/shale-gas-industry-harms-air-quality-public-health-shale-gas-outrage-news-bulletin-2/

673http://www.dispatch.com/content/stories/local/2014/10/22/fracking-well-monitors-find-high-pollution-levels.html

674*Ibid.*

675http://www.cdc.gov/chronicdisease/resources/publications/AAG/dcpc.htm

676http://eagleford.publicintegrity.org/

677http://www.fwweekly.com/index.php?option=com_content&view=article&id=5433:vapors-sicken-arlington&catid=76:metropolis&Itemid=377

678*Ibid.*

679*Ibid.*

680*Ibid.*

681http://www.fwweekly.com/2009/10/14/sacrificed-to-shale/

682http://townofdish.com/objects/DISH_-_final_report_revised.pdf

683*Ibid.*

684*Ibid.*

685http://www.alternet.org/water/150794/trailer_talk's_frack_talk%3A_why_a_mayor_was_forced_to_leave_his_town_because_of_gas_drilling_/?page=entire

686http://www.insideclimatenews.org/news/22032015/fracking-fumes-force-texas-resident-shutter-home-business-after-decade

687http://insideclimatenews.org/fracking-eagle-ford-shale-big-oil-and-bad-air-texas-prairie-press-release

688http://www.dallasnews.com/news/metro/20150302-studies-casting-light-on-natural-gas-production-and-health.ece

689http://stories.weather.com/fracking

690http://www.post-gazette.com/stories/local/marcellusshale/hunting-club-contends-with-spring-water-contaminations-from-gas-drilling-300746/?print=1

691http://www.democracynow.org/2012/8/8/chevron_oil_refinery_fire_in_richmond

692http://www.democracynow.org/2012/8/8/chevron_oil_refinery_fire_in_richmond

693http://www.mercurynews.com/breaking-news/ci_21302478/investigators-set-comb-through-chevron-richmond-fire-site

694http://www.democracynow.org/2012/8/8/chevron_oil_refinery_fire_in_richmond

695http://articles.latimes.com/2012/aug/14/opinion/la-oe-0814-juhasz-chevron-refinery-pollution-20120814

696http://sanfrancisco.cbslocal.com/2013/01/30/state-fines-chevron-1-million-for-richmond-refinery-fire/

697http://www.safetynewsalert.com/promoting-safety-vs-punishing-violators-a-tale-of-2-federal-agencies/

698http://www.contracostatimes.com/contra-costa-times/ci_26866574/butt-rogers-lead-early-numbers-richmond-mayor-council

699http://www.uticaod.com/news/x133071260/Environmental-group-says-pollution-is-streaming-from-fracking-sites

700http://www.alternet.org/environment/149760/oil_and_gas_companies_illegally_using_diesel_in_fracking_

701http://ecowatch.org/2012/fracking-children/

702http://www.dec.ny.gov/data/dmn/rdsgeisfull0911.pdf.

703http://www.health.ny.gov/press/reports/docs/high_volume_hydraulic_fracturing.pdf

704http://www.dep.state.pa.us/dep/deputate/airwaste/aq/cars/docs/Final_Act_124_Fact_Sheet.pdf

705http://www.thetrucker.com/News/Stories/2009/2/9/Pennsylvaniatruckidlinglawnowineffect.aspx

706http://www.epa.gov/oms/fuels/dieselfuels/index.htm

707http://www.dieselforum.org/news/advancements-in-clean-diesel-technology-and-fuel-to-continue-major-reductions-in-black-carbon-emissions

708http://www.sfgate.com/bayarea/article/Chevron-fire-truck-didn-t-spark-fire-3794857.php

[709]http://www.sustainableotsego.org/Risk Assessment Natural Gas Extraction-1.htm
[710]http://online.wsj.com/article/SB10001424127887323291704578199751783044798.html
[711]http://democrats.energycommerce.house.gov/index.php?q=news/reps-waxman-
markey-and-degette-report-updated-hydraulic-fracturing-statistics-to-epa
[712]*Ibid.*
[713]http://ecowatch.org/2012/still-not-regulated/
[714]http://www.ogj.com/articles/2013/10/congressmen-ask-omb-to-finalize-diesel-in-frac-
fluid-rule.html
[715]http://environmentalintegrity.org/wp-content/uploads/Fracking-Beyond-the-Law.pdf
[716]http://www.ogj.com/articles/2013/10/congressmen-ask-omb-to-finalize-diesel-in-frac-
fluid-rule.html.
[717]http://energyindepth.org/national/eip-diesel-fuel-report-lacks-data-integrity/
[718]http://bigstory.ap.org/article/39786bbf509e412a9feb9b58a6534a36/drilling-
boom-brings-rising-number-harmful-waste-spills
[719]http://ec.europa.eu/environment/integration/energy/unconventional_en.htm
[720]http://pennbpc.org/sites/pennbpc.org/files/CMSC-Final-Report.pdf
[721]http://www.energyindepth.org/wp-content/uploads/2012/05/myers-potential-
pathways-from-hydraulic-fracturing4.pdf
[722]http://www.eveningtribune.com/opinions/columnists/x912153418/New-fracking-
study-should-be-required-reading-for-local-leaders
[723]https://www.earthworksaction.org/files/pubs-
others/ACSA_Comprehensive_Analysis_of_Groundwater_Quality_in_the_Barnett_Shal
e_Region.pdf
[724]*Ibid.*
[725]http://www.scribd.com/doc/91428341/Delivered-Testimony-of-Professor-Margaret-
Rafferty-R-N-at-Hydrofracking-Forum-4-25-12
[726]http://old.post-gazette.com/pg/08322/928571-113.stm
[727]http://www.nytimes.com/2011/02/27/us/27gas.html?pagewanted=2&_r=1&ref=homep
age&src=me&
[728]http://www.propublica.org/article/wastewater-from-gas-drilling-boom-may-threaten-
monongahela-river
[729]http://www.enidnews.com/news/up-to-gallons-of-hydrochloric-acid-spilled-near-
hennessey/article_91b71349-670f-5840-88b4-9b804183465b.html
[730]http://pubs.acs.org/doi/abs/10.1021/es4011724
[731]http://today.duke.edu/2012/07/marcellus
[732]http://shale.sites.post-gazette.com/index.php/news/archives/24607-federal-agencies-
probe-range-resources-yeager-marcellus-shale-gas-drilling-site
[733]*Ibid.*
[734]http://protectingourwaters.wordpress.com/2012/09/17/methane-from-gas-drilling-
manning-family-told-dont-use-your-kitchen-stove/
[735]http://www.post-gazette.com/local/westmoreland/2014/07/05/Families-well-water-
disrupted-near-Ligonier/stories/201407050100
[736]http://news.nationalgeographic.com/news/2010/10/101022-energy-marcellus-shale-
gas-environment/
[737]http://pafaces.wordpress.com/2010/04/23/stephanie-hallowich-speaks-out/
[738]http://www.marcellus-shale.us/Stephanie-Hallowich.htm
[739]http://www.marcellus-shale.us/Stephanie-Hallowich.htm
[740]http://pafaces.wordpress.com/2010/04/23/stephanie-hallowich-speaks-out/
[741]http://pafaces.wordpress.com/copyright-and-disclosure-policies/
[742]http://news.nationalgeographic.com/news/2010/10/101022-energy-marcellus-shale-
gas-environment/
[743]*Ibid.*
[744]http://www.cnbc.com/id/45208498

[745]http://www.bloomberg.com/news/2013-02-19/texas-fracker-accused-of-bully-tactics-against-foes.html

[746]http://www.prnewswire.com/news-releases/dep-assesses-89-million-civil-penalty-against-range-resources-for-failure-to-repair-leaking-gas-well-300100011.html

[747]http://shale.sites.post-gazette.com/index.php/news/archives/24176-hallowich-family-files-court-action-against-range-resources

[748]http://citizenspeak.org/campaign/saynotofracking/epa-send-clean-water-families-impacted-fracking-butler-county-pa

[749]http://protectingourwaters.wordpress.com/2012/03/02/i-just-want-water-demonstrators-confront-rex-energy-in-butler-county/

[750]http://www.marcellusoutreachbutler.org/2/post/2012/03/the-plethora-of-excuses-and-explanations-disintegrates.html

[751]http://www.post-gazette.com/stories/local/marcellusshale/chamber-officials-lobby-to-keep-shale-costs-low-645818/

[752]http://citizensvoice.com/news/acid-spills-off-well-pad-1.1339831

[753]http://articles.philly.com/2012-08-06/business/33049766_1_pennenvironment-michael-krancer-shale-gas

[754]http://articles.philly.com/2012-08-06/business/33049766_1_pennenvironment-michael-krancer-shale-gas

[755]https://docs.google.com/file/d/0B4Y3VQLxjkxOWjM0QTlua2tnYkE/edit?pli=1

[756]http://www.ens-newswire.com/ens/feb2012/2012-02-16-02.html

[757]http://files.dep.state.pa.us/OilGas/BOGM/BOGMPortalFiles/RadiationProtection/rls-DEP-TENORM-01xx15AW.pdf

[758]   *See*—http://www.post-gazette.com/business/2014/05/27/Two-more-containers-found-with-Marcellus-Shale-sludge-radioactivity-in-Washington-County/stories/201405270162

[759]http://blog.shaleshockmedia.org/2013/07/12/orphaned-frackwaste-finds-foster-home-in-idaho/

[760]http://pubs.acs.org/doi/abs/10.1021/es402165b?prevSearch=%2528blacklick%2Bcreek%2529%2Band%2B%255BContrib%253A%2BVengosh%252CAvner%255D&searchHistoryKey=

[761]http://www.usatoday.com/story/news/nation/2013/10/02/fracking-radioactive-water-pennsylvania/2904829/

[762]http://www.delawareriverkeeper.org/resources/Reports/Decision%20RTK%20appeal%207.11.14.pdf

[763]*Ibid.*

[764]http://www.frackcheckwv.net/

[765]http://www.desmogblog.com/radionuclides-tied-shale-gas-fracking-can-t-be-ignored-possible-health-hazard

[766]http://www.epa.gov/radon/pubs/citguide.html

[767]http://ehp.niehs.nih.gov/wp-content/uploads/advpub/2015/4/ehp.1409014.acco.pdf

[768]http://ehp.niehs.nih.gov/wp-content/uploads/advpub/2015/4/ehp.1409014.acco.pdf

[769]http://www.nirs.org/radiation/radonmarcellus.pdf

[770]
http://news.psu.edu/story/143694/2012/12/17/analysis-marcellus-flowback-finds-high-levels-ancient-brines

[771]http://www.marcellus-shale.us/radioactive-shale.htm

[772]http://www.ohio.com/blogs/drilling/ohio-utica-shale-1.291290/study-says-pennsylvania-drilling-waste-high-in-radium-1.331703

[773]http://www.nytimes.com/2011/02/27/us/27gas.html?pagewanted=all

[774]http://www.energyforamerica.org/2012/11/29/hollywood-perpetuates-hydraulic-fracturing-myths-in-promised-land/

[775]http://www.dispatch.com/content/stories/local/2012/09/03/gas-well-waste-full-of-radium.html

[776]http://www.bloomberg.com/news/2013-10-02/radiation-in-pennsylvania-creek-seen-as-legacy-of-frackin.html

[777]*Ibid.*

[778]http://www.denverpost.com/breakingnews/ci_24106485/colorado-flood-rescue-operations-back-track-after-sundays

[779]http://www.denverpost.com/editorials/ci_24107674/colorado-floods-recede-will-rebuild

[780]http://www.thedenverchannel.com/news/local-news/13500-gallons-of-oil-spilled-along-st-vrain125-barrel-released-in-milliken

[781]http://www.coga.org/index.php/Events/ColoradoFloods#sthash.7kW2zBK7.dpbs

[782]http://www.counterpunch.org/2013/09/18/fracking-and-colorado-flooding-dont-mix/

[783]http://ecowatch.com/2014/01/08/shane-davis-fracking-colorado/

[784]http://www.denverpost.com/breakingnews/ci_24132296/oil-spill-along-st-vrain-river-near-platteville

[785]http://www.denverpost.com/breakingnews/ci_24107038/state-and-industry-struggle-assess-damage-flooded-oil

[786]http://thecoloradoobserver.com/2013/09/anti-fracking-activists-blasted-for-hyping-floods-to-push-political-agenda/

[787]http://www.coloradoindependent.com/144028/flood-damaged-andarko-tank-spilling-thousands-of-gallons-of-oil-into-south-platte-river/comment-page-1

[788]http://www.atsdr.cdc.gov/hac/pha/ChesapeakeATGASWellSite/ChesapeakeATGASWellSiteHC110411Final.pdf

[789]http://thetimes-tribune.com/news/after-blowout-most-evacuated-families-return-to-their-homes-in-bradford-county-1.1135253

[790]http://thetimes-tribune.com/news/after-blowout-most-evacuated-families-return-to-their-homes-in-bradford-county-1.1135253

[791]http://www.propublica.org/article/response-to-pa-gas-well-accident-took-13-hours-despite-state-plan-for-quick

[792]*Ibid.*

[793]http://m.thedailyreview.com/news/chesapeake-gets-dep-notice-of-violation-1.1136716

[794]http://www.oag.state.md.us/Press/2011/050211.html

[795]http://www.bloomberg.com/article/2012-06-14/a0r9O1j1J9Rk.html

[796]https://www.marcellusgas.org/drillco/drillco_details.php

[797]http://dl.dropbox.com/u/48182083/drilling/nov.pdf

[798]http://wnep.com/2012/07/06/acid-spill-at-gas-well/

[799]http://www.reuters.com/article/2013/03/31/us-exxon-pipeline-spill-idUSBRE92U00220130331

[800]http://www.upi.com/Business_News/Energy-Resources/2013/04/10/Exxon-faces-deadline-for-oil-spill/UPI-56081365598614/

[801]http://www.politico.com/story/2015/04/the-little-pipeline-agency-that-couldnt-117147_Page2.html

[802]http://www.denverpost.com/breakingnews/ci_22817087/parachute-creek-spill-continues-uncontained-cause-source-unknown

[803]http://fromthestyx.wordpress.com/2013/08/27/parachute-creek-spill-day-174/

[804]http://www.postindependent.com/news/7836384-113/creek-benzene-parachute-groundwater

[805]http://cogcc.state.co.us/Announcements/Hot_Topics/SpillsAndReleases/SpillAnalysisByYear2014Q3.pdf

[806]http://thetimes-tribune.com/news/wyoming-county-well-malfunction-causes-spill-evacuation-1.1458575

[807]http://thetimes-tribune.com/news/dep-spill-at-well-site-seeps-into-house-miniature-horse-farm-1.1481928

[808]http://www.marcellusgas.org/drillco/drillco_details.php

[809]http://yosemite.epa.gov/opa/admpress.nsf/d0cf6618525a9efb85257359003fb69d/dd0d219e038a8d4885257bac006b398c!OpenDocument

[810]http://www.courthousenews.com/2013/07/22/59559.htm
[811]http://in.reuters.com/article/2013/07/18/usa-energy-xto-idINL1N0FO27G20130718
[812]http://www.pennlive.com/midstate/index.ssf/2013/09/attorney_generals_criminal_cha.html
[813]http://www.xtoenergy.com/pressreleases/xto-energy-to-challenge-charges-by-pennsylvania-attorney-general
[814]http://www.nytimes.com/interactive/2014/11/23/us/north-dakota-oil-boom-downside.html?_r=1
[815]http://billingsgazette.com/news/local/a-year-after-spill-life-returns-to-normal-for-most/article_9b8347b5-a431-5736-8359-5a86fe08a472.html
[816]http://www.prairiebizmag.com/event/article/id/22444/
[817]http://www.dailykos.com/story/2015/01/19/1358683/-50-000-barrels-of-oil-spill-into-Yellowstone-River#
[818]http://edition.cnn.com/2015/01/20/us/yellowstone-river-spill/
[819]http://www.willistonherald.com/news/brine-spill-reaches-missouri-river/article_82d99970-9e83-11e4-9baf-f370ffb3b637.html
[820]http://www.willistonherald.com/news/brine-spill-reaches-missouri-river/article_82d99970-9e83-11e4-9baf-f370ffb3b637.html
[821]http://www.statesman.com/news/news/opinion/the-case-against-fracking/nRNQj/
[822]http://energyindepth.org/national/eip-diesel-fuel-report-lacks-data-integrity/
[823]http://www.propublica.org/article/scientific-study-links-flammable-drinking-water-to-fracking
[824]http://www.nicholas.duke.edu/cgc/pnas2011.pdf
[825]http://dukemagazine.duke.edu/article/fracking-findings
[826]http://www.pnas.org/content/early/2012/07/03/1121181109.full.pdf+html
[827]http://www.pnas.org/content/early/2013/06/19/1221635110.abstract
[828]http://bigstory.ap.org/article/studies-find-methane-pa-drinking-water
[829]http://s3.amazonaws.com/propublica/assets/methane/garfield_county_final2.pdf
[830]http://www.propublica.org/article/officials-in-three-states-pin-water-woes-on-gas-drilling-426
[831]http://www.dnr.state.oh.us/Portals/11/bainbridge/report.pdf
[832]*Ibid.*
[833]*Ibid.*
[834]http://solomonsads.blogspot.com/2011/03/editor-of-bradford-era-asks-how-dare.html
[835]http://www.casey.senate.gov/newsroom/releases/casey-calls-for-federal-help-with-gas-explosions-in-nw-pa
[836]http://stateimpact.npr.org/texas/2012/01/20/railroad-commission-responds-to-explosion-in-pearsall/
[837]http://news.nationalgeographic.com/news/2010/10/101022-energy-marcellus-shale-gas-environment/
[838]http://www.casey.senate.gov/issues/issue/?id=ce690501-bb77-4195-a6d1-87152b9ae298
[839]http://thedailyreview.com/dep-fines-chesapeake-900-000-for-bradford-county-pollution-1.1148038
[840]http://wcexaminer.com/index.php/archives/news/30232
[841]http://www.cleanair.org/program/outdoor_air_pollution/marcellus_shale/independent_study_finds_significant_fault_line_methane
[842]https://www.dropbox.com/s/kircmcdy7jdtfgw/DEP Letter to Minott 071212.pdf
[843]*Ibid.*
[844]http://www.boston.com/news/local/massachusetts/articles/2012/09/07/report_finds_methane_remains_issue_in_pa_township/
[845]http://thetimes-tribune.com/news/dep-secretary-methane-may-have-leaked-through-perforations-in-bradford-gas-well-1.1343005
[846]http://www.boston.com/news/local/massachusetts/articles/2012/09/07/report_finds_methane_remains_issue_in_pa_township/?page=2

[847]http://www2.epa.gov/enforcement/trans-energy-inc-clean-water-act-settlement#overview

[848]http://cleantechnica.com/2013/12/27/epa-hits-chesapeake-energy-with-record-fracking-fine/

[849]http://thehill.com/policy/energy-environment/227942-feds-fine-exxonmobil-23m-for-wva-fracking-violations

[850]http://theintelligencer.net/page/content.detail/id/575435/Chesapeake-to-Pay--600K-Fine-for-Filling-Wetzel-Co--Stream.html?nav=515

[851]https://stateimpact.npr.org/pennsylvania/2014/12/22/dep-slaps-vantage-energy-with-massive-fine-for-landslide-and-dumping/

[852]http://www.wfaa.com/story/news/local/tarrant-county/2015/06/16/arlington-officials-report-on-fracking-fluid-blowout/28844657/

[853]http://onlinelibrary.wiley.com/doi/10.1111/gwat.12316/abstract

[854]http://missoulian.com/news/local/usgs-study-links-montana-oil-spills-to-arsenic-releases-in/article_dc2f4fec-9bb4-5119-a556-8ae85c6af5ff.html

[855]http://stateimpact.npr.org/pennsylvania/2011/11/18/growing-tensions-within-the-delaware-river-basin-commission-halt-decision-on-gas-drilling/

[856]http://www.biologicaldiversity.org/campaigns/california_fracking/pdfs/20140915_State_Board_UIC_well_list_Category_1a.pdf
ANDhttp://www.biologicaldiversity.org/campaigns/california_fracking/pdfs/20140915_Bishop_letter_to_Blumenfeld_Responding_to_July_17_2014_UIC_Letter.pdf

[857]http://www.biologicaldiversity.org/news/press_releases/2014/fracking-10-06-2014.html

[858]http://www.cleveland.com/metro/index.ssf/2013/08/youngstown_man_admits_dumping.html

[859]http://www.cleveland.com/court-justice/index.ssf/2014/08/youngstown_contractor_sentence.html

[860]http://www.post-gazette.com/stories/local/state/state-charges-local-company-for-dumping-wastewater-and-sludge-287538/?print=1

[861]http://www.alternet.org/fracking/toxic-wastewater-dumped-streets-and-rivers-night-gas-profiteers-getting-away-shocking?page=0%2C12&paging=off

[862]http://old.post-gazette.com/pg/pdf/201103/20110317_shipman_awws_gjpresentment.pdf

[863]http://old.post-gazette.com/pg/pdf/201103/20110317_shipman_awws_gjpresentment.pdf

[864]http://www.post-gazette.com/stories/local/state/attorney-general-critical-of-pollution-sentence-wastewater-668583/

[865]http://www.alternet.org/fracking/toxic-wastewater-dumped-streets-and-rivers-night-gas-profiteers-getting-away-shocking?page=0%2C12&paging=off

[866]https://www.census.gov/prod/cen2000/phc-3-40.pdf

[867]http://www.timesleader.com/stories/Dimock-Twp-property-owners-sue-gas-driller-Cabot,106231

[868]http://www.propublica.org/article/officials-in-three-states-pin-water-woes-on-gas-drilling-426

[869]http://www.propublica.org/article/pennsylvania-tells-drilling-company-to-clean-up-its-act-1106

[870]http://weeklypress.com/shale-shame-cabot-fined-heavily-for-dimock-water-contamination-p1896-1.htm

[871]http://thetimes-tribune.com/news/gas-drilling/dep-drops-dimock-waterline-plans-cabot-agrees-to-pay-4-1m-to-residents-1.1077910

[872]http://marcellusmoney.org/candidate/white-donald-c

[873]http://thetimes-tribune.com/news/dimock-officials-reject-offer-of-water-deliveries-1.1241292

[874]http://dailyitem.com/0100_news/x431310713/Cabot-CEO-EPA-investigation-of-Dimock-water-wastes-taxpayer-money
[875]http://thetimes-tribune.com/news/dep-head-calls-epa-knowledge-of-dimock-rudimentary-1.1255658
[876]http://stateimpact.npr.org/pennsylvania/2012/06/01/krancer-once-again-tells-washington-to-back-off/
[877]http://www.ohio.com/news/break-news/pennsylvania-governor-says-drilling-opponents-are-unreasoning-1.336049
[878]http://energyindepth.org/national/video-dimock-residents-tell-binghamton-mayor-enough-is-enough/
[879]http://www.ipetitions.com/petition/enough-is-enough-dimock-residents/
[880]http://ecowatch.org/2012/epa-finds-water-safe-to-drink-despite-explose-levels-of-methane-and-other-toxins/
[881]http://www.propublica.org/article/so-is-dimocks-water-really-safe-to-drink
[882]http://waterdefense.org/blog/water-defense-cries-foul-epa-statement
[883]http://waterdefense.org/news/so-dimock%E2%80%99s-water-really-safe-drink
[884]http://citizensvoice.com/news/cabot-and-dimock-families-near-settlement-1.1358912
[885]http://citizensvoice.com/news/cabot-and-dimock-families-near-settlement-1.1358912
[886]http://wnep.com/2012/08/22/dep-cabot-allowed-to-frack-in-dimock/
[887]http://thetimes-tribune.com/news/dep-lets-cabot-resume-dimock-fracking-1.1361871
[888]http://www.latimes.com/news/nationworld/nation/la-na-epa-dimock-20130728,0,4847442.story
[889]http://www.epa.gov/aboutepa/states/dimock-atsdr.pdf
[890]*Ibid.*
[891]http://www.newpa.com/business/impact-awards
[892]http://www.epa.gov/region8/superfund/wy/pavillion/EPA_ReportOnPavillion_Dec-8-2011.pdf
[893]http://www.scientificamerican.com/article.cfm?id=fracking-linked-water-contamination-federal-agency
[894]http://docs.nrdc.org/energy/files/ene_12050101a.pdf
[895]*Ibid.*
[896]http://www.bloomberg.com/news/2012-09-26/diesel-compounds-found-in-water-near-wyoming-fracking-site-2-.html
[897]http://www.dallasnews.com/investigations/headlines/20110122-gas-in-parker-county-home%E2%80%99s-drinking-water-puts-drilling-epa-in-national-spotlight.ece
[898]http://www.dallasnews.com/investigations/headlines/20110122-gas-in-parker-county-home%E2%80%99s-drinking-water-puts-drilling-epa-in-national-spotlight.ece
[899]http://www.txcourts.gov/media/943997/130928.pdf
[900]http://www.dallasnews.com/investigations/headlines/20110122-gas-in-parker-county-home%E2%80%99s-drinking-water-puts-drilling-epa-in-national-spotlight.ece
[901]http://www.txcourts.gov/media/943997/130928.pdf
[902]http://yosemite.epa.gov/opa/admpress.nsf/ab2d81eb088f4a7e85257359003f5339/713f73b4bdceb126852577f3002cb6fb!OpenDocument
[903]http://www.oilandgaslawyerblog.com/Range Production Company Closing statement.pdf
[904]http://www.dallasnews.com/health/medicine/20101209-epa-2-parker-county-homes-at-risk-of-explosion-after-gas-from-fracked_well-contaminates-aquifer.ece
[905]http://www.dallasnews.com/investigations/headlines/20110122-gas-in-parker-county-home%E2%80%99s-drinking-water-puts-drilling-epa-in-national-spotlight.ece
[906]http://www.texaspolicy.com/pdf/Joint Stipulation of Dismissal.pdf
[907]http://www.cbn.com/cbnnews/healthscience/2012/April/EPA-Drops-Poison-Well-Case-against-Texas-Driller/?Print=true
[908]http://www.texastribune.org/2012/03/30/epa-withdraws-order-against-range-resources/

[909]http://ecowatch.com/2013/12/24/epa-drinking-water-fracking/
[910]http://www.txcourts.gov/media/943997/130928.pdf
[911]http://www.latimes.com/nation/nationnow/la-na-epa-dimock-20130728-m-story.html#page=1
[912]http://insideclimatenews.org/news/02032015/can-fracking-pollute-drinking-water-dont-ask-epa-hydraulic-fracturing-obama-chesapeake-energy
[913]http://www2.epa.gov/sites/production/files/2015-06/documents/hf_es_erd_jun2015.pdf
[914]See—https://www.documentcloud.org/documents/1183455-chk-prospective-study.html?key=932a33d9ba1eaac537a4;
andhttps://www.documentcloud.org/documents/808402-chk-revisions-to-case-study-qapp.html?key=932a33d9ba1eaac537a4#document/p24/a161974;
andhttps://www.documentcloud.org/documents/1183458-epa-hq-2012-001260-part-1-clean-part1.html?key=932a33d9ba1eaac537a4
[915] See—"Beyond the Headlines," by Dory Hippauf.
https://frackorporation.wordpress.com/2015/06/16/epa-2015-fracking-report-beyond-the-headlines/
[916]http://www.propublica.org/article/injection-wells-the-poison-beneath-us
[917]http://www.nrdc.org/energy/files/Fracking-Wastewater-FullReport.pdf
[918]http://www.pennfuture.org/userfiles12/EnvironmentalHearingBoardNOA2012-10-01.pdf
[919]http://switchboard.nrdc.org/blogs/amall/more_evidence_on_the_radioacti.html
[920]http://www.nytimes.com/2011/03/02/us/02gas.html?pagewanted=1
[921]http://www.cga.ct.gov/2013/rpt/2013-R-0469.htm
[922]http://www.post-gazette.com/stories/local/region/pennfuture-accuses-dep-of-permit-dishonesty-656236/
[923]http://www.post-gazette.com/stories/local/region/pennfuture-accuses-dep-of-permit-dishonesty-656236/
[924]http://www.post-gazette.com/stories/local/region/pennfuture-accuses-dep-of-permit-dishonesty-656236/
[925]*Ibid.*
[926]http://maps.fractracker.org/latest/?appid=eb1904df42c848ed967a48c52e873c91
[927]http://www.newsweek.com/oil-and-gas-wastewater-used-de-ice-roads-new-york-and-pennsylvania-little-310684
[928]http://www.newsweek.com/oil-and-gas-wastewater-used-de-ice-roads-new-york-and-pennsylvania-little-310684
[929]http://pennbpc.org/sites/pennbpc.org/files/CMSC-Final-Report.pdf
[930]http://thetimes-tribune.com/news/gas-company-whistle-blower-details-spills-errors-1.1234817
[931]http://www.latimes.com/local/lanow/la-me-ln-pits-oil-wastewater-20150226-story.html
[932]http://www.latimes.com/opinion/editorials/la-ed-fracking-wastewater-tainted-aquifers-california-20150213-story.html
[933]http://insideclimatenews.org/news/20120515/bureau-land-management-blm-fracking-regulations-natural-gas-chemical-disclosure?page=show
[934]http://newamericamedia.org/2012/05/feds-punt-on-leadership-over-fracking-rules-experts-say.php
[935]*Ibid.*
[936]http://ecowatch.com/2013/victory-in-fracking-wastewater-fight/
[937]http://www.portal.state.pa.us/portal/server.pt/community/newsroom/14287?id=%2017071%20&typeid=1
[938]http://www.epa.gov/region3/marcellus_shale/pdf/letter/krancer-letter5-12-11.pdf
[939]http://www.epa.gov/region03/marcellus_shale/pdf/ao/aocc-frs-hart-pbt-executed-5-8-13.pdf

[940]http://publicsource.org/investigations/briny-water-flows-into-southwestern-pa-streams

[941]http://www.waterworld.com/articles/2011/12/wastewater-plants-not-designed-for-fracking-water-says-robert-f-kennedy-jr.html

[942]http://www.toxicstargeting.com/MarcellusShale/documents/testimony/2011/12/12/canandaigua

[943]http://files.dep.state.pa.us/Mining/Abandoned%20Mine%20Reclamation/Abandoned MinePortalFiles/MIW/USGS_Dissolved_Metals_part1.pdf

[944]http://publicsource.org/investigations/briny-water-flows-into-southwestern-pa-streams

[945]http://oilandgas.ohiodnr.gov/industry/underground-injection-control

[946]http://water.epa.gov/type/groundwater/uic/class2/

[947]http://www.propublica.org/article/injection-wells-the-poison-beneath-us

[948]http://publicsource.org/investigations/briny-water-flows-into-southwestern-pa-streams.

[949]http://www.propublica.org/article/injection-wells-the-poison-beneath-us

[950]http://www.propublica.org/article/injection-wells-the-poison-beneath-us

[951]http://journalstar.com/news/opinion/editorial/editorial-a-trashy-reputation/article_eab59e83-2bdc-5fcd-8551-fe0931b8ab60.html

[952]http://www.hutchnews.com/print/Sun--explosions-reflection--1

[953]http://www.propublica.org/article/an-unseen-leak-then-boom

[954]*Ibid.*

[955]http://www.gao.gov/assets/670/665064.txt

[956]http://www.lcountyfracking.org/archives/465

[957]http://www.ohio.com/news/local/ohio-s-volume-of-drilling-waste-going-into-injection-wells-grows-by-15-percent-1.490607

[958]http://ecowatch.com/2012/cincinnati-becomes-first/

[959]http://www.ohio.com/blogs/drilling/ohio-utica-shale-1.291290/cincinnati-supports-proposed-statewide-ban-on-injection-wells-1.397947

[960]http://www.bcbr.com/article/20140725/EDITION/140729945

[961]http://www.usgs.gov/blogs/features/usgs_top_story/man-made-earthquakes/

[962]http://www.desmogblog.com/directory/vocabulary/6566

[963] Rubinstein, Justin L., *et al.*, "Present Triggered Seismicity Sequence in the Raton Basin of Southern Colorado/Northern New Mexico," professional paper presented to the American Geophysical Union Fall meeting, Dec. 3-7, 2012.

[964]http://www.cnn.com/2012/06/15/us/fracking-earthquakes/index.html

[965]http://www.rrc.state.tx.us/about/faqs/hydraulicfracturing.php

[966]http://www.utexas.edu/news/2012/08/06/correlation-injection-wells-small-earthquakes/

[967]http://www.texassharon.com/2012/07/13/three-earthquakes-last-night/

[968]http://www.ibtimes.com/articles/342886/20120518/earthquake-texas-hydraulic-fracturing-waste-water-injection.htm

[969]*Ibid.*

[970]http://redgreenandblue.org/2011/11/06/did-fracking-cause-the-oklahoma-earthquake/

[971]http://bizbeatblog.dallasnews.com/files/2014/10/Disposal-Well-Rule-Amendments-Oct-2014-1.pdf

[972]http://www.allgov.com/Controversies/ViewNews/Arkansas_Suspends_Drilling_of_Injection_Wells_after_Earthquake_Swarm_110302

[973]http://www.ibtimes.com/articles/342886/20120518/earthquake-texas-hydraulic-fracturing-waste-water-injection.htm

[974]http://www.aogc.state.ar.us/Hearing Orders/2011/July/180A-2-2011-07.pdf

[975]http://www.scientificamerican.com/article.cfm?id=did-fracking-cause-oklahomas-largest-recorded-earthquake

[976]http://www.okgeosurvey1.gov/pages/earthquakes/information.php

[977]http://www.dailykos.com/story/2015/04/03/1375464/-Oklahoma-s-Earthquakes-Are-Becoming-Too-Frequent-For-The-Oil-and-Gas-Industry-To-Hide?detail=facebook_sf#
[978]http://www.tulsaworld.com/news/local/state-orders-injection-well-shut-down-after-northwestern-oklahoma-earthquake/article_4184d155-d9b2-523d-a8d5-033174ff8584.html
[979]http://earthquake.usgs.gov/regional/ceus/products/newsrelease_05022014.php
[980]http://www.usgs.gov/newsroom/article.asp?ID=4132&from=rss_home#.VVEOhflViko
[981]http://www.tulsaworld.com/newshomepage1/quake-debate-science-questioned-while-state-s-earthquake-studies-go/article_eedb5ada-46eb-5550-8f56-04c5cc9966b4.html
[982]http://www.eenews.net/stories/1059988189
[983]http://www.cnn.com/2014/07/13/us/oklahoma-earthquakes/index.html
[984]http://www.tulsaworld.com/news/local/state-orders-injection-well-shut-down-after-northwestern-oklahoma-earthquake/article_4184d155-d9b2-523d-a8d5-033174ff8584.html
[985]http://www.eenews.net/stories/1060014342
[986]http://www.sciencemag.org/content/341/6142/164.abstract
[987]http://www.eenews.net/stories/1060014342
[988]http://www.reuters.com/article/2014/05/08/us-usa-oklahoma-earthquakes-idUSKBN0DO1FA20140508
[989]http://www.dailykos.com/story/2015/04/03/1375464/-Oklahoma-s-Earthquakes-Are-Becoming-Too-Frequent-For-The-Oil-and-Gas-Industry-To-Hide?detail=facebook_sf#
[990]http://www.nature.com/news/method-predicts-size-of-fracking-earthquakes-1.9608
[991]http://newsok.com/jumping-to-fracking-quake-conclusions-certainly-a-bad-idea/article/3665213
[992]http://newsok.com/mr.-president-welcome-to-the-town-that-fossil-fuel-built/article/3659626
[993]http://ksn.com/2015/01/19/multiple-earthquakes-felt-through-southern-kansas-monday-morning/
[994]http://earthquake.usgs.gov/earthquakes/eventpage/usc000tunu#general_summary
[995]http://www.occupydemocrats.com/fracking-disaster-kansas-went-from-1-earthquake-per-year-to-42-a-week/
[996]http://www2.ljworld.com/news/2015/jan/17/kansas-earthquakes-likely-caused-oil-and-gas-frack/?print
[997]http://www.science20.com/news_articles/fracking_linked_109_earthquakes_youngstown_ohio-118698
[998]http://ohiodnr.com/downloads/northstar/UICReport.pdf
[999]http://truth-out.org/news/item/10606-special-investigation-the-earthquakes-and-toxic-waste-of-ohios-fracking-boom
[1000]http://www.cnn.com/2012/06/15/us/fracking-earthquakes/index.html
[1001]http://truth-out.org/news/item/10606-special-investigation-the-earthquakes-and-toxic-waste-of-ohios-fracking-boom
[1002]http://truth-out.org/index.php?option=com_k2&view=item&id=7245:regulators-say-fracking-wastewater-well-caused-12-earthquakes-in-ohio
[1003]http://www.cbc.ca/news/canada/edmonton/fracking-linked-to-alberta-earthquakes-study-indicates-1.2829484?cmp=rss
[1004] ttp://i2.cdn.turner.com/cnn/2012/images/06/15/induced.seismicity.prepublication.pdf
[1005]*Ibid.*
[1006]http://www.thestar.com/business/article/1159854
[1007]http://www.ibtimes.com/articles/342886/20120518/earthquake-texas-hydraulic-fracturing-waste-water-injection.htm
[1008]http://www.seismosoc.org/society/press_releases/BSSA_105-1_Skoumal_et_al_Press_Release.pdf
[1009]http://www.ogs.ou.edu/pubsscanned/openfile/OF1_2011.pdf

[1010]http://blog.newsok.com/weather/2013/04/16/fyi-this-is-non-weather-but-in-case-you-missed-it-five-earthquakes-in-oklahoma-so-far-this-morning-national-earthquake-information-center-golden-colo/

[1011]http://www.cuadrillaresources.com/cms/wp-content/uploads/2011/11/Final_Report_Bowland_Seismicity_02-11-11.pdf

[1012]http://www.bcogc.ca/document.aspx?documentID=1270&type=.pdf

[1013]http://www.nrdc.org/media/2012/120509.asp

[1014]*Ibid.*

[1015]http://powersource.post-gazette.com/powersource/policy-powersource/2015/07/15/Study-finds-those-living-near-Marcellus-shale-wells-more-likely-to-be-hospitalized-Pennsylvania/stories/201507150215

[1016]http://wvwri.org/wp-content/uploads/2013/10/A-N-L-Final-Report-FOR-WEB.pdf.

[1017]http://www.ncbi.nlm.nih.gov/pubmed/23684268

[1018]http://ec.europa.eu/environment/integration/energy/pdf/fracking study.pdf

[1019]http://www.noiseandhealth.org/article.asp?issn=1463-1741;year=2004;volume=6;issue=23;spage=3;epage=20;aulast=Castelo

[1020]http://airportnoiselaw.org/dblevels.html

[1021]http://www.alternet.org/water/150794/trailer_talk's_frack_talk%3A_why_a_mayor_was_forced_to_leave_his_town_because_of_gas_drilling_/?page=entire

[1022]http://coloradoindependent.com/121266/fracking-operation-in-erie-begins-near-two-elementary-schools-wakes-up-neighborhood

[1023]*Ibid.*

[1024]http://news.nationalgeographic.com/news/2010/10/101022-energy-marcellus-shale-gas-environment/

[1025]http://www.pennlive.com/midstate/index.ssf/2014/01/us_judge_slaps_280000_in_penal.html

[1026]http://thinkprogress.org/climate/2015/02/08/3620627/oil-workers-strike-grows-to-include-ohio-indiana/

[1027]http://thinkprogress.org/climate/2015/02/02/3617967/oil-workers-strike-treatment-wages-safety/

[1028]http://www.bls.gov/ces/highlights052014.pdf

[1029]http://inthesetimes.com/working/entry/17267/building_trades_bctd_lauds_fracking_boom_laughs_off_environmental_concerns

[1030]http://www.usatoday.com/story/money/business/2014/04/20/fracking-foes-cringe-as-unions-back-drilling-boom/7938859/

[1031]http://www.liuna.org/keystone-xl-pipeline

[1032]https://www.uschamber.com/speech/state-american-business-2012-address-thomas-j-donohue-president-ceo-us-chamber-commerce

[1033]http://politicalcorrection.org/factcheck/201201120011

[1034]http://www.marketwired.com/press-release/-1604785.htm

[1035]http://www.ilr.cornell.edu/globallaborinstitute/research/upload/GLI_KeystoneXL_Reportpdf.pdf#page=6

[1036]http://www.dol.gov/opa/media/press/opa/OPA20131768.htm

[1037]http://data.bls.gov/cgi-bin/dsrv?fw

[1038]http://data.bls.gov/cgi-bin/dsrv?fi

[1039]http://www.houstonchronicle.com/news/special-reports/article/Texas-companies-with-fatalities-not-on-violator-5281494.php

[1040]http://article.wn.com/view/2014/05/01/AFLCIO_North_Dakotas_worker_death_rate_highest/

[1041]http://www.nytimes.com/2013/01/28/us/boom-in-north-dakota-weighs-heavily-on-health-care.html?pagewanted=all&_r=0

[1042]http://www.eenews.net/stories/1060007532

[1043]http://www.thedenverchannel.com/news/local-news/osha-fines-halliburton-7k-in-fracking-site-blast-that-killed-1-worker-injured-2-in-weld-county

[1044]http://www.thedenverchannel.com/news/local-news/1-killed-2-hurt-in-weld-county-fracking-site-accident

[1045]http://www.pittsburghlive.com/x/pittsburghtrib/news/regional/s_724557.html

[1046]http://www.victoriaadvocate.com/news/2013/aug/30/oil_rig_tceq_bm_083113_218486/

[1047]http://www.victoriaadvocate.com/news/2013/aug/29/oil_rig_fire_folo_bm_083013_218438/

[1048]http://www.nofrackingway.us/2014/02/17/fracker-leaves-pizza-coupons-near-shale-well-explosion/

[1049]http://www.marcellus-shale.us/pdf/Lanco-DEP-Ltr_3-18-14.pdf

[1050]http://s3.documentcloud.org/documents/1263238/boi-lanco-78.txt

[1051]http://www.naturalgasintel.com/articles/102530-chevron-to-pay-nearly-1m-for-fatal-pennsylvania-well-fire

[1052]http://marcellusdrilling.com/2014/07/parents-of-worker-killed-in-chevron-greene-co-fire-sue/

[1053]http://www.currentargus.com/ci_25664078

[1054]http://www.lawfirmnewswire.com/2014/08/west-texas-oil-field-explosion-kills-two/

[1055]http://www.thedickinsonpress.com/energy/bakken/3717918-nabors-hit-97k-fine-over-oilfield-death

[1056]http://america.aljazeera.com/tools/pressreleases/al-jazeera-americasfaultlinespresentsdeathonthebakkenshale.html

[1057]http://www.thedickinsonpress.com/energy/bakken/3717918-nabors-hit-97k-fine-over-oilfield-death

[1058]http://www.reuters.com/article/2013/07/08/us-antero-fire-idUSBRE9670Q420130708

[1059]http://pipeline.post-gazette.com/news/archives/25234-eight-injured-in-west-virginia-gas-well-explosion

[1060]http://www.nbclosangeles.com/news/local/Torrance-Refinery-Exxon-Mobil-Explosion-292413021.html

[1061]http://www.latimes.com/local/lanow/la-me-ln-exxon-mobil-refinery-blast-20150223-story.html

[1062]http://thinkprogress.org/climate/2015/02/20/3624959/torrance-refinery-explosion-health-concerns/

[1063]http://thinkprogress.org/climate/2015/02/20/3624959/torrance-refinery-explosion-health-concerns/

[1064]http://www.mrt.com/business/oil/article_871a0f22-c856-11e4-844c-13913ca130a9.html

[1065]https://www.osha.gov/pls/imis/establishment.inspection_detail?id=958669.015

[1066]http://www.tandfonline.com/doi/abs/10.1080/.VBDknKOuRas#.VBbOrpRdW1o

[1067]http://www.krextv.com/news/around-the-region/NC5-INVESTIGATION-Deadly-Gas-Cover-Up-Revealed-126869973.html

[1068]*Ibid.*

[1069]http://www.postindependent.com/article/20110708/VALLEYNEWS/110709930

[1070]http://www.stopthefrackattack.org/wp-content/uploads/2012/07/STFA-BREATHE_Fact-Sheet1.pdf

[1071]http://polis.house.gov/news/documentsingle.aspx?DocumentID=229905

[1072]http://www.denverpost.com/breakingnews/ci_23637022/osha-fines-3-firms-finds-workers-at-parachute

[1073]http://www.youtube.com/watch?v=N0on1TiO4DU&feature=youtu.be

[1074]http://protectingourwaters.wordpress.com/2012/07/18/whistle-blowing-truck-driver-on-law-flouting-fracking-companies/

[1075]http://www.twincities.com/localnews/ci_27525671/proposed-frac-sand-mining-operation-could-be-biggest

[1076]http://shalenow.com/the-marcellus-shale/

[1077]http://grist.org/list/frackers-are-strip-mining-the-midwest-for-sand/

[1078]http://www.trefis.com/stock/nsc/articles/137803/how-railroad-companies-could-benefit-from-shale-gas-boom/2012-08-09

[1079]http://marcellusdrilling.com/2012/02/csx-other-railroads-get-boost-from-shale-gas-shipments/

[1080]*Ibid.*

[1081]http://www.progressiverailroading.com/csx_transportation/article/Marcellus-and-Utica-shale-drilling-keeps-railroads-busy--34779

[1082]http://www.progressiverailroading.com/mechanical/news/ACF-Industries-reopens-Pennsylvania-railcar-plant-due-to-booming-energy-sector--36411

[1083]http://stateimpact.npr.org/pennsylvania/2013/06/05/touring-wall-street-moguls-company-corbett-denies-politics-influences-his-visits/

[1084]http://dnr.wi.gov/topic/Mines/documents/SilicaSandMiningFinal.pdf

[1085]https://www.osha.gov/silica/factsheets/OSHA_FS-3682_Silica_GIM.html

[1086]http://www.osha.gov/dts/hazardalerts/hydraulic_frac_hazard_alert.html

[1087]http://www.npr.org/blogs/health/2013/03/29/175042708/Sand-From-Fracking-Operations-Poses-Silicosis-Risk

[1088]http://ecowatch.org/2012/mining-companies-invade-wisconsin-for-frac-sand/

[1089]http://truth-out.org/news/item/16095-the-mines-that-fracking-built

[1090]http://chippewa.com/search/?l=25&sd=desc&s=start_time&f=html&q=silica%20sand%20moratorium

[1091]http://thegazette.com/2013/02/04/allamakee-supervisors-approve-frac-sand-mining-moratorium/

[1092]http://www.iowadnr.gov/Portals/idnr/uploads/water/wse/jordan_aquifer.pdf

[1093]http://www.aflcio.org/content/download/.../1/.../safetyhealth_05222012.pdf

[1094]http://www.mintpressnews.com/aflcio-demands-government-agencies-protect-fracking-workers-dangerous-exposure/

[1095]http://inthesetimes.com/working/entry/13286/fracking/

[1096]http://www.osha.gov/dts/hazardalerts/hydraulic_frac_hazard_alert.html

[1097]https://www.osha.gov/silica/index.html

[1098]http://permianshale.com/news/id/15/osha-calls-silica-limit-proposed-silica-limit-hurt-fracking-industry-group-says-2/

[1099]https://www.osha.gov/Publications/OSHA3763.pdf

[1100]http://documents.foodandwaterwatch.org/doc/Social_Costs_of_Fracking.pdf

[1101]http://documents.foodandwaterwatch.org/doc/Social_Costs_of_Fracking.pdf

[1102]http://www.motherjones.com/environment/2012/11/fracking-safety-north-dakota

[1103]http://bigstory.ap.org/article/ap-impact-deadly-side-effect-fracking-boom-0

[1104]http://www.nytimes.com/2012/05/15/us/for-oil-workers-deadliest-danger-is-driving.html?_r=2&pagewanted=2

[1105]http://thedailyreview.com/news/fire-chiefs-traffic-congestion-is-delaying-emergency-response-times-in-bradford-county-1.1095692

[1106]http://documents.foodandwaterwatch.org/doc/Social_Costs_of_Fracking.pdf

[1107]http://documents.foodandwaterwatch.org/doc/Social_Costs_of_Fracking.pdf

[1108]http://keystoneresearch.org/sites/default/files/20141217report.pdf

[1109]https://drive.google.com/viewerng/viewer?a=v&pid=sites&srcid=ZGVmYXVsdGRvbWFpbnxtdWx0aXN0YXRlc2hhbGVZZ3g6NTEyYzY5OWZjYzY4NWM0

[1110]https://drive.google.com/viewerng/viewer?a=v&pid=sites&srcid=ZGVmYXVsdGRvbWFpbnxtdWx0aXN0YXRlc2hhbGVZZ3g6NTEyYzY5OWZjYzY4NWM0

[1111]https://drive.google.com/viewerng/viewer?a=v&pid=sites&srcid=ZGVmYXVsdGRvbWFpbnxtdWx0aXN0YXRlc2hhbGVZZ3g6NTEyYzY5OWZjYzY4NWM0

[1112]http://www.npr.org/2011/09/25/140784004/new-boom-reshapes-oil-world-rocks-north-dakota

[1113]http://data.bls.gov/timeseries/LNS14000000

[1114]http://www.bls.gov/web/laus/laumstrk.htm

[1115]http://ecofriendlydevelopment.net/investment_files/Brian%20Williams%20broadcast,%20N.%20Dakota%27s%20Housing%20Shortage.pdf

[1116]http://www.nber.org/papers/w21359

[1117]http://money.cnn.com/2011/09/28/pf/north_dakota_jobs/index.htm
[1118]http://www.npr.org/2011/09/25/140784004/new-boom-reshapes-oil-world-rocks-north-dakota
[1119]http://www.willistonherald.com/news/williston-rent-highest-in-nation/article_b0d5b4b4-9699-11e3-8b68-001a4bcf887a.html
[1120]http://www.npr.org/2011/12/02/142695152/oil-boom-puts-strain-on-north-dakota-towns
[1121]http://www.ag.nd.gov/Reports/BCIReports/CrimeHomicide/Crime12.pdf
[1122]http://bismarcktribune.com/bakken/crime-up-percent-last-year-in-north-dakota/article_f0a23ec4-f940-11e2-ad25-0019bb2963f4.html
[1123]http://www.heitkamp.senate.gov/public/index.cfm/2013/11/after-heitkamp-brought-white-house-drug-czar-to-state-administration-takes-crucial-steps-to-address-western-nd-drug-problems
[1124]http://www.heitkamp.senate.gov/public/index.cfm/press-releases?ID=8ea1711b-8010-4c64-84cd-a87b1b426b9d
[1125]http://indiancountrytodaymedianetwork.com/2014/11/15/rising-crime-bakken-region-leads-fbi-open-north-dakota-office-157862
[1126]http://www.traffickedreport.com/
[1127]https://lakotalawproject.wordpress.com/2015/03/10/bakken-oil-boom-brings-grave-issues-to-native-communities/
[1128]http://indiancountrytodaymedianetwork.com/2013/08/28/brave-heart-women-fight-ban-man-camps-which-bring-rape-and-abuse-151070
[1129]http://www.nytimes.com/aponline/2014/11/14/us/ap-us-oil-patch-fbi-office.html?_r=0
[1130]http://protectingourwaters.wordpress.com/2012/08/01/shale-gas-industry-puts-workers-at-risk-in-rush-to-frack/
[1131]http://www.motherjones.com/environment/2012/11/fracking-safety-north-dakota
[1132]http://www.eenews.net/stories/1059976658
[1133]http://www.eenews.net/stories/1059976658
[1134]http://www.environmentamerica.org/sites/environment/files/reports/EA_FrackingNumbers_scrn.pdf
[1135]http://www.undeerc.org/bakken/bakkenformation.aspx
[1136]http://www.energyfromshale.org/bakken-shale-gas
[1137]http://www.undeerc.org/bakken/developmenthistory.aspx
[1138]http://www.eia.gov/energyexplained/index.cfm?page=US_energy_home
[1139]https://www.dmr.nd.gov/oilgas/riglist.asp
[1140]http://newsok.com/regulator-nd-oil-boom-like-nascar-race/article/feed/627570
[1141]http://www.exxonmobilperspectives.com/2012/09/19/a-hundred-fold-increase-in-oil-production/
[1142]http://oilshalegas.com/bakkenshale.html
[1143]http://www.reuters.com/assets/print?aid=USBRE9BT0OV20131231
[1144]http://thebakken.com/articles/1140/u-s-out-produces-rest-of-world-in-oil-and-gas-from-shale
[1145]http://www.ogj.com/unconventional-resources/bakken-shale.html
[1146]http://oilshalegas.com/bakkenshale.html
[1147]http://www.investmentu.com/2011/September/natural-gas-flaring.html
[1148]*Ibid.*
[1149]http://www.eia.gov/todayinenergy/detail.cfm?id=18451
[1150]https://www.dmr.nd.gov/oilgas/or24665.pdf
[1151]http://www.beaumontenterprise.com/news/texas/article/ND-farmer-finds-oil-spill-while-harvesting-wheat-4888671.php
[1152]http://www.beaumontenterprise.com/news/texas/article/ND-farmer-finds-oil-spill-while-harvesting-wheat-4888671.php
[1153]http://www.theguardian.com/environment/2013/oct/10/north-dakota-tioga-tesoro-oil-spill

[1154]http://minnesota.cbslocal.com/2015/05/24/cleanup-of-oil-spill-at-nd-farm-to-take-2-more-years/

[1155]http://bismarcktribune.com/news/state-and-regional/cleanup-of-oil-spill-at-nd-farm-to-take-more/article_8bdb5f7f-95d8-5a79-a193-21dc07ca72d0.html

[1156]http://www.theguardian.com/environment/2013/oct/25/north-dakota-oil-pipeline-spills-secrecy

[1157]http://www.ndwheat.com/buyers/default.asp?ID=295

[1158]http://www.minotmilling.com/durum/durum.html

[1159]http://www.ndwheat.com/buyers/default.asp?ID=295

[1160]http://www.newstimes.com/news/us/article/Poor-durum-crop-likely-to-impact-pasta-consumers-5924329.php

[1161]http://seattletimes.com/html/businesstechnology/2024172633_columbiatrainsxml.html

[1162]*Ibid.*

[1163]http://marcelluseffect.blogspot.com/2012/02/canadian-farmers-call-for-fracking.html

[1164]http://www.telegraph.co.uk/earth/energy/fracking/11141913/Farmers-fear-fracking-could-spell-financial-ruin.html

[1165]http://www.laprogressive.com/fracking-in-california/

[1166]http://gen.usc.edu/assets/001/84787.pdf

[1167]http://www.bloomberg.com/news/2014-05-21/eia-cuts-monterey-shale-estimates-on-extraction-challenges-1-.html

[1168]http://www.latimes.com/business/la-fi-oil-20140521-story.html

[1169]http://www.nytimes.com/2000/09/06/dining/national-origins-california-s-central-valley-where-the-mountains-are-almonds.html

[1170]http://www.fas.usda.gov/agx/ISMG/Global%20Wine%20Report%20Final%20Aug2006.pdf

[1171]http://ca.water.usgs.gov/projects/central-valley/san-joaquin-basin.html

[1172]http://www.bakersfieldcalifornian.com/business/kern-gusher/x634489929/State-poised-to-shut-down-11-local-oil-injection-wells

[1173]http://www.marcellusfieldguide.org/index.php/guide/pre_development_issues/effects_on_agriculture/

[1174]*Ibid.*

[1175]http://pubs.usgs.gov/of/2012/1154/of2012-1154.pdf

[1176]http://pubs.cas.psu.edu/FreePubs/PDFs/ee0020.pdf

[1177]http://www.marcellusfieldguide.org/index.php/guide/pre_development_issues/effects_on_agriculture/

[1178]http://www.thenation.com/article/171504/fracking-our-food-supply

[1179]http://www.thenation.com/article/171504/fracking-our-food-supply

[1180]http://www.dec.ny.gov/energy/46288.html

[1181]http://www.dec.ny.gov/energy/46288.html

[1182]http://pubs.usgs.gov/of/2012/1154/of2012-1154.pdf

[1183]http://www.thenation.com/article/171504/fracking-our-food-supply

[1184]http://www.sustainableotsego.org/Risk Assessment Natural Gas Extraction-1.htm

[1185]http://www.dailygazette.com/Standard/ShowStoryTemplate.asp?Path=SCH/2013/03/21&ID=Ar02302&Section=Sports

[1186]http://www.pennfuture.org/UserFiles/Daze/20120416_Letter_DEPFBC_SusqSmallmouthBass.pdf

[1187]*Ibid.*

[1188]*Ibid.*

[1189]http://63.134.196.109/documents/mpactsofGasDrillingonHumanandAnimalHealth.pdf

[1190]http://truth-out.org/news/item/13058-why-are-cows-tails-dropping-off

[1191]http://www.thenation.com/article/171504/fracking-our-food-supply

[1192]http://www.fwweekly.com/2009/10/14/sacrificed-to-shale/

[1193]*Ibid.*

[1194]http://www.alternet.org/water/150794/trailer_talk's_frack_talk%3A_why_a_mayor_was_forced_to_leave_his_town_because_of_gas_drilling_/?page=entire

[1195]http://www.usgs.gov/newsroom/article.asp?ID=3677
[1196]http://www.theintelligencer.net/page/content.detail/id/561195/Consol-Sued-for-Dunkard-Creek-Fish-Kill.html?nav=515
[1197]http://www.tpwd.state.tx.us/landwater/water/environconcerns/hab/ga/
[1198]http://www.theintelligencer.net/page/content.detail/id/561195/Consol-Sued-for-Dunkard-Creek-Fish-Kill.html?nav=515
[1199]http://www.post-gazette.com/stories/news/us/consol-to-pay-55m-for-clean-water-act-violations-286950/
[1200]http://www.dispatch.com/content/stories/local/2014/07/01/state-agency-fracking-fire-likely-fouled-creek.html
[1201]http://www.dispatch.com/content/stories/local/2014/06/29/glitchsparks-smoky-fire-at-gas-well.html
[1202]http://www.dispatch.com/content/stories/local/2014/07/21/details-on-chemicals-trickle-in-after-spill.html
[1203]http://www.dispatch.com/content/stories/local/2014/07/21/details-on-chemicals-trickle-in-after-spill.html
[1204]http://www.eenews.net/public/Landletter/2010/10/21/1
[1205]http://www.wcs.org/press/press-releases/natural-gas-development-linked-to-habitat-loss.aspx
[1206]http://www.propublica.org/article/science-lags-as-health-problems-emerge-near-gas-fields/single
[1207]http://thetimes-tribune.com/news/gas-drilling/after-blowout-most-evacuated-families-return-to-their-homes-in-bradford-county-1.1135253
[1208]http://eagleford.publicintegrity.org/?_ga=1.61509482.71452413.1402680016
[1209]Ibid.
[1210]http://www.water-contamination-from-shale.com/louisiana/louisiana-fracking-operations-suspect-in-cow-deaths/
[1211]http://www.facebook.com/#!/Realpromisedland/info
[1212]http://www.usatoday.com/news/nation/2010-12-14-1Alouisiana14_CV_N.htm
[1213]http://lancasterfarming.com/news/northeedition/Couple-Reeling-From-Natural-Gas-Mess-
[1214]Ibid.
[1215]Ibid.
[1216]http://www.ahs.dep.pa.gov/eFACTSWeb/searchResults_singleViol.aspx?InspectionID=2115722
[1217]http://www.ahs.dep.pa.gov/eFACTSWeb/searchResults_singleViol.aspx?InspectionID=2076777
[1218]http://www.ahs.dep.pa.gov/eFACTSWeb/searchResults_singleViol.aspx?InspectionID=2095088.
[1219]http://www.observer-reporter.com/article/20121218/NEWS04/121219324
[1220]http://www.naturalgaswatch.org/?cat=8
[1221]http://en.wikipedia.org/wiki/Transportation_in_the_United_States
[1222]http://www.rita.dot.gov/bts/sites/rita.dot.gov.bts/files/publications/national_transportation_statistics/html/table_01_04.html
[1223]http://www.trb.org/HMCRP/HMCRPOverview.aspx
[1224]http://www.alternet.org/fracking/toxic-wastewater-dumped-streets-and-rivers-night-gas-profiteers-getting-away-shocking?paging=off
[1225]http://dailyitem.com/0100_news/x685074962/People-flee-as-chemical-cloud-hovers-near-New-Columbia
[1226]Ibid.
[1227]http://wnep.com/2012/09/26/crash-spills-fluid-from-gas-drilling-into-creek/
[1228]http://www.post-gazette.com/local/marcellusshale/2014/04/22/Washington-County-crash-sends-fracking-water-in-stream/stories/201404220103
[1229]http://www.rt.com/usa/160604-texas-fracking-wastewater-dumping/

601

1230http://www.pagoppolicy.com/Display/SiteFiles/112/Hearings/6_10_10/Christie_Testi
mony_6_10_10.pdf
1231http://repository.cmu.edu/cgi/viewcontent.cgi?article=1065&context=cee
1232http://eagleford.publicintegrity.org/?_ga=1.61509482.71452413.1402680016
1233http://www.chron.com/news/houston-texas/houston/article/Shale-play-turns-Karnes-
County-around-but-not-4303201.php
1234http://www.capitol.state.tx.us/Search/DocViewer.aspx?ID=83RSJR000013B&Query
Text=%242.9%2bOR%2bbillion&DocType=B
1235http://fuelfix.com/blog/2012/09/21/study-7b-bill-for-nd-road-upkeep-over-20-years/
1236http://fuelfix.com/blog/2012/09/21/study-7b-bill-for-nd-road-upkeep-over-20-years/
1237http://un-naturalgas.org/NYSDOT%20Transportation%20Impacts%20Paper.pdf
1238http://news.nationalgeographic.com/news/energy/2013/07/130711-oil-train-crash-
five-key-issues/
1239http://www.charlestondailymail.com/article/20150216/DM01/150219449
1240http://marcellusdrilling.com/2012/02/railamerica-buys-wellsboro-corning-short-line-
railroad/
1241http://www.progressiverailroading.com/rail_industry_trends/article/Gas-exploration-
and-drilling-in-the-Marcellus-Shale-means-more-traffic-for-Class-Is-short-lines--29103
1242http://www.progressiverailroading.com/short_lines_regionals/article/Marcellus-
shalerelated-activity-has-been-a-growth-driver-for-North-Shore-Railroad-Co-but-it-
hasn39t-been-the-only-one--30587
1243http://truth-out.org/buzzflash/commentary/fracking-train-shipments-regularly-
delay-amtrak-trains-and-passengers-for-hours/18720-fracking-train-shipments-
regularly-delay-amtrak-trains-and-passengers-for-hours
1244http://www.eia.gov/petroleum/transportation/
1245http://www.eia.gov/todayinenergy/detail.cfm?id=20592&src=email
1246http://www.phmsa.dot.gov/staticfiles/PHMSA/DownloadableFiles/1_2_14%20Rail_Sa
fety_Alert.pdf
1247https://www.federalregister.gov/articles/2013/11/20/2013-27785/safety-and-security-
plans-for-class-3-hazardous-materials-transported-by-rail
1248http://www.desmogblog.com/2014/01/05/exclusive-permit-shows-bakken-oil-
casselton-train-contained-high-levels-volatile-chemicals
1249http://truth-out.org/buzzflash/commentary/item/18350-chemical-and-oil-companies-
accused-of-using-unsafe-rail-tank-cars
1250http://www.mcclatchydc.com/2014/01/20/215143/more-oil-spilled-from-trains-in.html
1251http://abcnews.go.com/US/wireStory/ap-exclusive-fuel-hauling-trains-derail-10-year-
29139529
1252http://www.usatoday.com/story/news/world/2013/07/08/40-still-missing-in-deadly-
canada-oil-train-crash/2497875/
1253http://news.nationalgeographic.com/news/energy/2013/07/130711-oil-train-crash-
five-key-issues/
1254http://www.cbsnews.com/news/officials-30-missing-in-quebec-train-crash-presumed-dead/
1255http://www.claimsjournal.com/news/international/2014/07/07/251345.htm
1256http://www.dispatch.com/content/stories/local/2012/07/12/speed-not-cause-of-
derailment-officials-say.html
1257http://www.theguardian.com/world/2013/oct/19/canada-rail-fire-derailment/print
1258http://articles.latimes.com/2013/nov/09/nation/la-na-nn-train-crash-alabama-oil-
20131109
1259http://www.mcclatchydc.com/2014/01/20/215143/more-oil-spilled-from-trains-in.html
1260http://www.huffingtonpost.com/2013/11/08/crude-oil-train-derails-alabama-
_n_4240799.html
1261http://www.inforum.com/event/article/id/422351/
1262http://bismarcktribune.com/bakken/north-dakota-discloses-oil-train-shipment-
details/article_a2f3d6c4-fceb-11e3-b648-001a4bcf887a.html

[1263]http://www.reuters.com/article/2014/01/08/us-train-idUSBREA0703S20140108
[1264]http://www.reuters.com/article/2014/04/30/us-railways-accident-virginia-idUSBREA3T0YW20140430?feedType=RSS&feedName=topNews&utm_source=dlvr.it&utm_medium=twitter&dlvrit=992637
[1265]http://www.wset.com/story/25392359/continuing-coverage-at-least-50000-gallons-of-crude-oil-missing-after-train-derailment
[1266]http://www.mcclatchydc.com/news/nation-world/national/economy/article31832745.html
[1267]http://www.deq.virginia.gov/Portals/0/DEQ/Enforcement/Public%20Notice%20RP%20Signed%20Orders/CSX%20order%20signed%203Feb2015.pdf
[1268]http://stateimpact.npr.org/pennsylvania/2014/01/20/train-carrying-crude-oil-derails-on-philadelphia-bridge/
[1269]http://www.philly.com/philly/news/breaking/CSX_freight_train_partial_derailment_South_Philadelphia.html
[1270]http://www.cbc.ca/news/canada/calgary/cp-rail-train-derails-in-southwestern-alberta-1.2957823
[1271]https://www.mail-archive.com/sustainablelorgbiofuel@lists.sustainablelists.org/msg80803.html
[1272]http://globalnews.ca/news/1833353/west-virginia-train-derailment-sends-oil-tanker-into-river/
[1273]http://www.thestate.com/2015/02/17/3992089_west-virginia-train-derailment.html?rh=1
[1274]http://popist.com/s/7112041/
[1275]http://rt.com/usa/232899-evacuation-train-derail-virginia/
[1276]http://www.reuters.com/article/2015/02/17/us-usa-train-derailment-csx-idUSKBN0LK1ST20150217
[1277]http://www.thonline.com/news/breaking/article_fb191072-c373-11e4-9c86-9bc87b7364e4.html
[1278]http://www.dispatch.com/content/stories/national_world/2015/03/07/oil-train-derails-leaks-into-waterway.html
[1279]http://www.toledoblade.com/local/2014/07/03/Train-derailment-blocks-crossing-in-western-Lucas-Co.html
[1280]https://www.facebook.com/media/set/?set=a.331450470295077.64568.263861137054011&type=3
[1281]http://www.wtae.com/news/train-derails-near-homes-in-uniontown/30857328
[1282]http://thedailyreview.com/news/section-of-train-derails-in-bradford-county-1.1859315
[1283]http://www.amtrak.com/servlet/ContentServer?c=AM_Alert_C&pagename=am/AM_Alert_C/Alerts_Popup&cid=1251626267188
[1284]http://www.npr.org/2014/02/06/272150321/oil-transports-gum-up-passenger-route-across-northern-plains
[1285]http://www.truth-out.org/buzzflash/commentary/fracking-train-shipments-regularly-delay-amtrak-trains-and-passengers-for-hours/18720-fracking-train-shipments-regularly-delay-amtrak-trains-and-passengers-for-hours
[1286]http://www.phmsa.dot.gov/pv_obj_cache/pv_obj_id_D9E224C13963CAF0AE4F15A8B3C4465BAEAF0100/filename/Final_EO_on_Transport_of_Bakken_Crude_Oi_05_07_2014.pdf
[1287]
See:http://desmogblog.com/sites/beta.desmogblog.com/files/BNSF%20June%206%20Letter%20to%20ID%20BHS.pdf
[1288]http://www.desmogblog.com/2014/06/18/white-house-meeting-logs-big-rail-lobbying-bomb-trains-regulations-touts-publicly
[1289]http://www.desmogblog.com/2014/08/13/rail-ceos-investors-bomb-trains-safe-at-almost-any-speed

[1290]http://desmogblog.com/sites/beta.desmogblog.com/files/BNSF%20June%206%20Lett
er%20to%20ID%20BHS.pdf
[1291]http://www.desmogblog.com/2014/07/02/oil-rail-battle-between-right-know-and-
need-know
[1292]http://www.sej.org/sites/default/files/local-DAVIS-7-0.pdf
[1293]http://www.post-gazette.com/news/state/2014/07/07/Pa-won-t-release-info-on-trains-
carrying-crude-oil-Pa-refuses-to-release-route-info-on-oil-
trains/stories/201407070048#ixzz36qh0Co6y
[1294]http://www.post-gazette.com/local/city/2014/10/03/Pennsylvania-Office-of-Open-
Record-orders-to-release-oil-train-records/stories/201410030206
[1295]http://www.fra.dot.gov/eLib/details/L05222
[1296]http://www.buffalonews.com/city-region/government-orders-rail-companies-to-warn-
communities-about-hazardous-train-cargo-20140507
[1297]http://www.reuters.com/article/2014/04/23/railways-regulations-canada-
idUSL2N0NF0LS20140423
[1298]http://www.fra.dot.gov/eLib/details/L16355#p1_z5_gD
[1299]http://abcnews.go.com/US/fiery-scene-train-derails-north-dakota/story?id=30845073
[1300]http://washpost.bloomberg.com/Story?docId=1376-NQ21JF6JIJVI01-
1D5N4OU0DCLSS6F8U7MEENEE2J
[1301]http://www.startribune.com/business/285687401.html
[1302]http://www.propublica.org/article/pipelines-explained-how-safe-are-americas-2.5-
million-miles-of-pipelines
[1303]Ibid.
[1304]http://www.politico.com/story/2015/04/the-little-pipeline-agency-that-couldnt-
117147_Page2.html
[1305]http://www.nofrackingway.us/2015/03/11/pipelines-or-rail-for-fossil-fuels-are-false-
choices/
[1306]http://www.corrosion-doctors.org/Pipeline/Carlsbad-explosion.htm
[1307]http://lubbockonline.com/stories/082100/loc_082100695.shtml
[1308]http://www.bizjournals.com/albuquerque/stories/2002/09/16/story3.html?page=all
[1309]http://www.justice.gov/enrd/4431.htm
[1310]http://www.mercurynews.com/ci_16051111?source=most_emailed&nclick_check=1
[1311]http://www.huffingtonpost.com/2010/09/12/san-bruno-explosion-
photos_n_713976.html#138569
[1312]http://www.mercurynews.com/ci_16051111?source=most_emailed&nclick_check=1
[1313]http://www.mercurynews.com/ci_16045798?IADID=Search-www.mercurynews.com-
www.mercurynews.com
[1314]http://www.sfgate.com/bayarea/article/PG-E-diverted-safety-money-for-profit-
bonuses-2500175.php
[1315]http://abclocal.go.com/kgo/story?section=news%2Flocal%2Fpeninsula&id=9258863
[1316]http://insideclimatenews.org/news/20140804/san-bruno-gas-blast-case-cozy-emails-
between-judge-and-defendant
[1317]http://www.latimes.com/business/la-fi-puc-fine-20150409-story.html
[1318]http://www.nbcbayarea.com/news/local/CPUC-Fines-PGE-16B-After-Deadly-2010-
San-Bruno-Explosion-299218111.html
[1319]http://www.huffingtonpost.com/2012/12/11/sissonville-west-virginia-
explosion_n_2279577.html?ir=Green
[1320]http://www.wvpubcast.org/newsarticle.aspx?id=27911
[1321]http://newsok.com/wv-gas-explosion-comes-amid-line-shut-off-
debate/article/feed/476071
[1322]http://www.puc.state.pa.us/naturalgas/Act_127_Info.aspx
[1323]http://www.philly.com/philly/news/special_packages/inquirer/marcellus-
shale/135273768.html
[1324]http://jurist.org/dateline/2012/07/garrett-eisenhour-pipeline-regulation.php

604

[1325]http://www.postgazette.com/stories/local/marcellusshale/wva-blast-heightens-concerns-over-natural-gas-pipelines-666131/
[1326]http://articles.philly.com/2012-04-08/news/31308559_1_gas-safety-gas-explosion-natural-gas
[1327]http://www.gao.gov/assets/590/589514.pdf
[1328]*Ibid.*
[1329]http://www.philly.com/philly/news/special_packages/inquirer/marcellus-shale/20111208_Gas_lines_proliferating_in_Pa__are_lightly_regulated.html
[1330]http://www.peer.org/news/news-releases/2013/07/30/federal-safety-agency-plays-pipeline-rupture-roulette/
[1331]http://articles.mcall.com/2014-05-10/news/mc-pa-federal-pipeline-regulation-20140510_1_pipeline-safety-pipeline-replacement-phmsa
[1332]http://insideclimatenews.org/news/20130911/exclusive-pipeline-safety-chief-says-his-regulatory-process-kind-dying
[1333]http://www.politico.com/story/2015/04/the-little-pipeline-agency-that-couldnt-117147_Page2.html
[1334]http://www.politico.com/story/2015/04/the-little-pipeline-agency-that-couldnt-117147_Page3.html#ixzz3YbkQXdIh
[1335]http://www.papipelinesafety.org/news/archive/2012/01/
[1336]http://shale.sites.post-gazette.com/index.php/news/archives/24437-compressor-station-explosion-shuts-down-at-least-10-wells
[1337]http://citizensvoice.com/news/drilling/susquehanna-county-compression-station-up-and-running-without-state-permission-1.1295296
[1338]http://thetimes-tribune.com/news/gas-drilling/no-fine-for-compressor-station-operator-1.1467447
[1339]http://www.mycentraljersey.com/article/20130530/NJNEWS/305300063/
[1340]http://www.usnews.com/news/business/articles/2014/03/31/4-hurt-200-evacuated-after-washington-plant-fire
[1341]http://www.naturalgaswatch.org/?p=1305
[1342]http://www.washingtonpost.com/wp-dyn/content/article/2010/10/29/AR2010102900044.html
[1343]http://nyfriendsofcleanairandwater.blogspot.com/2014/01/williams-compressor-station-windsor-ny.html
[1344]http://www.reuters.com/article/2014/04/24/usa-blast-wyoming-idUSL2N0NF2FZ20140424
[1345]http://trib.com/news/state-and-regional/fire-still-burns-at-damaged-wyoming-gas-plant/article_67b11cdf-0b76-53b9-8050-a2504d3e2e3d.html
[1346]http://trib.com/news/state-and-regional/two-suffer-injuries-due-to-fire-at-wyoming-natural-gas/article_1e7e73ac-e3a0-563a-bff1-0b6a716457da.html
[1347]http://www.pinedaleroundup.com/v2_news_articles.php?heading=0&story_id=2294&page=72
[1348]http://sunad.com/index.php?tier=1&article_id=26501
[1349]http://www.pennenergy.com/articles/pennenergy/2012/06/worker-killed-in-explosion.html
[1350]http://www.wboy.com/story/21949055/firefighters-working-to-subdue-gas-well-explosion-in-tyler-county-eureka-triad-hunter-magnum-hunter-resources-pipeline
[1351]http://alerts.skytruth.org/report/05f2642d-8a93-3b6f-bbac-9c57a37b
[1352]http://www.cnn.com/2015/10/08/us/louisiana-deadly-gas-explosion/
[1353]http://www.gmanetwork.com/news/story/350457/pinoyabroad/news/2-pinoys-among-12-dead-in-qatar-restaurant-gas-explosion
[1354]http://www.dailymail.co.uk/wires/ap/article-2712486/Evacuees-return-Taiwan-gas-explosions.html
[1355]http://www.nytimes.com/2015/01/30/world/americas/mexico-city-hospital-gas-explosion.html?_r=0

[1356]http://www.scientificamerican.com/article.cfm?id=drill-for-natural-gas-pollute-water&print=true

[1357]http://cogcc.state.co.us/forms/PDF_Forms/form35.pdf

[1358]http://www.denverpost.com/environment/ci_22827696/colorado-docs-chafe-at-secrecy-oath-needed-access

[1359]*Ibid.*

[1360]http://archives.legislature.state.oh.us/bills.cfm?ID=129_SB_315

[1361]http://archive.longislandpress.com/2012/03/06/cracks-emerge-in-gop-over-hydraulic-fracturing/

[1362]http://www.arktimes.com/arkansas/health-consequences-post-spill/Content?oid=2860352

[1363]http://www.chicagobusiness.com/article/20140829/NEWS07/140829764/long-awaited-fracking-rules-released#

[1364]http://www.fayobserver.com/news/local/article_c6a307c4-a794-5b44-9e7f-77e7f1adbdb8.html

[1365]http://www.ncleg.net/Sessions/2013/Bills/Senate/PDF/S786v7.pdf

[1366]http://www.ama-assn.org/ama/pub/news/news/2015/2015-06-09-new-policies-annual-meeting-afternoon.page

[1367]http://www.americanlegislator.org/2012/03/alec-encourages-responsible-resource-production/

[1368]http://concernedhealthny.org/letter-from-dr-jerome-a-paulson-to-the-pennsylvania-department-of-environmental-protection/

[1369]http://www.npr.org/2012/05/17/152268501/pennsylvania-doctors-worry-over-fracking-gag-rule

[1370]http://blog.shaleshockmedia.org/2012/09/02/marcellus-shale-coalition-in-the-lobby/

[1371]http://hosted2.ap.org/PAWIC/APUSnews/Article_2012-04-19-Gas Drilling-Health/id-d48d0a70cde6443c9f105279abfac99f

[1372]http://www.samsmithpahouse.com/NewsItem.aspx?NewsID=14132

[1373]http://www.paenvironmentdigest.com/newsletter/default.asp?NewsletterArticleID=22066&SubjectID=

[1374]http://nurses.3cdn.net/39c3056f1d418b5a7f_xfm6bkbib.pdf

[1375]http://www.courthousenews.com/2012/07/31/48847.htm

[1376]*Ibid.*

[1377]http://www.law360.com/articles/482868

[1378]http://thetimes-tribune.com/news/doctor-loses-lawsuit-that-challenged-secrecy-of-fracking-fluid-1.1850073

[1379]http://www.nejm.org/doi/full/10.1056/NEJMsb1209858

[1380]http://www.denverpost.com/opinion/ci_18436002

[1381]http://truth-out.org/news/item/7323:fracking-pennsylvania-gags-physicians

[1382]http://closup.umich.edu/files/nsee-fracking-fall-2012.pdf

[1383]http://www.whitehouse.gov/state-of-the-union-2012

[1384]http://www.doi.gov/news/pressreleases/loader.cfm?csModule=security/getfile&pageid=293916

[1385]http://www.chron.com/business/article/Q-amp-A-Openness-on-fracturing-ured-3797370.php

[1386]http://www.counterpunch.org/2012/03/19/the-perils-of-fracking/

[1387]http://www.upstreamonline.com/live/article1330207.ece

[1388]http://www.epa.gov/oecaagct/lcra.html

[1389]http://www.ombwatch.org/node/12130

[1390]http://switchboard.nrdc.org/blogs/amall/new_nrdc_analysis_state_fracki.html

[1391]*Ibid.*

[1392]http://www.rangeresources.com/getdoc/50e3bc03-3bf6-4517-a29b-e2b8ef0afe4f/Well-Completion-Reports.aspx

[1393]http://www.bizjournals.com/dallas/news/2013/11/11/exxon-range-resources-get-an-f-in.html?page=all
[1394]http://www.scribd.com/doc/169829862/Range-Resources-Docket-2011-149-R#scribd
[1395]http://www.ragingchickenpress.org/2013/09/23/lawsuit-shows-range-resources-doesnt-know-chemical-composition-of-fracking-fluid-used-in-2011-spill/
[1396]http://www.observereporter.com/article/20131105/NEWS01/131109670#.UnuFt3CTjT8
[1397]http://law.justia.com/cases/pennsylvania/superior-court/2015/1130-wda-2014.html
[1398]http://comptroller.nyc.gov/wp-content/uploads/2014/04/NYC-Comptroller-Stringer-and-As-You-Sow-Reach-Agreement-with-ExxonMobil-on-Fracking-Disclosure.pdf
[1399]http://www.desmogblog.com/2014/04/06/exxonmobil-agrees-disclose-fracking-risks-investors
[1400]http://public.bakerhughes.com/shalegas/disclosure.html
[1401]http://www.forbes.com/sites/greatspeculations/2014/11/18/why-halliburton-is-buying-baker-hughes/
[1402]http://billingsgazette.com/news/state-and-regional/wyoming/wyoming-halliburton-agree-to-greater-fracking-disclosure/article_6f238843-8854-5cc1-8737-af8fbd60e1ee.html
[1403]http://fracfocus.org/welcome
[1404]http://www.bloomberg.com/news/2012-08-14/fracking-hazards-obscured-in-failure-to-disclose-wells.html
[1405]http://www.bloomberg.com/news/2012-08-14/fracking-hazards-obscured-in-failure-to-disclose-wells.html
[1406]http://www.eia.gov/neic/experts/natgastop10.htm
[1407]http://ecowatch.org/2012/water-for-fracking/
[1408]http://stateimpact.npr.org/texas/2012/12/14/the-number-of-fracking-trade-secrets-in-texas-will-likely-surprise-you/
[1409]http://www.fracfocus.org/major-improvements-fracfocus-announced
[1410]http://www.fracfocus.org/
[1411]http://www.bloomberg.com/news/2012-08-14/fracking-hazards-obscured-in-failure-to-disclose-wells.html
[1412]http://blogs.law.harvard.edu/environmentallawprogram/files/2013/04/4-23-2013-LEGAL-FRACTURES.pdf
[1413]*Ibid.*
[1414]*Ibid.*
[1415]http://www.bloomberg.com/news/2012-08-14/fracking-hazards-obscured-in-failure-to-disclose-wells.html
[1416]*Ibid.*
[1417]http://www.fracfocus.org/node/344
[1418]http://stateimpact.npr.org/pennsylvania/2015/06/18/dep-plans-new-fracking-chemical-disclosure-site-promises-more-transparency/
[1419]*Ibid.*
[1420]http://www2.canada.com/story.html?id=6763527
[1421]http://earthjustice.org/sites/default/files/Hallowich_Brief.pdf
[1422]http://s3.documentcloud.org/documents/1204483/nuisance-easement.pdf
[1423]http://www.moopdf.net/file/civil-division-stephanie-hallowich-and-no-c-63-cv-201003954.html
[1424]http://earthjustice.org/sites/default/files/Hallowich_Brief.pdf
[1425]http://earthjustice.org/
[1426]http://stateimpact.npr.org/pennsylvania/2012/12/07/appeals-court-agrees-with-newspapers-in-sealed-fracking-case/
[1427]http://earthjustice.org/sites/default/files/Hallowich-Opinion-Order.pdf
[1428]http://www.alternet.org/fracking/pennsylvania-court-deals-blow-secrecy-obsessed-fracking-industry-corporations-not-same

[1429]http://pipeline.post-gazette.com/news/archives/25271-confidential-agreement-should-have-been-part-of-washington-county-marcellus-shale-case-record
[1430]http://www.post-gazette.com/stories/local/washington/confidential-agreement-should-have-been-part-of-washington-county-marcellus-shale-case-record-697530/
[1431]http://catholicclimatecovenant.org/catholic-teachings/pope-francis/
[1432]http://www.theguardian.com/world/2015/jun/15/pope-francis-destruction-ecosystem-leaked-encyclical
[1433]http://www.reuters.com/article/2015/06/16/us-pope-environment-leak-idUSKBN0OV2JC20150616
[1434]http://www.cruxnow.com/church/2015/06/18/laudato-si-chapter-three-the-human-roots-of-the-ecological-crisis/
[1435]http://environment.yale.edu/climate-communication/article/american-catholics-worry-about-global-warming-and-support-u.s.-action/
[1436]http://www.salon.com/2015/06/17/fox_news_greg_gutfeld_is_not_happy_with_the_pope_the_most_dangerous_person_on_the_planet/
[1437]http://www.rushlimbaugh.com/daily/2015/06/16/the_pope_s_leaked_marxist_climate_rant
[1438]http://www.theguardian.com/us-news/2015/jun/17/jeb-bush-joins-republican-backlash-pope-climate-change
[1439]http://www.masslive.com/news/index.ssf/2015/06/politicians_prefer_pope_franci.html
[1440]http://www.masslive.com/news/index.ssf/2015/06/politicians_prefer_pope_franci.html
[1441]http://www.cornwallalliance.org/about/
[1442]http://www.cornwallalliance.org/2009/05/01/evangelical-declaration-on-global-warming/
[1443]http://en.radiovaticana.va/news/2015/04/28/un_leader_meets_pope_and_addresses_climate_change_conference/1140143
[1444]http://www.huffingtonpost.co.uk/2015/06/18/pope-issues-climate-change-warning_n_7609690.html
[1445]http://www.asiantribune.com/node/6210
[1446]http://www.islamset.com/env/pref2.html
[1447]http://www.nationalcouncilofchurches.us/about/
[1448]http://www.pelicanweb.org/solisustv08n06page1.html
[1449]http://www.interfaithpowerandlight.org/wp-content/uploads/2013/06/IPL-Fact-Sheet-2014-forprinter2.pdf
[1450]http://www.interfaithpowerandlight.org/about/mission-history/
[1451]http://www.paipl.us/2012/09/27/effects-on-poverty-and-social-injustice/
[1452]http://www.uua.org/statements/statements/14257.shtml
[1453]http://www.ucc.org/news/commentary-earth-day-04222014.html
[1454]http://www.quaker.org.uk/quaker-response-crisis-climate-change
[1455]http://downloads.elca.org/html/jle/www.elca.org/what-we-believe/social-issues/environment.aspx.htm#read
[1456]http://usatoday30.usatoday.com/news/religion/2011-07-09-gas-drilling-debate-religion_n.htm
[1457]http://forward.com/articles/139831/fracking-comes-to-jewish-summer-camp/
[1458]*Ibid.*
[1459]http://blueandgreenblog.wordpress.com/2011/07/15/frack-you-from-rabbi-arthur-waskow-the-shalom-center/
[1460]http://jewsagainsthydrofracking.org/rabbis-liebling-waskow-at-multifaith-blessing-of-the-waters/566
[1461]http://ncronline.org/blogs/eco-catholic/pope-francis-inspires-300-rabbis-sign-rabbinic-letter-climate
[1462]http://ncronline.org/blogs/eco-catholic/pope-francis-inspires-300-rabbis-sign-rabbinic-letter-climate
[1463]http://www.coejl.org/2012/05/jewish-community-position-on-hydrofracking-adopted-2/
[1464]http://ccarnet.org/rabbis-speak/resolutions/all/national-energy-policy/

[1465]http://urj.org/about/union/governance/reso/?syspage=article&item_id=107281&print able=1

[1466]*Ibid.*

[1467]http://www.coejl.org/aboutus/history/

[1468]http://www.coejl.org/2012/05/jewish-community-position-on-hydrofracking-adopted-2/

[1469]http://www.motherjones.com/environment/2014/09/hillary-clinton-fracking-shale-state-department-chevron

[1470]http://www2.canada.com/calgaryherald/iphone/business/calgary/story.html?id=6107167

[1471]http://www.upi.com/Science_News/2011/04/21/South-Africa-halts-gas-fracking-plan/UPI-23521303431387/

[1472]http://gdacc.org/2012/06/15/daily-maverick-confessions-of-a-fracking-defector/

[1473]http://af.reuters.com/article/investingNews/idAFJOE88601L20120907

[1474]http://www.perthnow.com.au/news/western-australia/gas-fracking-will-cause-irreversible-damge-says-conservation-council-of-wa/story-fnhocxo3-1226660869144

[1475]http://www.health.ny.gov/press/reports/docs/high_volume_hydraulic_fracturing.pdf

[1476]http://www.reuters.com/article/2009/12/23/energy-fracking-newyork-idUSN2220711920091223?type=marketsNews

[1477]http://www.dailykos.com/story/2010/11/30/924263/-Fracking-A-New-York

[1478]http://rac.org/advocacy/rjv/issues/fracking/index.cfm?&printable=1

[1479]http://eidmarcellus.org/marcellus-shale/artists-against-fracking-no-artists-looking-for-relevance-part-2/12534/

[1480]http://eidmarcellus.org/marcellus-shale/artists-against-fracking-no-artists-looking-for-relevance-part-i/12509/

[1481]http://www.cnn.com/2012/08/29/us/new-york-fracking-artists-protest/index.html

[1482]http://artistsagainstfracking.com/our-trip/

[1483]http://bigstory.ap.org/article/yoko-ono-sean-lennon-tour-pa-gas-drilling-sites

[1484]http://susquehannacounty.wnep.com/news/news/141213-celebrity-fracking-bus-tour-causes-commotion

[1485]http://bigstory.ap.org/article/yoko-ono-sean-lennon-tour-pa-gas-drilling-sites

[1486]http://susquehannacounty.wnep.com/news/news/141213-celebrity-fracking-bus-tour-causes-commotion

[1487]http://www.earthisland.org/journal/index.php/elist/eListRead/robert_kennedy_jr._on_fracking_media_and_changing_light_bulbs

[1488]http://www.energyindepth.org/tag/pa/

[1489]http://pennsylvaniafrack.com/2014/12/21/a-very-brief-history-of-the-new-york-frack-ban-movement/

[1490]http://www.nytimes.com/2012/10/01/nyregion/with-new-delays-a-growing-sense-that-gov-andrew-cuomo-will-not-approve-gas-drilling.html?pagewanted=all&_r=0

[1491]http://newyork.newsday.com/news/region-state/cuomo-fracking-regulations-likely-delayed-into-2013-1.4246369

[1492]http://www.syracuse.com/news/index.ssf/2013/06/attorney_general_eric_schneide_2.html

[1493]http://www.wlf.org/org/support.asp

[1494]http://www.sourcewatch.org/index.php?title=Washington_Legal_Foundation

[1495]http://www.wlf.org/litigating/projects.asp

[1496]http://blog.shaleshockmedia.org/2013/01/27/washington-legal-foundation-wlf-frackin-joe-camel-with-law-books/

[1497]http://blog.timesunion.com/capitol/archives/200828/moreau-we-shall-not-be-moved-from-promoting-fracking/

[1498]http://www.nydailynews.com/opinion/yes-fracking-n-y-article-1.1299789

[1499]http://www.propublica.org/article/ed-rendell-new-york-fracking-op-ed-disclosure

[1500]http://www.pressconnects.com/story/opinion/editorial/2014/08/09/editorial-ny-can-learn-pas-fracking-problems/13796943/

[1501]http://www.timesunion.com/opinion/article/Editorial-The-gas-industry-s-hot-air-4572748.php [reprinted with permission.]

[1502]http://www.health.ny.gov/press/reports/docs/high_volume_hydraulic_fracturing.pdf

[1503]http://www.timesunion.com/local/article/Citing-perils-state-bans-fracking-5964402.php#photo-7285238

[1504]*Ibid.*

[1505]http://earthjustice.org/news/press/2014/earthjustice-hails-governor-andrew-cuomo-s-historic-energy-leadership

[1506]http://www.washingtonpost.com/opinions/new-york-takes-the-wrong-approach-to-fracking/2014/12/27/c2ef0006-87c5-11e4-b9b7-b8632ae73d25_story.html?wprss=rss_opinions

[1507]http://www.timesunion.com/local/article/Citing-perils-state-bans-fracking-5964402.php

[1508]http://www.nytimes.com/2014/12/18/nyregion/cuomo-to-ban-fracking-in-new-york-state-citing-health-risks.html?_r=0

[1509]http://observer.com/2014/12/david-paterson-goes-to-bat-for-andrew-cuomo-after-gop-rips-fracking-ban/

[1510]http://observer.com/2014/12/david-paterson-goes-to-bat-for-andrew-cuomo-after-gop-rips-fracking-ban/

[1511]http://www.csmonitor.com/Environment/2014/1217/Is-New-York-governor-s-ban-on-fracking-grounded-in-science-video

[1512]http://www.democratandchronicle.com/story/news/local/2014/12/17/new-york-andrew-cuomo-fracking-casinos/20527181/

[1513]http://www.ithacajournal.com/story/news/local/2014/12/17/dryden-lawsuit-new-york-fracking-ban/20554997/

[1514]http://www.nrdc.org/media/2014/141007.asp

[1515]http://documents.foodandwaterwatch.org/doc/Frack_Actions_IthacaCityNY.pdf#_ga=1.1911473.128869311.1419185802

[1516]http://www.washingtonpost.com/blogs/govbeat/wp/2014/12/17/cuomo-administration-rules-against-fracking/

[1517]http://www.siena.edu/uploadedfiles/home/parents_and_community/community_page/sri/sny_poll/SNY%20June%202013%20Poll%20Release%20--%20FINAL.pdf

[1518]http://www.nrdc.org/media/2014/141007.asp

[1519]http://www.scribd.com/doc/250787268/122214-NY-GOV-BP-doc

[1520]http://www.poughkeepsiejournal.com/story/opinion/editorials/2015/02/18/fracking-waste-must-dealt-responsibly/23642879/

[1521]http://www.baltimoresun.com/news/opinion/editorial/bs-ed-fracking-20141007-story.html

[1522]http://www.baltimoresun.com/news/opinion/editorial/bs-ed-fracking-20141127-story.html

[1523]http://www.fredericknewspost.com/opinion/editorial/ban-don-t-delay-fracking-in-maryland/article_960c17fb-1354-5ebb-8816-8027970a10a0.html

[1524]https://soundcloud.com/foodandwaterwatch/edward-norton-radio-ad-dont-frack-maryland

[1525]http://mgaleg.maryland.gov/webmga/frmMain.aspx?pid=billpage&stab=03&id=hb0449&tab=subject3&ys=2015RS

[1526]http://abclocal.go.com/wpvi/story?section=news/politics&id=8819877

[1527]http://insideclimatenews.org/news/20141203/boulder-ups-anti-fracking-game

[1528]http://www.dailycamera.com/erie-news/ci_20126684/noaa-study-erie-gas-drilling-moratorium-fracking-propane-butane

[1529]http://www.dailycamera.com/erie-news/ci_20126684/noaa-study-erie-gas-drilling-moratorium-fracking-propane-butane

[1530]http://ecowatch.org/2012/fighting-fracking-gold-rush/

[1531]http://denverijournal.com/article.php?id=8176

610

[1532]http://www.post-gazette.com/stories/local/marcellusshale/pittsburgh-inspired-colo-towns-fracking-ban-665637/

[1533]http://www.nytimes.com/2012/11/26/us/with-ban-on-fracking-colorado-town-lands-in-thick-of-dispute.html?_r=0

[1534]http://ballotpedia.org/wiki/index.php/Longmont_City_Fracking_Ban_Question_(November_2012)

[1535]http://www.bizjournals.com/denver/news/2012/12/06/hickenlooper-colorado-wont-sue.html

[1536]*Ibid.*

[1537]http://www.huffingtonpost.com/2013/07/11/longmont-fracking-ban_n_3581819.html

[1538]http://media.bizj.us/view/img/3280941/longmont-fracking-ruling-072414.pdf

[1539]http://www.boulderijournal.com/article.php?id=8252

[1540]http://blog.cleanwateraction.org/2013/02/20/fort-collins-bans-fracking-as-democracy-comes-alive-in-colorado/

[1541]http://www.boulderijournal.com/article.php?id=8665

[1542]http://www.coloradoan.com/article/20130531/NEWS01/305310040/Fort-Collins-group-seeks-fracking-moratorium

[1543]http://www.coloradoan.com/story/news/2015/04/07/pro-fracking-group-spends-another-help-martinez/25412783/

[1544]http://www.dailycamera.com/news/boulder/ci_23390239/boulder-approves-one-year-fracking-moratorium

[1545]http://www.dailycamera.com/boulder-county-news/ci_24323588/colorado-oil-and-gas-association-gives-600k-fight

[1546]http://www.denverpost.com/breakingnews/ci_24460947/900-000-spent-four-colorado-anti-fracking-measures

[1547]http://www.scientificamerican.com/article.cfm?id=three-of-four-colorado-communities-fracking-gas

[1548]http://news.nationalgeographic.com/news/energy/2013/11/131105-colorado-ohio-fracking-ban-election-results/

[1549]http://www.reporterherald.com/news/loveland-local-news/ci_26028522/voters-reject-loveland-fracking-moratorium

[1550]http://ecowatch.com/2014/06/25/oil-gas-industry-buys-colorado-fracking-election/

[1551]http://www.scribd.com/doc/234995530/13CV63-Order-Granting-Motions-for-Summary-Judgment

[1552]http://www.leagle.com/decision/19921875830P2d1045_11856.xml/COUNTY%20CO M'RS%20v.%20BOWEN/EDWARDS%20ASSOC.

[1553]http://www.leagle.com/decision/19921891830P2d1061_11871

[1554]http://insideclimatenews.org/news/20141203/boulder-ups-anti-fracking-game

[1555]http://www.dailycamera.com/editorials/ci_26994523/editorial-question-leadership

[1556]http://en.wikipedia.org/wiki/Denton,_Texas

[1557]http://www.stateoftheair.org/2015/states/texas/denton.html

[1558]http://www.dentonrc.com/local-news/local-news-headlines/20140703-petition-seeks-fracking-support.ece

[1559]http://www.texassharon.com/2014/07/13/caught-on-video-petition-circulators-deceive-denton-residents/

[1560]http://www.dallasnews.com/news/metro/20140715-denton-city-council-fracking-meeting-goes-late.ece

[1561]http://www.rrc.state.tx.us/media/21731/denton-ltr-7-10-14.pdf

[1562]http://energyindepth.org/wp-content/uploads/2014/07/Perryman-Denton-Fracking-Ban-Impact.pdf

[1563]http://rodenfordenton.com/2014/07/fracking-and-dentons-economy-a-quick-response-to-the-perryman-study/

[1564]http://www.texastribune.org/2014/11/05/denton-bans-fracking-spurring-bigger-clashes/

[1565]http://rt.com/usa/202543-texas-fracking-ban-denton/
[1566]http://txoga.org/assets/doc/TXOGA_Petition_Against_the_City_of_Denton.pdf
[1567]http://www.desmogblog.com/sites/beta.desmogblog.com/files/denton%20filed%20copy.pdf
[1568]http://georgepfortexas.org/george-p-bush-campaign-announces-endorsement-of-the-texas-oil-and-gas-pac/
[1569]http://www.texasmonthly.com/daily-post/what-other-texas-municipalities-are-considering-new-fracking-laws
[1570]http://earthjustice.org/news/press/2014/groups-move-to-intervene-in-defense-of-denton-s-fracking-ban
[1571]http://www.dallasnews.com/opinion/editorials/20150322-editorial-overreaching-bill-could-bring-fracking-to-your-doorstep.ece
[1572]http://www.texastribune.org/2014/11/05/denton-bans-fracking-spurring-bigger-clashes/
[1573]http://www.christicraddick.com/we-must-work-together-to-find-a-solution/
[1574]http://trailblazersblog.dallasnews.com/2014/11/craddick-railroad-commission-will-continue-permitting-in-denton-not-ruling-out-action-against-ban.html/
[1575]*Ibid.*
[1576]http://www.houstonchronicle.com/news/politics/texas/article/The-Victory-chair-A-daughter-of-West-Texas-rises-5920275.php
[1577]http://eagleford.publicintegrity.org/
[1578]http://www.christicraddick.com/houston-chronicle-a-daughter-of-west-texas-rises-in-austin/
[1579]http://www.thedailybeast.com/articles/2014/12/09/two-texas-regulators-tried-to-enforce-the-rules-they-were-fired.html
[1580]http://eagleford.publicintegrity.org/
[1581]http://www.bizjournals.com/dallas/news/2015/04/17/bill-that-would-ban-fracking-bans-breezes-through.html
[1582]http://www.texastribune.org/2015/04/17/texas-house-drill-denton-fracking-bill/
[1583]http://www.earthworksaction.org/media/detail/texas_governor_signs_hb40_into_law_guts_community_rights#.VVtySflViko
[1584]https://www.texastribune.org/bidness/explore/myra-crownover/
[1585]http://kut.org/post/denton-repeals-city-ban-fracking
[1586]http://www.star-telegram.com/news/business/barnett-shale/article24627469.html
[1587]http://newsok.com/state-agency-to-continue-drill-permits-in-denton/article/feed/756980
[1588]http://reuters_th.adam.com/content.aspx?productId=16&pid=16&gid=52790
[1589]http://www.huffingtonpost.com/2012/07/02/north-carolina-governor-fracking-perdue_n_1642004.html
[1590]http://www.arnoldporter.com/resources/documents/Center%20for%20Biological%20Diversity%20v%20BLM.pdf
[1591]http://switchboard.nrdc.org/blogs/amall/theyre_not_willing_to_do_it_ri.html
[1592]http://westhawaiitoday.com/sections/news/local-news/council-says-no-fracking-way.html
[1593]http://documents.foodandwaterwatch.org/pressreleases/californians-against-fracking-launch-statewide-campaign-to-pressure-gov-brown-to-ban-fracking/
[1594]https://www.indybay.org/newsitems/2015/02/04/18768050.php
[1595]http://www.capitolweekly.net/article.php?_c=116j1gfoevttuuo&xid=116j0pkc6zsdpel&done=.116j1gfoevu2uuo
[1596]https://www.indybay.org/newsitems/2015/02/04/18768050.php
[1597]http://maplight.org/california/bill/2013-sb-1132/4077296/total-contributions?start=09-01-2009&end=10-31-2009
[1598]http://www.huffingtonpost.com/2013/09/21/california-fracking-bill_n_3965069.html
[1599]http://www.santacruzsentinel.com/santacruz/ci_25801873/santa-cruz-county-first-ban-fracking

[1600]http://ballotpedia.org/San_Benito_County_Fracking_Ban_Initiative,_Measure_J_(November_2014)

[1601]http://www.celdf.org/downloads/Nov_2014_election_PR_CA.pdf

[1602]http://www.depreportingservices.state.pa.us/ReportServer/Pages/ReportViewer.aspx?/Oil_Gas/Permits_Issued_Detail

[1603]http://www.usatoday.com/money/industries/energy/2010-11-16-nat-gas-ban-pittsburgh_N.htm

[1604]http://2politicaljunkies.blogspot.com/2011/06/what-frack-catholic-cemeteries-gas.html

[1605]http://blogs.philadelphiaweekly.com/phillynow/2011/01/25/city-council-to-vote-on-drilling-moratorium-green-party-response/

[1606]http://www.portal.state.pa.us/portal/server.pt?open=514&objID=1072223&parentname=ObjMgr&parentid=396&mode=2

[1607]http://www.state.nj.us/drbc/library/documents/naturalgas/WayneCtyCommrs-ltr-to-DRBC061313.pdf

[1608]http://m.citizensvoice.com/news/corbett-throws-prose-punch-at-agency-over-gas-ban-1.1512888

[1609]http://www.poconorecord.com/apps/pbcs.dll/article?AID=/20130707/NEWS04/307070312&cid=sitesearch

[1610]http://thetimes-tribune.com/news/state-budget-cuts-river-agency-aid-1.1721100

[1611]http://closup.umich.edu/files/nsee-fracking-fall-2012.pdf

[1612]http://www.pennenvironment.org/news/pae/100000-signatures-moratorium-fracking-delivered-gov-tom-corbett

[1613]http://www.senatorferlo.com/senator-ferlo-announces-legislation-for-shale-drilling-moratorium

[1614]http://blog.shaleshockmedia.org/2013/06/16/6-15-13-pa-state-democratic-committee-passes-resolution-to-support-fracking-moratorium/

[1615]http://blog.shaleshockmedia.org/2013/06/17/natural-gas-media-seep/

[1616]http://www.pennlive.com/midstate/index.ssf/2013/06/rendell_pa_democrats_call_for.html

[1617]http://www.eenews.net/stories/1059975812

[1618]http://www.slideshare.net/MarcellusDN/letter-todemstatecommitteeopposingfrackingmoratorium

[1619]http://timesleader.com/news/news/776294/Ahead-of-Obamas-visit-GOP-urges-against-anti-fracking-action

[1620]http://triblive.com/opinion/editorials/4198306-74/stop-india-teachers#axzz2kumMznrx

[1621]http://www.pennlive.com/midstate/index.ssf/2013/06/corbett_challengers_sidestep_d.html

[1622]http://www.edf.org/energy/getting-natural-gas-right

[1623]http://m.npr.org/story/140872251

[1624]http://63.134.196.109/documents/mpactsofGasDrillingonHumanandAnimalHealth.pdf

[1625]http://www.post-gazette.com/local/marcellusshale/2012/05/25/Shale-drilling-contaminated-water-families-say-in-lawsuit/stories/201205250177

[1626]*Ibid.*

[1627]*Ibid.*

[1628]http://canon-mcmillan.patch.com/blog_posts/the-pennsylvania-dep-another-red-herring-2

[1629]http://www.sec.gov/Archives/edgar/data/858470/000104746912001751/a2207418z10-k.htm

[1630]https://www.google.com/finance?q=NYSE:XOM&fstype=ii&ei=61EXU7iCB8r_kAX7-wE

[1631]http://www.cfr.org/united-states/new-north-american-energy-paradigm-reshaping-future/p28630

[1632]http://www.dummies.com/how-to/content/true-conspiracy-the-ford-pinto-memorandum.html
[1633]http://www.motherjones.com/politics/1977/09/pinto-madness?page=1
[1634]http://archive.sltrib.com/article.php?id=11274601&itype=NGPSID
[1635]http://thereaganyears.tripod.com/environment.htm
[1636]http://archive.sltrib.com/article.php?id=11330541&itype=NGPSID
[1637]http://www.nytimes.com/2009/02/05/us/05leases.html?_r=0
[1638]http://www.bidder70.org/files/64701_64800/64778/indictment.pdf
[1639]http://music.peacefuluprising.org/album/midnight-vigil-for-tim-dechristopher
[1640]http://www.nbcnews.com/id/41735935/ns/business-press_releases/t/activist-supporters-organize-response-federal-trial-intimidation/#.UYJe1KLEWM8
[1641]http://www.peacefuluprising.org/tims-official-statement-at-his-sentencing-hearing-20110726
[1642]http://www.deseretnews.com/article/705388228/Activist-Timothy-DeChristopher-sentenced-to-2-years-in-prison.html?pg=all2012
[1643]http://www.sltrib.com/sltrib/news/52263987-78/dechristopher-federal-leases-trial.html.csp
[1644] http://www.ca10.uscourts.gov/opinions/11/11-4151.pdf
[1645]http://www.sltrib.com/sltrib/news/52263987-78/dechristopher-federal-leases-trial.html.csp.
[1646]http://www.moviespirit.com/2013/02/06/bidder-70-climate-activist-tim-dechristopher-acts-for-humanity/
[1647]http://www.sltrib.com/sltrib/news/53133186-78/utah-utahn-named-whose.html.csp
[1648]http://www.ca10.uscourts.gov/opinions/11/11-4151.pdf
[1649]http://www.ca10.uscourts.gov/opinions/11/11-4151.pdf
[1650]http://ecowatch.com/2012/four-eco-documentaries-that-will-fuel-your-fire/
[1651]http://www.alternet.org/story/155069/earth_day%3A_9_films_that_will_change_the_way_you_think_about_the_world
[1652]http://www.bidder70film.com/#!press/cjnx
[1653]http://ecowatch.com/2013/michigan-common-ground-against-fracking/
[1654]http://www.woodtv.com/news/michigan/fracking-auction-brings-protesters
[1655]http://www.woodtv.com/news/michigan/fracking-auction-brings-protesters
[1656]http://inthesetimes.com/uprising/entry/14116/michiganers_fight_auction_of_state_land_for_fracking/
[1657]http://www.reuters.com/article/2012/06/25/us-chesapeake-land-deals-idUSBRE85OOEI20120625
[1658]http://www.reuters.com/article/2013/02/25/us-chesapeake-encana-antitrust-idUSBRE91O15D20130225
[1659]http://issuu.com/nrdcnaturesvoice/docs/naturesvoice-summer15/2
[1660]http://www.boiseweekly.com/boise/you-have-the-right-to-remain-silent-dont-you/Content?oid=3352777
[1661]http://www.boiseweekly.com/boise/you-have-the-right-to-remain-silent-dont-you/Content?oid=3352777
[1662]http://environews.tv/040815-fracking-activist-alma-hasse-files-1-5-million-lawsuit-for-wrongful-arrest-after-all-charges-dropped/
[1663]http://www.boiseweekly.com/boise/charges-dropped-against-activist-alma-hasse/Content?oid=3447165
[1664]http://money.cnn.com/quote/quote.html?symb=COG
[1665]http://independentweekender.com/index.php/2013/10/23/fractivist-barred-from-cabot-sites/#.UmkgmXCTjT8
[1666] *See:* Instument 200713681 Susquehanna County
[1667] *See:* MJ-34301-CR-0000052-2015, Susquehanna County, Pa.
[1668] *See:* 18 Pa. Cons. Stat. §5702

[1669]http://www.post-gazette.com/local/breaking/2010/09/14/Rendell-terminates-contract-with-anti-terrorism-firm.print

[1670]http://www.post-gazette.com/breaking/2010/09/14/Rendell-terminates-contract-with-anti-terrorism-firm/stories/201009140202#ixzz0zYg5qLyB

[1671]http://www.earthisland.org/journal/index.php/eij/article/we_are_being_watched/

[1672]http://www.propublica.org/documents/item/pennsylvania-intelligence-bulletin-no.-131-aug.-30-2010

[1673]http://www.post-gazette.com/breaking/2010/09/14/Rendell-terminates-contract-with-anti-terrorism-firm/stories/201009140202#ixzz0zYg5qLyB

[1674]https://legiscan.com/WY/text/SF0012/id/1151882

[1675]http://www.tn.gov/attorneygeneral/op/2013/op13-39.pdf

[1676]http://www.chicagotribune.com/news/nationworld/ct-idaho-secret-filming-animal-abuse-law-20150804-story.html

[1677]http://www.onearth.org/blog/town-board-to-residents-shut-up-about-fracking

[1678]http://www.post-gazette.com/stories/news/environment/drilling-on-campus-marcellus-shale-boom-puts-colleges-at-crossroads-322630/?print=1

[1679]http://eaglefordtexas.com/news/id/4911/cant-stop-fracking-says-isu-president-2/

[1680]http://uchicagogate.com/2014/01/16/cracks-in-the-basin-part-1-illinois-a-national-model/

[1681]http://blogs.knoxnews.com/humphrey/university-of-tennessee/

[1682]http://www.nytimes.com/2013/03/16/science/earth/university-of-tennessees-fracking-plan-wins-approval-and-stirs-protests.html?_r=0

[1683]http://meadvilletribune.com/local/x2094920399/Local-group-protest-against-oil-and-natural-gas-well-fracking

[1684]http://epw.senate.gov/public/index.cfm?FuseAction=Minority.PressReleases&ContentRecord_id=fd161ae5-802a-23ad-410b-549dd302203b

[1685]http://fuelfix.com/blog/2011/12/08/epa-says-hydraulic-fracturing-polluted-groundwater/

[1686]http://www.opensecrets.org/politicians/industries.php?cycle=2012&type=C&cid=N00005582&newMem=N&recs=20

[1687]http://www.opensecrets.org/politicians/industries.php?cycle=2014&type=C&cid=N00005582&newMem=N&recs=20

[1688]http://trib.com/opinion/editorial/epa-s-silence-does-a-disservice-to-wyoming/article_0921b4ec-3d86-5a6e-bd5b-67bb739c138a.html

[1689]http://trib.com/opinion/columns/finding-answers-for-pavillion-residents/article_2d4e5b10-4399-5e45-9f5e-ee75d8949de8.html

[1690]http://science.house.gov/sites/republicans.science.house.gov/files/documents/hearings/HHRG-112-SY20-20120201-SD001.pdf

[1691]http://www.huffingtonpost.com/2013/06/21/pavillion-wyoming-fracking-study-epa_n_3478685.html?utm_hp_ref=green

[1692]http://www.propublica.org/article/epas-abandoned-wyoming-fracking-study-one-retreat-of-many

[1693]http://www.gastruth.org/?cat=7

[1694]http://files.dep.state.pa.us/AboutDEP/AboutDEPPortalFiles/RemarksAndTestimonies/MLK-Testimony-111611.pdf

[1695]http://www.nicholas.duke.edu/cgc/pnas2011.pdf

[1696]http://articles.philly.com/2011-12-02/news/30467569_1_drinking-water-water-resources-methane

[1697]http://public-accountability.org/2015/02/frackademia-in-depth/

[1698]http://public-accountability.org/2015/02/frackademia-in-depth/

[1699]http://public-accountability.org/2015/02/frackademia-in-depth/

[1700]http://www.nofrackingway.us/2015/02/11/study-confirms-frackers-use-fracademics-to-promote-fracking/

[1701]http://www.bloomberg.com/news/articles/2015-05-15/oil-tycoon-harold-hamm-wanted-scientists-dismissed-dean-s-e-mail-says

[1702]http://www.eenews.net/assets/2015/02/27/document_ew_03.pdf

[1703]http://www.eenews.net/assets/2015/02/27/document_ew_03.pdf

[1704]http://www.eenews.net/stories/1060018280

[1705]http://washpost.bloomberg.com/Story?docId=1376-NOEQS46KLVR801-2FGERR8VOHQN8SE2AJ3OHSIVID

[1706]http://www.bloomberg.com/news/articles/2015-03-30/big-oil-pressured-scientists-over-fracking-wastewater-s-link-to-quakes

[1707]http://www.eenews.net/stories/1060020697

[1708]http://www.eenews.net/stories/1060020697

[1709]http://www.bloomberg.com/news/articles/2015-03-30/big-oil-pressured-scientists-over-fracking-wastewater-s-link-to-quakes

[1710]http://www.bloomberg.com/news/articles/2015-05-15/oil-tycoon-harold-hamm-wanted-scientists-dismissed-dean-s-e-mail-says

[1711]http://washpost.bloomberg.com/Story?docId=1376-NOEQS46KLVR801-2FGERR8VOHQN8SE2AJ3OHSIVID

[1712]http://www.eenews.net/stories/1060014342

[1713]http://inside.mines.edu/UserFiles/Image/EconGeol/Mhitzman_CV.pdf

[1714]http://www.mines.edu/Bio

[1715]http://bizwest.com/special-report-campus-payrolls-defy-recession/

[1716]http://public-accountability.org/2015/06/freedom-fracked/

[1717]http://public-accountability.org/2015/06/freedom-fracked/

[1718]http://inside.mines.edu/UserFiles/File/BeyondMines/bio_nigel4.pdf

[1719]http://www.coloradoindependent.com/32990/school-of-mines-claims-former-prof-merely-warned-about-public-comments

[1720]http://www.boulderweekly.com/article-9467-fracking-and-academic-freedom.html

[1721]https://www.youtube.com/watch?v=Y_7FK2q2eJ4

[1722]http://commonsensepolicyroundtable.com/wp-content/uploads/2010/11/Economic-Impact-of-Fracking-Moratorium-Final-Report-March-2014-032414.pdf

[1723]http://www.colorado.edu/business/sites/default/files/attached-files/colorado_oil_and_gas_update_-_prices_082015.pdf).

[1724]http://www.boulderweekly.com/article-14970-behind-the-curtain.html

[1725]http://www.metrodenver.org/about/staff/

[1726]http://denversouthedp.org/about-us/board-members/

[1727]http://www.greenpeace.org/usa/frackademia-investigation-reveals-colorado-university-involved-in-pro-fracking-pr-scheme/

[1728]http://www.boulderweekly.com/article-14970-behind-the-curtain.html

[1729]http://blogs.artvoice.com/avdaily/2012/07/20/ub-the-buffalo-news-bamboozled-by-natural-gas-industry/

[1730]*Ibid.*

[1731]http://www.srsi.buffalo.edu/wp-content/uploads/2012/05/UBSRSI-Environmental-Impacts-Single-Page.pdf

[1732]http://www.desmogblog.com/2012/09/19/frackademia-the-brewing-suny-buffalo-shale-resources-society-institute-storm

[1733]http://www.timeshighereducation.co.uk/features/fracking-research-playing-with-fire/2/2007351.article

[1734]http://public-accountability.org/2012/05/ub-shale-play/

[1735]http://www.buffalonews.com/apps/pbcs.dll/article?AID=/20121119/CITYANDREGION/121119113/1010

[1736]http://media.syracuse.com/news/other/2015/04/10/sigel%20paper%20on%20methane.pdf

[1737]http://insideclimatenews.org/news/06042015/fracking-study-water-contamination-under-ethics-review

[1738]http://www.syracuse.com/news/index.ssf/2015/04/fracking_study_syracuse_university_chesapeake_energy_methane_wells.html

[1739]http://pubs.acs.org/doi/abs/10.1021/acs.est.5b01800

[1740]http://webapp.und.edu/dept/our/uletter/?p=32501

[1741]http://webapp.und.edu/dept/our/uletter/?p=32501

[1742]http://investors.clr.com/phoenix.zhtml?c=197380&p=irol-newsArticle&ID=2019668

[1743]http://www.fool.com/investing/general/2014/02/11/americas-oil-boom-is-fueling-growth-at-continental.aspx

[1744]http://www.engr.utexas.edu/features/mulva-eerc-gift

[1745]http://www.utexas.edu/news/2014/04/15/tillerson-pledge-cockrell-school/

[1746]http://energy.utexas.edu/images/ei_shale_gas_regulation120215.pdf

[1747]*Ibid.*

[1748]http://public-accountability.org/wp-content/uploads/ContaminatedInquiry.pdf

[1749]http://www.bloomberg.com/news/2012-07-23/frackers-fund-university-research-that-proves-their-case.html

[1750]http://stateimpact.npr.org/texas/2012/07/24/texas-professor-on-the-defensive-over-fracking-money/

[1751]*Ibid.*

[1752]http://www.bizjournals.com/houston/morning_call/2012/12/university-of-texas-making-changes-in.html

[1753]http://www.chron.com/news/houston-texas/houston/article/Panel-criticizes-fracking-study-UT-ethics-rules-4098257.php

[1754]http://www.bloomberg.com/news/print/2012-12-06/texas-energy-institute-head-quits-amid-fracking-study-conflicts.html

[1755]http://www.pnas.org/content/early/2013/10/09/1304880110

[1756]http://energyindepth.org/national/bombshell-study-confirms-low-methane-leakage-from-shale-gas/

[1757]http://www.desmogblog.com/print/7546

[1758]http://www.scribd.com/doc/168661042/PSE-Study-on-EDF-Greenwashing-of-Methane

[1759]http://articles.philly.com/2012-02-14/news/31058972_1_ann-weaver-hart-higher-education-cuts

[1760]http://www.mainlinemedianews.com/articles/2011/04/29/main_line_suburban_life/news/doc4db9eae2a7f7c181410251.txt

[1761]http://www.marcellusmoney.org/candidates

[1762]http://www.legis.state.pa.us/cfdocs/billinfo/billinfo.cfm?syear=2011&sind=0&body=S&type=B&bn=367

[1763]http://www.ragingchickenpress.org/2013/09/24/education-matters-more-than-money-the-apscuf-anti-extraction-resolution/

[1764]http://www.post-gazette.com/stories/news/environment/drilling-on-campus-marcellus-shale-boom-puts-colleges-at-crossroads-322630/?print=1

[1765]*Ibid.*

[1766]http://www.scribd.com/doc/208975640/2013-14-Undergraduate-Catalog#scribd

[1767]http://www.marcellusgas.org/graphs/PA-Tioga

[1768]http://mansfield.edu/marcellus-institute/events/marcellus-summer-camp/

[1769]http://www.ccp.edu/site/news_room/press_releases/2012/111412.php

[1770]http://protectingourwaters.wordpress.com/2012/11/28/breaking-faculty-group-stands-up-to-marcellus-shale-coalition-at-community-college-of-philadelphia/#comments

[1771]http://protectingourwaters.wordpress.com/2012/12/06/aft-union-calls-fracking-risks-dangers-unacceptable-urges-clean-energy-jobs-training/

[1772] http://www.opednews.com/articles/The-Fracking-Prostitutes-o-by-Walter-Brasch-College_Energy_Energy_Energy-140427-490.html

[1773] http://www.bloomberg.com/news/2012-07-23/frackers-fund-university-research-that-proves-their-case.html

[1774] http://marcelluscoalition.org/about/

[1775] http://www.bloomberg.com/news/2012-07-23/frackers-fund-university-research-that-proves-their-case.html

[1776] http://www.rural.palegislature.us/documents/reports/Marcellus_and_drinking_water_2012.pdf

[1777] *Ibid.*

[1778] http://www.pennlive.com/midstate/index.ssf/2012/03/gas_drilling_industry_paying_p.html

[1779] http://onwardstate.com/2010/09/18/pegula-marcellus-shale-development-good-for-students/

[1780] http://www.bizjournals.com/buffalo/blog/morning_roundup/2012/04/pegula-donation-to-penn-state-hockey.html

[1781] http://www.marcellus.psu.edu/about/What_we_do.php

[1782] http://www.shaletec.org/

[1783] http://www.shaletec.org/

[1784] http://www.shaletec.org/whatis.htm

[1785] http://news.smeal.psu.edu/news-release-archives/2014/september/smeal-to-contribute-to-ge-supported-center-to-study-natural-gas-supply-chains

[1786] http://sustainability.psu.edu/external-spotlight/boom-how-penn-state-students-are-being-educated-about-fracking

[1787] http://sustainability.psu.edu/external-spotlight/boom-how-penn-state-students-are-being-educated-about-fracking

[1788] http://news.psu.edu/story/331716/2014/10/27/campus-life/%E2%80%98-boom%E2%80%99-chosen-penn-state-reads-2015-16-common-text

[1789] http://www.cps.org.uk/files/reports/original/131202135150-WhyEverySeriousEnvironmentalistShouldFavourFracking.pdf

[1790] http://www.nationofchange.org/print/42096

[1791] http://nrcce.wvu.edu/wp-content/uploads/FINAL_UTICA_REPORT_07012015.pdf

[1792] http://www.westernenergyalliance.org/sites/default/files/2013%20Energy%20Dev%20Impact.pdf

[1793] http://public-accountability.org/2015/09/frackademia-update/

[1794] https://www.heartland.org/sites/default/files/east_coast_energy_study.pdf

[1795] http://www.ocregister.com/articles/oil-614459-economic-production.html

[1796] http://www.ocregister.com/articles/oil-614459-economic-production.html

[1797] http://www.latimes.com/business/la-fi-oil-20140521-story.html

[1798] http://www.api.org/~/media/files/policy/taxes/2015/economic-impacts-of-the-proposed-natural-gas-severance-tax-in-pennsylvania.pdf

[1799] http://www.hbs.edu/competitiveness/Documents/america-unconventional-energy-opportunity.pdf

[1800] http://public-accountability.org/2015/09/frackademia-update/

[1801] http://www.frackingjobs.co/which-shale-plays-produce-the-most-oil-and-gas/

[1802] http://www.blankrome.com/index.cfm?contentID=31&itemID=2730

[1803] http://www.prnewswire.com/news-releases/pennsylvania-environmental-protection-secretary-michael-krancer-joins-blank-rome-llp-199531021.html

[1804] http://www.portal.state.pa.us/portal/server.pt/community/about_dep/13464/office_of_the_secretary/585259

[1805] http://stateimpact.npr.org/pennsylvania/2013/12/04/corbetts-pick-to-run-dep-does-not-view-climate-change-as-harmful/

[1806] http://stateimpact.npr.org/pennsylvania/2013/12/04/corbetts-pick-to-run-dep-does-

not-view-climate-change-as-harmful/
[1807]http://stateimpact.npr.org/pennsylvania/2013/12/04/corbetts-pick-to-run-dep-does-not-view-climate-change-as-harmful/
[1808]http://articles.philly.com/2014-10-03/news/54603988_1_explicit-messages-corbett-e-mail-account
http://www.mcall.com/news/breaking/mc-pa-kane-pornography-emails-1001-20141002-story.html
[1810]http://www.businesswire.com/news/home/20130821005883/en/Kevin-Sunday-DEP-spokesman-joins-Quantum-Communications
[1811]http://stateimpact.npr.org/pennsylvania/2015/01/23/corbetts-top-energy-adviser-hired-by-gas-industry/
[1812]*Ibid.*
[1813]http://paenvironmentdaily.blogspot.com/2015/01/gov-corbett-appoints-john-t-hines-to.html
[1814]http://newsok.com/residents-pa.-ignoring-their-health-complaints/article/feed/381606
[1815]http://www.dcnr.state.pa.us/discoverdcnr/ataglance/secretary/index.htm
[1816]http://paenvironmentdaily.blogspot.com/2013/10/brian-grove-appointed-dcnr-deputy-for.html
[1817]http://stateimpact.npr.org/pennsylvania/2013/10/28/former-chesapeake-energy-executive-hired-as-deputy-secretary-for-dcnr/
[1818]http://public-accountability.org/wp-content/uploads/Fracking-and-the-Revolving-Door-in-Pennsylvania.pdf
[1819]http://public-accountability.org/wp-content/uploads/Fracking-and-the-Revolving-Door-in-Pennsylvania.pdf
[1820]http://www.vnf.com/van-ness-feldman-welcomes-senator-mary-landrieu
[1821]http://www.desmogblog.com/sites/beta.desmogblog.com/files/John%20Krohn%20CV%20(1).pdf
[1822]https://twitter.com/Krohn_DC
[1823]http://www.eia.gov/todayinenergy/detail.cfm?id=15351#
[1824]http://www.desmogblog.com/2014/05/23/energy-in-depth-john-krohn-us-energy-information-administration
[1825]http://shalebubble.org/drill-baby-drill/
[1826]http://www.desmogblog.com/2014/05/23/energy-in-depth-john-krohn-us-energy-information-administration
[1827]http://goldwynstrategies.com/Default.aspx
[1828]http://www.desmogblog.com/2014/11/19/natural-gas-exports-washington-revolving-door-fuels-climate-threat
[1829]https://www.republicreport.org/wp-content/uploads/2014/11/LNG_Report_Final_Nov_2015.pdf
[1830]http://www.desmogblog.com/natural-gas-exports-washington-s-revolving-door-fuels-climate-threat-democrats-and-obama-administration
[1831]http://www.desmogblog.com/natural-gas-exports-washington-s-revolving-door-fuels-climate-threat-democrats-and-obama-administration
[1832]https://www.republicreport.org/2014/natural-gas-exports-bush-administration-aides-and-lng/
[1833]http://www.republicreport.org/2014/natural-gas-exports-democrats-and-the-obama-administration/
[1834]http://www.eenews.net/assets/2014/07/28/document_gw_07.pdf
[1835]http://www.politico.com/story/2013/10/heather-zichal-stepping-down-97937.html
[1836]http://www.cheniere.com/corporate/directors.shtml
[1837]http://www.desmogblog.com/2014/06/27/heather-zichal-met-with-cheniere-executives-as-obama-energy-aide-before-board-nomination
[1838]http://www.americanrhetoric.com/speeches/mariosaviosproulhallsitin.htm

[1839]http://ecowatch.com/2013/shut-down-fracking-wastewater-site/
[1840]http://www.ohio.com/blogs/drilling/ohio-utica-shale-1.291290/ohio-anti-fracking-activists-rally-today-in-columbus-1.378910
[1841]http://www.collegegreenmag.com/question-and-anwser-with-madeline-ffitch
[1842]http://www.winonadailynews.com/news/opinion/columnists/local/article_35109a46-b991-11e2-965e-0019bb2963f4.html
[1843]http://insideclimatenews.org/news/20131111/tiny-minnesota-city-draws-line-frac-sand-boom
[1844]http://earthfirstjournal.org/newswire/2013/07/08/breaking-hundreds-of-protesters-shut-down-oil-gas-chemical-supplier-to-protest-fracking/
[1845]http://www.ohio.com/blogs/drilling/ohio-utica-shale-1.291290/twelve-activists-arrested-at-north-carolina-chemical-plant-1.411820
[1846]http://economie.hotnews.ro/stiri-energie-13971682-cum-schimbat-pozitia-usl-fata-gazele-sist-motiune-cenzura-pina-idee-care-trebuie-tratata-serios-pozitiv.htm
[1847]http://www.motherjones.com/environment/2014/09/hillary-clinton-fracking-shale-state-department-chevron
[1848]http://www.naturalgaseurope.com/romania-balances-shale-gas-investment-and-environment
[1849]http://www.reuters.com/article/2014/11/10/romania-shalegas-idUSL6N0SZ1A720141110
[1850]http://www.foodandwaterwatch.org/blogs/highlights-the-second-international-day-to-ban-fracking/
[1851]http://www.nytimes.com/2014/09/22/nyregion/new-york-city-climate-change-march.html?_r=0
[1852]http://peoplesclimate.org/wrap-up/
[1853]http://peoplesclimate.org/wrap-up/
[1854]https://madmimi.com/p/fd6c45?fe=1&pact=25139966606
[1855]http://peoplesclimate.org/wrap-up/
[1856]http://www.washingtonexaminer.com/the-peoples-march-against-common-sense/article/2553859
[1857]http://canon-mcmillan.patch.com/articles/white-still-no-word-on-range-resources-worstell-impondment-from-dep
[1858]http://www.forbes.com/sites/jeffmcmahon/2015/03/12/russians-financed-the-u-s-anti-fracking-movement-oil-tycoon/
[1859]http://www.kxii.com/fox/home/headlines/2-arrested-at-work-site-for-Keystone-XL-pipeline-205268161.html
[1860]http://www.democracynow.org/2014/3/3/xl_dissent_398_youth_arrested_at
[1861]http://www.huffingtonpost.com/jamie-henn/keystone-xl-protest_b_4886208.html
[1862]http://350.org/press-release/keystone-xl-protest-at-the-white-house-leads-to-mass-arrests/
[1863]http://www.omaha.com/news/metro/couple-hosting-anti-keystone-xl-concert-didn-t-expect-to/article_9247cc6f-abee-5bdb-ae7f-64d4ccd22fb5.html
[1864]http://www.omaha.com/news/metro/couple-hosting-anti-keystone-xl-concert-didn-t-expect-to/article_9247cc6f-abee-5bdb-ae7f-64d4ccd22fb5.html
[1865]http://www.omaha.com/go/willie-nelson-neil-young-lend-their-talents-to-keystone-xl/article_c1ac3d34-4676-11e4-8524-001a4bcf6878.html
[1866]http://www.omaha.com/go/willie-nelson-neil-young-lend-their-talents-to-keystone-xl/article_c1ac3d34-4676-11e4-8524-001a4bcf6878.html
[1867]http://www.omaha.com/go/willie-nelson-neil-young-lend-their-talents-to-keystone-xl/article_c1ac3d34-4676-11e4-8524-001a4bcf6878.html
[1868]http://fortune.com/fortune500/exxon-mobil-corporation-2/
[1869]http://reclaimdemocracy.org/exxon-ceo-fracking-lawsuit/
[1870]http://wagingnonviolence.org/feature/forest-service-seeks-to-silence-smokey-the-bear-over-fracking/

[1871]http://www.damascuscitizensforsustainability.org/dcs-programs/atsdr/
[1872]http://www.damascuscitizensforsustainability.org/
[1873]http://www.pressconnects.com/article/20110925/NEWS01/109250340/Gas-face-off-set-DEC-public-hearing
[1874]http://saneenergyproject.org/2014/07/07/ferc-and-the-regulatory-trap/
[1875]http://gasfreeseneca.com/?page_id=123
[1876]http://www.dcbureau.org/201206157397/natural-resources-news-service/epa-refuses-to-release-seismic-data.html
[1877]http://www.dcbureau.org/201410031011/natural-resources-news-service/ferc-approves-ny-methane-storage-project.html
[1878]http://ecowatch.com/2013/03/18/12-arrested-blockading-fracking-infrastructure/
[1879]https://www.commondreams.org/headline/2013/04/18-7
[1880]http://ourfutureisunfractured.wordpress.com/participants/
[1881]http://www.the-leader.com/x1416816020/Marcellus-Watch-Budgeting-orange-jumpsuits-in-Schuyler
[1882]http://www.stargazette.com/story/news/local/2014/10/29/crestwood-protest-arrests/18114435/
[1883]http://org.salsalabs.com/o/423/p/salsa/web/blog/public/?blog_entry_KEY=23522
[1884]http://www.chesapeakeclimate.org/index.php?option=com_k2&view=itemlist&layout=category&task=category&id=7&Itemid=11#sthash.9cgEG60o.dpuf
[1885]http://chesapeakeclimate.org/press-releases/stop-cove-point-coalition-collects-over-150000-public-comments-urging-federal-regulators-to-reject-lng-export-plan/
[1886]http://www.wearecovepoint.org/about/
[1887]http://www.washingtonpost.com/business/community-divided-over-cove-point-natural-gas-terminal/2014/12/05/8f4e7300-7003-11e4-8808-afaa1e3a33ef_story.html
[1888]http://mgaleg.maryland.gov/2013RS/fnotes/bil_0006/sb0656.pdf
[1889]http://www.desmogblog.com/sites/beta.desmogblog.com/files/Dominion-Calvert%20County%20NDA.pdf
[1890]http://articles.baltimoresun.com/2014-08-12/features/bs-md-cove-point-ruling-20140812_1_cove-point-county-officials-county-attorney
[1891]http://chesapeakeclimate.org/press-releases/groups-condemn-federal-ruling-on-cove-point-lng-export-facility/
[1892]http://www.dcmediagroup.us/2015/05/26/gas-export-opponents-stage-blockade-of-ferc/
[1893]http://www.pennfuture.org/content.aspx?MenuID=10&SubSectionID=381&SectionID=148#LoyalsockSF
[1894]http://www.responsibledrillingalliance.org/
[1895]http://www.openrecordspa.org/sunshine.html#public_participation
[1896]http://stateimpact.npr.org/pennsylvania/2011/09/07/ridge-shale-drilling-makes-america-safe/
[1897]http://www.post-gazette.com/stories/local/marcellusshale/pa-gov-tom-corbett-gives-warm-welcome-to-marcellus-shale-coalition-654316/?print=1
[1898]http://thehill.com/blogs/e2-wire/e2-wire/143279-gingrich-calls-for-elimnating-epa-expanding-domestic-energy-production
[1899]http://stateimpact.npr.org/pennsylvania/2013/09/26/shale-insight-2013-newt-gingrich-and-the-r-word/
[1900]http://shaleinsight.com/schedule/
[1901]http://shaleinsight.com/registration/
[1902]http://shaleinsight.com/exhibitors/
[1903]http://shaleinsight.com/wp-content/uploads/2011/02/Sponsorship-Opportunities-and-Application.pdf
[1904]http://shaleinsight.com/pdf/17936-MSC-Mailer-Single-Pages.pdf
[1905]http://stateimpact.npr.org/pennsylvania/2013/09/27/philadelphia-the-grassroots-anti-fracking-movements-new-old-front/
[1906]http://fuelfix.com/blog/2013/09/26/americans-support-for-keystone-oppose-fracking/

[1907]http://www.oikoumene.org/en

[1908]http://www.ccp.edu/site/news_room/press_releases/2012/111412.php

[1908]http://protectingourwaters.wordpress.com/2012/11/28/breaking-faculty-group-stands-up-to-marcellus-shale-coalition-at-community-college-of-philadelphia/#comments

[1908]http://protectingourwaters.wordpress.com/2012/12/06/aft-union-calls-fracking-risks-dangers-unacceptable-urges-clean-energy-jobs-training/

[1908]http://www.opednews.com/articles/The-Fracking-Prostitutes-o-by-Walter-Brasch-College_Energy_Energy_Energy-140427-490.html

[1908]http://www.bloomberg.com/news/2012-07-23/frackers-fund-university-research-that-proves-their-case.html

[1908]http://marcelluscoalition.org/about/

[1908]http://www.bloomberg.com/news/2012-07-23/frackers-fund-university-research-that-proves-their-case.html

[1908]http://www.westernenergyalliance.org/sites/default/files/2013%20Energy%20Dev%20Impact.pdf

[1908]http://www.ocregister.com/articles/oil-614459-economic-production.html

[1908]http://www.ocregister.com/articles/oil-614459-economic-production.html

[1908]http://www.latimes.com/business/la-fi-oil-20140521-story.html

[1908]http://www.rural.palegislature.us/documents/reports/Marcellus_and_drinking_water_2012.pdf

[1908]*Ibid.*

[1908]http://www.pennlive.com/midstate/index.ssf/2012/03/gas_drilling_industry_paying_p.html

[1908]http://onwardstate.com/2010/09/18/pegula-marcellus-shale-development-good-for-students/

[1908]http://www.bizjournals.com/buffalo/blog/morning_roundup/2012/04/pegula-donation-to-penn-state-hockey.html

[1908]http://www.marcellus.psu.edu/about/What_we_do.php

[1908]http://www.shaletec.org/

[1908]http://www.shaletec.org/

[1908]http://www.shaletec.org/whatis.htm

[1908]http://news.smeal.psu.edu/news-release-archives/2014/september/smeal-to-contribute-to-ge-supported-center-to-study-natural-gas-supply-chains

[1908]http://www.frackingjobs.co/which-shale-plays-produce-the-most-oil-and-gas/

[1908]http://www.blankrome.com/index.cfm?contentID=31&itemID=2730

[1908]http://www.prnewswire.com/news-releases/pennsylvania-environmental-protection-secretary-michael-krancer-joins-blank-rome-llp-199531021.html

[1908]http://www.portal.state.pa.us/portal/server.pt/community/about_dep/13464/office_of_the_secretary/585259

[1908]http://stateimpact.npr.org/pennsylvania/2013/12/04/corbetts-pick-to-run-dep-does-not-view-climate-change-as-harmful/

[1908]http://stateimpact.npr.org/pennsylvania/2013/12/04/corbetts-pick-to-run-dep-does-not-view-climate-change-as-harmful/

[1908]http://stateimpact.npr.org/pennsylvania/2013/12/04/corbetts-pick-to-run-dep-does-not-view-climate-change-as-harmful/

[1908]http://articles.philly.com/2014-10-03/news/54603988_1_explicit-messages-corbett-e-mail-account

[1908]http://www.mcall.com/news/breaking/mc-pa-kane-pornography-emails-1001-20141002-story.html

[1908]http://www.businesswire.com/news/home/20130821005883/en/Kevin-Sunday-DEP-spokesman-joins-Quantum-Communications

[1908]http://stateimpact.npr.org/pennsylvania/2015/01/23/corbetts-top-energy-adviser-hired-by-gas-industry/

[1908]*Ibid.*
[1908]http://paenvironmentdaily.blogspot.com/2015/01/gov-corbett-appoints-john-t-hines-to.html
[1908]http://newsok.com/residents-pa.-ignoring-their-health-complaints/article/feed/381606
[1908]http://www.dcnr.state.pa.us/discoverdcnr/ataglance/secretary/index.htm
[1908]http://paenvironmentdaily.blogspot.com/2013/10/brian-grove-appointed-dcnr-deputy-for.html
[1908]http://stateimpact.npr.org/pennsylvania/2013/10/28/former-chesapeake-energy-executive-hired-as-deputy-secretary-for-dcnr/
[1908]http://public-accountability.org/wp-content/uploads/Fracking-and-the-Revolving-Door-in-Pennsylvania.pdf
[1908]http://public-accountability.org/wp-content/uploads/Fracking-and-the-Revolving-Door-in-Pennsylvania.pdf
[1908]http://www.vnf.com/van-ness-feldman-welcomes-senator-mary-landrieu
[1908]http://www.desmogblog.com/sites/beta.desmogblog.com/files/John%20Krohn%20CV%20(1).pdf
[1908]https://twitter.com/Krohn_DC
[1908]http://www.eia.gov/todayinenergy/detail.cfm?id=15351#
[1908]http://www.desmogblog.com/2014/05/23/energy-in-depth-john-krohn-us-energy-information-administration
[1908]http://shalebubble.org/drill-baby-drill/
[1908]http://www.desmogblog.com/2014/05/23/energy-in-depth-john-krohn-us-energy-information-administration
[1908]http://goldwynstrategies.com/Default.aspx
[1908]http://www.desmogblog.com/2014/11/19/natural-gas-exports-washington-revolving-door-fuels-climate-threat
[1908]https://www.republicreport.org/wp-content/uploads/2014/11/LNG_Report_Final_Nov_2015.pdf
[1908]http://www.desmogblog.com/natural-gas-exports-washington-s-revolving-door-fuels-climate-threat-democrats-and-obama-administration
[1908]http://www.desmogblog.com/natural-gas-exports-washington-s-revolving-door-fuels-climate-threat-democrats-and-obama-administration
[1908]https://www.republicreport.org/2014/natural-gas-exports-bush-administration-aides-and-lng/
[1908]http://www.republicreport.org/2014/natural-gas-exports-democrats-and-the-obama-administration/
[1908]http://www.eenews.net/assets/2014/07/28/document_gw_07.pdf
[1908]http://www.politico.com/story/2013/10/heather-zichal-stepping-down-97937.html
[1908]http://www.cheniere.com/corporate/directors.shtml
[1908]http://www.desmogblog.com/2014/06/27/heather-zichal-met-with-cheniere-executives-as-obama-energy-aide-before-board-nomination
[1908]http://www.americanrhetoric.com/speeches/mariosaviosproulhallsitin.htm
[1908]http://ecowatch.com/2013/shut-down-fracking-wastewater-site/
[1908]http://www.ohio.com/blogs/drilling/ohio-utica-shale-1.291290/ohio-anti-fracking-activists-rally-today-in-columbus-1.378910
[1908]http://www.collegegreenmag.com/question-and-anwser-with-madeline-ffitch
[1908]http://www.winonadailynews.com/news/opinion/columnists/local/article_35109a46-b991-11e2-965e-0019bb2963f4.html
[1908]http://insideclimatenews.org/news/20131111/tiny-minnesota-city-draws-line-frac-sand-boom
[1908]http://earthfirstjournal.org/newswire/2013/07/08/breaking-hundreds-of-protesters-shut-down-oil-gas-chemical-supplier-to-protest-fracking/

[1908]http://www.ohio.com/blogs/drilling/ohio-utica-shale-1.291290/twelve-activists-arrested-at-north-carolina-chemical-plant-1.411820

[1908]http://economie.hotnews.ro/stiri-energie-13971682-cum-schimbat-pozitia-usl-fata-gazele-sist-motiune-cenzura-pina-idee-care-trebuie-tratata-serios-pozitiv.htm

[1908]http://www.motherjones.com/environment/2014/09/hillary-clinton-fracking-shale-state-department-chevron

[1908]http://www.naturalgaseurope.com/romania-balances-shale-gas-investment-and-environment

[1908]http://www.reuters.com/article/2014/11/10/romania-shalegas-idUSL6N0SZ1A720141110

[1908]http://www.foodandwaterwatch.org/blogs/highlights-the-second-international-day-to-ban-fracking/

[1908]http://www.nytimes.com/2014/09/22/nyregion/new-york-city-climate-change-march.html?_r=0

[1908]http://peoplesclimate.org/wrap-up/

[1908]http://peoplesclimate.org/wrap-up/

[1908]https://madmimi.com/p/fd6c45?fe=1&pact=25139966606

[1908]http://peoplesclimate.org/wrap-up/

[1908]http://www.washingtonexaminer.com/the-peoples-march-against-common-sense/article/2553859

[1908]http://canon-mcmillan.patch.com/articles/white-still-no-word-on-range-resources-worstell-impondment-from-dep

[1908]http://www.forbes.com/sites/jeffmcmahon/2015/03/12/russians-financed-the-u-s-anti-fracking-movement-oil-tycoon/

[1908]http://www.kxii.com/fox/home/headlines/2-arrested-at-work-site-for-Keystone-XL-pipeline-205268161.html

[1908]http://www.democracynow.org/2014/3/3/xl_dissent_398_youth_arrested_at

[1908]http://www.huffingtonpost.com/jamie-henn/keystone-xl-protest_b_4886208.html

[1908]http://350.org/press-release/keystone-xl-protest-at-the-white-house-leads-to-mass-arrests/

[1908]http://www.omaha.com/news/metro/couple-hosting-anti-keystone-xl-concert-didn-t-expect-to/article_9247cc6f-abee-5bdb-ae7f-64d4ccd22fb5.html

[1908]http://www.omaha.com/news/metro/couple-hosting-anti-keystone-xl-concert-didn-t-expect-to/article_9247cc6f-abee-5bdb-ae7f-64d4ccd22fb5.html

[1908]http://www.omaha.com/go/willie-nelson-neil-young-lend-their-talents-to-keystone-xl/article_c1ac3d34-4676-11e4-8524-001a4bcf6878.html

[1908]http://www.omaha.com/go/willie-nelson-neil-young-lend-their-talents-to-keystone-xl/article_c1ac3d34-4676-11e4-8524-001a4bcf6878.html

[1908]http://www.omaha.com/go/willie-nelson-neil-young-lend-their-talents-to-keystone-xl/article_c1ac3d34-4676-11e4-8524-001a4bcf6878.html

[1908]http://fortune.com/fortune500/exxon-mobil-corporation-2/

[1908]http://reclaimdemocracy.org/exxon-ceo-fracking-lawsuit/

[1908]http://wagingnonviolence.org/feature/forest-service-seeks-to-silence-smokey-the-bear-over-fracking/

[1908]http://www.damascuscitizensforsustainability.org/dcs-programs/atsdr/

[1908]http://www.damascuscitizensforsustainability.org/

[1908]http://www.pressconnects.com/article/20110925/NEWS01/109250340/Gas-face-off-set-DEC-public-hearing

[1908]http://saneenergyproject.org/2014/07/07/ferc-and-the-regulatory-trap/

[1908]http://gasfreeseneca.com/?page_id=123

[1908]http://www.dcbureau.org/201206157397/natural-resources-news-service/epa-refuses-to-release-seismic-data.html

[1908]http://www.dcbureau.org/2014100310011/natural-resources-news-service/ferc-approves-ny-methane-storage-project.html

[1908]http://ecowatch.com/2013/03/18/12-arrested-blockading-fracking-infrastructure/
[1908]https://www.commondreams.org/headline/2013/04/18-7
[1908]http://ourfutureisunfractured.wordpress.com/participants/
[1908]http://www.the-leader.com/x1416816020/Marcellus-Watch-Budgeting-orange-jumpsuits-in-Schuyler
[1908]http://www.stargazette.com/story/news/local/2014/10/29/crestwood-protest-arrests/18114435/
[1908]http://org.salsalabs.com/o/423/p/salsa/web/blog/public/?blog_entry_KEY=23522
[1908]http://www.chesapeakeclimate.org/index.php?option=com_k2&view=itemlist&layout=category&task=category&id=7&Itemid=11#sthash.9cgEG60o.dpuf
[1908]http://chesapeakeclimate.org/press-releases/stop-cove-point-coalition-collects-over-150000-public-comments-urging-federal-regulators-to-reject-lng-export-plan/
[1908]http://www.wearecovepoint.org/about/
[1908]http://www.washingtonpost.com/business/community-divided-over-cove-point-natural-gas-terminal/2014/12/05/8f4e7300-7003-11e4-8808-afaa1e3a33ef_story.html
[1908]http://mgaleg.maryland.gov/2013RS/fnotes/bil_0006/sb0656.pdf
[1908]http://www.desmogblog.com/sites/beta.desmogblog.com/files/Dominion-Calvert%20County%20NDA.pdf
[1908]http://articles.baltimoresun.com/2014-08-12/features/bs-md-cove-point-ruling-20140812_1_cove-point-county-officials-county-attorney
[1908]http://chesapeakeclimate.org/press-releases/groups-condemn-federal-ruling-on-cove-point-lng-export-facility/
[1908]http://www.dcmediagroup.us/2015/05/26/gas-export-opponents-stage-blockade-of-ferc/
[1908]http://www.pennfuture.org/content.aspx?MenuID=10&SubSectionID=381&SectionID=148#LoyalsockSF
[1908]http://www.responsibledrillingalliance.org/
[1908]http://www.openrecordspa.org/sunshine.html#public_participation
[1908]http://stateimpact.npr.org/pennsylvania/2011/09/07/ridge-shale-drilling-makes-america-safe/
[1908]http://www.post-gazette.com/stories/local/marcellusshale/pa-gov-tom-corbett-gives-warm-welcome-to-marcellus-shale-coalition-654316/?print=1
[1908]http://thehill.com/blogs/e2-wire/e2-wire/143279-gingrich-calls-for-elimnating-epa-expanding-domestic-energy-production
[1908]http://stateimpact.npr.org/pennsylvania/2013/09/26/shale-insight-2013-newt-gingrich-and-the-r-word/
[1908]http://shaleinsight.com/schedule/
[1908]http://shaleinsight.com/registration/
[1908]http://shaleinsight.com/exhibitors/
[1908]http://shaleinsight.com/wp-content/uploads/2011/02/Sponsorship-Opportunities-and-Application.pdf
[1908]http://shaleinsight.com/pdf/17936-MSC-Mailer-Single-Pages.pdf
[1908]http://stateimpact.npr.org/pennsylvania/2013/09/27/philadelphia-the-grassroots-anti-fracking-movements-new-old-front/
[1908]http://fuelfix.com/blog/2013/09/26/americans-support-for-keystone-oppose-fracking/
[1908]http://www.oikoumene.org/en
[1908]http://www.theguardian.com/environment/2015/apr/30/church-of-england-ends-investments-in-heavily-polluting-fossil-fuels
[1909]http://gofossilfree.org/commitments/
[1910]http://www.quaker.org.uk/news/quakers-disinvest-fossil-fuels
[1911]http://www.post-gazette.com/news/environment/2014/07/06/Religious-and-liberal-groups-target-fossil-fuel-divestiture/stories/201407060217)
[1912]http://episcopaldigitalnetwork.com/ens/2015/06/30/fossil-fuels-climate-advisory-committee-resolutions-move-to-house-of-deputies/

625

[1913]*See:*http://www.msn.com/en-au/sport/afl/episcopal-church-votes-to-divest-from-fossil-fuels-this-is-a-moral-issue/vp-AAcwJbI

[1914]http://www.post-gazette.com/news/environment/2014/07/06/Religious-and-liberal-groups-target-fossil-fuel-divestiture/stories/201407060217#ixzz3C4LDNHyU

[1915]http://insideclimatenews.org/news/15052015/pennsylvania-high-school-students-convince-school-divest-coal

[1916]http://usatoday30.usatoday.com/news/religion/2011-07-09-gas-drilling-debate-religion_n.htm

[1917]http://www.nola.com/business/index.ssf/2015/05/exxon_mobil_vote_climate_chang.html

[1918]*Ibid.*

[1919]http://time.com/2853203/union-fossil-fuels/.

[1920]http://dealbook.nytimes.com/2013/09/05/a-new-divestment-focus-fossil-fuels/?module=Search&mabReward=relbias%3Ar%2C{%221%22%3A%22RI%3A10%22}

[1921]http://www.unity.edu

[1922]http://gofossilfree.org/commitments/

[1923]http://chronicle.com/article/Pitzer-College-to-Take/145905/

[1924]http://www.nytimes.com/2014/09/22/us/heirs-to-an-oil-fortune-join-the-divestment-drive.html?_r=0

[1925]http://www.stanforddaily.com/2014/05/07/university-divests-from-coal-mining-companies/

[1926]http://www.commondreams.org/news/2015/01/12/something-momentous-happening-hundreds-stanford-professors-call-full-fossil-fuel

[1927]http://dailyorange.com/2015/03/syracuse-university-announces-it-will-divest-from-fossil-fuels/

[1928]http://www.cnbc.com/id/102020458

[1929]http://www.harvardfacultydivest.com/open-letter-new

[1930]http://www.nytimes.com/2014/11/21/business/energy-environment/nrg-sets-goals-to-cut-carbon-emissions.html?_r=0

[1931]http://www.greenpeace.org/usa/en/media-center/news-releases/Norways-No-to-Coal-Marks-Major-Victory-for-Climate-Movement-Divests-from-Major-US-Utilities/

[1932]http://blog.sfgate.com/energy/2013/04/23/san-francisco-supervisors-vote-to-divest-from-big-oil/

[1933]http://mayormcginn.seattle.gov/an-update-on-fossil-fuel-divestment/

[1934]https://www.sustainablebusiness.com/index.cfm/go/news.display/id/24815

[1935]https://www.aquaamerica.com/our-states/pennsylvania.aspx

[1936]http://www.businesswire.com/news/home/20120430006113/en/CORRECTING-REPLACING-2000-Truck-Trips-Removed-PA

[1937]http://aspe.hhs.gov/poverty/11poverty.shtml

[1938]https://commerce.us.reuters.com/purchase/showReportDetail.do?docid=55145489

[1939]http://www.pvresource.com/News/Press-Release-Details/2012/PVR-Partners-Announces-Completion-Of-Marcellus-Shale-Midstream-Pipeline-System-Acquisition-From-Chief-and-Announces-New-Bsoard/default.aspx

[1940]http://eidmarcellus.org/marcellus-shale/property-rights-where-do-we-draw-the-line/7918/

[1941]http://huffmaster.com/

[1942]https://www.aquaamerica.com/about-aqua/news/view-article.aspx?id=1570

[1943]http://www.stallionoilfield.com/index.php?id=34

[1944]http://truth-out.org/news/item/17605-former-mobil-vp-warns-of-fracking-and-climate-change

[1945]http://www.controlrisks.com/OurThinking/Pages/Global-anti-fracking-movement.aspx

[1946]http://www.people-press.org/2014/11/12/little-enthusiasm-familiar-divisions-after-the-gops-big-midterm-victory/attachment/17/

[1947]http://www.variety.com/review/VE1117941971/

[1948]http://www.momscleanairforce.org/2013/01/29/interview-with-angela-monti-fox-founder-of-the-mothers-project/

[1949]http://www.eenews.net/assets/2011/02/01/document_gw_02.pdf

[1950]http://www.energyindepth.org/wp-content/uploads/2011/11/Debunking-Gasland.pdf

[1951]http://fractoids.blogspot.com/

[1952]http://www.cfr.org/united-states/new-north-american-energy-paradigm-reshaping-future/p28630

[1953]http://triblive.com/x/pittsburghtrib/opinion/s_786914.html#axzz2kumMznrx

[1954]http://lancasteronline.com/article/local/361158_State-geologist-discusses-Marcellus-Shale-in-talk-here.html

[1955]http://lancasteronline.com/article/local/362603_Pa--official-apologizes-for-Nazi-propaganda-comment-made-here.html

[1956]http://johnhanger.blogspot.com/2011/02/gasland-and-oscars.html

[1957]http://www.media-stakeholder-relations-hydraulic-fracturing.com/5/speakers/

[1958]http://commonsense2.com/2012/09/naturalgasdrilling/connecting-the-dots-part-10-let%e2%80%99s-go-to-the-movies/

[1959]http://www.youtube.com/watch?v=-eWhdEVDHPQ

[1960]http://sourcewatch.org/index.php/Energy_for_America

[1961]http://www.rollingstone.com/movies/news/q-a-gasland-part-ii-director-josh-fox-on-the-fight-against-fracking-20130423

[1962]*Ibid.*

[1963]http://www.energyindepth.org/eid-to-fox-how-about-including-facts-in-gasland-2/

[1964]*Ibid.*

[1965]http://www.energyindepth.org/debunking-gasland-part-ii/

[1966]http://www.indiewire.com/article/tribeca-review-gasland-gets-a-sequel-but-does-it-offer-anything-new-looking-at-gasland-part-ii

[1967]http://www.imdb.com/title/tt1558250/

[1968]http://www.boxofficemojo.com/movies/?id=promisedland2012.htm

[1969]http://www.boxofficemojo.com/movies/?id=promisedland2012.htm

[1970]http://www.movieinsider.com/m9828/promised-land/production/

[1971]http://www.huffingtonpost.com/robert-f-kennedy-jr/fracking-movie-promised-land_b_2251339.html

[1972]http://www.foxnews.com/opinion/2012/10/01/hollywood-hypocrisy-and-matt-damon-anti-fracking-film/

[1973]http://www.nationalreview.com/planet-gore/328561/ithe-bourne-stupidityi-matt-damon-takes-fracking-greg-pollowitz

[1974]http://www.breitbart.com/Big-Hollywood/2012/09/27/Matt-Damon-s-Anti-Fracking-Movie-Goes-Full-Conspiracy-Theory

[1975]http://www.cnn.com/2013/01/02/opinion/bennett-fracking-movie/index.html

[1976]http://www.facebook.com/#!/Realpromisedland/info

[1977]http://www.politico.com/story/2013/01/matt-damon-called-liar-by-pro-fracking-filmmaker-85733.html

[1978]http://blog.heritage.org/2012/09/28/matt-damons-anti-fracking-movie-financed-by-oil-rich-arab-nation/

[1979]http://johnhanger.blogspot.com/2013/01/promised-land-fracks-or-blows-it-my.html

[1980]http://www.participantmedia.com/wp-content/uploads/2011/05/Participant-Media-Our-History.pdf

[1981]http://www.businessinsider.com/matt-damon-movie-blasts-fracking-backed-by-uae-2012-9#ixzz2Ha60FQfw

[1982]http://cnsnews.com/node/616420

[1983]http://marcelluscoalition.org/2013/01/msc-engages-public-with-natural-gas-facts-counters-promised-land/

[1984]http://www.desmogblog.com/sites/beta.desmogblog.com/files/John%20Krohn%20CV%20(1).pdf

[1985]http://www.truthlandmovie.com/

[1986]http://blog.littlesis.org/2012/06/13/fracking-industrys-answer-to-gasland-devised-by-astroturf-lobbying-group-and-political-ad-agency/

[1987]http://www.desmogblog.com/sites/beta.desmogblog.com/files/ANGA%202012%20990.pdf

[1988]http://commonsense2.com/2012/09/naturalgasdrilling/connecting-the-dots-part-10-let%e2%80%99s-go-to-the-movies/

[1989]http://blog.littlesis.org/2012/06/13/fracking-industrys-answer-to-gasland-devised-by-astroturf-lobbying-group-and-political-ad-agency/

[1990]http://www.desmogblog.com/sites/beta.desmogblog.com/files/John%20Krohn%20CV%20(1).pdf

[1991]http://fracktoids.blogspot.com/2012/07/energy-in-depth-dots.html?showComment=1341985781003

[1992]http://commonsense2.com/2012/05/unsung-heroes/meet-the-1st-lady-of-fractivism-the-woman-who-connects-the-dots/

[1993]http://www.prnewswire.com/news-releases/fracknation-premieres-january-22-on-axs-tv-183812371.html

[1994]http://www.reuters.com/article/2013/01/03/ca-fracknation-film-idUSnPnPH36379+160+PRN20130103

[1995]http://www.post-gazette.com/stories/business/news/industry-gets-cast-in-fracknation-the-latest-documentary-on-the-drilling-debate-220871/

[1996] See:http://www.youtube.com/watch?v=dEB_Wwe-uBM

[1997]http://www.marcellusgas.org/graphs/PA-Potter

[1998]http://www.groundswellrising.com/the-film/

[1999]http://www.groundswellrising.com/

[2000]http://www.hollywoodreporter.com/review/groundswell-rising-film-review-796898

[2001]http://www.dear-governor-cuomo.com/

[2002]http://www.switchenergyproject.com/

[2003]http://variety.com/2012/film/reviews/switch-1117948696/

[2004]http://www.washingtonpost.com/gog/movies/switch,1225679/critic-review.html

[2005]http://www.boston.com/ae/movies/2012/09/11/switch-drill-maybe-drill/UBjxp448y66L2zE5XQ6khP/story.html

[2006]http://www.switchenergyproject.com/about.php

[2007]http://www.boston.com/ae/movies/2012/09/11/switch-drill-maybe-drill/UBjxp448y66L2zE5XQ6khP/story.html

[2008]http://www.stoparkansasfracking.org/2004-08-17.pd

[2009]http://www.commondreams.org/news/2014/09/05/disruption-film-offers-grassroots-global-revolt-key-answer-climate-crisis

[2010]http://www.nytimes.com/2014/10/10/movies/the-overnighters-looks-at-migrant-workers.html?_r=0

[2011]http://www.latimes.com/entertainment/movies/la-et-mn-overnighters-movie-review-20141031-story.html

[2012]http://www.washingtonpost.com/goingoutguide/movies/the-overnighters-movie-review/2014/11/12/6bef2490-66a3-11e4-9fdc-d43b053ecb4d_story.html

[2013]http://www.whitewaterblackgold.com/about

[2014]http://www.whitewaterblackgold.com/sites/default/files/WWBG_Screening%20Guide.pdf

[2015]http://www.capp.ca/aboutUs/mediaCentre/CAPPCommentary/Pages/ShatteredGroundignoresthefacts.aspx

[2016]http://www.cbc.ca/natureofthings/episode/tipping-point.html

[2017]http://www.cbc.ca/natureofthings/episode/shattered-ground.html

[2018]http://www.bullfrogfilms.com/catalog/split.html

[2019]*Ibid.*

[2020]http://thegazette.com/2012/10/24/northeast-iowa-residents-speak-out-against-frac-sand-mining-proposal/

[2021]http://thepriceofsand.com/

[2022]http://www.wkpublishing.com/general/fees.html

[2023]http://www.frackusa.com/?page_id=2

[2024]http://www.amazon.com/Just-Fracks-Maam-hydrofracking-American/dp/159932332X/ref=pd_sim_b_10

[2025]http://www.seamusmcgraw.com/

[2026]http://www.rti.org/publications/rtipress.cfm?pubid=19220

[2027]*Ibid.*

[2028]http://www.amazon.com/Environmental-Impacts-Hydraulic-Fracturing-Spellman/dp/1466514671/ref=sr_1_sc_1?ie=UTF8&qid=1373980864&sr=8-1-spell&keywords=Environmental+Impacts+of+Hydraulic+Fracturin

[2029]http://www.earthworksaction.org/library/detail/blackout_summary#.VQLriI7F_-s

[2030]https://www.kirkusreviews.com/book-reviews/business-economics/books/?availability=recent-reviews&stars=na

[2031]http://www.marcellusgas.org/graphs/PA#pcnty

[2032]http://www.sungazette.com/page/content.detail/id/585875/EPA-study-on-fracking-could-prove-valuable-commodity.html?nav=5004

[2033]http://www.sungazette.com/page/content.detail/id/590610/Drilling-boom-environment-alliance-takes-step-forward.html?nav=5004

[2034]http://www.sungazette.com/page/content.detail/id/598724/New-numbers-show-natural-gas-drilling-s-robust-production.html?nav=5004

[2035]http://triblive.com/x/pittsburghtrib/opinion/s_786914.html#axzz2kumMznrx

[2036]http://triblive.com/opinion/editorials/2949293-74/energy-oil-obama-administration-america-independence-blow-dramatic-extraction-fracking#axzz2kumMznrx

[2037]http://triblive.com/opinion/editorials/4556807-74/gas-shale-boom?printerfriendly=true#axzz2kumMznrx

[2038]http://www.timesleader.com/news/editorial/50401834/OUR-VIEW-Do-we-know-health-impact-of-drilling

[2039]http://www.sltrib.com/opinion/1751351-155/county-infrastructure-projects-coalition-counties-future

[2040]http://www.sltrib.com/opinion/1867952-155/editorial-utah-should-be-less-willing

[2041]http://www.fayobserver.com/opinion/editorials/our-view-fracking-can-begin-but-safeguards-are-lacking/article_942c7a6a-20e7-521e-a0b2-f2c61ea9e63f.html

[2042]http://www.starnewsonline.com/article/20150129/ARTICLES/150129657?p=2&tc=pg

[2043]http://www.fredericknewspost.com/opinion/editorial/ban-don-t-delay-fracking-in-maryland/article_960c17fb-1354-5ebb-8816-8027970a10a0.html

[2044]http://www.dailycamera.com/editorials/ci_28283321/editorial-is-colorados-fracking-battle-over

[2045]http://www.sfchronicle.com/opinion/editorials/article/As-new-study-shows-we-don-t-know-how-dangerous-6380670.php

[2046]http://www.news-gazette.com/opinion/editorials/2014-08-31/ardor-lacking-fracking.html

[2047]http://www.chron.com/opinion/article/Fracking-s-fire-5997842.php

[2048]http://www.reviewjournal.com/opinion/editorial-nevada-prepares-arrival-fracking

[2049]https://www.thestateofthegulf.com/media/84690/Remarks-by-Geoff-Morrell-at-the-Society-of-Environmental-Journalists-Conference-9-3-14.pdf

[2050]http://www.reuters.com/article/2014/09/04/bp-gulfmexico-ruling-idUSL1N0R51H120140904

[2051]http://www.newsweek.com/former-army-general-lambasts-oil-industry-hijacking-democracy-269086

[2052]http://pipeline.post-gazette.com/about

[2053]http://www.timesleader.com/news/editorial-ouropinion/151628230/OUR-VIEW-Even-an-energy-revolution-needs-leaders

[2054]http://insideclimatenews.org/news/20141211/health-worries-pervade-north-texas-fracking-zone

[2055]http://energyindepth.org/national/fracking-insideclimate-news-public-integrity/

[2056]http://insideclimatenews.org/news/20140729/insideclimate-news-responds-steve-everley-energy-depth

[2057]http://insideclimatenews.org/news/20140729/insideclimate-news-responds-steve-everley-energy-depth

[2058]http://www.sej.org/initiatives/winners-sej-14th-annual-awards-reporting-environment

[2059]http://whirlmagazine.com/cbs-radios-marcellus-shale-festival/

[2060]http://pittsburgh.cbslocal.com/2013/07/31/marcellus-shale-festival/

[2061]http://www.pghcitypaper.com/pittsburgh/all-gassed-up-cbs-radio-hands-over-the-mic-to-natural-gas-boosters/Content?oid=1685087

[2062]http://www.alleghenyfront.org/2015-awards/full

[2063]http://www.journalism.columbia.edu/page/924-2013-winners/707

[2064]https://www.youtube.com/watch?v=XGEc8KBKO8I

[2065]http://tvbythenumbers.zap2it.com/2014/11/04/sunday-final-ratings-the-good-wife-brooklyn-nine-nine-once-upon-a-time-60-minutes-adjusted-up-final-football-ratings/323716/

[2066]http://www.theguardian.com/environment/climate-consensus-97-per-cent/2013/may/28/global-warming-consensus-climate-denialism-characteristics

[2067]http://www.nytimes.com/1998/04/26/us/industrial-group-plans-to-battle-climate-treaty.html

[2068]http://www.adweek.com/news/advertising-branding/grey-energizes-americas-natural-gas-alliance-100371

[2069]http://archive.stargazette.com/article/20130330/NEWS11/303300030/Exxon-Mobil-spent-2-million-pro-fracking-ad-campaign

[2070]http://blogs.democratandchronicle.com/voteup/2012/10/03/new-ads-%E2%80%9Cwe%E2%80%99ve-waited-long-enough%E2%80%9D-for-hydrofracking/

[2071]https://stations.fcc.gov/collect/files/47904/Political%20File/2015/Non-Candidate%20Issue%20Ads/American%20Petroleum%20Institute/API%20Superbowl53 1661--1%20(14223876057116)_.pdf

[2072]http://ecowatch.com/2013/11/01/misleading-fracking-ads-on-npr/

[2073]http://www.frackcheckwv.net/2013/11/01/national-public-radio-carries-misleading-ads-of-frackers/

[2074]http://thedailystar.com/letters/x1912998904/Local-station-doesnt-support-fracking/print

[2075]http://www.ftc.gov/sites/default/files/documents/statutes/federal-trade-commission-act/ftc_act_incorporatingus_safe_web_act.pdf

[2076]http://www.lawpublish.com/edenfield-fane.html

[2077]http://www.theguardian.com/environment/2014/sep/03/pro-fracking-advert-banned-asa-breitling-energy-shale-gas-daily-telegraph

[2078]https://www.wbur.org/npr/316552595/no-matter-how-you-spell-it-fracking-stirs-controversy?ft=3&f=316552595

[2079]http://blog.gregoryfca.com/2011/02/are-we-winning-or-losing-public.html

[2080]http://www.oxforddictionaries.com/us/definition/american_english/fracking

[2081]http://www.merriam-webster.com/dictionary/fracking

[2082]http://articles.philly.com/2014-09-26/business/54322631_1_gas-industry-marcellus-shale-coalition-largest-gas-producer

[2083]http://www.nofrackingway.us/2014/09/28/putting-lipstick-on-the-fracking-pig/

[2084]http://s3.amazonaws.com/ownlocal.adforge.production/ads/1269209/original_images.jpg?1411660579

[2085]http://unitedshaleadvocates.com/rocksolidfacts/?sp=1
[2086]https://www.youtube.com/watch?v=VY34PQUiwOQ
[2087]http://www.rollingstone.com/politics/blogs/national-affairs/mark-ruffalo-on-the-fracking-fight-20120516
[2088]https://www.youtube.com/watch?v=VfymhAEe-TM
[2089]http://www.theclaycenter.org/education/Power-Your-Future.aspx
[2090]http://www.wvgazettemail.com/article/20150615/GZ04/150619636
[2091]http://www.wvgazettemail.com/article/20150615/GZ04/150619636
[2092]http://pipeline.post-gazette.com/index.php/news/business/23997
[2093]http://www.foxnews.com/us/2011/07/14/fracking-coloring-book-by-talisman-energy-blasted-by-critics/
[2094]http://www.colbertnation.com/the-colbert-report-videos/391552/july-11-2011/anti-frack-attack
[2095]http://www.adweek.com/news/advertising-branding/grey-energizes-americas-natural-gas-alliance-100371
[2096]http://www.stargazette.com/article/20130330/NEWS11/303300030/Exxon-Mobil-spent-2-million-pro-fracking-ad-campaign
[2097]https://www.commondreams.org/headline/2013/12/24-2
[2098]http://www.burlingtonfreepress.com/story/news/local/vermont/2014/07/02/woman-arrested-knit-protest-pipeline/12118239/
[2099]http://www.biggreenradicals.com/
[2100]http://www.bermanco.com/
[2101]http://www.nytimes.com/2014/10/31/us/politics/pr-executives-western-energy-alliance-speech-taped.html?_r=0
[2102]http://www.sltrib.com/sltrib/opinion/58061863-82/lands-blm-federal-energy.html.csp
[2103]http://shalegasoutrage.org/wp-content/uploads/2011/09/Craig-and-Julie-Sautners-story-in-press-packet.pdf
[2104]http://thetimes-tribune.com/news/controversial-billboard-depicting-contaminated-water-comes-down-1.1184658
[2105]http://www.citypaper.net/article.php?No-free-speech-in-fracking-country-10140
[2106]http://www.nofrackingway.us/2013/07/06/censored-fractavist-billboards/
[2107]http://www.youtube.com/watch?v=pLGnbYYnuzU
[2108]http://dailyitem.com/0100_news/x1059018339/Turnpike-billboards-slam-gas-drilling-critics
[2109]http://www.nogreenslime.com/what-is-green-slime/
[2110]http://www.nytimes.com/2012/10/21/business/energy-environment/in-a-natural-gas-glut-big-winners-and-losers.html?pagewanted=all&_r=3&
[2111]http://www.smh.com.au/business/mining-and-resources/plunging-oil-price-could-trigger-next-gfc-20141127-11ut7p.html
[2112]http://www.alternet.org/economy/97-fracking-now-operating-loss-current-oil-prices
[2113]http://www.ft.com/cms/s/0/e964a8a6-2c38-11e3-8b20-00144feab7de.html
[2114]http://www.carbontracker.org/wastedcapital
[2115]http://www.bloomberg.com/news/2014-05-26/shakeout-threatens-shale-patch-as-frackers-go-for-broke.html
[2116]http://money.cnn.com/2015/01/07/investing/big-oil-price-energy-stocks/
[2117]http://www.reuters.com/article/2015/01/08/wbh-bankruptcy-idUSL1N0UN01Z20150108
[2118]http://fuelfix.com/blog/2015/07/15/sabine-files-for-bankruptcy-in-new-york-amid-falling-oil-prices/#31510101=0
[2119]http://wolfstreet.com/2015/08/11/big-natural-gas-driller-samson-resources-chapter-11-bankruptcy-smart-money-kkr-crushed/
[2120]http://www.naturalgasintel.com/articles/104182-range-sticking-with-marcellus-in-2016-projected-utica-eurs-cant-compete

631

[2121]http://fuelfix.com/blog/2015/07/31/chevron-profits-fall-90-percent-on-oil-impairments/#34624101=0

[2122]http://fuelfix.com/blog/2015/08/03/moodys-more-oil-driller-defaults-coming-as-banks-restrict-lending/#34624101=0

[2123]http://www.investorvillage.com/uploads/77263/files/OXFORD19CODEBTHITLIST.pdf

[2124]http://www.theguardian.com/environment/2014/oct/13/mark-carney-fossil-fuel-reserves-burned-carbon-bubble

[2125]http://en.wikipedia.org/wiki/HSBC

[2126]http://europe.newsweek.com/hsbc-warns-clients-fossil-fuel-investment-risks-323886

[2127]http://www.nytimes.com/2012/10/21/business/energy-environment/in-a-natural-gas-glut-big-winners-and-losers.html?pagewanted=all&_r=0

[2128]http://www.nytimes.com/2012/10/21/business/energy-environment/in-a-natural-gas-glut-big-winners-and-losers.html?pagewanted=all&_r=0

[2129]http://www.nytimes.com/2012/10/21/business/energy-environment/in-a-natural-gas-glut-big-winners-and-losers.html?pagewanted=all&_r=0

[2130]http://www.theguardian.com/environment/2013/apr/19/carbon-bubble-financial-crash-crisis

[2131]http://shalebubble.org/wp-content/uploads/2013/02/SWS-report-FINAL.pdf

[2132]http://www.theguardian.com/environment/2013/apr/19/carbon-bubble-financial-crash-crisis

[2133]http://www.postcarbon.org/drill-baby-drill/es

[2134]http://www.examiner.com/article/drilling-permits-decline-sharply-for-the-pennsylvania-marcellus-formation

[2135]http://articles.philly.com/2012-07-08/business/32589447_1_natural-gas-prices-drilling-natural-gas/2

[2136]http://www.thereporteronline.com/article/20121001/NEWS03/121009986/next-cold-war-gas-drilling-boom-rattles-russia&pager=full_story

[2137]http://www.bizjournals.com/pittsburgh/blog/energy/2013/02/more-than-half-of-excos-warrendale.html

[2138]http://wvmetronews.com/2015/01/06/natural-gas-drilling-company-lays-off-contract-workers/

[2139]http://www.bloomberg.com/news/articles/2015-09-29/chesapeake-cuts-15-of-workforce-as-oil-rout-deepens

[2140]http://stateimpact.npr.org/pennsylvania/2013/10/08/chesapeake-energy-lays-off-800-employees-nationwide/

[2141]http://www.naturalgasintel.com/articles/104153-consol-looking-to-sell-up-to-23b-in-coal-natgas-assets

[2142]http://fuelfix.com/blog/2015/07/31/chevron-profits-fall-90-percent-on-oil-impairments/#34624101=0

[2143]http://www.forbes.com/sites/christopherhelman/2015/10/22/as-oil-bust-kills-200000-jobs-a-headhunter-looks-at-the-bright-side/

[2144]http://www.reuters.com/article/2015/01/16/usa-oil-drilling-idUSL1N0UV2AA20150116

[2145]http://www.forbes.com/sites/christopherhelman/2015/10/22/as-oil-bust-kills-200000-jobs-a-headhunter-looks-at-the-bright-side/

[2146]http://fuelfix.com/blog/2015/10/19/halliburton-posts-54-million-loss-amid-slumping-oil-field-activity/#36089101=0

[2147]http://www.forbes.com/sites/christopherhelman/2015/10/22/as-oil-bust-kills-200000-jobs-a-headhunter-looks-at-the-bright-side/

[2148]http://www.philly.com/philly/business/20150125_Falling_oil_prices_bring_slowdown_in_shale_drilling.html

[2149]http://www.businesswire.com/news/home/20141222005922/en/Chesapeake-Energy-Corporation-Closes-Southern-Marcellus-Utica#.VL0yMUfF_-s

[2150]http://fortune.com/2015/01/09/oil-prices-shale-fracking/

[2151]http://investors.clr.com/phoenix.zhtml?c=197380&p=irol-newsArticle&ID=2001684

[2152]http://powersource.post-gazette.com/powersource/companies-powersource/2015/02/24/More-risks-today-in-oil-and-gas-CEO-survey-finds/stories/01502240013

[2153]http://www.naturalgasintel.com/articles/104153-consol-looking-to-sell-up-to-23b-in-coal-natgas-assets

[2154]http://www.alternet.org/fracking/anti-fracking-activists-celebrate-cancellation-gas-leases-and-drilling-plans-ne

[2155]http://www.sfgate.com/business/energy/article/Hess-Corp-plans-18-percent-cut-in-Bakken-6045360.php

[2156]http://bakken.com/news/id/244490/again-continental-resources-cuts-budget-bakken-rig-count/

[2157]https://finance.yahoo.com/q/is?s=COG

[2158]http://gis.bakerhughesdirect.com/Reports/StdRptTotals.aspx

[2159] file:///C:/Users/Walt/Downloads/RigCnt2015.pdf

[2160]http://gis.bakerhughesdirect.com/RigCounts/default2.aspx

[2161]http://www.aogr.com/web-exclusives/us-rig-count/2015

[2162]http://www.aogr.com/web-exclusives/us-rig-count/2015

[2163]http://bakkenshale.com/drilling-rig-count/

[2164] file:///C:/Users/Walt/Downloads/Rig%20Count%20Summary_120415%20(1).pdf

[2165]http://business.financialpost.com/2013/10/18/u-s-has-overfracked-and-overdrilled-shell-director-says/?__lsa=d589-16d8

[2166]http://gomarcellusshale.com/group/tiogacounty/forum/topics/talisman-to-trim-capital-spending?commentId=2274639%3AComment%3A222446&groupId=2274639%3AGroup%3A12579

[2167]http://money.cnn.com/magazines/fortune/global500/2013/full_list/?iid=G500_sp_full

[2168]http://online.wsj.com/article/SB10001424127887323681904578641183049294110.html

[2169]http://thetimes-tribune.com/news/wpx-to-divest-from-marcellus-shale-as-profiting-from-gas-proving-more-difficult-1.1789797

[2170]http://www.wpxenergy.com/news-and-media/press-releases/pressrelease.aspx?id=7e96bcb1-6d83-46f3-8a75-9f60d313901a

[2171]http://www.marketwatch.com/story/wpx-energy-completes-200-million-asset-sale-2015-05-27

[2172]http://knappap.blogspot.com/2012/01/wet-gas-vs-dry-gas-do-you-know-what.html

[2173]*Ibid.*

[2174]http://www.frackcheckwv.net/2014/05/14/oil-gas-drillers-exploring-utica-shale-in-west-virginia/

[2175]http://www.reuters.com/article/2013/10/09/us-americanenergy-utica-idUSBRE9980N020131009?feedType=RSS&feedName=businessNews

[2176]http://www.bloomberg.com/news/2013-04-15/ohio-s-500-billion-oil-dream-fades-as-utica-turns-gassy.html

[2177]http://www.canada.com/business/Asian+market+looking+natural+executive+director+says/7095235/story.html

[2178]http://www.canada.com/business/Asian+market+looking+natural+executive+director+says/7095235/story.html

[2179]http://www.afr.com/p/australia2-0/panama_canal_expansion_to_draw_us_pcKvGanRKDqJNYGZgTPVcK

[2180]http://uk.reuters.com/article/2011/11/16/natgas-export-chesapeake-idUKN1E7AF1L720111116

[2181]http://seekingalpha.com/instablog/310523-expstocktrader/161531-cheniere-energy-a-tsunami-about-to-happen-in-the-worldwide-lng-market

[2182]http://www.businessweek.com/articles/2014-11-06/u-dot-s-dot-natural-gas-exports-will-fire-up-in-2015

[2183]http://www.bg-group.com/OurBusiness/WhereWeOperate/Pages/GEMS.aspx

[2184]http://www.miamiherald.com/2013/10/14/3689151/pascagoula-lng-terminal-awaits.html

[2185]http://www.utsandiego.com/news/2012/Apr/17/sempra-wins-backing-natural-gas-export-terminal/

[2186]http://articles.philly.com/2013-04-03/business/38221532_1_cove-point-dominion-resources-lng

[2187]http://www.thenation.com/article/180645/battling-cove-point-capital

[2188]http://www.prnewswire.com/news-releases/cabot-oil--gas-corporation-provides-corporate-update-announces-agreement-to-provide-natural-gas-to-the-dominion-cove-point-lng-terminal-236538531.html

[2189]http://lowersusquehannariverkeeper.org/publications/EarthReports%20et%20al%20Comments%20on%20EA%20final.pdf

[2190]http://fuelfix.com/blog/2013/09/09/freeport-lng-signs-deal-with-toshiba-and-south-korean-utility/

[2191]http://www.philly.com/philly/business/20141107_Sunoco_Logistics_annouces__2_5B_pipeline_project.html

[2192]http://fuelfix.com/blog/2013/08/07/feds-approve-lake-charles-lng-export-project/

[2193]http://ecowatch.com/2013/10/31/new-yorkers-rally-against-fracking-flawed-regulations/

[2194]http://pubs.acs.org/doi/abs/10.1021/es063031o;https://www.sciencemag.org/content/343/6172/733.summary

[2195]http://ntl.bts.gov/lib/000/400/422/20021101_alt_fuel.pdf

[2196]http://www.gao.gov/products/GAO-07-316

[2197]https://www.hsdl.org/?view&did=486464

[2198]http://www.psehealthyenergy.org/data/LNG_SignOnLetterPDF.pdf

[2199]http://www.psehealthyenergy.org/events/view/144

[2200]http://www.reuters.com/article/2012/05/15/us-excelerateenergy-idUSBRE84E0WO20120515

[2201]http://www.reuters.com/article/2014/12/30/us-usa-lng-excelerate-idUSKBN0K81CP20141230

[2202]http://www.reuters.com/article/2014/12/30/us-usa-lng-excelerate-idUSKBN0K81CP20141230

[2203]http://nationalacademies.org/onpi/06072005.pdf

[2204]http://www.pnas.org/content/106/6/1704.short

[2205]http://www.eia.gov/forecasts/ieo/index.cfm

[2206]https://news.vice.com/article/the-end-of-fracking-is-closer-than-you-think

[2207]http://www.postcarbon.org/publications/drill-baby-drill/

[2208]http://www.postcarbon.org/wp-content/uploads/2014/10/DBD-Exec-Summary.pdf

[2209]http://www.postcarbon.org/drill-baby-drill/

[2210]http://science.howstuffworks.com/environmental/green-science/clean-coal.htm

[2211]http://energy-facts.org/?utm_source

[2212]http://www.environmentalhealthnews.org/ehs/news/2015/jun/10-reasons-clean-coal-is-a-marketing-myth

[2213]http://sierraclub.typepad.com/compass/2012/02/the-sierra-club-and-natural-gas.html

[2214]http://www.pelicanweb.org/solisustv08n06page1.html

[2215]http://lakotavoice.com/2014/11/15/house-vote-in-favor-of-the-keystone-xl-pipeline-an-act-of-war/

[2216]http://www.freep.com/story/money/business/michigan/2014/10/12/fracking-oil-michigan-shale/17119337/

[2217]http://www.ren21.net/Portals/0/documents/Resources/GSR/2013/GSR2013_lowres.pdf

[2218]http://www.ren21.net/Portals/0/documents/Resources/GSR/2013/GSR2013_lowres.pdf

[2219]https://en.wikipedia.org/wiki/Three_Gorges_Dam

[2220]http://www.nzherald.co.nz/business/news/article.cfm?c_id=3&objectid=10826025

[2221]http://www.klimanetvaerk.dk/forum/topics/350-local-gruppe?commentId=3477612%3AComment%3A15313&xg_source=activity
[2222]http://www.klimanetvaerk.dk/forum/topics/350-local-gruppe?commentId=3477612%3AComment%3A15313&xg_source=activity
[2223]http://web.archive.org/web/20121020171232/https://www.cia.gov/library/publications/the-world-factbook/rankorder/2178rank.html
[2224]http://www.energy-enviro.com/index.php?PAGE=5&PRINT=yes&ID=4456
[2225]http://www.bloomberg.com/news/2012-05-10/saudi-arabia-plans-109-billion-boost-for-solar-power.html
[2226]http://www.upi.com/Business_News/Energy-Resources/2013/02/14/Sign-of-the-times-Saudis-go-solar/UPI-44571360875054/
[2227]http://web.archive.org/web/20121020171232/https://www.cia.gov/library/publications/the-world-factbook/rankorder/2178rank.html
[2228]http://www.bbc.co.uk/news/world-middle-east-21830846
[2229]http://www.npr.org/blogs/thetwo-way/2013/03/19/174745557/flush-with-oil-abu-dhabi-opens-worlds-largest-solar-plant
[2230]http://www.theguardian.com/environment/2015/oct/26/morocco-poised-to-become-a-solar-superpower-with-launch-of-desert-mega-project
[2231]http://www.afdb.org/fileadmin/uploads/afdb/Documents/Environmental-and-Social-Assessments/Morocco_-_Ouarzazate_Solar_Power_Station_Project_II_-_ESIA_Summary.pdf
[2232]https://ecowatch.com/2015/09/09/worlds-largest-solar-farm/
[2233]http://www.israel21c.org/environment/here-comes-the-sun-israel-launches-a-solar-power-field/
[2234]http://www.israel21c.org/environment/brightsource-takes-solar-thermal-energy-home-to-israel/print/
[2235]http://en.wikipedia.org/wiki/Wind_power_in_India
[2236]http://timesofindia.indiatimes.com/business/india-business/Bhel-to-build-worlds-biggest-solar-plant-near-Jaipur/articleshow/22829879.cms
[2237]http://reneweconomy.com.au/2013/india-reveals-plans-for-worlds-largest-solar-plant-79438
[2238]http://www.reuters.com/article/2009/07/28/idUSDEL104230
[2239]http://en.twwtn.com/Bignews/75906.html
[2240]http://www.solarimpulse.com/leg-8-from-Nagoya-to-Hawaii
[2241]http://rt.com/news/solar-energy-record-break-332/
[2242]http://www.seia.org/research-resources/solar-industry-data
[2243]http://www.seia.org/news/new-report-shows-growth-clean-solar-energy-benefitting-environment
[2244]http://www.bloomberg.com/news/2014-10-29/while-you-were-getting-worked-up-over-oil-prices-this-just-happened-to-solar.html
[2245]http://www.nytimes.com/2014/11/24/business/energy-environment/solar-and-wind-energy-start-to-win-on-price-vs-conventional-fuels.html?_r=0
[2246]http://cleantechnica.com/2014/02/04/current-cost-solar-panels/
[2247]http://energyinformative.org/solar-panels-cost/
[2248]http://www.nrel.gov/learning/re_photovoltaics.html
[2249]http://www.huffingtonpost.com/entry/solar-jobs-
[2250]http://www.heraldonline.com/2015/01/15/6704562_solar-industry-creating-jobs-nearly.html?rh=1
[2251]http://www.whitehouse.gov/the-press-office/2014/09/18/fact-sheet-white-house-announces-executive-actions-and-commitments-acros
[2252]http://yosemite.epa.gov/opa/admpress.nsf/0/B3E0D71E7D0988D285257E440009ADFF
[2253]http://www.usda.gov/wps/portal/usda/usdahome?contentid=2014/09/0202.xml&navid=NEWS_AUSUMS&navtype=RT&parentnav=ENERGY&edeployment_action=retrievecontent
[2254]http://www.brightsourceenergy.com/hidden-hills#.VETsg_nF9js

635

[2255]http://www.brightsourceenergy.com/ivanpah-solar-project#.VETsrfnF9js
[2256]http://www.abengoasolar.com/export/sites/abengoasolar/resources/pdf/Solana_factsh eet_09092013.pdf
[2257]http://lpo.energy.gov/projects/abengoa-solar-inc/
[2258]http://www.abengoasolar.com/export/sites/abengoasolar/resources/pdf/Solana_factsh eet_09092013.pdf
[2259]http://www.firstsolar.com/en/about-us/projects/topaz-solar-farm
[2260]http://www.usatoday.com/story/tech/2015/02/10/worlds-largest-solar-plant-california-riverside-county/23159235/
[2261]http://www.brightsourceenergy.com/palen#.VETsz_nF9js
[2262]http://www.energy.ca.gov/sitingcases/blythe_solar/
[2263]http://www.midamericanrenewablesllc.com/solarstar_solar.aspx
[2264]http://www.solarreserve.com/what-we-do/csp-projects/crescent-dunes/
[2265]http://money.cnn.com/2014/06/17/news/companies/solarcity-silevo/
[2266]http://www.buffalonews.com/city-region/riverbend/historic-day-for-buffalo-zemsky-says-of-solarcity-riverbend-plans-20140923
[2267]http://www.baltimoresun.com/business/bs-bz-ccbc-solar-project-20150505-story.html
[2268]http://www.seia.org/research-resources/us-solar-market-insight
[2269]http://hms.harvard.edu/news/bionic-leaf
[2270]http://www.hbs.edu/faculty/Publication%20Files/12-105.pdf
[2271]http://query.nytimes.com/mem/archive-free/pdf?res=990CE7DF1E3FE233A25751C0A9619C946796D6CF
[2272]http://uk.reuters.com/article/2014/09/29/us-solar-iea-electricity-idUKKCN0HO11K20140929
[2273]http://barnardonwind.com/2013/03/13/have-five-minutes-correct-anti-wind-myths-in-articles-and-comments/comment-page-1/
[2274]http://www.sibleyguides.com/conservation/causes-of-bird-mortality/
[2275]http://www.newscientist.com/article/dn23697-up-to-half-of-all-birds-threatened-by-climate-change.html#.VDpopvldVe_
[2276]http://qz.com/357332/chinas-wind-farms-can-now-produce-more-energy-than-all-of-americas-nuclear-plants/
[2277]http://www.takepart.com/article/2014/08/18/forget-fracking-wind-energy-prices-have-hit-record-low
[2278]http://www.forbes.com/sites/peterdetwiler/2013/03/26/denmark-1000-megawatts-of-offshore-wind-and-no-signs-of-slowing-down/
[2279]http://elpais.com/elpais/2014/01/15/inenglish/1389798670_862500.html
[2280]http://www.scotsman.com/news/environment/green-energy-is-scotland-s-leading-source-of-power-1-3617247?utm_source=twitter&utm_medium=post&utm_term=renewable,scotland&utm_campaign=Climate&__surl__=IgHe1&__ots__=1417099345993&__step__=1
[2281]http://www.ewea.org/fileadmin/files/library/publications/reports/Deep_Water.pdf
[2282]http://www.ewea.org/fileadmin/files/library/publications/reports/Deep_Water.pdf
[2283]http://www.ikeafoundation.org/1-billion-for-climate-action/
[2284]http://www.ikeafoundation.org/1-billion-for-climate-action/
[2285]http://www.desmogblog.com/2015/03/10/booming-u-s-renewable-energy-sector-growing-faster-expected
[2286]http://www.renewableenergyworld.com/articles/2010/07/worlds-largest-wind-project-is-underway.html
[2287]http://en.wikipedia.org/wiki/Alta_Wind_Energy_Center#cite_note-alta-2
[2288]http://www.sustainablebusiness.com/index.cfm/go/news.display/id/21699
[2289]http://www.sierraclub.org/sierra/201207/pennsylvania-fracking-shale-gas-200-sidebar.aspx

[2290]http://www.takepart.com/article/2014/08/18/forget-fracking-wind-energy-prices-have-hit-record-low

[2291]https://www.dropbox.com/sh/qml781qhmw431wm/FqB7lGX3Zx/20130612_VolturnUSGridConnectPressRelease.pdf

[2292]http://www.bostonglobe.com/metro/2014/09/14/vermont-milestone-green-energy-efforts/fsLHJl4eoqv6QoFNewRYBK/story.html

[2293]https://ecowatch.com/2015/05/20/kodiak-island-renewable-energy/

[2294]http://www.usnews.com/news/us/articles/2015/07/12/apnewsbreak-south-getting-its-first-big-wind-farm-soon

[2295]http://iberdrolarenewables.us/rel_15.06.11.html

[2296]http://iberdrolarenewables.us/rel_15.06.11.html

[2297]http://oceana.org/sites/default/files/reports/offshore_energy_by_the_numbers_report_final.pdf

[2298]http://theenergycollective.com/globalwarmingisreal/2188916/renewable-energy-beats-natural-gas-new-capacity-2014

[2299]http://www.gallup.com/poll/161519/americans-emphasis-solar-wind-natural-gas.aspx

[2300]http://www.truth-out.org/news/item/17605-former-mobil-vp-warns-of-fracking-and-climate-change?tmpl=component&print=1

[2301]http://www.economist.com/blogs/schumpeter/2013/08/interview-george-mitchell

[2302]http://edition.cnn.com/2010/OPINION/02/22/jacobson.nuclear.power.con/index.html

[2303]http://edition.cnn.com/2010/OPINION/02/22/jacobson.nuclear.power.con/index.html

[2304]http://www.stanford.edu/group/efmh/jacobson/Articles/I/sad1109Jaco5p.indd.pdf

[2305]http://www.stanford.edu/group/efmh/jacobson/Articles/I/sad1109Jaco5p.indd.pdf

[2306]http://www.nrel.gov/analysis/re_futures/

[2307]http://www.worldenergyoutlook.org/media/weowebsite/2012/factsheets.pdf

[2308]https://www.whitehouse.gov/the-press-office/2015/11/06/statement-president-keystone-xl-pipeline

[2309]http://www.speaker.gov/press-release/statement-rejection-keystone-xl-pipeline

[2310]See:http://www.usatoday.com/story/news/nation/2014/01/28/keystone-pipeline-poll/4935083/;http://www.cbsnews.com/news/americans-view-of-the-economy-most-positive-in-eight-years.

[2311]http://www.villagevoice.com/2012-09-19/news/boom-or-doom-fracking-environment

[2312]http://www.pewinternet.org/2015/01/29/chapter-3-attitudes-and-beliefs-on-science-and-technology-topics/

Several thousand anti-fracking activists from around the world met in Paris in December 2015 to protest the destruction of the environment and public health, and to present alternate solutions to fossil fuel energy.

PHOTO: Karen Feridun

INDEX

647

Government Accountability Office
18, 209, 280-281
GPX/GXP 225-226
Graves, John 489, 490-491
Great Sioux Nation 551
Green Party 416, 419
Greenberg, Steven 326
Greenman, Josh 321
Grewal, Paul 339
Grillot, Larry 379-371
Grimshaw, Thomas W. 388-389
Groat, Charles G. 388-390
Grosbard, Miles 394
*Groundswell Rising: Protecting
Our Children's Air and Water*
414, 478-479
Groundwater Protection Council
297
Grove, Brian 177, 408
Gu, Jeff 217
Guesman, Michael 189
Guggenheim, David 476
Gutfeld, Greg 305
Guthrie Health 143-144

H
Hall, James 3
Halliburton 161, 162, 177, 188,
228, 257, 267-268, 285-287, 295-
296, 383, 397, 492, 536
Halliburton Loophole 162, 188,
383
*Hallowich v. Range Resources, et
al.*, 167-170, 171, 300-302
Hallowich, Chris 167-170, 171,
300-302
Hallowich, Stephanie 167-170,
171, 300-302
Halpin, Barbara 434
Haluszczak, Lara O. 175
Hamm, Harold 379-380, 387, 388,
423
Hammer, Rebecca 201, 204, 218-
219
Hammond, Wendi 58
Hang, Walter 206
Hanger, John 97, 160, 177, 191,
410, 468-469, 474
Hannah, Daryl 317, 424

Hanson, Rolf 339
Harbert, Karen 170
Hardrock Excavating 189
Harold Hamm School of Geology
and Geological Engineering 387
Harris, Barb 159
Harris, Bill 57
Harris, Darlene 342
Harrisburg Patriot-News 99, 101,
118, 345
Hart Resources Technology 204-
205
Harvard University 33, 298, 359,
403-404, 405, 450-451
Hasemeyer, David 501-506 *loc.
cit.*
Hasse, Alma 64, 362-363
Hatch, Orin 69
Hawaii 125, 340
Haynesville Shale 2, 15, 212, 549
HB 1950 (See: Act 13)
HBO 467, 470-471
Health Effects (primary section)
147-162
Heartland Institute 2
Heinberg Richard 489
Heinz Foundation 511
Henderson, Patrick 86, 208, 290,
408
Hense, Debora 60
Hercik, James 108
Heritage Foundation 423, 474
Hess Co. 48, 312, 537
Hess, Kevin 114
Hetzler, Paul 12-13
Hickenlooper, John 176-177, 330-
333
Hilcorp Energy 65
Hill + Knowlton 525
Hill, Anne 520
Hill, Michael C. 50-51
Hinchey, Maurice 32, 232-233
Hines, John T. 97
Hippauf, Dory xvi, 38, 39, 89, 95,
103, 278, 321, 476, 506, 508
History of Fracking (*See:
Fracking—History*)
Hitzman, Murray 211, 383
Hoffman, Gary P. 17-18

651

Pritchard, Meighan 309
Promised Land 472-475, 476, 490
Proppants (*See*: Fracking—
  Proppants)
ProPublica 96, 97, 151, 177-178,
  181-182, 193, 194, 196, 200,
  207, 258, 277, 321, 369, 377-
  378, 507, 521
Prospect Energy 330
Protecting Our Waters 242, 442,
  445-446
Psychological Effects from
  Fracking (*See*: Fracking—
  Psychological Effects)
Public Accountability Initiative
  378-379, 385, 389, 405, 410, 475
Public Employees for
  Environmental Response 281
Public Health Issues (*See*:
  Fracking—Public Health Issues)
Pyle, Thomas J. 11

Q

Qualls, Rozanne 210
Quantum Communications 409
Quicksilver Resources 531, 532
Quigley, John 299
Quigley, Keith 113
Quisenberry, Kevin 456

R

Rabinowitz, Peter M. 151
Rabobank 49
Rachael Carson Bridge 439
Rachel Carson State Office
  Building (Pa.) 93
Radford, Tom 484-485
Radioactive Elements (*See:*
  Strontium, TENORM, Thorium,
  Uranium)
Radioactive Gases (*See*: Radium,
  Radon)
Radioactive Waste Alert 530
Radium 172-176 *loc. cit.*, 202, 289
Radon 11, 72, 174
Radow, Elisabeth 35, 48, 50
Railroad Commission of Texas 56-
  57, 59, 98, 129, 154, 181, 186,

197-198, 212-213, 334, 336-337,
  502-504, 506
Railroads 270-277, 234-235 (*See*:
  BNSF; CSX ; Genesee &
  Wyoming Railroad, Montreal,
  Maine & Atlantic Railway,
  Norfolk Southern Railroad,
  North Shore Railroad)
Railroad Disasters 271-277
Regulations 275-277
Rain for Rent 232
Range Production 197-198
Range Resources 3, 16, 46, 53-54,
  82, 84, 168-170, 171, 197, 198,
  223, 294-295, 300-302, 311, 345,
  350, 376, 444, 461, 508, 533, 537
Rao, Vikram 490, 492
Raton Basin 211
Raulerson, Josh 508
Reagan, Ronald 29, 353, 373
Reaven, Sheldon 164
Redford, Robert 357, 526
Reed, Cyrus 336-337
Reed, Dave 84
Reed, Tom 81, 326
Reibsome, MaryJayne vi, xvii
Reid, Harry 73
Reitzel, Rick 210
Religious Action Center of Reform
  Judaism 312, 317
Rendell, Ed 65, 94, 97, 99, 100,
  106, 108, 180, 191, 320-321,
  344-345, 370, 410, 443, 468
Renewable Energy 2-3, 6, 39, 90-
  91, 226, 304, 307-308, 324, 397,
  402, 419, 420, 422, 423, 443,
  448, 449, 451, 479, 480, 523,
  535, 543, 546-565
  Bioenergy and Biogenic
    Energy 1, 552
  Geothermal Energy 62, 551-
    553, 554, 563, 564, 551
  Hydropower Energy 3, 526,
    552
  Solar energy 1, 4, 33, 38-39,
    133, 308, 526, 546, 547,
    550-551, 553-559, 560, 562,
    563, 564
  Thermogenic Energy 1

660

Wind energy 510, 552, 553, 559-561, 562

Renwand, Thomas 295

Republican Party xiii, 8, 26, 34, 37, 39, 40, 67-121 *loc. cit.*, 131, 141, 192, 233, 281, 287-288, 304-305, 320, 325, 326, 328, 332, 338, 343, 344-346, 360, 378, 385, 410, 464, 476, 497, 556, 562 (*Also See*: George H.W. Bush, George W. Bush, Gerald Ford, Richard M. Nixon, Ronald Reagan)

Research and Policy Center of Environment America 129, 245

Resnikoff, Marvin 174, 175

Resource Conservation and Recovery Act 28, 178, 207

Responsible Drilling Alliance 298, 436-438

Restuccia, Andrew 282

Reuters News 18, 59, 60, 231,316, 341, 360, 540, 544

Revolving Door Politics 407-413

Rex Energy 170

Rhoads, Stephen W. 165

Richardson, Bill 486

Richardson, Janie 437

Richmond, Matt 508-509

Richter, Wolf 532

Ridge, Tom 410, 443

Rippel, Raina 221

Riverdale Mobile Home Village 452-465

Road Damage from Fracking (*See*: Fracking—Road Damage)

Robinson, Leonard 311

Rockefeller Brothers Fund 451

Roden, Paul R. xvi

Rogers, Deborah 41, 44, 446, 534, 565

Romney, Mitt 67

Roosevelt, Franklin D. 26

Rose, Arthur W. 175

Rose, Patti xvi, 344

Rosenbaum, Liz xvi

Roter, Rebecca xvi, 318, 528

Rove, Karl 439

Rovicks, David 520

Rowland, Marcus C. 61

Royal Dutch Shell 44, 66, 69, 89-90, 135, 239, 399, 532, 538

Royal Society of Great Britain 135

Royte, Elizabeth 255

Rozell, Daniel 164

Ruffalo, Mark 317, 417, 519

Rugh, Peter 428

Russell, Barry 36, 181, 491

Russia 6, 33, 423, 474, 527, 543

Ryan, Joseph 143

Ryan, Matthew T. 192

Ryder, Andrea 97-98

S

Saavedra, Diana xvii

Saberi, Poune 36 242

Sabine Oil & Gas 532-533

Sacrifice Zone 49, 418, 441

Safe Water Drinking Act 28, 31, 32, 70, 161, 204

Salazar, John 32

Salazar, Ken 354, 358

Salazar, Mario 200

Salt Lake City Tribune 354, 356-357, 495

"Same River" 466, 521-522

Samson Resources 532-533

San Antonio Express–News 129, 500

San Juan Basin 2, 486

San Juan Citizens Alliance 129

Sandberg, Claire 193-194

Sanders, Bernie 38

Sanford, N.Y. 372-373

Sanford Basin 287

Sanko, David 111, 113

Sanserino, Michael 500

Santandr 49

Santoro, Renee 126-127

Santorum, Rick 67, 305

Sarandon, Susan 317, 318

Sassoon, David 501-506 *loc. cit.*

Saudi Arabia 306, 527, 531-532, 553-554

Savage, Heather 214

Savio, Mario 415

Sawyer, Hall 257, 258

Scarnati, Joseph B. 84, 119

664

665

(Index compiled by
Rosemary and Walter Brasch)

ABOUT THE AUTHOR

Walter M. Brasch, Ph.D., is an award-winning social issues journalist and the author of 20 other books, most of which fuse historical and contemporary social issues. Among his books is *Before the First Snow*, a critically-acclaimed novel that looks at what happens when government and energy companies form a symbiotic relationship, using "cheaper, cleaner" fuel and the lure of jobs in a depressed economy but at the expense of significant health and environmental impact.

He is professor emeritus of mass communications and journalism, and a former newspaper and magazine reporter and editor, and multi-screen multimedia writer–producer.

He is vice-president of the Central Susquehanna chapter of the ACLU, vice-president and co-founder of the Northeast Pennsylvania Homeless Alliance, a member of the board of the Keystone Beacon Community for healthcare coordination, and is active in numerous social causes. He was a Commonwealth Speaker for the Pennsylvania Humanities Council, and was active in emergency management.

He was president of the Pennsylvania Press Club and the Keystone State professional chapter of the Society of Professional Journalists, vice-president of the Pennsylvania Women's Press Association, and founding coordinator of Pennsylvania Journalism Educators. He is a featured columnist for *Liberal Opinion Week*, senior correspondent for the *American Reporter*, senior editor for *OpEdNews*, and a former editorial board member of *Journalism History* and the *Journal of Media Law and Ethics*.

He is a member of the National Society of Newspaper Columnists, Authors Guild, The Newspaper Guild (CWA/AFL-CIO), and the Society of Environmental Journalists. He is listed in *Who's Who in America, Contemporary Authors, Who's Who in the Media,* and *Who's Who in Education.*

He was recognized in 2012 by the Pennsylvania Press Club with the Communicator of Achievement award for lifetime achievement in journalism and public service. Among recent writing awards are multiple awards from the National Society of Newspaper Columnists, Society of Professional Journalists, National Federation of Press Women, USA Book News, Independent Book Publishing Professionals Group, Pennsylvania Press Club, Pennsylvania Women's Press Association, Pennsylvania Associated Press Broadcasters Association, Penn-Writers, Pacific Coast Press Club, Press Club of Southern California, and the International Association of Business Communicators.

He was honored by San Diego State University as a Points of Excellence winner in 1997. At Bloomsburg University, he earned the Creative Arts Award, the Creative Teaching Award, and was named an Outstanding Student Advisor. He received the first annual Dean's Salute to Excellence in 2002, a second award in 2007, and the Maroon and Gold Quill Award for nonfiction. He is the 2004 recipient of the Martin Luther King Jr. Humanitarian Service Award.

Dr. Brasch earned an A.B. in sociology/social welfare from San Diego State College, an M.A. in journalism from Ball State University, and a Ph.D. in mass communication/journalism, with cognate areas in both American government/public policy and language and culture studies, from The Ohio University.

He is married, has two children, and is surrounded by animals and the rural beauty of northeastern Pennsylvania.

To learn more about Dr. Brasch, visit
http://www.walterbrasch.com

CPSIA information can be obtained at www.ICGtesting.com
Printed in the USA
LVOW11s0825120316

478896LV00003B/219/P